Microwave Processing
of Materials V

MATERIALS RESEARCH SOCIETY
SYMPOSIUM PROCEEDINGS VOLUME 430

Microwave Processing of Materials V

Symposium held April 8-12, 1996, San Francisco, California, U.S.A.

EDITORS:

Magdy F. Iskander
University of Utah
Salt Lake City, Utah, U.S.A.

James O. Kiggans, Jr.
Oak Ridge National Laboratory
Oak Ridge, Tennessee, U.S.A.

Jean-Charles Bolomey
Supelec
Gif-Sur-Yvette, France

MATERIALS
RESEARCH
SOCIETY

PITTSBURGH, PENNSYLVANIA

Single article reprints from this publication are available through
University Microfilms Inc., 300 North Zeeb Road, Ann Arbor, Michigan 48106

CODEN: MRSPDH

Published by:

Materials Research Society
9800 McKnight Road
Pittsburgh, Pennsylvania 15237
Telephone (412) 367-3003
Fax (412) 367-4373
Website: http://www.mrs.org/

Library of Congress Cataloging in Publication Data

Microwave processing of materials V ; symposium held April 8-12,1996,
    San Francisco, California, U.S.A. / editors: Magdy F. Iskander,
    James O. Kiggans, Jr., Jean-Charles Bolomey
        p. cm—(Materials Research Society symposium proceedings ; v. 430)
        Includes bibliographical references and index.
        ISBN 1-55899-333-9
        I. Iskander, Magdy F. II. Kiggans, Jr., James O. III. Bolomey, Jean-Charles
    IV. Series: Materials Research Society symposium proceedings ; v. 430.

Manufactured in the United States of America

CONTENTS

PART IV: <u>MICROWAVE SYSTEM DESIGN</u>

PART V: <u>DIELECTRIC PROPERTIES MEASUREMENTS AND ANALYSIS</u>

PART XIII: PLASMA PROCESSING

PREFACE

The fifth MRS symposium on microwave processing of materials was held in conjunction with the 1996 MRS Spring Meeting in San Francisco, California. The symposium was organized to continue the thrust of the four previous symposia held in 1988, 1990, 1992, and 1994. The theme for the 1996 symposium was "Science and Applications of the Microwave Processing Technology". It was highlighted by the presence of an excellent international gathering of scientists and engineers who reported results of their recent work in the growing area of microwave processing technology. Specifically, more than 120 papers from 17 countries were accepted for presentation in the 1996 conference. Represented countries include: Argentina, Australia, Belgium, Canada, China, France, Germany, India, Italy, Japan, Mexico, Poland, Russia, Spain, Ukraine, United Kingdom, and the United States.

The symposium was organized in 21 sessions, including a poster session to provide a more casual but effective forum for interaction and discussions between authors and participants. In addition, a luncheon plenary session on "the microwave effect" was organized to help update attendees on activities and recent results in this area of fundamental research.

This volume provides an excellent reference to the continued progress made since the 1988 conference and is a complement to the four earlier volumes in the MRS Proceedings Series on Microwave Processing—Volumes 124, 189, 269, and 347. Work reported in this book is divided into 13 parts including:

- scale-up and commercialization
- microwave nondestructive testing
- microwave processing
- dielectric properties measurements and analysis
- modeling of microwave heating
- microwave interaction and mechanisms
- microwave processing using variable frequency sources
- alternate microwave sources
- temperature modeling and measurements
- microwave processing of polymers
- plasma processing
- microwave system design
- remediation of hazardous waste

The papers included in this volume were reviewed and carefully selected from among the over 120 papers presented at the meeting. They show significant advances in understanding and control of microwave energy and its use in the processing and testing of materials. Development in modeling and numerical simulation, microwave processing techniques, nondestructive testing, use of variable frequency and alternative sources in designing microwave processing systems and advances in the area of high-temperature dielectric properties measurements were particularly notable.

It is generally felt that this area of research has continued to attract talented researchers who are focusing their activities on promising ventures and contributing to the ultimate success and the possible commercialization of the emerging microwave processing technology. It is expected that this technology will continue to make significant progress, and research efforts in areas such as computer modeling and simulation, process monitoring and control, and the innovative design of microwave processing systems, will result in major advances in the near future.

Magdy F. Iskander
James O. Kiggans, Jr.
Jean-Charles Bolomey

July, 1996

ACKNOWLEDGMENTS

We wish to thank all of the speakers, session chairs, and participants who made this symposium a great success. We are especially grateful to the authors and referees who assisted in the timely publication of the proceedings. The session chairs, who did an excellent job of managing the program and who were also responsible for refereeing the papers from the respective sessions, were:

D. Blackham	Z. Fathi	D.A. Lewis
J. Ch. Bolomey	A. Gourdenne	B. Maestrali
J.H. Bookse	M.F. Iskander	A.D. McMillan
S. Bringhurst	H.W. Jackson	D. Rytting
R.W. Bruce	D.L. Johnson	M.J White
C. Buffler	J.O. Kiggans, Jr.	T.L. White
Y. Bykov	H.D. Kimrey	M. Willert-Porada

We extend special thanks to Holly Cox, who spent many hours assisting in organizing the symposium, managing the collection and handling of manuscripts, handling of correspondence, repairing manuscripts, and compiling the final manuscript; and to Cathy Cheverton, for her assistance in the collection and handling of manuscripts. We are also grateful to Shane Bringhurst and Mikel J White for their invaluable assistance in organizing sessions and for their help in compiling the indices.

We acknowledge the financial support provided for this symposium by:

The Heavy Vehicle Propulsions System Materials Program, DOE Office of Transportation Technologies, under contract with Lockheed Martin Energy Research Corporation

Energy Research Laboratory, Technology Transfer Program, DOE Office of Energy Research

Oak Ridge National Laboratory, Advanced Industrial Concepts Program, DOE Office of Energy Efficiency and Renewable Energy

Hewlett-Packard Company, Microwave Instruments Division

Electricité de France (EDF)

This symposium was endorsed by the American Ceramic Society and the International Microwave Power Institute.

MATERIALS RESEARCH SOCIETY SYMPOSIUM PROCEEDINGS

MATERIALS RESEARCH SOCIETY SYMPOSIUM PROCEEDINGS

Prior Materials Research Society Symposium Proceedings available by contacting Materials Research Society

Part I

Scale-up and Commercialization

SCALE-UP OF THE NITRIDATION AND SINTERING OF SILICON PREFORMS USING MICROWAVE HEATING

J. O. Kiggans, Jr.*, T. N. Tiegs*, C. C. Davisson*, M. S. Morrow** and G. J. Garvey***
*Oak Ridge National Laboratory, Oak Ridge, Tenn., 37831-6087
**Y-12 Development Division, Oak Ridge, Tenn., 37831-8096
***Golden Technologies, Inc., Golden, Colo.; Presently, Ceradyne, Inc. Costa Mesa, CA.

ABSTRACT:

Scale-up studies were performed in which microwave heating was used to fabricate reaction-bonded silicon nitride and sintered reaction-bonded silicon nitride (SRBSN). Tests were performed in both a 2.45 GHz, 500 liter and a 2.45 GHz, 4000 liter multimode cavities. A variety of sizes, shapes, and compositions of silicon preforms were processed in the studies, including bucket tappets and clevis pins for diesel engines. Up to 230 samples were processed in a single microwave furnace run. Data were collected which included weight gains for nitridation experiments, and final densities for nitridation and sintering experiments. For comparison, nitridation and sintering studies were performed using a conventional resistance-heated furnace.

INTRODUCTION

Silicon nitride materials represent a class of materials having a wide range of compositions where silicon nitride (Si_3N_4) is the major phase. These materials are of interest in numerous applications for such diverse items as cutting tools, rotors and stator vanes for advanced gas turbines, valves and cam roller followers for gasoline and diesel engines, and radomes on missiles.[1] However, these materials tend to be very expensive and are not competitive on a cost basis with metal parts.[2]

SRBSN, made from silicon, has been identified as a cost-effective alternative to Si_3N_4 made from high cost Si_3N_4 powders. Silicon is economical compared to high purity Si_3N_4 powders (approximately 1/4 the cost). In addition, SRBSN materials are attractive in that they exhibit improved control over the dimensional tolerances due to less shrinkage during sintering.[3,4]

In the last few years, microwave heating has been investigated extensively for use in the thermal processing of ceramics.[5-9] In addition, the microwave processing of Si_3N_4 has been the focus of considerable research.[10-16]

EXPERIMENTAL PROCEDURES

Nitridation Scale-Up Study

Silicon preforms of a clevis pin geometry were supplied by Golden Technologies, Inc., Golden, Co., as part of a Cooperative Research and Development Agreement (CRADA) sponsored by Department of Energy with the Oak Ridge National Laboratory. The composition of the clevis pins is proprietary information. The pins are a cylindrical shape, 3.84 cm long by 1.18 cm diameter, with a 0.24 cm diameter hole through center and parallel to the longitudinal axis.

Scale-up microwave nitridation processing of the preforms was conducted in an ORNL, 500 liter cylindrical multimode cavity, equipped with dual 6 kW, 2.45 GHz power generators. The clevis pins were placed inside one of two rectangular-shaped sample crucibles having walls composed of 40, 60, or 80 wt % SiC with a balance of Si_3N_4 and Y_2O_3, and with the crucible top and bottom composed of hot-pressed boron nitride. One crucible measured 9.5 cm by 10.8 cm by 4.4 cm, and the larger crucible measured 15 cm by 16.8 cm by 4.4 cm. Each sample crucible was placed inside a 2.5 cm thick alumina fiberboard insulation casket inside the microwave furnace. The furnace was evacuated twice to 800 millitorr and backfilled with nitrogen, and then evacuated and backfilled with N_2 - 4 vol % H_2. Nitridation was performed at a final temperature of 1300°C. Temperatures were measured continuously during processing using a molybdenum-sheathed type "C" thermocouple. Complete details of experiments have been reported elsewhere.[17]

Mat. Res. Soc. Symp. Proc. Vol. 430 © 1996 Materials Research Society

Clevis pins were weighed before and after experiments, and weight gain values calculated to determine the extent of nitridation.

<u>Nitridation and Sintering Scale-Up Study</u>

Three types of silicon preforms were used for the nitridation and sintering scale-up study: gelcast-isopressed preforms fabricated by ORNL; die-pressed silicon preforms purchased from Coors Ceramics Co.; and injection-molded bucket tappet preforms purchased from the Cremer Forschungs Institut Gmbh and Co., Rödental Germany. Further details concerning preparation, sizes, and compositions of the various preforms have been reported.[12-14]

Scale-up microwave nitridation and sintering experiments were conducted in the ORNL, 500 liter cavity and in a 4000 liter multimode cavity, which is equipped with eight, 6 kW, 2.45 GHz power generators. Eight to 230 samples, with masses ranging from 9 to 32 g each, were processed in each experiment. Microwave furnaces were evacuated and backfilled with N_2 and then N_2 - 4 vol % H_2 prior to start of the heating cycle. Each microwave experiment was conducted in a single heating cycle. Conventional nitridation and sintering experiments were conducted in a graphite-resistance heated furnace in two separate furnace runs using flowing N_2 - 4 vol % H_2 (nitridation run) or N_2 (sintering run). The final sintering temperature for the microwave and conventional experiments ranged from 1750 to 1800°C. Samples were weighed before and after experiments, and the densities of sintered pieces were measured using the Archimedes method. Further details concerning experimental conditions for individual conventional and microwave heating experiments have been reported.[13-15,18]

Results

Nitridation Scale-up Study

This study was conducted in order to determine the proper crucible set-up for the scale-up of the microwave nitridation of silicon preforms. Table 1 summarizes the results obtained from a number of experiments. The first experiment involved the nitridation of 36 clevis pins, with 0.5 cm separating each pin, in the small 40 wt % SiC crucible. SiC was included in this crucible as a surrogate material to absorb microwave energy and heat the crucible walls. The average nitridation weight gain was 43.5 wt %, which is comparable to results obtained from a conventional nitridation. A closer look at data in experiment 1 indicates that the weight gain of samples along the inside edge of the crucible, "Edge CP," were comparable to the weight gain of samples completely surrounded by other samples, "Inside CP." However, some samples in the center of the crucible showed minor shrinkage due to sintering. To gain a further insight into the partial sintering of these samples, experiment 2 was performed. In this experiment, 60 clevis pins were in the same crucible with no separation between pins. Once again the weight gains were good, however the unwanted sintering of the center-most samples in the crucible increased. This result pointed out the need for more space between clevis pins to prevent spot sintering. In a larger scale-up experiment, experiment 3, eighty-one clevis pins were spaced at a greater distance, 0.65 cm from each other, in the large 40 wt % SiC crucible. Table 1 shows that good nitridation was obtained with no shrinkage of "inside" samples, however, the statistical deviation of average weight gain rose slightly. Although the weight gain of the "Edge CP" samples appeared to be normal, a slight silicon color on the side of pins facing the crucible wall suggested that the temperature along the edge was lower than the center of the crucible.

Tests were conducted in crucibles of the same size as experiment 3, but having walls made of 60 and 80 wt % SiC. The higher SiC additions to these crucibles were made to increase microwave heating of the crucibles. The results in Table 1 indicate that the standard deviation of the average densities improved by increasing the SiC content. Overall, the last three experiments showed that proper sample spacing eliminated undesirable localized sintering of samples, and that matching the absorption of microwaves by crucibles and samples resulted in a more uniform temperature distribution within the crucibles. Figure 1, visual evidence of the success of the microwave nitridation scale-up, shows the crucible arrangement of a microwave nitridation with 90 clevis pins using the 80 wt % SiC crucible set-up.

Table 1. Summary of microwave nitridation scale-up experiments

Exp #	Sample #	Crucible Type	Average % Wt Gain "Total CP[1]"	Average % Wt Gain "Inside CP[2]"	Average % Wt Gain "Edge CP[3]"	St. Dev. Average % Wt Gain "Total CP"
1	36	40 wt % SiC	43.49	43.49	43.50	0.14
2	60	40 wt % SiC	43.85	43.80	43.91	0.08
3	81	40 wt % SiC	43.83	43.73	43.98	0.16
4	81	60 wt % SiC	43.46	43.46	43.55	0.12
5	81	80 wt % SiC	43.82	43.78	43.88	0.09

[1]CP = Clevis Pin
[2]Inside CP = clevis pins surrounded only by other samples
[3]Edge CP = clevis pins with the crucible wall on one side and samples on the opposite side

Fig. 1. Insulation package containing 90 clevis pins after nitridation
using microwave heating.

Nitridation and Sintering Scale-up Study

The second study involved the scale-up of microwave nitridation and sintering of silicon preforms.
Table 2 summarizes important details and results concerning individual experiments. In
experiment 1, eight bucket tappets were nitrided and sintered. A high density and a low standard
deviation of the average density was obtained. Samples processed in experiment 7, using
conventional heating, showed a similar sintered density and standard deviation. Experiments 2
through 6 were microwave scale-up runs with progressively larger numbers of samples. Varied

average sintered densities were obtained due to sintering conditions chosen for the experiments. Slight warping of bucket tappet samples located in the corners of the crucible occurred in experiment 5, due the higher heat losses from the crucible corners. This problem could probably be solved through the use of a cylindrical crucible (no corners). In general, the standard deviation of the densities for these experiments increased as the sample number increased. Likewise, results in experiment 8 indicate that the same trend was true for scale-up runs using conventional heating. An examination of the microwave power data for experiments 2 through 6 shows that increasing power was required as the number of sample was increased. However, general conclusions relating power and sample number are not possible, since different sample types and insulation containers were used in various experiments. Figure 2 is a photo showing the arrangement of one layer of the total of three layers of bucket tappets and rectangular silicon preforms, as packaged inside a crucible for sample processing in microwave experiments 5 and 6. Figure 3 is a photograph of the same samples after microwave processing. Figure 3 gives a good indication of the size of samples and the extent of the scale-up.

Table 2. Summary of conventional and microwave nitridation and sintering scale-up experiments

Exp #	Furnace Type	Furnace Size	Sample #	Total Starting Sample Mass (kg)	Peak Power (kW)	Final Sample Density (g/cm³)	Standard Deviation Densities (g/cm³)
1	microwave	500 L	8[a]	0.3	1.8	3.27	0.005
2	microwave	500 L	21[a]	0.7	3.7	3.23	0.002
3	microwave	500 L	72[b]	1.0	2.3	3.30	0.018
4	microwave	500 L	108[c]	1.1	5.2	3.25	0.008
5	microwave	4000 L	90[a]	2.9	2.6	3.18	0.020
6	microwave	4000 L	230[c]	2.1	2.9	3.19[f]	0.019
7	resistance	70 L	9[a]	0.3	--	3.26	0.003
8	resistance	70 L	210[d]	2.9	--	3.22[f]	0.052

[a]Cremer Forschung Institut bucket tappets, proprietary composition
[b]ORNL gelcast - isopressed material, Si_3N_4 - 9 wt % La_2O_3 - 3 wt % Al_2O_3
[c]ORNL gelcast - isopressed material, Si_3N_4 - 9 wt % Y_2O_3 - 3 wt % Al_2O_3
[d]Coors die pressed samples , proprietary composition
[e]ND = not determined
[f]Average density value, based on measurement of density of every fourth sample

Fig. 2 Photos of silicon bucket tappet (left) and silicon gelcast rectangular bar preforms (right),as packaged for microwave scale-up experiments 5 and 6, respectively.

Fig. 3. Bucket tappet (left) and rectangular bar (right) SRBSN samples fabricated using microwave in experiments 5 and 6, respectively.

Conclusion

Scale-up was successfully demonstrated for microwave fabrication of reaction-bonded silicon nitride and sintered reaction-bonded silicon nitride. A variety of sizes, shapes, and compositions of silicon preforms were processed in the studies, including bucket tappets and clevis pins for diesel engines, with up to 230 samples processed at one time. Different sample packaging requirements were needed for microwave nitridation versus microwave sintering, since nitridation was done at a peak temperature 1300°C, whereas nitridation and then sintering required peak temperatures of 1750 to 1800°C.

Critical factors for success in both the microwave nitridation and sintering scale-up included proper sample spacing and selection of hybrid-heating crucible materials. These studies were meant to prove capability only. Additional refinements in the processes, such as the development of cylindrical hybrid-heating processing crucibles and the use of continuously moving microwave platforms, will be necessary for transfer of the microwave process to industry. It is believed that the lower microwave power requirements for microwave nitridation, 1/2 the amount required for sintering, make it the more attractive of the two processes.

REFERENCES

1. R. N. Katz, Nitrogen Ceramics 1976-1981, pp. 3-20 in Progress in Nitrogen Ceramics, ed. F. L. Riley, Martinus Nijhoff Pub., The Hague, Netherlands (1983).
2. L. M. Sheppard, "Cost-Effective Manufacturing of Advanced Ceramics," pp. 692-707 in Am. Ceram. Soc. Bull., 70 [4] (1991).
3. T. Quadir, R. W. Rice, J. C. Chakraverty, J. A. Breindel, and C. C. Wu, "Development of Lower Cost Si_3N_4," pp. 9-10 in Ceram. Eng. Sci. Proc., 12, 1952-1957 (1991).

4. A. J. Moulson, "Reaction-Bonded Silicon Nitride: Its Formation and Properties," pp. 1017-1051 in J. Mater. Sci., Vol. 14, (1979).
5. W. H. Sutton, M. H. Brooks, and I. J. Chabinsky, eds., Microwave Processing of Materials, Vol. 124, Materials Research Soc., Pittsburgh, PA 1988.
6. W. B. Snyder, W. H. Sutton, D. L. Johnson, and M. F. Iskander, eds., Microwave Processing of Materials-II, Vol. 189 Materials Research Soc., Pittsburgh, PA 1991.
7. R. L. Beatty, W. H. Sutton, and M. F. Iskander, eds., Microwave Processing of Materials-III, Vol. 269, Materials Research Soc., Pittsburgh, PA 1992.
8. D. E. Clark, F. D. Gac, and W. H. Sutton, eds., Ceramic Transactions Microwaves: Theory and Application in Materials Processing, American Ceramic Society, Westerville, OH, 1988.
9. M. F. Iskander, R. J. Lauf, and W. H. Sutton, eds., Microwave Processing of Materials IV, Vol. 347, Materials Research Soc., Pittsburgh, PA, 1994.
10. T. N. Tiegs and J. O. Kiggans, jr., "Fabrication of Silicon Nitride Ceramics by Microwave Heating," pp. 665-671 in Proc. 4th International Symp. Ceram. Mater. & Compon. for Engines, Elsevier Applied Sci., New York (1992).
11. M. L. C. Patterson, P. S. Ape, R. M. Kimber, and R. Roy, "Batch Process For Microwave Sintering of Silicon Nitride," pp. 291-300 in Materials Research Society Proceedings, Vol. 269.
12. T. N. Tiegs and J. O. Kiggans, and H. D. Kimrey, "Microwave Sintering of Silicon Nitride," pp. 9-10, Ceram. Eng. Sci. Proc., 12 (1991).
13. T. N. Tiegs, J. O. Kiggans, and K. L. Ploetz, "Sintered Reaction-Bonded Silicon Nitride By Microwave Heating" pp. 283-288 in Materials Research Society Proceedings, Vol. 287.
14. T. N. Tiegs, J. O. Kiggans, jr., H. T. Lin, and C. A. Willkens, "Comparison of Properties of Sintered and Sintered Reaction-Bonded Silicon Nitride Fabrication by Microwave and Conventional Heating," pp. 501-506 in Materials Research Society Proceedings, Vol. 347.
15. J. O. Kiggans, jr., T. N. Tiegs, H. D. Kimrey, and J-P. Maria, "Studies on the Scale-Up of the Microwave-Assisted Nitridation and Sintering of Reaction-Bonded Silicon Nitride," pp. 71-76 in Materials Research Society Proceedings, Vol. 347.
16. D. R. Clarke and W. W. Ho, "Effect of Intergranular Phases on the High-Frequency Dielectric Losses of Silicon Nitride Ceramics," pp. 246-252, in Additives and Interfaces in Electronic Ceramics, Advances in Ceramics, Vol. 7, ed. M. F. Yan and A. H. Heuer, Amer. Ceram. Soc., Westerville, OH, 1983.
17. ORNL Ceramic Technology Project Semi-annual Progress Report, April - Sept. 1995, ORNL/TM (in publication).
18. ORNL Ceramic Technology Project Semi-annual Progress Report, Oct. - Mar. 1996, ORNL/TM (in progress).

ACKNOWLEDGMENT

Research sponsored by the U. S. Department of Energy, Assistant Secretary for Energy Efficiency and Renewable Energy, Office of Transportation Technologies, as part of the Propulsion System Materials Program, under contract DE-AC05-96OR 22464 with Lockheed Martin Energy Research Corporation.

MICROWAVE FIRING AT 915 MHz - EFFICIENCY AND IMPLICATIONS

N.G. EVANS, M.G. HAMLYN
School of Engineering, Staffordshire University,
Stafford, UK, ST18 0AD

ABSTRACT

The conversion efficiency of electrical to microwave power by magnetrons operating at 896 / 915 MHz is significantly greater than that of magnetrons operating at 2.45 GHz. The increased efficiency at this frequency improves the prospects of microwave firing being adopted as a commercially viable process but the change in frequency alters the microwave-material interactions that take place.

This paper discusses the influence and implications of using 896 / 915 MHz for firing ceramics and compares dielectric property data at both frequencies. The results suggest that using the lower frequency will result in more efficient heating.

INTRODUCTION

Previous research work has shown that, for small batches, microwave firing uses more energy per mass of product than conventional firing [1]. As larger batches are used, however, the efficiency approaches that of conventional firing (FIG. 1) [1]. The main contribution to the energy consumption is the conversion of electrical to microwave power [1]. Magnetrons operating at 2.45 GHz are typically 50 - 55% efficient. When control systems are used to vary the output power the overall efficiency falls to 40 - 50%. Magnetrons working at the lower frequency of 896 / 915 MHz typically have conversion efficiencies of the order of 80 - 90% [2], with control the overall efficiency is about 80%. Thus, switching to the lower frequency has the potential to raise the overall efficiency by 60 - 100%.

The equations describing the interactions of microwaves with a dielectric material are well described [3]. Two important parameters are the reflection coefficient, which depends on the permittivity, and the penetration depth, which depends also on the frequency of the electromagnetic wave. The Microwave Processing Group at Staffordshire University is currently carrying out development aimed at assessing the commercial viability of microwave firing.

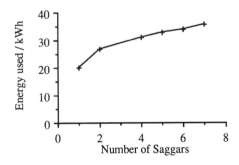

Figure 1. Graph of Energy Consumption for Increasing Numbers of Saggars (Refractory Containers) being Fired.

Mat. Res. Soc. Symp. Proc. Vol. 430 © 1996 Materials Research Society

In support of this work a dielectric property measurement apparatus is used to measure the permittivity and loss factor of products, kilnware and insulation materials. Measurements have routinely been performed up to 1400°C at 2.45 GHz for several years but the apparatus has been expanded to allow measurements at six discrete frequencies from about 400 MHz to nearly 3 GHz. Samples can be measured as powders, unfired cylinders or core drilled dense material. The expanded measurement apparatus is being used to assess materials and kilnware for use at 896 / 915 MHz.

EXPERIMENTAL

Samples of a high alumina (99% alumina), a debased alumina (86% alumina) and a vitreous pottery body were core drilled from fired samples. The dielectric properties of each sample were determined by the cavity perturbation technique [4, 5] (FIGS. 2 - 7). The dielectric properties of the high alumina have been reported previously at 2.45 GHz as part of a comparison between various research establishments [6].

The measurements at the two frequencies were carried out during the same heating run. Due to the nature of the technique the sample cools between the two measurements and the actual sample temperature is calculated by a heat transfer equation in the software.

The measured dielectric properties were inserted in a spreadsheet so that the reflection coefficient and the penetration depth at each frequency could be calculated for each product over the temperature range of the measurement. The results are shown graphically below (FIGS. 8 - 13). The resonant frequency of the cavity is about 911 MHz so the dielectric property data is quoted at this frequency. The penetration depth was calculated using 915 MHz as the microwave frequency which will be used during heating, using the dielectric property data at 911 MHz since it is not expected that there will be any significant alteration between the two frequencies. The calculated values are quoted, therefore, at 915 MHz.

RESULTS

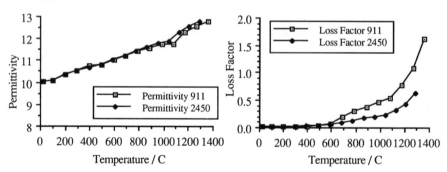

Figure 2. Variation of Permittivity with Temperature for a high purity alumina at 911 MHz and 2.45 GHz

Figure 3. Variation of Loss Factor with Temperature for a high purity alumina at 911 MHz and 2.45 GHz

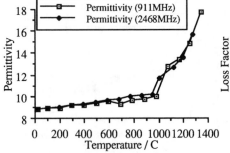

Figure 4. Variation of Permittivity with Temperature for a debased alumina at 911 MHz and 2.45 GHz

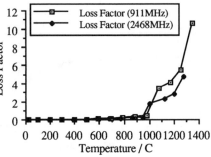

Figure 5. Variation of Loss Factor with Temperature for a debased alumina at 911 MHz and 2.45 GHz

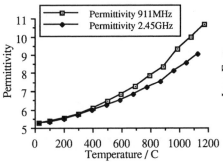

Figure 6. Variation of Permittivity with Temperature for a vitreous pottery body at 911 MHz and 2.45 GHz

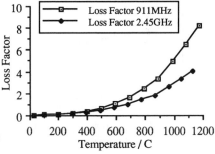

Figure 7. Variation of Loss Factor with Temperature for a vitreous pottery body at 911 MHz and 2.45 GHz

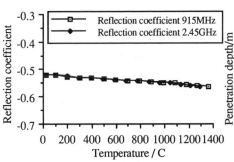

Figure 8. Variation of Reflection coefficient with Temperature for a high purity alumina at 915 MHz and 2.45 GHz

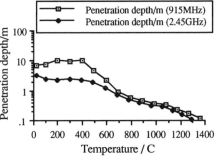

Figure 9. Variation of Penetration Depth with Temperature for a high purity alumina at 915 MHz and 2.45 GHz

11

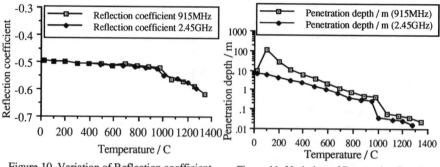

Figure 10. Variation of Reflection coefficient with Temperature for a debased alumina at 915 MHz and 2.45 GHz

Figure 11. Variation of Penetration Depth with Temperature for a debased alumina at 915 MHz and 2.45 GHz

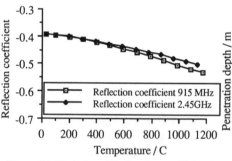

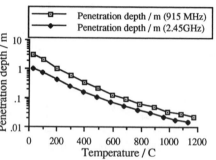

Figure 12. Variation of Reflection coefficient with Temperature for a vitreous pottery body at 915 MHz and 2.45 GHz

Figure 13. Variation of Penetration Depth with Temperature for a vitreous pottery body at 915 MHz and 2.45 GHz

DISCUSSION

Calculations based on the measured dielectric properties show that the penetration depth is greater at the lower frequency by a factor of about two. The reflection coefficients show little change with frequency. Importantly, the shape of the dielectric property curves as a function of temperature are basically the same, indicating that similar heating behaviour can be expected.

The results show that microwave firing at 896 / 915 MHz should produce similar results to 2.45 GHz. The two fold increase in penetration depth will be beneficial for the uniformity of individual products but will also allow an eight fold increase in the volume of product which can be fired without the same attenuation problems that may occur with 2.45 GHz. We are currently building a large scale-up furnace with a 75 kW, 896 / 915 MHz supply.

Although microwave-only firings have been successfully carried out at 2.45 GHz, up to the limits of the 17.6 kW currently available, temperature gradient control has been achieved by insulating the products in the cavity [1]. On a commercial scale surrounding individual pieces or groups of pieces with insulation becomes too costly in terms of material costs, kiln loading time and energy efficiency. For these reasons, therefore, a secondary heating system will be incorporated in the kiln.

Because 896 / 915 MHz is already commercially used in a number of drying and curing applications high power microwave generators are available at reasonable cost. The use of a correctly designed cavity with a suitable secondary heating system should provide a similar degree of uniformity to that achieved already.

The increase in loss factor at the lower frequency only partly compensates for the lowering of the frequency so the electric field strength will be increased to maintain the power absorbed per unit volume.

CONCLUSION

Microwave firing, which has been shown to be feasible at 2.45 GHz, should not only be feasible at 896 / 915 MHz but considerably more efficient. Only experimental trials will determine if the process will be successful but initial indications are positive and dielectric property measurements show that materials should behave similarly at the two frequencies.

ACKNOWLEDGEMENTS

The authors gratefully acknowledge the financial support of Midlands Electricity plc, UK, for their help with the work

REFERENCES

1. M.J. Bratt, Process Technology for Firing Ceramic Products with Microwave Energy, PhD thesis, April 1995, Staffordshire University.

2. Dielectric Heating for Industrial Processes, p24. Dielectric Heating Working Group, Chairman P. Hulls, Secretary De Hoe J-M. Published by U.I.E. 1992

3. A.C. Metaxas, R. Meredith, Industrial Microwave Heating. IEE Power Engineering Series 4. Peter Peregrinus Ltd 1988.

4. R. Hutcheon, M. de Jong, F.. Adams, J. Microwave Power and Electromagnetic Energy vol 27(2) pp 93 - 102, 1992.

5. N.G. Evans, The Dielectric Properties of Ceramics at 2.45 GHz and their Influence on Microwave Firing, PhD thesis, April 1995, Staffordshire University.

6. M. Arai, J.G.P. Binner, A.L. Bowden, T.E. Cross, N.G. Evans, M.G. Hamlyn, R. Hutcheon, G. Morin, B. Smith, Ceramic Transactions Volume 36. Proc of the Symp 'Microwaves: Theory and Application in Materials Processing II' 1993 pp539 - 546.

MICROWAVE SINTERING OF LARGE PRODUCTS

TASABURO SAJI
Fujidempa Kogyo Co., LTD. Tsukuba Laboratory
Ibaraki Japan

ABSTRACT

A practical microwave ceramic sintering technology has been developed. To achieve the highest merit from microwave sintering, susceptors were not used. In this paper, the ceramic sintering process using the 28GHz mm-wave energy generated by a Gyrotron oscillator is described. Large sized (ϕ 200 × 200) and complex shaped material of Si_3N_4, Al_2O_3, and ZrO_2 were successfully and reproducibly sintered. The sintered bodies exhibited high density and fine crystalline structure. The processed materials were found to have exceptional strength and toughness.

The critical processing technology developed dealt with the means by which uniform and defect free sintered bodies are obtained. The processing stressed the importance of achieving a uniform electric field within the applicator and to decrease the sintering dependence on the materials' temperature based dielectric loss properties. With the use of innovative support and thermal insulation structures, the processing system allowed for very precise temperature controls.

INTRODUCTION

Some merits [1] of microwave sintering of ceramics are rapid heating, selective heating and high densification achieved at relatively low temperature. However, the microwave sintering process has been inherently unusable due to difficulty in sufficiently heating the ceramic material which have low dielectric loss properties. In typical processing with microwave, thermal runaway and sample cracking are very common and difficult to control.

Because the problem of sintering ceramics with microwave is not readily solvable, industrial use of such processing has not been common. It is well known that much problems with the microwave sintering of ceramics have been experienced at 2.45GHz frequency. By using a shorter wavelength (28GHz mm-wave frequency), the problems encountered with 2.45GHz frequency have been effectively resolved. As well, with the use of higher frequency, here-to-fore not well understood ceramic sintering interaction characteristics with mm-wave frequency was more closely studied. The author was able to obtain consistently, defect free ceramic sintered bodies with the mm-wave processing.

EXPERIMENT

Si_3N_4 (Ube Grade E-10) with 5 wt% Al_2O_3 and 5 wt% Y_2O_3 were milled for 48 hours and casted in a plaster mold. The water content of the slurry was 37.5%. The size of sample was ϕ 250 × 250. Al_2O_3 (Iwatani Grade UA-5155) with 0.05 wt% MgO and 0.2wt% Cr_2O_3 were milled for 20 hours and formed by slip casting. The water content of the slurry was 29.2%. The size was same as Si_3N_4. ZrO_2 with 3wt% Y_2O_3 (Toso Grade TZ-3YS) were milled 40 hours and casted. The water content of the slurry was 20.5%. The sizes were ϕ 250 × 250 and ϕ 90 × 100.

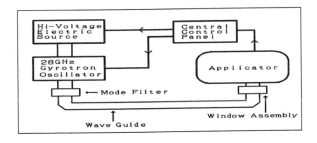

Fig. 1 A diagram of experimental equipment.

A high power Gyrotron tube is used to generate 28GHz millimeter-wave energy.

The output power is computer controlled 0.1 ~ 10kW continuously and oscillating mode is TE_{02}.

Mat. Res. Soc. Symp. Proc. Vol. 430 © 1996 Materials Research Society

A high power Gyrotron was used to generate 28GHz mm-wave energy. The output power was computer controlled from 0.1 to 10 kW and oscillating mode was TE_{02}. A wave guide system consisting of mode filter, two benders, a 2.5 inch wave guide, and output window assembly was used. The applicator was ϕ 650 × 1100 with water cooling jacket. In the applicator, the mode was converted from TE_{02} to multi-mode by a mode converter. A diagram of the experimental equipment is presented in Fig. 1

RESULT AND DISCUSSION

In case of sintering ϕ 250 × 250 Si_3N_4, the temperature rise rate was 30℃/min for obtaining defect free sintered bodies. The density of the sintered sample was found to be uniformly densified (98.5~99.2%) when sintered at 1650℃ for 30 minutes. Because of high sintering uniformity, the sintered sample exhibited no stress cracks.

The ϕ 250 × 250 Al_2O_3 greens were sintered at 1400℃ for 60 minutes in the temperature rise rate 15℃/min. The density of the sample was very uniform (99.3~99.8%)

The ϕ 250 × 250 ZrO_2 were sintered, but the sintering condition for obtaining crack free sintered bodies were not found out. The ϕ 90 × 100 ZrO_2 greens were tried to sinter, and succeeded in obtaining crack free bodies. The sintering condition was heating at 1400℃ for 60 minutes in average rate of temperature rise 2℃/min. The density was 99.5~99.9%.

Uniform density is the result of even temperature distribution within the sample body. This uniform temperature is the result of well controlled distribution of the electric field in the applicator. Millimeter wave frequency made it possible to generate the uniform electric field within the applicator.

Millimeter wave Sintering

It is not possible to sinter defect free practically sized and shaped ceramics with 2.45GHz multi-mode microwave. If the work piece is very small (ϕ 10), it is possible to sinter by single mode 2.45GHz microwave. By hybrid heating, it is also possible to sinter, but the benefits of microwave heating will be spoiled. Defect free sintered body can be obtained by millimeter wave heating without using any susceptors.

Uniformity of Electric Field

A relational parameter was developed to help define the appropriate size of applicator for specific frequency to be used for sintering.

Parameter $= (L/\lambda)^3$ where L is the length of applicator and λ is the wave length

When 28GHz, 0~15kW continuos wave was introduced within a stainless steel applicator (ϕ 250 × 300) containing ϕ 25 × 10 Alumina greens, the sample sintering result was not satisfactory due to local heating and thermal runaway. In this case, the value of $(L/\lambda)^3$ was 2.19×10^4. When the applicator size was increased to ϕ 600 × 900, an disk shaped Alumina of ϕ 300 × 10t was uniformly sintered. Here the $(L/\lambda)^3$ was 5.95×10^5. Another Alumina sample of ϕ 250 × 250 was uniformly sintered in ϕ 800 × 1200 size applicator. The $(L/\lambda)^3$ was 1.40×106. When ϕ 25 × 10t Alumina sample was sintered in ϕ 800 × 2000 size applicator with 2.45GHz CW microwave, the sample ruptured. Here the $(L/\lambda)^3$ was 4.41×10^4. From these results, it was found that following applicator length and wave length relationship hold true.

$(L/\lambda)^3 < 4.41 \times 10^4$, lack electric field uniformity for sintering
$(L/\lambda)^3 > 5.95 \times 10^5$, sufficient uniform electric field can be attained to sinter.

It can be postulated that for 2.45GHz microwave, applicator size of $L > 1.03 \times 10^4$ would be required to obtain uniform electric field within the applicator. Unfortunately, this size applicator is not practical due to large loss associated with the wall loss and low heating efficiency resulting from the applicator size.

The information presented in the remainder of this paper is based on the experimental results attained with the use of ϕ 650 × 1100 applicator. The distribution of electric field in the applicator was measured and is presented in Fig. 2.

Temperature dependency of dielectric loss factor

The temperature dependency of dielectric loss factor (ε'') is smaller when the frequency of microwave is higher. This is very important to prevent thermal runaway. It is necessary to satisfy following formula to prevent thermal runaway.

$$Pt_1 - Pt_2 < \{(t_1 - t_2)/l\} AR \qquad (1)$$

where Pt_1, Pt_2 are the absorption energy at the temperature t_1, t_2, $t_1 - t_2/l$ is the gradient of temperature, A is the section area of thermal conduction, R is the thermal conductivity.

In other words $Pt_1 - Pt_2$ is temperature dependency of dielectric loss factor. It is understood that greens, which have low thermal conductivity, should be heated under the condition of low temperature dependency of dielectric loss factor for obtaining uniform sintered bodies. Millimeter wave make the condition. This is remarkable benefit of millimeter wave for the sintering technology.

Efficiency of heating
Absorption energy P is as following

$$P = 2 \pi f \varepsilon_0 \varepsilon_r \tan \delta E^2 V \Theta \qquad (2)$$

where f is the frequency, ε_0 is the dielectric constant of vacuum, ε_r is the relative dielectric constant, $\tan \delta$ is the loss tangent, E is the strength of electric field, V is the volume of sample, Θ is the shape factor.
When E, V and Θ are constant, the absorption energy P depends on $f \times \varepsilon'' (\varepsilon'' = \varepsilon_0 \varepsilon_r \tan \delta)$ Comparing 28GHz wave to 2.45GHz wave, f is about 11 (28/2.45)times. Since ε'' of Alumina is about 5 times, it means that Alumina absorbs 28GHz wave energy 55 times compare to 2.45GHz wave.
When ω / Q is assumed to be the attenuation constant, wall loss can be expressed as

$$Q = 3V / 2A \delta \qquad (3)$$

where V is the volume of applicator, A is the surface area of applicator, δ is the skin depth, and the skin depth is as following

$$\delta = 1 / \sqrt{\pi f \mu \sigma} \qquad (4)$$

where f is the frequency, μ is the electric conductance, σ is the magnetic permeability.
When the wall material is the same, the wall loss is inverse proportion to the square root of frequency. Therefore, the wall loss of 28GHz wave $\sqrt{2.45} / \sqrt{28} = 1 / 3.38$ to 2.45GHz wave.
It is found that energy at 28GHz frequency work, on heating ceramics in much higher efficiency than 2.45GHz microwave, because of high absorption energy and low wall loss.

Nonthermal effect
While many investigators have commented about apparent non thermal effects of microwave energy, a clear understanding of the mechanisms causing the reported phenomenon has not been agreed upon. The observed reduced dencification temperature when ceramics are sintered with microwave energy is one of the probable non thermal effects reported by many investigators. [2,3] The author and others have observed that as the sintering energy's frequency is increased the densification temperature decreases. This reduction in densification temperature, one of the most important benefits of millimeter wave sintering, is not a major factor in the technology of sintering large products.

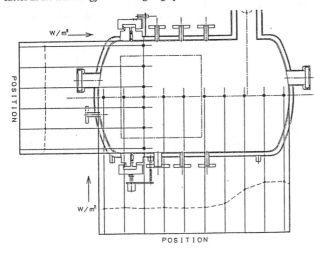

Fig. 2   The distribution of electric field in ϕ 650 $\times$ 1100 applicator.

The distribution is sufficient uniform electric field for attaining to sinter ceramics.

Quasi-optical behaviour

Because of its short wavelength, it is convenient to use quasi-optical techniques to distribute millimeter wave energy can be concentrated at the focal point of mirror or lens. The minimum diameter of the concentrated area is approximately twice the wavelength ; i.e., when 28GHz energy (wavelength = 10.7mm) was focused by mirror, over 80% of the energy was concentrated with in a 20mm diameter circle. There are many other ways to concentrate the energy. The author is attempting to apply quasi-optical technology to the sintering large, complex shaped ceramics.

It is also possible to make a gentle grade electric field. Furthermore, it is possible to adapt special sintering technology like free transfer of electric field distribution. Such technique as controlling direction of grain growth can be investigated.

Control of Heating

Studying the mechanism of microwave sintering is important for obtaining defect free sintered ceramics. Heating ceramics should be controlled suitably for the mechanism.

Si_3N_4

In case of Si_3N_4 sintering [4] , generally the dielectric loss factor of additives such as Al_2O_3, and Y_2O_3, are lower than Si_3N_4. But when they become oxinitride as $(Si,Al,Y)_2ON_2$, the loss factors increase and become much higher than Si_3N_4. Namely that, the additives are heated selectively from around 1350℃, which is the assumed temperature to start forming oxinitride, and the temperature of grain boundary rises selectively.

The temperature of Si_3N_4, which is the majority component, is the average temperature indicated on the thermometer. The temperature rises slowly but the temperature of grain boundary rises quickly, and liquid phase is formed in the boundary. Then the Si_3N_4 ceramics densify in low temperature with rearrangement of grains and transformation of α to β phase.

For confirming the above, an experiment was done. A green of Si_3N_4 with 5 wt% Al_2O_3 and 5 wt% Y_2O_3 was heated by fixed microwave power. As shown in Fig. 4, rising speed of the temperature was accelerated from around 1350℃. Dielectric loss factor of Si_3N_4 from 1200℃ to 1500℃ is monotonous increase. It does not increase suddenly, so the acceleration of the speed must be caused by another reason. The reason is the sudden change of loss factor of the additives, and the sudden change is caused by the formation of oxinitride.

Therefore, the temperature should be controlled accurately and carefully at the temperature around 1350℃.

Al_2O_3

In case of Al_2O_3, surface activation by microwave is the major factor of the mechanism. Electric field concentrate in pores, so that the surface of grains are exposed to high electric field, then apparent surface energy becomes high.

$$\varepsilon_p \ E_p = \varepsilon_g \ E_g \tag{5}$$

where ε_p is the dielectric constant of pore, E_p is the electric field of pore, ε_g is the dielectric constant of grain and E_g is the electric field of grain.

Generally, the dielectric constant of pure Al_2O_3 is about 8, and that of air is about 1. Then $\varepsilon_g / \varepsilon_p = 8$, so $E_p / E_g = 8$. When Al_2O_3 compacts are heated in atmosphere by microwave, electric field of pore is 8 times that of grains.

The electric field is the accumulating area of energy. With a given electric field

$$Wp = \varepsilon_p \ E_p^2 / 2 \quad [J/m^3] \tag{6}$$

where Wp is the accumulated energy in pores.

When Al_2O_3 compacts are heated in atmosphere by microwave, the accumulated energy in pores is 8 times that of grains. In the pores, high potential energy is accumulated, so pores are always forced to be smaller. When the electric field in pores is very high, sometimes plasma breaks out. Then the temperature around pores rises by microscopic local heating and the mobility of substance increases locally. The three elements which are high electric field, high potential energy and the plasma, join as a " new driving force " for densification.

Dielectric loss factor of Al_2O_3 increases suddenly from around 300℃, and the " new driving force " has an effect on the densification from around 950℃. Therefore, the temperature should be controlled carefully around 300℃ and 950℃,

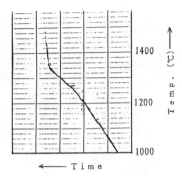

Fig. 3   The curve of temperature rise vs
time for Si_3N_4 heated by mm-wave..

Fig. 4   A schematic of surface activation.

ZrO_2

   With ZrO_2, the sintering mechanism is similar to that of Al_2O_3. Because the dielectric loss factor of ZrO_2 change suddenly, so it is difficult to obtain defect free sintered bodies. The control of heating was discussed with consideration of the uniformity of electric field, and the shape of the thermal insulating structure was decided. The technology is still being developed, it has been found that the maximum rate of temperature rise should be only 5℃/min. In some ranges of the temperature, the rate should be controlled accurately not more than 0.5℃/min

Simulation of Temperature Distribution

   When the distribution of electric field is perfectly uniform, a sample absorbs microwave energy uniformly. But the energy radiates from its surface. Therefore, the temperature of parts where ratio of surface to volume is large, become lower, and in case of small ratio, the temperature become higher. This temperature differential causes the dielectric loss factor become variable. The higher temperature part is heated easier because its dielectric loss factor is high. In opposite side, the lower temperature part is difficult to heat. Then the microwave energy concentrates in the higher temperature part. As a result, the work piece is cracked by thermal stress.

   Before the heating , an attempt was made to analyze the temperature distribution of the heating structure consisting of work piece and thermal insulating materials. And the suitable conditions for making uniform distribution were found. Even sintering simple shaped products, it was necessary to analyze the temperature distribution. The analysis was done by finite element method (FEM). Thermal conduction was calculated by following

$$c\,\rho\,\frac{\partial T}{\partial t} = \frac{\partial}{\partial x}\left(R\frac{\partial T}{\partial x}\right) + \frac{\partial}{\partial y}\left(R\frac{\partial T}{\partial y}\right) + \frac{\partial}{\partial z}\left(R\frac{\partial T}{\partial z}\right) \qquad (7)$$

where c is the specific heat, ρ is the density, R is the thermal conductivity, T is the temperature, t is the time and x, y, z are coordinates.

   R-T curves and c-T curves of all materials in the heating structure were found. For example, the R-T curve $(200 \sim 1300℃)$ of Si_3N_4 is indicated as following

$$R=0.125+1.512\times10^{-4}\,T+1.029\times10^{-7}\,T^2 \quad [W/cm\cdot℃] \qquad (8)$$

and c-T curve $(100 \sim 1200℃)$ is

$$c=2.291+3.419\times10^{-5}\,T+1.652\times10^{3}\,T^{-3} \quad [cal/g\cdot℃] \qquad (9)$$

The data were inputted to the vertexes of coordinate system of FEM.

19

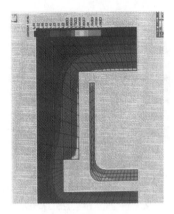

Fig. 5 An example of simulation of the temperature distribution of heating structure.
The work piece is ϕ 250×250 Si$_3$N$_4$ and the thermal insulating material is high purity Al$_2$O$_3$ fiber.

Fig. 6 Examples of mm-wave sintered bodies.
The largest one is ϕ 200 ×200 (sintered size) Si$_3$N$_4$, and others are pure Al$_2$O$_3$, Al$_2$O$_3$ with Cr$_2$O$_3$(Ruby) and ZrO$_2$.

The absorption energy P should be calculated by formula (1), but it is difficult to know electric field E in materials. So relative P was found and inputted as the heating energy to the vertexes. The relative P has temperature dependency. Therefore, P-T curve was investigated.

For an example, a simulation of the temperature distribution of heating structure is shown in Fig. 5. The work piece is ϕ 250×250 crucible shaped Si$_3$N$_4$, and is covered with thermal insulating materials made of high purity Al$_2$O$_3$ fibre. Examples of mm-wave sintered Si$_3$N$_4$, Al$_2$O$_3$ and ZrO$_2$ are shown in Fig. 6.

CONCLUSION

Microwave sintering of large size ceramics is possible, when following conditions are satisfied.
1. The electric field has sufficient uniform distribution.
2. Temperature dependency of the dielectric loss factor is low.
3. Control of heating is suitable for the mechanism of microwave sintering ceramics.
4. Heating structure consisting of the work piece and insulating materials is structured for making uniform temperature of the work piece.

1. and 2. can be actualized by using mm-wave.

REFERENCES

1. W.H.Sutton, " Microwave processing of Ceramics-an Overview " Materials Research Society Symposium   Proceedings Vol. 269, Microwave Processing of Materials (1992) 2-3

2. T.N.Tiegs, J.O.Kiggans and H.D.Kimrey, " Microwave Sintering of Silicon Nitride" Ceramic Engineering Society proceedings 12 (1991)

3. H.D.Kimrey, J.O.Kiggans, M.A.Janney, R.L.Beatty, " Microwave Sintering of Zirconia-Toughened   Alumina Composites " Materials Research Society Symposium proceedings Vol. 189   P 243-255 (1991)

4. T.Saji, " Mechanism of Microwave Sintering, and Millimeter Wave Sintering "   New Ceramics Vol. 8, No, 5   P 21-30 (1995)

INDUSTRIAL APPLICATIONS OF VARIABLE FREQUENCY MICROWAVE ENERGY IN MATERIALS PROCESSING

Z. Fathi, D. A. Tucker, W.A. Lewis and J. B. Wei

Lambda Technologies, Inc., 8600 Jersey Ct, Ste C, Raleigh, NC 27613

ABSTRACT

A review of some market-driven research, process applications and systems development is provided. The variable frequency microwave processing concepts are briefly described. Industrial processing using variable frequency microwave energy in the areas of polymerization, composite processing, bonding and plasma is discussed. Analytical applications inherent in the use of variable frequency and its control are demonstrated in the areas of materials signature analysis for volumetric cure monitoring.

INTRODUCTION

Microwave energy interacts with materials at the molecular level. The ability of the microwave electric field to polarize molecules and the ability of these molecules to follow the rapid reversal of the electric field results in the conversion of electromagnetic energy into heat within the irradiated material. The microwave energy absorbed within materials depends, among other things, on the incident microwave frequency, the dielectric constant and the effective dielectric loss of the material and the distribution of electric fields within the material.

Microwave processing is generally performed within a metallic applicator. There are three categories of applicators: resonant applicators, traveling wave applicators and multimode applicators. The traveling wave applicator and the single mode applicator (resonant cavity) are successful in processing simple material geometries such as fibers. However, the multimode applicator has the capability of forcing electromagnetic energy onto large and complex shaped materials. Thus, the multimode applicator is compatible with the production of large and complex functional material components.

Multimode applicators can be powered with fixed and/or variable frequency sources. The fixed frequency microwave oven is similar to home kitchen models. When a fixed frequency microwave signal is launched within the microwave cavity (the metallic box), it suffers multiple reflections and results in the establishment of several modal patterns. The overall distribution of electromagnetic energy is not uniform throughout the microwave cavity resulting in high and low energy field areas, i.e., hot and cold spots. Heating using multimode applicators has been successful in various food and rubber related applications. The thermal gradients established due to non-uniform electromagnetic energy distribution can be adjusted for to some extent and are tolerated for food applications, but this is not the case for advanced materials.

Multimode microwave furnaces powered with variable frequency sources are better suited for advanced materials applications related to polymerization, composites processing, bonding and others. The electric field distribution, on a time averaged basis, is uniform throughout the entire cavity volume leading to uniform exposure of the processed material(s) to the microwave energy [1-3]. Furthermore, the incident microwave frequencies can be changed to optimize the microwave energy absorption by the material(s) of interest. Figure 1 provides a simplistic illustration of how a series of fixed frequencies can be synchronized and launched inside a

Mat. Res. Soc. Symp. Proc. Vol. 430 © 1996 Materials Research Society

multimode cavity to obtain a uniform electromagnetic energy distribution on a time averaged basis, leading to uniform processing of materials.

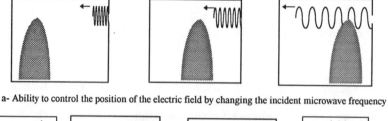

a- Ability to control the position of the electric field by changing the incident microwave frequency

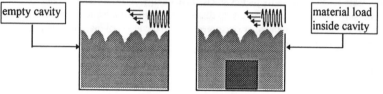

b- Uniform electromagnetic energy distribution through broadband frequency irradiation inside empty and/or loaded cavities ⇒ Uniform Processing

Figure 1. A simplistic illustration of the variable frequency microwave process in uniformly heating materials. The frequency bandwidth necessary for uniform heating is material dependent. Different frequency bandwidths are required for different materials.

The frequency agile sources include tunable magnetrons, tunable klystrons, and Traveling Wave Tubes (TWT's). Unlike magnetrons, the TWT is a broadband microwave amplifier. Typically, a microwave YIG oscillator generates a microwave signal (fixed and/or variable frequency) and the signal is then fed to the TWT for high power amplification [4]. The microwave energy is then fed to a multimode applicator for processing materials. This relatively new technology has been applied to the processing of various materials ranging from high temperature sintering of ceramics to low temperature curing of polymers [5-13].

The variable frequency microwave furnaces are computer controlled, taking into account several key process parameters such as temperature, operating frequencies, pressure, and depending on the application, vacuum. The closed loop system is adjusted to pulse and/or continuously feed microwave energy, increase or decrease incident microwave power, change incident frequencies, and control the sweep rate - the rate with which microwave frequencies are changed. Figure 2 illustrates the generic closed loop control and microwave generation used in variable frequency microwave processing. These controllable/adjustable processing parameters make the variable frequency microwave furnaces the systems of choice for many processing applications where production reliability and product consistency are of utmost importance. Furthermore, design changes for an existing product can be accommodated for by taking advantage of the flexibility of this technology; i.e., changing operating parameters including frequencies.

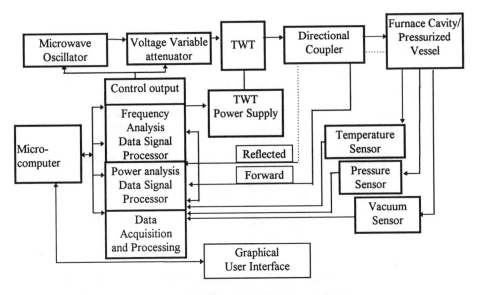

Figure 2. The variable frequency microwave control and DATA acquisition.

In addition to uniformity, the operating frequencies are typically selected to maximize the heating efficiency of the process. This is achieved by avoiding the frequencies at which the reflected power is high. A technique referred to as cavity/ material characterization [15] is typically utilized to obtain a broadband spectrum of the microwave absorption as a function of frequency. The frequencies that lead to strong coupling of energy inside the materials are therefore selected. Heating using variable frequency microwave energy can be achieved by frequency hopping among a series of fixed frequencies or among a series of broadband frequencies. The sweep rate, the rate with which frequencies are changed within each frequency band, is independently controlled, and can range from milliseconds to two minutes.

Lambda Technologies is working on a new computer control that will take into account the material changes in dielectric properties in that the frequency bands that lead to high absorption of energy shift during the processing will be accommodated for in real time using the closed loop control. Figure 3 is a schematic illustration of the intelligent process provided through the use of variable frequency microwave energy.

EXPERIMENTAL, PROCESSING AND SYSTEM DESIGN

There are several advantages resulting from the use of microwave energy. Some of these advantages include selective and volumetric heating, rapid curing, elimination of temperature excursions, and controllable energy deposition (through pulsing of the incident electromagnetic energy). These advantages make microwave processing an appealing processing method provided the process is scaleable to full production. The variable frequency microwave systems meet these scale-up requirements.

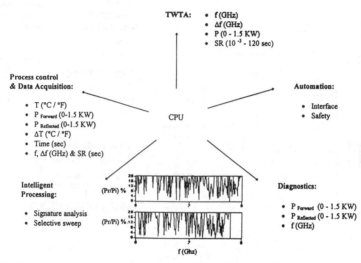

TWTA:
- f (GHz)
- Δf (GHz)
- P (0 - 1.5 KW)
- SR (10⁻³ - 120 sec)

Figure 3. Illustration of the computer control and closed loop feedback for intelligent processing of materials. The spectra displayed in the figure refer to the microwave signature collected during the cure of a polymeric material at two different stages of heating.

Polymerization

Variable frequency microwave energy has been applied to the curing of several generic systems: DGEBA/mpda, DGEBA/dds, ERL 2258/mpda. An illustration of the typical microwave and conventional thermal profiles required for full curing of ERL 2258 / mpda is presented in Figure 4. In addition to fast curing rates, the use of microwave energy leads to comparable if not greater glass transition temperatures of the processed samples. Microwave energy has been applied to process large samples and multiple small samples inside the cavity.

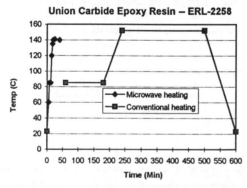

Figure 4. The thermal profiles used in microwave and conventional heating for curing large samples and/or multiple small samples with equivalent extent of cure.

Rapid processing rates are needed for industrial applications such as optical and contact lenses manufacturing. Quality cannot be sacrificed for an increased processing speed. Increased processing speeds of contact lenses with equivalent optical and mechanical properties have been obtained using variable frequency microwave energy. The processing speed has been reduced by 75%, while the tensile strength, modulus (Mdynes/cm^2), elongation (%),image, hydration and extractables (%) are comparable between the two processing methods. The microwave system design for the continuous / batch processing of contact lenses is described in Figure 5.

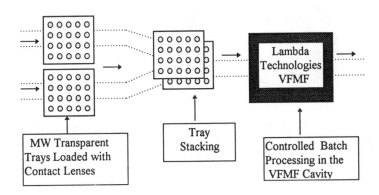

Figure 5. A schematic illustration of a continuous/batch microwave system for processing optical articles.

Bonding

Various bonding applications have been investigated using variable frequency microwave energy, some of which have been implemented on an industrial scale. These applications include bonding Polymer Matrix Composites (PMC's) to PMC's, PMC's to metals, polymers to metals, and semiconductors to PMC's as in Die-Attach applications.

In general, bonding is achieved with direct fusing of the components of interest or through the use of hot melts or structural adhesives. The variable frequency process has been applied to curing hot melts and structural adhesives. In curing structural adhesives, two approaches have been taken to assist the microwave process. In some cases, the chemistry of some adhesives has been tailored to couple to microwave energy, leading to selective absorption of microwave energy in the joint area. In other cases, a sacrificial conductive closed loop has been inserted at the joint area to assist the absorption of microwave energy and no tailoring of the adhesive's chemistry is needed[16].

Figure 6 provides a schematic illustration of a bonding process using a conductive closed loop insert that absorbs microwave energy and provides joule heating leading to direct fusing and/or curing of structural adhesives at the joint area of components to be bonded.

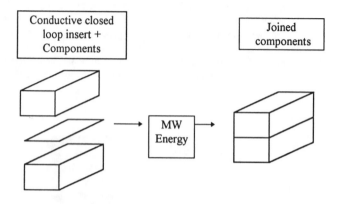

Figure 6. A schematic illustration of a bonding application using a sacrificial conductive closed loop insert for direct fusing applications and/or curing structural adhesives at the joint area.

In the area of surface mount technology, variable frequency microwave energy has been successfully applied as an alternative means by which die attach and glob top curing can be achieved rapidly without arcing and/or damaging electronic components included within circuit boards [17].

In addition to heating uniformity, variable frequency applicators eliminate the arcing problems experienced in microwave ovens when a metal or a semiconducting material is irradiated. Arcing is the result of excessive charge build-up in metallic materials in the presence of standing waves patterns. In the variable frequency microwave technique, the electric fields are electronically stirred and the microwave energy is not focused at any given location more than a fraction of a second. The dynamics of charge build-up that lead to arcing are never achieved, hence leading to no arcing and defect- free processing. Furthermore, the ability to select the incident frequencies to tune to a given material, such as the adhesive at the interface joint, makes this generation of microwave applicators suitable for surface mount technology.

A schematic diagram of a die-attach process is provided in Figure 7. The sweep rates used for surface mount in general, and die attach in particular, are in the range of milliseconds leading to defect free curing of adhesives and liquid encapsulants. The typical time reductions obtained in the variable frequency microwave process range from 75% to 90% as compared to conventional processes.

Other applications of variable frequency microwave energy are microwave-assisted chemical vapor deposition [12-14] and non-destructive evaluation [15]. The ability to raster plasmas by changing incident frequencies in multimode applicators has been demonstrated. The application of this technology in diamond coating of complex shapes is now being investigated.

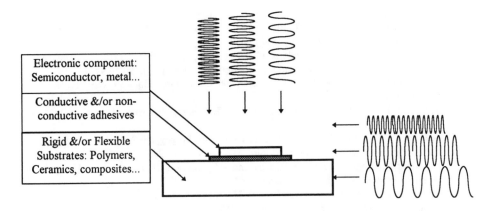

Electronic component: Semiconductor, metal...	
Conductive &/or non-conductive adhesives	
Rigid &/or Flexible Substrates: Polymers, Ceramics, composites...	

Figure 7. Arc-free bonding application in the electronic
industry using variable frequency microwave energy.

CONCLUSIONS

The frequency agile microwave applicators offer considerable advantages when compared
to fixed frequency applicators. Electronic tuning of frequency, frequency bandwidth and sweep
rates are integral features of variable frequency microwave systems. A closed loop computer
control system allows for intelligent processing. In addition to heating uniformity and frequency
tuning capability, the variable frequency microwave processing technology leads to successful
processing in areas otherwise unexplored in the microwave community: die attach and surface
mount applications in the electronic packaging industry.

A multidisciplinary approach combining materials science, chemistry, plasma physics,
microwave engineering and design, has been taken to assist end users with process and systems
design for industrial implementation of microwave related processes.

REFERENCES

1. A. C. Johnson, R.J. Lauf, A. D. Surret, in Microwave Processing of Materials IV, edited
   by M.F. Iskander, R.J. Lauf and W.H. Sutton (Materials Research Society Proceedings
   347, Pittsburgh, PA, 1994) pp. 453-458.
2. A.C. Johnson, R.J. Espinosa, W.A. Lewis, L.T. Thigpen, C.A. Everleigh, and R.S. Garard
   in Microwaves: Theory and Applications in Materials Processing II, edited by D.E. Clark,
   W.R. Tinga, and J.R. Laia (American Ceramic Society Transactions **36**, Westerville, OH,
   1993) pp. 563-570.
3. M.J. White, S.F. Dillon, M.F. Iskander, and S. Bringhurst "FDTD Simulation of Variable
   Frequencies in a large multimode cavity used for microwave sintering," elsewhere in these
   proceedings.
4. C.A. Everleigh, A.C. Johnson, R.J. Espinosa, and R.S. Garard in Microwave Processing
   of Materials IV, edited by M.F. Iskander, R.J. Lauf and W.H. Sutton (Materials Research
   Society Proceedings **347**, Pittsburgh, PA, 1994) pp. 79-89.

5. D.W. Bible, R.J. Lauf, and C.A. Everleigh in Microwave Processing of Materials III, edited by R.L. Beatty, W.H. Sutton, and M.F. Iskander (Materials Research Society Proceedings 269, Pittsburgh, PA, 1992) pp. 77-81.

6. R.J. Lauf, D.W. Bible, S.R. Maddox, C.A. Everleigh, R.J. Espinosa, and A.C. Johnson in Microwaves: Theory and Applications in Materials Processing II, edited by D.E. Clark, W.R. Tinga, and J.R. Laia (American Ceramic Society Transactions 36, Westerville, OH, 1993) pp. 571-579.

7. R.J. Lauf, D.W. Bible, A.C. Johnson, and C.A. Everleigh, Microwave Journal 36(11), 24 (1993).

8. R.J. Lauf, F.L. Paulauskas, and A.C. Johnson in 28th Microwave Symposium Proceedings (International Microwave Power Institute, Manassas, VA, 1993), pp. 150-155.

9. R.J. Lauf, A.D. Surrett, F.L. Paulauskas, and A.C. Johnson, "Polymer Curing Using Variable Frequency Microwave Processing," in Microwave Processing of Materials IV, edited by M.F. Iskander, R.J. Lauf and W.H. Sutton (Materials Research Society Proceedings 347, Pittsburgh, PA, 1994) pp. 453-458.

10. Z. Fathi, R.S. Garard, J.Clemons. C. Saltiel, R.M. Hutcheon, M.T. DeMeuse, in Microwaves Theory and Application in Materials Processing III, edited by D.E. Clark, D.C. Folz, S.J. Oda, and R. Silberglitt (American Ceramic Society Transactions 59, Westerville, OH, 1995) pp. 441-448.

11. R.S. Garard, Z. Fathi, and J.B. Wei, in Microwaves Theory and Application in Materials Processing III, edited by D.E. Clark, D.C. Folz, S.J. Oda, and R. Silberglitt (American Ceramic Society Transactions 59, Westerville, OH, 1995) pp. 117-124.

12. R.A. Rudder, R.C. Hendry, G.C. Hudson, R.J. Markunas, A.C. Johnson, L.T. Thigpen, R.S. Garard, and C.A. Everleigh in Microwaves: Theory and Application in Materials Processing II, edited by D.E. Clark, W.R. Tinga, and J.R. Laia (American Ceramic Society Transactions 36, Westerville, OH, 1993) pp. 377-384.

13. A.C. Johnson, R.A. Rudder, W.A. Lewis, and R.C. Hendry, "Use of Variable Frequency Microwave Energy as a Flexible Plasma Tool," in Microwave Processing of Materials IV, edited by M.F. Iskander, R.J. Lauf and W.H. Sutton (Materials Research Society Proceedings 347, Pittsburgh, PA, 1994) pp. 617-622.

14. D.A. Tucker, M.T. McClure, Z. Fathi, Z. Sitar, B. Walden, W.H. Sutton, W.A. Lewis, and J.B. Wei, "Microwave Plasma Assisted CVD of Diamond On Titanium and Ti-6Al-4V", elsewhere in these proceedings.

15. J.B. Wei, Z. Fathi, D.A. Tucker, M.L. Hampton, R.S. Garard, and R.J. Lauf, " Materials Characterization and Diagnosis using Variable Frequency Microwaves", elsewhere in these proceedings.

16. Patent Pending.

17. Patents Pending.

FDTD SIMULATION OF MICROWAVE SINTERING IN LARGE (500/4000 liter) MULTIMODE CAVITIES

MARTA SUBIRATS[1], MAGDY F. ISKANDER[1], MIKEL J WHITE[1], AND JIM KIGGANS[2]
[1]Electrical Engineering Department, University of Utah, Salt Lake City, UT 84112
[2]Oak Ridge National Laboratory, Oak Ridge, TN 37831-6087

ABSTRACT

To help develop large-scale microwave-sintering processes and to explore the feasibility of the commercial utilization of this technology, we used the recently developed multi-grid 3D Finite-Difference Time-Domain (FDTD) code and the 3D Finite-Difference Heat-Transfer (FDHT) code to determine the electromagnetic (EM) fields, the microwave power deposition, and temperature-distribution patterns in layers of samples processed in large-scale multimode microwave cavities.

This paper presents results obtained from the simulation of realistic sintering experiments carried out in both 500 and 4000 liter furnaces operating at 2.45 GHz. The ceramic ware being sintered is placed inside a cubical crucible box made of rectangular plates of various ceramic materials with various electrical and thermal properties. The crucible box can accommodate up to 5 layers of ceramic samples with 16 to 20 cup-like samples per layer. Simulation results provided guidelines regarding selection of crucible-box materials, crucible-box geometry, number of layers, shelf material between layers, and the fraction volume of the load vs. that of the furnace.

Results from the FDTD and FDHT simulations will be presented and various tradeoffs involved in designing an effective microwave-processing system will be compared graphically.

INTRODUCTION

Microwave processing of materials has been under laboratory investigation for several years. Some benefits in specific applications have been identified and have resulted in exploring the scale-up and commercialization of the microwave-sintering technology. With the significant interest in this technology, there has been a need to better model and simulate realistic sintering experiments to help guide the commercialization and scale-up efforts.

Several papers have already been published, and simulation results have been presented based on a uniform grid 3D FDTD code [1]. In order to help simulate electrically large sintering processes, the University of Utah developed a multi-grid 3D FDTD code [2]. The multi-grid FDTD code allows a smaller mesh size only in the region of interest (near the ceramic ware), and makes it possible to simulate large models with localized fine structures. The multi-grid FDTD code is used to calculate and predict the EM fields and the microwave power deposition patterns in the large 500 and 4000 liter sintering systems available at the Oak Ridge National Laboratory. The temperature distribution of microwave-sintering systems is calculated using a 3D FDHT code [3]. A unique feature of the developed code is the utilization of realistic microwave power deposition values from the FDTD simulations.

To improve the computational time and accuracy in the simulation of large-scale sintering experiments, the multi-grid FDTD and the FDHT codes have been integrated to provide realistic heating patterns during the microwave-sintering process. In this paper, the simulation models and the solution procedure will be briefly described, and both microwave absorption and heating patterns in large multimode cavities will be presented.

Mat. Res. Soc. Symp. Proc. Vol. 430 © 1996 Materials Research Society

MODEL DESCRIPTION

The multi-grid FDTD and the FDHT codes were used to simulate two large-scale microwave-sintering cavities. Figure 1 shows the schematic of the 500 liter, cylindrical, multimode cavity (73.66 cm in diameter and 111.76 cm long) equipped with a 6 kW, 2.45 GHz source. The 4000 liter furnace, on the other hand, is a rectangular, multimode cavity (111.76 cm in width and 279.4 cm long) with eight feed waveguides operating at 2.45 GHz.

The ceramic ware being sintered is placed inside a cubical crucible box in both cavities. The cubical crucible box is made of 18.3 cm x 18.3 cm identical rectangular plates, which are built from 0.25-1 cm thick material. The composition of the crucible plates used in these simulations include the following: 1) Hot-pressed boron nitride, 2) silicon nitride containing 40 wt. % silicon carbide, and 5 wt. % sintering aids, 3) silicon nitride containing 50 wt. % silicon carbide, and 5 wt. % sintering aids, and 4) silicon carbide. Crucibles formed from these plates are referred to as BN, UBE-40, UBE-50, and SiC, respectively. The crucible box can accommodate up to 5 layers with up to 20 samples per layer. The samples are cup-like structures of 4.08 cm diameter and 3.38 cm height. The ceramic cups have a nominal composition of silicon containing 3 wt. % Si_3N_4 and 5 wt. % each of Al_2O_3 and Y_2O_3 sintering material. The plates between the layers of ceramic ware are composed of silicon nitride (Si_3N_4). The samples and the crucible box are insulated by a 3.8 cm layer of insulation powder to help contain the heat. The crucible box and powder are held in a 2.5 cm alumina fiberboard box to further help simulate the heating process.

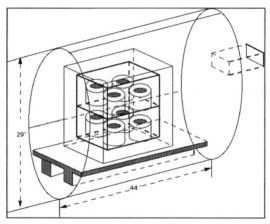

Figure 1 Schematic of the model used in the simulation of the 500 liter cylindrical cavity. Up to five layers of ware (cup-like structures) are placed inside the crucible box. Up to 20 (4x5 array) samples per layer have been simulated.

SOLUTION PROCEDURE

Computer modeling and simulation of microwave-sintering systems involves several steps. The first step is to calculate the EM fields and the microwave power deposition patterns using the multi-grid FDTD code. These values are then used by the FDHT program to calculate the temperature distribution and heating patterns in samples and the surrounding insulation. Finally, the dielectric properties of materials are updated as a function of temperature, and the calculation process is repeated.

To help in the ongoing numerical simulations and microwave-sintering research, high-temperature dielectric-properties measurements have been made in our laboratories [4] for various materials. Thermal properties were either obtained from Oak Ridge National Laboratory, or found in the available literature.

SIMULATION RESULTS

The following are samples of the obtained results from simulating the 500 liter, multimode, microwave cavity. Specific results will be presented to examine the effect of the crucible-box material, geometry of the crucible box, the possibility of using rings of highly conducting materials instead of the crucible box, and the effect of the material type to be used as shelves between the layers of the ceramic samples.

Crucible box

a) Composition of the crucible box

The microwave power deposition and the temperature-distribution patterns in samples were calculated for different suscepter crucible types made of silicon nitride-based materials, silicon carbide, and boron nitride materials. Results were used to analyze the optimal design of the crucible box.

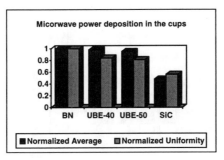

Figure 2 Average and uniformity of microwave power deposition per cell in the ceramic samples for different crucible materials (increasing the electrical conductivity of the crucible box from left to right). ε^* of UBE-40 and UBE-50 are 6.46-j0.064 and 8.79-j0.1198 respectively.

Figure 2 shows the average and the uniformity of microwave power deposition in the ceramic cups for different crucible-box materials. The average is calculated by adding the power-deposition value of each cell of ceramic sample, and dividing it by the total number of such cells. The uniformity is obtained by dividing the average by the maximum power-deposition value. The obtained results have been normalized by the maximum values. As it can be seen, crucible boxes made of SiC (ε^*=29-j5.88) resulted in the lowest power deposition and the worst uniformity in the ceramic ware. BN (ε^*=3.44-j0.0025) resulted in the most uniformity of microwave power deposition in the ceramic samples.

Graphical illustrations of temperature-distribution patterns were obtained after running the FDTD code for several cycles and the FDHT code for 30 minutes. Figure 3 and 4 show cross-sectional views of the temperature-distribution patterns in materials for BN and SiC crucible boxes, respectively.

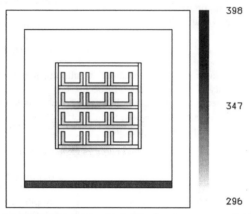

Figure 3 Cross-sectional view of the temperature-distribution pattern when the crucible box is made of BN, and using the microwave power deposition values provided by the FDTD code.

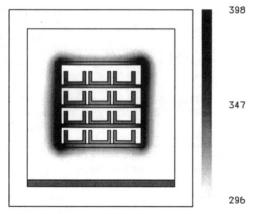

Figure 4 Cross-sectional view of the temperature-distribution pattern when the crucible box is made of SiC, and using the microwave power deposition values provided by the FDTD code.

The power-deposition and temperature-distribution patterns obtained from these simulations show that in the region close to the SiC walls, the temperature of the ceramics increases due to the quick heating of the crucible box. This phenomenon is known as hybrid heating and explains why crucibles containing SiC may present a non-uniform pattern due to the excessive heating in the crucible walls. Results also show that while the SiC crucible stimulates the heating process by absorbing significant amounts of the microwave power, it shields the samples from the microwave energy and results in considerable non-uniformities in the heating of the samples.

Several solutions are proposed to increase the uniformity and efficiency of the heating when SiC crucible boxes are used. The first one is to use materials with less weight percentage of SiC. Another possible suggestion is to reduce the thickness of the SiC crucible wall. The third recommendation is to use of SiC rings instead of solid crucible boxes. The use of highly conducting rings will be analyzed with more detail in the following sections.

b) Geometry of the crucible box

The shape of the crucible box is another important design issue. Cubical and cylindrical crucible boxes have been simulated and results of the heating patterns were compared.

Figure 5 shows the difference of average and uniformity of microwave power deposition in the samples for both cases. From figure 5 it may be seen that the cylindrical crucible box creates a higher average and uniformity of microwave power deposition in the ceramic samples. A tradeoff between difficulty of construction of the cylindrical crucible box and the amount of increase in the power deposition in the ceramic ware should hence be made.

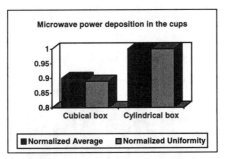

Figure 5 Average and uniformity of microwave power deposition per cell in the ceramic samples for a cubical and cylindrical SiC crucibles.

c) The use of highly conducting rings

As may be observed from previous results, solid crucible plates of high-loss materials such as SiC, stimulate the heating process and, at the same time, shield the samples from the microwave energy. To overcome this difficulty and in particular to facilitate an increased penetration of microwave energy in samples, it is suggested that rings of highly conducting materials be used instead of the solid crucible plates. To help guide the experimental development in this area, the number of rings, the separation between rings, and the diameter of rings have been analyzed in various groups of simulations to evaluate their effect on microwave power deposition.

Figure 6 shows a comparison of the obtained simulation results for a solid SiC crucible and various SiC ring arrangements. It may be seen from these results that rings resulted in increased microwave power deposition and more uniform deposition patterns than the solid crucible boxes. The number of rings and separation of rings were found to have less effect on the microwave power deposition until the density of rings approaches that of a solid crucible. Figure 7 shows the average and the uniformity of microwave power deposition in the samples as a function of the ring diameter (relative to wavelength). Maximum power deposition is reach when the ring diameter is 4 times the wavelength. The uniformity of heating, on the other hand, oscillates with the increase in the ring diameter.

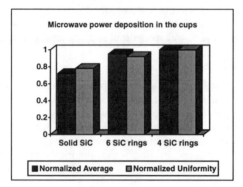

Figure 6 Average and uniformity of microwave power deposition per cell in the ceramic samples for solid SiC crucible box, 6 SiC rings, and 4 SiC rings.

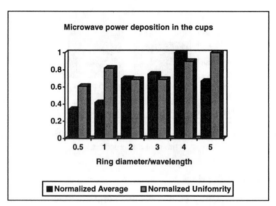

Figure 7 Average and uniformity of microwave power deposition per cell in the ceramic samples as a function of the ring diameter (increasing ring diameter/wavelength from left to right).

Ceramic load

a) Shelf material type

The composition of the plates between the layers of the ceramic ware critically influences the field distribution in the samples. Simulations were performed using SiC, alumina fiber, and Si_3N_4 shelves.

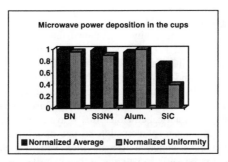

Figure 8 Average and uniformity of microwave power deposition per cell in the ceramic samples for different shelf materials.

Figure 8 shows the average and the uniformity of microwave power deposition in the ceramic ware when different shelf materials were used. Based on these results, it may be observed that SiC shelves considerably reduce both the amount of power penetration to the samples, and the uniformity.

b) Number of layers of ceramic ware

In a typical experimental arrangement, it was observed that different number of layers provides different field distribution inside the multimode cavity. Simulations were performed with 1, 2, 3, and 4 layers of ceramic ware.

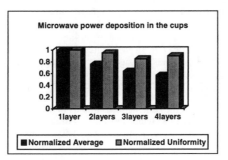

Figure 9 Average and uniformity of microwave power deposition per cell in the ceramic samples for 1, 2, 3, and 4 layers (increasing the number of layers from left to right).

Figure 9 shows that the average and the uniformity of microwave power deposition in the ceramic samples decrease as the number of layers increases. The overall efficiency of the heating process, however, increases with the increase in the number of layers.

CONCLUSIONS

The multi-grid 3D FDTD and the FDHT codes have been integrated to simulate microwave-sintering processes in large multimode cavities. Results concerning the simulation of realistic experiments carried out in the 500 liter cavity operating at 2.45 GHz are presented. Similar results are available for the 4000 liter cavity.

For the 500 liter multimode cavity it is shown that ceramics processed in the BN crucible have more uniform and higher microwave power deposition than when SiC crucibles are used. Crucibles containing SiC may present a non-uniform pattern due to the excessive heating in the crucible walls, and the microwave shielding effect of the SiC. A cylindrical crucible box creates more uniformity and higher microwave power deposition in the ceramic samples than a cubical crucible. An increase in uniformity of 20% is observed when 4 layers of 4x4 cups were heated using a cylindrical crucible. Microwave power deposition and uniformity may also be improved by using a non-solid crucible, created from highly electrically conducting rings, rather than a solid crucible. Increasing the diameter of the rings (relative to wavelength) also resulted in higher and more uniform patterns in the ceramic ware. A diameter of 4 times the wavelength is suggested. Shelves composed of SiC allow less electric-field penetration and decrease the uniformity of heating in the ceramic load. The increase of the number of layers results in lower fields and poor uniformity in each layer although the overall efficiency of the furnace increases.

These as well as other available results will certainly help in identifying some trends that are important in optimizing the microwave-sintering processes of advanced materials and ceramics.

ACKNOWLEDGMENTS

Research sponsored by the U.S. Department of energy, Assistant Secretary for Energy Efficiency and Renewable Energy, Office of Transportation Technologies, as part of the Ceramic technology Project of the Propulsion System Materials Program, under contract DE-AC05-84OR21400 with Martin Marietta Energy Systems, Inc.

REFERENCES

1. M. F. Iskander, A. Octavio, M. Andrae, H. D. Kimrey, and Lee M. Walsh, FDTD Simulation of Microwave Sintering of Ceramics in Multimode Cavities, IEEE Trans. on Microwave Theory and Techniques, Vol. 42, NO.5, MAY 1994.

2. M. J. White, M. F. Iskander, Z. Huang, and H. D. Kimrey, Development of a Multi-Grid FDTD Code for Three Dimensional Applications, IEEE Trans. Ant. and Prop., submitted for publication.

3. J. Tucker, M. F. Iskander, and Z. Huang, Calculation of Heating Patterns in Microwave Sintering Using a Finite-Difference Code, Mat. Res. Soc. Symp. Proc. Vol. 347, 1994.

4. S. Bringhust, M. F. Iskander, and P. Gartside, FDTD simulation of an open-ended metallized ceramic probe for broadband high-temperature dielectric-properties measurements, Mat. Res. Soc. Symp. Proc. Vol. 347, 1994.

APPLICATION OF THE 50 Ω RADIO FREQUENCY TECHNOLOGY IN THE AUTOMOTIVE INDUSTRY :

FAST BONDING OF COMPOSITE MATERIALS :

REAR DOORS OF THE CITROEN ZX AND CITROEN XANTIA CARS

J.P. BERNARD*, Mr. SABRAN**, L. COLLET***
*SAIREM, 24 rue Louis Saillant,, 69120 Vaulx en Velin , France
**MANDUCHER, rue Ampère, 01100 Oyonnax, France
***Electricité de France, DER/ADEI, BP1, 77250 Morêt sur Loing, France

ABSTRACT

In the field of plastic and composite materials the radio frequency dielectric heating is more and more used. Compared to traditional techniques such as conduction and convection heating, the radio frequency technology is interesting, because it allows fast heating of thick materials and heat insulation materials.
As bonding techniques are more and more integrated in production lines, the polymerization of glues must be realized in a very short time. The 50 use of the Ω radio frequency technology makes this heating process possible.
The authors describe the industrial application of this technology to the CITROEN ZX and CITROEN XANTIA cars. Steps involved in implementing this industrial process (laboratory - pilot -industrial equipment) are presented and analysis the technical and economic results of this application.

INTRODUCTION

The plastic materials are generally poor thermic conductors which makes their heating very difficult by conduction or convection. Dielectric heating (here at 27 MHz), by producing the heat directly into the material, offers in this case an efficient and often advantageous solution. CITROËN has chosen such a solution for the gluing of the two parts of the back door of its ZX and XANTIA models. Being more easily heated than the two parts of the back door (dielectric losses are greater), the glue joint alone will be heated (140 to 180 °C). Therefore, the processing time is shorter, and the power to supply is lower (10 to 12 kW).
The PVC seal gelling around a sunroof window pane yields to the same principles. The Radio Frequency heating selectivity allows the energy to be brought solely into the PVC. This process has numerous advantages. These two applications are completely automated and integrated in assembly lines.
The automotive industry uses more and more plastic materials in its products. The addition of reinforcing fibbers improves their mechanical characteristics. With this kind of plastic materials traditional assembling technics such as welding are replaced by bonding technics.
Composite materials are used for doors, engine hoods, rear doors, etc... These pieces are usually composed of 2 parts in polyester - glass fibber, such as BMC (bulk moulding compound).
The traditional bonding operation is realised with a metallic applicator with hot air or heated by a resistance. The operation takes 3 minutes : this is the time needed for rising the glue temperature. This temperature must be limited in order to prevent a deformation of the parts to be glued. Today production rates require bonding times below 1 minute.

Mat. Res. Soc. Symp. Proc. Vol. 430 © 1996 Materials Research Society

Consequently in 1988 PEUGEOT has turned to the radio frequency technology, in order to examine the possible acceleration of bonding for the production of the rear door for its new ZX model. The radio frequency technology makes possible the direct heating of glue through the 2 parts in composite materials.

PRINCIPLE OF RADIO FREQUENCY HEATING

In the field of radio frequency heating, the more commonly used frequencies for ISM applications (Industrial, Scientific and Medical) are 13,56 MHz and 27,12 MHz. In these frequencies waves are reflected by metals and can cross insulating materials (dielectric materials). Heating is then caused by the absorption of the energy, which is carried by the wave. Energy will then decrease progressively while transforming into heat inside the material.

PRINCIPLE OF THE 50 Ω RADIO FREQUENCY TECHNOLOGY

Only a perfect control of the energy absorbed by the glue joint makes it possible to fully appreciate the reaction kinetics of the glue and guarantees the bonding quality. The traditional radio frequency technologies do not meet these quality requirements. The 50 Ω techniques initiated by Electricité de France, developed and manufactured by SAIREM has successfully addressed this problem. Appropriate use of the technology allows a perfect matching of the emitter onto the load all along the operation.
The principle is a modular system of specialised and independent components as shown in figure 1.

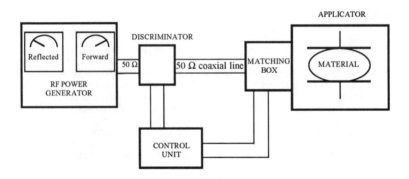

Fig.1 : R.F. 50 OHM TECHNOLOGY

MAIN STEPS TO THE MANUFACTURE OF THE 50 Ω RADIO FREQUENCY EQUIPMENT

Physico-Chemical Study of Glue Formulation

The problem was the following : how to heat 1 mm of glue between 2 thicknesses of composite materials ? The thickness of each composite part was 3 mm. This material was aimed at the rear door of the ZX car.
The purpose of the study was to determine a glue formulation. The glue had to be polymerizable by radio frequency and to meet the following requirements :
- polymerisation time below 30 s,
- mechanical quality in accordance with the standards imposed by PEUGEOT,
- easy deposition by a robot.

The glue is a 2-component polyurethane, which has been chosen for its good adherence on BMC without surface treatment.
This study has also helped to determine :
- the energitical density (in kWh kg^{-1}) required for the treatment
- the power density (in kW kg^{-1}) supported.

The study has been carried out with standard test pieces in an instrumented equipment. It has shown the absolute necessity to control energy all along the polymerisation cycle.

Technological Study of the Applicator

The aim of the applicator is to maintain in position the 2 parts to be glued, in order to avoid deformation and to confine the radio frequency field in the glue area. An experimental model (1/4 of the industrial size) has been realised.

Description of the 50 Ω Radio Frequency Industrial Installation at MANDUCHER for the Bonding of the Rear Door of the ZX Car.

The production line of the rear door is completely integrated and automated. The 50 Ω radio frequency equipment was installed in September 1989. From March 90 to June 91 a pilot production has been launched for improving the process. Since June 91 the machine produces more than 1 000 pieces per day without noticeable problem.
The 50 Ω radio frequency equipment is composed as follows :
- two 50 Ω radio frequency generators, adjustable output power : 20 kW, quartz driven, amplifier type, frequency : 27.12 MHz,
- 2 applicators,
- 2 automatic matching systems.

A set is treated in 80 s with a radio frequency heating time of 35 s. The working radio frequency power is 12 kW.

Economic comparison between the traditional process and the 50 Ω radio frequency process.

After more than 2 years of production MANDUCHER, which is the manufacturer of the rear door for the ZX car, has completely integrated the radio frequency process because of its economic advantages and reliability. A comparison provided by MANDUCHER between the traditional process and the 50 Ω radio frequency process is given in tables 1 and 2.
This table is made for a production of 1,000 pieces per day of identical rear doors in composite materials for 2 European cars, using 2 different gluing technologies : by hot air in a tunnel and by radio frequency. The economic advantages of the 50 Ω radio frequency technology clearly stand out as well for operating costs as for capital costs.

	HOT AIR PROCESS	RADIO FREQUENCY PROCESS
Number of generators	1 tunnel	2 RF generators
Installed power	360 kW gas + elec.	2 x 20 kW elect.
Consumption	8640 kWh	120 kWh
Safety/breakdowns	no	yes
Number of applicators	20	2
Cooling of applicators	yes	no
Handling	robots + conveyor	robots
Occupied place	5 times	1 time

OPERATING COST		
Energy	approx 1 000 FF/day	approx 60 FF/day
CAPITAL COST		
Civil engineering	> 1 MFF	
Oven/RF generator	2 MFF	2 MFF
Applicators	3 MFF	1,5 MFF
Total	> 6 MFF	3,5 MFF

TABLE 1 : TECHNICAL COMPARISONS BETWEEN HOT AIR AND RADIO
FREQUENCY

TABLE 2 : OPERATING COST

CONCLUSION

Today the use of the 50 Ω radio frequency technology for the structural bonding of composite materials is well accepted in the automotive industry due to its efficiency and reliability. The new equipments at INERGA in Spain for the rear door of the ZX car and at CITROEN in Rennes for the rear door of the XANTIA car are the concrete examples of this acceptance. Thanks to this technology the "on line" process is fully automated. This solution makes it possible to have a good production quality in real time. Other applications seem promising in this field using the bonding principle.

WINDOW ENCAPSULATION IN CAR INDUSTRY
BY USING THE 50 Ω RF TECHNOLOGY

J.P. BERNARD*, Mr. BARBOTEAU**, L. COLLET***

*SAIREM, 24 rue Louis Saillant, 69120 Vaulx-en-Velin, France
**WEBASTO HEULIEZ, 85700 Les Chatelliers Chateaumur, France
*** Electricité de France, DER/ADEI, BP1, 77250 Morêt sur Loing, France

ABSTRACT

Throughout the world car industry has been using window encapsulation for a few years now. This technology is mainly used in production lines and is called RIM for polyurethane reaction injection moulding.
This technology, however brings about some problems such as :
- glass breaking during mould closure,
- high production cost,
- systematic rough edges.

The PSA Group (Peugeot-Citroën), a pioneer in this field, in collaboration with SAIREM has launched a new innovating process for window encapsulation by using the 50 Ω RF technology for gelling PVC Plastisol.
The study was followed by an industrial prototype. Industrial equipment was then installed at WEBASTO HEULIEZ for window encapsulation of the sunshine roof for the Citroën Xantia.
The authors describe the principle of window encapsulation and the different existing processes. He describes the 50 Ω RF technology, an industrial installation and the constraints of this technology in order to get maximum efficiency.
In the conclusion they present a technical and economical analysis of the different solutions for window encapsulation. They also present the advantages of the 50 Ω RF technology and the new opportunities it offers.

PRINCIPLE

It consists in manufacturing a joint around a window placed in a mould. A coating material is then injected in the mould.

THE DIFFERENT PROCESSES

RIM

So far the most commonly used process is the RIM. It is a low pressure process that offers some advantages compared to a high pressure process. But it brings about some problems such as :

- polyurethane is becoming yellow due to the action of UV and requires a protection against them,
- glass breaking during mould closing,
- high production cost.

RF process

The principle consists in polymerising a single-component injected liquid by using the 50 Ω RF technology. In this case the liquid is PVC Plastisol. It is injected in a mould at ambient temperature. The mould is composed of 2 electrodes made of aluminium and of a membrane that determines the shape of the piece to be produced. After having filled up the mould, a RF voltage is applied to the clamps of the electrodes in order to rise temperature in the material to be polymerised. The following diagram details the different parts of the system :

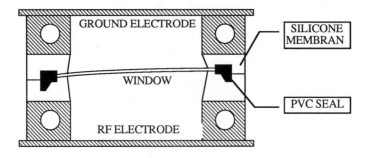

Fig.1 : Principle of Moulding Tool

On this diagram you can see that the membrane ensures to lightness of the system. This principle also enables to hold the glass without breaking it.

ADVANTAGES OF THE PVC PLASTISOL PROCESS COMPARED TO RUBBER AND RIM

- a single component instead of 2 for the RIM,
- reduced size of the press due to the injection of PVC Plastisol at low pressure,
- very short cycle,
- high energy saving,
- better quality of the final product,
- lower cost price compared to traditional technics.

CONCLUSION

The 50 Ω RF technology offers a real industrial interest for window encapsulation.
Its advantages are the following :
- homogeneous and quick heating inside the material,
- no thermic inertia,
- low cost price of the product,
- creation of various shapes.

FUTURE DEVELOPMENTS

The new process is very competitive compared to the other technics and offers a possible development of new markets by creating new materials.

MODELLING RF DIELECTRIC HEATING APPLICATIONS

In the field of dielectric heating the two important parameters are the distributions of electric field and temperatures. In order to succeed in building high-performance applications it is necessary to get a good understanding of electrical and thermic phenomenons. Such tools currently exist and must be used. The wellknown modelling software FLUX 2D using finite elements, developed and marketed by CEDRAT RECHERCHE FRANCE, in its latest module provides this ability.
EDF is carrying out studies to prove the improvements in the treatment quality and to extend the process to the manufacture of more complex pieces (more complex geometry, inclusion of metallic parts...)

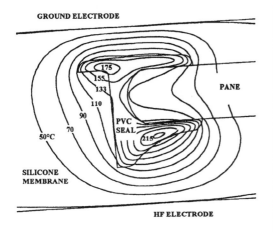

One study carried out by EDF in the process of moulding PVC seals around window panes, in the automotive industry :
here are the isotherms at the end of the Radio Frequency heating time (about 60 s).
FLUX 2D modelling software enables to improve the treatment's homogeneity.

Fig.2 : modelling of the temperature in the PVC seals moulding process

REFERENCES

J.J. OLIVIER : 'Process for Gelling Viny Resin by Means of Electromagnetic Waves', CFE Conference in Nice, 1991, Vol. 2, 11-14

C. MARCHAND and T. MEUNIER : 'Recent Developments In Industrial Radio Frequency Technology', Journal of Microwave Power, 1990, 25 (1) 39-46

B. MEYER, C. DEBARD, M. JACQUIN : 'Installation d'Expérimentation et de Caractérisation de Traitement Thermique par Haute Fréquence', CFE conference in Nice, 1991, Vol 2, 101-104.

J.F. ROCHAS : 'Intérêt de la haute fréquence 50 Ω dans les procédés de collage', CFE Conference in Nice, 1991, Vol 2, 7-10.

B. MEYER : 'Fast Bonding Technics Using 50 Ω Radio Frequency Technology to Assemble the Composite Parts of Automobiles', International Microwave Power Institute, 1992, Washington.

MICROWAVE SINTERING OF TUNGSTEN CARBIDE COBALT HARDMETALS

T. GERDES*, M. WILLERT-PORADA*, K. RÖDIGER**

*   University of Dortmund, Dept. Chem. Eng., Dortmund, FRG
**   WIDIA GmbH, Essen, FRG

ABSTRACT

The variety of possible microstructures obtained upon microwave sintering of hardmetals is described with regard to microwave specific heating mechanisms. While the reduction of processing time and temperature is evident, the influence of the microwave sintering process on the microstructure is not fully understood. From microwave sintering experiments at different power levels processes that depends on the local field strength are identified. The mechanical properties of microwave sintered tool bits are compared with commercial WC-Co hardmetals. An incising wear resistance of MW-sintered hardmetals for speed turning of cast iron is found.

INTRODUCTION

Among ceramic products made by powder metallurgical methods hardmetals are one of the most successful class of materials for nearly 70 years [1]. This long-term commercial success has been made possible by a continuous development of improved sintering technologies. Such sintering techniques like induction heating, hot isostatic pressing (HIP) and sinter-HIP were first implemented into commercial processes in the area of hardmetal tools. Significantly improved mechanical properties were achieved by using fine grained, highly sinteractive powders. However, such powders can not be densified by pressureless sintering without excessive grain growth. Both the improvement of the mechanical properties of hardmetal cutting tools by advanced powder development, e.g. nanosized powders, and the improvement of the HIP-technology does not resolve the technological disadvantage connected with high pressure sintering, e.g. expensive furnace technology, batch process, limitation of batch size and tool-geometry. Furthermore, the sintering mechanism of cemented carbides requires a certain sequence of dissolution and precipitation processes that do not take place at very low sintering temperatures and high pressure.

Therefore, research activity devoted to the development of a new sintering technology with the capability of pressureless fast sintering at moderate temperatures and because of the action of additional forces accelerating dissolution and liquid phase distribution in the porous compact started a few years ago. Despite the high electrical conductivity of hardmetal green parts, microwave sintering turned out to be a feasible method to heat and sinter these materials to full density, at temperatures that are approximately 100°C lower than in a conventional sintering process [2,3]. Due to the high electrical conductivity of cemented carbides heating by a microwave field can be used to achieve quite different microstructures of the sintered product by a particular choice of the electric field strength and distribution within a batch.

EXPERIMENTAL

Green parts were made from attritor milled graded commercial tungsten carbide powder with metallic binder content varying from 0-25 wt.% (mainly Co) by isostatic or axial pressing at 300 MPa. The compacts contained 2 wt.% of a paraffin binder.

Mat. Res. Soc. Symp. Proc. Vol. 430 © 1996 Materials Research Society

Microwave sintering experiments were performed in a 2.5 kW multi-mode cylindrical cavity (Figure 1) operating at 2.45 GHz frequency, at atmospheric pressure in Ar. The field strength within a batch is adjusted by movable walls, as shown in Figure 1. Thermal insulation is provided with fibrous alumina (84% Al_2O_3, 16% SiO_2) casket. Temperature measurements were taken by a thermocouple (20 - 600°C) and by an optical pyrometer (450 - 1600°C).

Sintered cutting tools were characterized by density-, Vickers-Hardness- (H_V 20) and magnetic measurements. Cutting performance for cast iron is characterized by wear-resistance tests on as sintered tool bits.

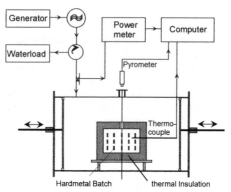

Figure 1.   Schematic of the variable size microwave applicator

RESULTS AND DISCUSSION

Numerous processing parameters control the densification and sintering behavior of hardmetals in a microwave field. Most important among these parameters are the local field strength and the homogeneity of the MW-field distribution.

Cutting tools made from WC-6Co as well as from WC-25Co can be completely densified by microwave sintering, as shown in Figure 2. However, the heating rate that mainly depends upon the local field strength turns out to govern the amount of open porosity.

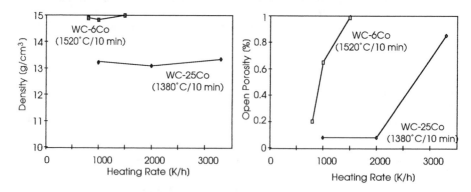

Figure 2.   Density and porosity of MW sintered hardmetals as a function of heating rate

As shown in Figure 3 a significant increase of open porosity is observed with an increasing heating rate. This porosity is located in the near surface area of the tool bits, indicating Co-evaporation due to local overheating.

Local overheating may result from limited penetration depth of the microwave radiation into a hardmetal green part. As discussed earlier [3], volume heating of WC-Co-green parts is only possible at high lubricant contents. With lubricant concentration of 2 wt.% a penetration depth for microwaves with 2.45 GHz frequency of < 1 mm has been measured [3]. Therefore, at a certain external microwave field strength an internally very high filed strength might develop in lubricant free pores within the penetrated depth of the sample, causing plasma ignition which leads to local overheating combined with Co-evaporation, as shown schematically in Figure 3. This process will also occur upon sintering with constant field strength, because the increased density of the hardmetal will further reduce the penetration depth. In this case a porous surface layer is observed in a fully densified hardmetal tool bit, as indicated in Figure 3.

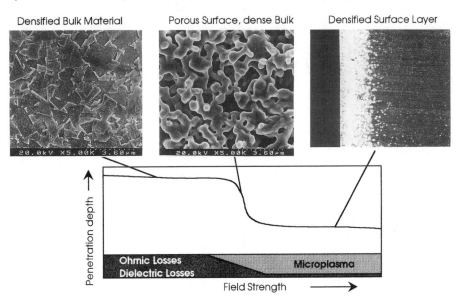

Figure 3.   Influence of the field strength on the dissipation process for microwave radiation and the resulting microstructure of hardmetals

If the microplasma extends over the whole surface layer of a hardmetal cutting tool, the penetration depth for microwave radiation will immediately been reduced. Consequently densification proceeds only within a 200 μm surface layer, with the interior of the cutting tool left porous, as shown in Figure 3.

Therefore, in order to fully densified hardmetals by microwave sintering the microwave field strength has to be adjusted to the changing penetration depth during a sintering cycle. With optimized processing parameters accelerated densification of hardmetals can be achieved by microwave sintering as compared to conventional sintering, as shown in Figure 4 for WC-6Co.

Further conclusions about the densification behavior upon microwave sintering can be drawn , when comparing the amount of closed and open porosity corresponding to a certain density. As shown in Figure 5 elimination of closed porosity below 1 % is already completed at 1150 °C upon microwave sintering as compared to > 5% closed porosity for hardmetals sintered conventionally at 1300°C for 10 min. A similar result is obtained for the open porosity. Microwave sintered hardmetals contain less than 1% open porosity after a 10 min. halt at 1150°C as compared to > 25% closed porosity for conventionally sintered hardmetals.

Particularly the significant reduction of closed as well as open porosity at rather a low sintering temperature indicates an influence of the microwave heating on the mechanism of liquid phase sintering. Usually it is assumed that the main part of the densification process during sintering of cemented carbides occurs after the liquid phase formation. However, the ternary W-C-Co eutectic is formed above 1280 °C. Therefore, for microwave sintering either a thermal non equilibrium has to be assumed, with local formation of the eutectic liquid, or an accelerated distribution of carbon and Co by evaporation and condensation throughout the sintering body due to ohmic and local plasma heating, followed by dissolution of C and W within the parent WC-crystals.

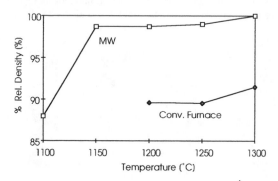

Figure 4.   Densification behavior of MW and conventional sintered WC-6Co Hardmetals

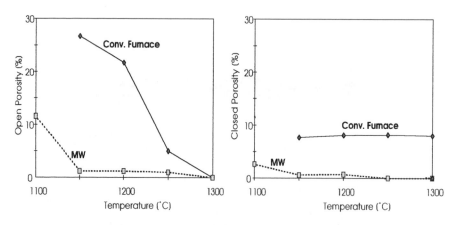

Figure 5.   Residual porosity of MW and conventional sintered WC-6Co at different sintering temperatures

The important role of Co for the difference in the efficiency of microwave sintering as compared to conventional sintering is evident from microwave sintering results on carbides with a decreased metallic binder content. As shown in Figure 6 complete densification is no longer possible with metallic binder contents below 1.0 wt.%. However, the amount of open porosity left after microwave sintering is low enough to enable sinter-HIP and achieve complete densification. Carbides with a very low metallic binder content are expected to show excellent corrosion resistance and increased hardness.

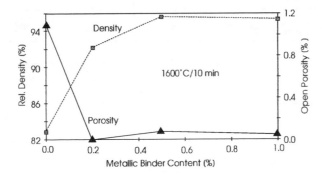

Figure 6.   Densification behavior of MW-sintered hardmetals with low metallic binder content

For commercial implementation of a microwave sintering process into the cutting tool production the reproducibility of mechanical properties will play a crucial role. As shown in Figure 7 excellent reproducibility of the magnetic properties is achieved upon microwave sintering of WC-6Co batches with 25 tool bits /batch. The magnetic properties indirectly characterize the hardness and grain size of the cemented carbide tools.

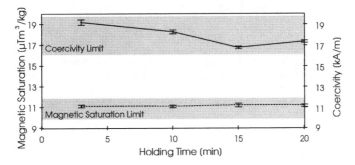

Figure 7.   Magnetic properties of MW-sintered WC-6Co cutting tools (25 tools each)

In comparison to the scattering values known for conventionally sintered hardmetals microwave sintering has the potential for a significant increase in reproducibility of mechanical properties for cutting tools at an improved hardness level, as shown in Table 1.

Furthermore, an increased flank wear resistance for speed turning of cast iron is found for microwave sintered WC-Co6 as shown in Figure 8.

	WC-Co6		WC-Co25	
	MW-sintering	conventional sintering	MW-sintering	conventional sintering
% theo. Density	> 99.9	> 99.9	100	> 99.9
Hardness HV30	1600 - 1700	1580	800 - 850	700

Table 1. Density and hardness of microwave sintered hardmetals as compared to commercial materials

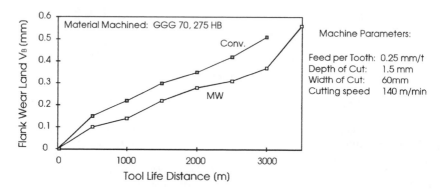

Figure 8. Turning of speroidal cast iron with MW- and commercial sintered WC-6Co Tools

CONCLUSION

Because of the relative high electrical conductivity of powder metallurgical products like hardmetals the penetration depth for microwave radiation of a WC-Co green parts is determined by the powder connectivity and the pore structure. At increasing field strengths plasma based loss mechanism can decrease the penetration depth, causing overheating of the surface layer of the sintered hardmetal.

Microwave field strength adjustment to the heating behavior of a WC-Co batch results in enhanced densification behavior in the early and middle sintering stage yielding hardmetal tool bits with excellent mechanical properties.

REFERENCES

1. **H. Kolaska**, Pulvermetallurgie der Hartmetalle, Fachverband Pulvermetallurgie, Hagen, 1992
2. **T. Gerdes, M. Willert-Porada**, Microwave Sintering of Metal-Ceramic and Ceramic-Ceramic Composites, MRS Symp. Proc. Vol. 347, p. 531-537, 1994
3. **T. Gerdes, M. Willert-Porada, H. Kolaska, K. Rödiger**, Guidelines for Large Scale MW-Processing of Hardmetals, Ceram. Trans. 59, p. 423-431, 1995

Part II

Microwave Nondestructive Testing

SOME ASPECTS RELATED TO THE TRANSFER
OF MICROWAVE SENSING TECHNOLOGY

J.Ch.BOLOMEY
Supélec, Plateau de Moulon, 91192 Gif-sur-Yvette, France, bolomey@supelec.fr

ABSTRACT

This paper reviews the major concerns related to the development of microwave sensing techniques. For their practical importance, testing and measurement of conveyed products via microwave multipoint sensors are more particularly considered. After reporting some promising sensing configurations and applications, technical and non technical aspects involved in the development of microwave sensors are analysed. Some of these aspects are encountered in any technology transfer process, while some others are more particularly specific to microwave sensors. In the conclusion, some general guidelines are provided for a successfull development of microwave sensing technology.

INTRODUCTION

The improvement of existing material processing techniques as well as the development of new materials are demanding for additional control and testing modalities. Beside well established modalities such as ultrasound, infrared, X or gamma rays, etc... microwaves have an increasing and specific role to play. Their recognized sensibility with respect to many physical and chemical factors of practical relevance, make that they have been considered for many years as a possible sensing agent for noninvasive testing and measurement of industrial materials [1]. Moisture content is probably one of the most significant areas for which microwaves constitute a priviligiated candidate [2]. But other factors such as temperature, chemical composition or structural changes, faults, etc... are also accessible via microwave sensing in most of the materials, except conductors. This paper adresses the development of microwave sensing modalities.

MICROWAVE SENSING INDICATIONS

From an economical point of view, the investment in new sensing technologies is only justified by the corresponding, direct or indirect, expected pay-back. Consequently, the most promising applications of microwave sensors are those for which they will efficiently contribute to reduce fabrication costs and/or to improve product quality. For instance, cost reduction can be achieved via a better control of the fabrication process resulting in energy saving or by means of minimizing wasted materials. Maintaining reliable quality is also another way for improving market acceptance and recognition. For its importance, the case of conveyed products or high rate batch processes will be more particularly considered. It corresponds, indeed, to a very broad area of applications, extending from mass production to high tech materials. In the first case, quantity effects are expected to be gained from an appropriate sensing device, while significant added value considerations may apply to the second case. Even if microwave sensing can be considered for any suitable material, whatever the processing technique, two special configurations are more particularly interesting, namely microwave processing and microwave products. In both configurations, the dielectric constant of the material to be processed plays a crucial role. In microwave processing, the dielectric constant has a direct impact on the processing efficiency. Indeed, the heat delivery in the material is directly dependent on the imaginary part of the dielectric constant. Often, the dielectric constant changes during the process, and testing the material dielectric constant during or after the process provides some indication on the actual or past process evolution. Thin film or glue polymerisation, rubber vulcanization, textile drying, food defrozing, etc... constitue some representative examples of

Mat. Res. Soc. Symp. Proc. Vol. 430 © 1996 Materials Research Society

such situations. For microwave products, i.e. products to be used in the microwave frequency range, the final properties to be obtained can be directly derived from microwave measurements on the dielectric constant or transmission/reflexion coefficients. Microwave products and materials have been initially almost exclusively developped for military applications. For instance, radar furtivity has stimulated the development of several kinds of microwave absorbers (coatings for aerospace industry, vehicles, clothes, camouflages, etc...) requiring low to extremely low reflectivity. Oppositely, electromagnetic windows for radomes require the maximum transparency in the operating frequency band. Actually, microwave products are increasingly extended to civilian applications due mainly to the explosive development of microwave cellular communications, wireless local area networks and, more generally, due to electromagnetic compatibility constraints. However, beside these major indications, microwave sensing techniques can also be considered for non microwave processes and/or non microwave products. Some examples are provided by wood, pulp and paper industries, plastics, food industry, etc...

MICROWAVE SENSING TECHNOLOGY

Until now, most of the microwave sensors have been single port sensors, only providing spatially averaged testing or measurement capabilities [1]. Since a few years, multiport sensors have been reported to avoid any mechanical movement and to result in improved imaging potentials in terms of rapidity and spatial resolution. Both aspects are of prime importance for wide laminated products and/or for rapid processing techniques. The possibility of determining real time transverse profiles is essential for achieving an efficient control of the fabrication process or for on-line testing of the products. For such objectives, microwaves undoubtly offer specific advantages. As a matter of fact two major kinds of applications can be identified, differing from the spatial resolution point of view. High spatial resolution requirements concern the detection of faults or structural defects from their corresponding effect on the local dielectric constant. Detecting small defects or determining accurately the boundary between two parts of a material needs high spatial resolution. In the microwave range, high spatial resolution means the millimeter/centimeter range. Such resolutions can be usefull for an optimized cut of the product aiming to minimize the lost material. An example is provided by the optimal sawing of wooden boards for obtaining knotless boards with standard dimensions. Such an optimized cutting process can also be very profitable for very high added value technical materials. In these applications, the discrimination capability is more important than the accuracy of the measurement. On the contrary, when the objective of the microwave sensing is to control the fabrication/transformation process, the required spatial resolution is directly dictated by the the fineness with which this process can be effectively controlled. For instance, during a drying process, the required spatial resolution is fixed by the dimensions of the fine heating modules designed for adjusting locally the transverse moisture content profile. Typical dimensions approximately extend from 10 cm to 50 cm and then result in moderate spatial resolution. However, the required accuracy on the spatially averaged measurement is usually high, about 1 % or less.

Two rather distinct technologies are available to fit the high resolution/qualitative and mid resolution/quantitative requirements. They just reflect the two major approaches to the microwave dielectric characterization of materials. As well known, they consist either in cavity or free-space measurements. In cavity measurements, the material is considered as a part of a resonator and the measurement of the resonance frequency and of the Q factor allows to determine the real and imaginary parts of the dielectric permittivity, via perturbation formulas [1,3]. On the contrary, the material can be coupled more loosely to a microwave beam for which the reflexion and/or the transmission coefficients are measured [1,4,5]. These coefficient can be correlated to the material quantity of practical relevance, with or without explicit reference to the complex permittivity. From a technical point of view, these two approaches are quite different. Cavity techniques imply phaseless but frequency swept measurements. On the contrary, free-space measurements do not necessarily require multifrequency but, usually, phase measurements, for instance, for wavefront processing or calibration purposes. Globally and despite possible exceptions, it seems that cavity and free-space measurements are better suited for high resolution/qualitative and mid reslution/quantitative applications, respectively. Time resolution is in favour of single frequency operation, on the contrary to cavity methods which need frequency sweeping. This general trend may have to be balanced because, when using high quality factors, cavity methods provide better sensitivity than free space methods. The lack of sensitivity for free space techniques can be compensated via time integration, and hence loss in rapidity. Coupling between adjacent elements of sensor arrays constitutes a limiting factor for spatial resolution. It seems that cavity elements are generally larger than antennas (e.g. slots, printed circuit patches, etc...) used for free space measurements. Moreover, for these antennas, coupling effects can be more easily compensated by introducing a linear coupling matrix.

The challenge in developping a microwave sensor is to obtain, simultaneously, a convenient sensitivity with respect to the quantity (ies) of practical relevance while maintaining sufficient unsensitivity with respect to other quantities or environmental constraints encountered in the operational situation (dust, vibrations, high temperature, etc...). A determinant parameter in optimizing these requirements is the operating frequency, or the operating frequency band [6]. Roughly speaking, increasing the frequency increases the sensitivity with respect to both the desired and parasitic quantities. In addition, increasing the frequency is usually beneficial from the spatial resolution point of view, but reduces the penetration in lossy materials. Consequently, the highest operating frequency will be often determined by taking into account the maximum tolerable noise floor (thermal noise plus parasitic signals) for a required investigation depth in a given material. Incidentally, the frequency has also a direct impact on the dimensions of the sensor and on its cost. Decreasing the frequency lowers quite rapidly the fabrication cost while the sensor dimensions are becoming larger and can pose mechanical compatibility difficulties for online integration.

DEVELOPMENT SCHEMES AND ASSOCIATED DIFFICULTIES

The successfull development of a microwave sensor is strongly influenced by several factors. The most important factors are now analysed and discussed. Some of them are common to any technology transfer process, and are shared by any sensing modality. Some others are more specifically related to microwaves.

Multidisciplinary aspects

Clearly, the developement of a microwave sensor requires more than just microwave know how. This is particularly evident for non microwave products for which the quantity of practical relevance is obtained via its influence on the dielectric constant. The major problem is to relate this quantity to the measured data provided by the microwave sensor (dielectric constant, reflexion/transmission coefficient, ...). The difficulty mainly results from the already mentioned fact that the measurement will be also influenced by other "parasitic parameters", the dependence of which must be removed or at least reduced at an acceptable level. This task has to be performed by an adequate model which incorporates more or less additional data stemming from other sensors or from the available a priori information on the product. The model complexity results, partly, on the material complexity itself and on its dielectric characteristics which govern the interaction scattering mechanism. The so-called inverse scattering problem consists of retrieving the material properties from its measured scattered field [7]. This inverse problem is greatly stabilized by any a priori available information. Building the model is a difficult task; validating it is really a need before online integration of the sensor. To summarize, the development of an efficient sensor requires the fusion of microwave and material know-how's. An efficient sensor is a dedicated sensor which has been designed and optimized for given products in a given environment. Evidently, the situation is much more simple in the case of microwave products, just because the measurement performed by the microwave sensor is very directly related to the desired performance of the material under inspection.

The length of the development process

As a direct consequence of the dedicated aspect of microwave sensors, the extension of the existing microwave technology to another application requires specific development. The complete development scheme extends from preliminary feasibility studies to industrialization and integration on the production unit. At least four steps are typically involved. Firstly, basic feasibility must be demonstrated from the material properties. These properties have generally to be obtained via dielectric characterization on small calibrated samples. The dielectric characterization allows to conduct preliminary numerical modelling and experiments for estimating the penetration depth in the material, the spatial resolution as well as the sensitivity to the quantity to be measured or the defect to be detected. All these aspects constitute key factors in the selection of the operating frequency. Dynamic testing conducted by means of existing equipment can, when possible, provide a valuable assistance in optimizing the overall sensing configuration. This first step is expected to result in the specifications of a dedicated sensor, taking into account the available technology as well as the environmental constraints. Electromagnetic and mechanical compatibilities constitute two aspects to be carefully addressed. The second step aims to fabricate a sensor prototype satifying all the previous requirements. Indeed, this prototype will be engaged in in situ assessments, preferably on a pilot unit on which different parameters of the process can be quantitatively changed. Such a step is essential to 1) more completely analyse the sensor behaviour under dynamic conditions, 2) to calibrate it, by using all the available information, 3) to realistically estimate the global accuracy and the performances of the microwave sensing approach and 4) to assess the economical aspects. Fusion with other sensor modalities can be also considered. This second step usually results in a revision of the initial prototype. In case of positive results, the third step will be initiated aiming to fabricate an advanced prototype for experimental integration on the production line. This advanced prototype will be modified and improved according to the experience gained. Then the final fourth step will start with the industrialization of the sensor.

Non stabilized requirements

One difficulty encountered in the development of microwave sensors has been the lack of fixed requirements. Some misunderstanding of the true problem may result in oversized expectations on microwave potentials, especially if existing techniques have failed and if, consequently, microwaves are considered as the ultimate miraculous means to solve the problem. Microwave images can either provide a partial coverage of the requirements or show more, and in some cases much more, than it was expected. For instance, all the defects to be detected may not be detected: in this case, the question is to determine what are the global defect detection probability and resulting ambiguities. Another case is that both undesirable and tolerable defects are visible: the question is then to discriminate between them. Note that the distinction between acceptable and not acceptable defects may be subject to some more or less strict criteria: a kind of defect, resulting from a given fault in the processing evolution, may be acceptable below a certain degree of importance (spatial extent, shape, intensity, ...) and unacceptable above this level. A determination of this level is sometimes very critical. The result is that during the development of the sensor, the requirements may be subject to some evolution which makes not so simple the question to know if the sensor fits or not the requirements. The solution is to fabricate as soon as possible calibrated samples representative of the configurations of typical faults/changes to be detected or measured. Closely related to this problem, it is important to estimate the "realty" of the need, whatever this need is already existing or has to be created.

Integration difficulties

The integration of a sensor on a production line has to be carefully prepared. More particularly, electromagnetic and mechanical compatibilities must be accounted for. In the contrary case, the installation of the sensor could have disastrous consequences on the production process, via unproper control resulting in an aggravation of the initial dysfunctionning to be corrected or via sorting errors of the finished products. At high conveying speeds, the least problem can generate drastic losses in materials and disorganization of the fabrication plans. For instance, a convenient adjustment of critical thresholds must prevent against false alarms while maintaining high probability detection. Non expected effects from preliminary assessments can appear and perturb the identification / measurement schemes. Beside such direct effects, this could also seriously affect the acceptance of the new sensing equipment. A convenient training is a necessary way to facilitate the acceptance, especially at the beginning of the installation.

<u>The role of existing competing technologies</u>

Of course, the introduction of microwave sensors in production lines must account for existing sensing technologies. Very often, a user thinks to microwave just because all the available techniques do not solve his sensing problem. Indeed, why trying to use microwaves if the existing techniques already provide a convenient solution ? Only clear and definite specific advantages (cost/performance ratio) can overcome such a positioning handicape. Consequently, either microwaves have to be introduced in a priori difficult operational situations, or they have to be much better than other techniques to replace them. This is the price to pay which results from the recent development of microwave sensors. The recognized specific advantages of microwaves are 1) non-contact penetration, 2) centimeter or less spatial resolution over large dimensions, 3) high sensitivity, especially versus water content, 4) no harmfulness versus ionizing and thermal effects, while disadvantages are mainly image calibration and costs, from technical and financial points of view, respectively.

The situation of microwave testing techniques can be, to some extent, compared to the situation of microwave processing techniques. As now well known, microwave processing techniques have required some time for being recognized as uncontournable for some applications concerning thermal insulators (e.g. vulcanization), food defrozing or pasteurization (e.g. meat, fish), etc...). For such applications, microwaves provide definite advantages such as rapidity, uniformity, material homogeneity/robustness, biological immunity, etc... in the fabrication/transformation process. In some other applications, microwave processing techniques have been shown to be efficiently combined with other previously existing techniques. Examples are provided by microwave/conventional heating combinations. Finally, other applications will definitely stay out of the scope of microwaves techniques. Under the testing point of view, the same situations hold. But, as compared to processing techniques, testing techniques can not yet justify the same experience.

<u>Some additional non technical difficulties</u>

In addition to technical difficulties, some others of non technical order must not be underestimated. Cultural gaps belongs to this category. They can be observed between the different technical areas involved in the sensor development. But they can also exist, more surprisingly, between different divisions of the same company, for instance between R&D management and production units. Such cultural gaps may result in misunderstanding and seriuously affect the development of a microwave sensor. Another difficulty is related to the newness of microwave sensing modalities. There is an evident lack of references corresponding to recognized success of microwave sensors. Indeed, even in case of success, confidentiality is often an obstacle for advertizing purposes. Finally, habits and past investments have a direct impact on the decision to install a microwave sensor. This effect is reinforced by high development and unit fabrication costs. These costs mainly result from the need to customize the available microwave technology for a given application.

CONCLUSIONS

Microwave sensing exhibits undiscutable technical potentialities. Their dissemination is mainly delayed by the fact that there is generally no "on the shelve" solution for new applications involving materials with different dielectric properties and shapes. Such a situation results from complex interaction mechanisms between microwave radiations and materials, which is well known for microwave processing techniques. As already explained, any new application requires a customization of the existing technology. As a direct result, the development cycle is particularly long between preliminary feasibility studies and on-line integration. The cost of unit fabrication is too often much higher than possibly accepted by the market, despite convenient performances. However, even if a systematic equipment of fabrication lines with microwave sensors is not acceptable, the use of a portable microwave sensor for expertise purposes may be very useful for optimizing the key parameters of the process under investigation. Furthermore, the development of microwave integrated circuits, for instance for cellular communications, will probably offer, in the near future, efficient and cost effective solutions for industrial sensing applications. More immediately, the selection of an appropriate modular technology selection is expected to reduce both development length and fabrication cost for dedicated sensors.

Finally, the efforts must be focused on the most profitable areas. Such areas can be identified from existing or expected "true" needs by means of market surveys and by taking into account the specific advantages of microwaves with respect to all other available techniques. In addition, it appears that an effective partnership between microwave and material companies is essential to reach the ultimate goal of sensor installation on a production site. This partnership may be also considered with an integrator of sensing machines.

To conclude, microwave sensing has evident capabilities which are not yet sufficiently exploited. It can be reasonnably expected that these capabilities will be offered soon at acceptable costs.

ACKNOWLEDGMENTS

This study has been conducted within the frame of Contract N°7E 2288/AEE 2015 from Electricité de France, Direction des Etudes et Recherches, Department ADE-Industrie.

REFERENCES

1. E.Nyförs and P.Vainikainen, Industrial Microwave Sensors, Artech House, 1989
2. A.Kraszewski, IEEE Trans. MTT-39,5,828 (1991)
3. M.Fischer, P.Vainikainen, E.Nyfors, IEEE Trans.IM-36,4,1036 (1987)
4. J.Ch.Bolomey, G.Cottard, B.J.Cown, Mat.Res.Soc.,189, 49 (1991)
5. J.Ch.Bolomey, G.Cottard, P.Berthaud, A.Lemaitre, J.F.Portala, Mat.Res.Soc.,347, 161 (1994)
6. J.Ch.Bolomey, Frontiers in Industrial Process Tomography, Chap.6, Engineering Foundation, 1995
7. J.Ch.Bolomey, SPIE Proc.Series, 2275, 2 (1994)

USE OF MICROWAVES FOR NONDESTRUCTIVE MATERIAL INSPECTION

B.R. HOKKANEN *, J.F. LINDSEY III **
Department of Technology, Southern Illinois University at Carbondale
Carbondale, IL 62901-6603
*brhokk@siu.edu
**lindsey@engr.siu.edu

ABSTRACT

Wideband S-parameter measurements have been used to investigate the properties of electrically non-conductive materials using a vector network analyzer. Using S_{21} scattering parameters, amplitude and phase shift measurements were taken for a variety of both homogeneous and heterogeneous materials. With these initial S-parameter measurements as a reference, surface defects were modeled on the test samples and the S-parameter measurements were reevaluated. The effect of varying levels of moisture absorption in certain materials was also investigated. The results of this study indicate that there is indeed potential for the use of a vector network analyzer as a nondestructive testing tool to evaluate the structural integrity and/or moisture content of electrically non-conductive materials through the transmission of microwave energy.

INTRODUCTION

One hundred percent inspection of products or raw materials is often required in a manufacturing environment. This need arises in situations where a small physical defect could affect the capability of the product to adequately perform the function for which it was designed. Destructive testing methods are obviously suitable only for inspection of samples; thus, a non-destructive testing (NDT) method is required. Those methods which are non-invasive allow inspection without the need to physically disturb the product or raw material under test, and so are preferred in industry. Compared to the number of non-invasive NDT methods which utilize other forms of energy, the use of microwaves is far less common.[1]

The specific type of NDT method required in a given situation depends upon the type of material being studied and upon the types of defects each method is capable of detecting. Microwave NDT methods are gaining importance due to the emergence of new and advanced materials such as polymeric and ceramic composites which pose new challenges for NDT and require new solutions.[2] When testing electrically nonconducting materials (dielectrics), microwave methods are used to sense changes in the complex permittivity of the material. The technique of microwave tomography has been shown to be well suited for the dielectric characterization of materials.[3] Microwave NDT methods have been used to detect and evaluate the presence of voids in dielectric media, to accurately measure the thickness of various dielectric composites and ceramics, and to evaluate properties of materials consisting of a mixture of constituents.[4] Work has been done in the investigation of the complex dielectric tensors of wood for determination of moisture content, density and grain angle.[5] Also, the process of active microwave imaging is beginning to be realized as a viable approach in certain applications, namely in the biomedical and geophysical areas. A specific geophysical application which has been researched is the detection and identification of buried inhomogeneities.[6] This paper investigates the feasibility of a nondestructive material inspection technique involving the use of a vector

Mat. Res. Soc. Symp. Proc. Vol. 430 © 1996 Materials Research Society

network analyzer for microwave transmission measurements and the subsequent statistical analysis of the results.

EXPERIMENT

Source of Data

The devices used to gather the data for this study were a Hewlett-Packard Model HP 8530A vector network analyzer (VNA)/microwave receiver, HP 83651A synthesized sweeper and an HP 8517A S-parameter test set. The test setup included a Microbot Teachmover robot for adjustment of test sample position. The robot arm was lined with microwave absorbent material to minimize the likelihood of interference from fields reflected from the arm. Two separate pairs of monopole antennas were constructed for the study. The first pair was an identical set of conical monopole antennas (CMAs) designed as per Funk and Saddow.[7] These antennas were found to be suitable for testing over the frequency range of approximately 3 to 25 GHz. The second pair was an identical set of sleeve monopole antennas designed using specifications presented by Rudge et al.[8] These antennas were found to be suitable for testing over the frequency range of 21 to 50 GHz. The list of materials tested in this study is as follows: Hexcel HRH-10 Honeycomb (with 1/8" cells and 1/16" cells, respectively), unidirectional fiberglass (S-glass loaded with polyphenylene sulfide (PPS)), polyimide glass, a sandwich formed of two polycarbonate sheets bonded to a core of Plascore polycarbonate honeycomb with a polyurethane adhesive, Milliken Research Corporation Contex™ fabrics (denoted C120 and C160, respectively), shoddy material (jean fabric composed of 50% polyester/50% cotton and 100% cotton, respectively) and microwave absorbing material samples denoted UI80, SL22 and SL24. The UI80 sample consists of a urethane resin loaded with 80% by weight iron carbonyl powder. The SL22 and SL24 samples consist of silicone resin filled with microballoons and carbon black powder. The thickness of every test material was measured using a micrometer, the results are shown in Table I.

Table I
Thicknesses of all test samples as measured with a micrometer

Material	Thickness
Hexcel Honeycomb (1/8" cell)	0.51 in. (1.3 cm)
Hexcel Honeycomb (1/16" cell)	0.59 in. (1.5 cm)
Plascore Sandwich	0.316 in. (8.03 mm)
Polyimide Glass	0.07 in. (1.78 mm)
SL22 Microwave Absorber	0.243 in. (6.17 mm)
SL24 Microwave Absorber	0.297 in. (7.54 mm)
UI80 Microwave Absorber	0.117 in. (2.97 mm)
Unidirectional Fiberglas	0.09 in. (2.29 mm)
Contex™ C120 Fabric	0.005 in. (0.13 mm)
Contex™ C160 Fabric	0.016 in. (0.41 mm)
50% polyester/50% cotton jean fabric	0.027 in. (0.69 mm)
100% cotton jean fabric	0.024 in. (0.61 mm)

Procedure

The basis of this study was to use the VNA to investigate the effect of various material samples upon the amplitude and phase of microwave signals transmitted between a pair of antennas when the material sample is placed between them. The VNA can be operated in one of four distinct measurement modes at any given time. The mode to be used (denoted by the S-parameter) is determined by the usage of the device I/O ports, labeled port 1 and port 2, which transmit and/or receive the microwave energy. The S-parameters S_{21} and S_{12} enable transmission measurements to be taken by using one port to transmit and the other to receive. The other two S-parameters, S_{11} and S_{22}, enable reflection measurements by using the same port both to transmit and receive. The parameter used for this study was strictly S_{21}.

The VNA has two separate channels for input/output of measurement data. The data present on these channels can be viewed in real time, either one channel-at-a-time or simultaneously on the CRT display. The number of frequency points in the sweep at which data is to be collected is also an option. In these experiments, one channel was used for amplitude measurement and the other for phase; data was taken at 201 points in the frequency sweep.

The experimental setup was virtually the same for both the CMA antennas and the sleeve monopole antennas. The bases of the antennas were fixed to a section of particle board such that the antennas themselves protruded from the edge of the board and were oriented parallel to the ground, to the surface of the board and each other. The distance of separation between the centers of the cones of the two CMAs was measured to be $5^1/_8$". The distance separating the centers of the two sleeve monopole tips was $2^1/_2$".

The overall experiment was divided into two stages. The first stage was to investigate the possibility of using the microwave measurements for detection of a modeled defect on the surface of the test material. The second stage was to investigate the sensitivity of the microwave measurements to varying degrees of moisture absorption in the material. In both stages, the material samples would be affixed to one end of a rectangular sheet of styrofoam using masking tape. The styrofoam sheet would be supported at the other end by the gripper of the Teachmover robot. The styrofoam and masking tape were used because of their transparency to the microwave transmission. The base of the robot remained in a fixed position for the entire study. The distance between the robot arm and the antennas had been determined experimentally to be sufficient in preventing the movement of the arm from interfering with the measurement calibration.

For each material tested in the first experimental stage, a 5/8" square section was cut from one end of the sample and affixed to the sample to serve as a defect model. The modeled defect was centered vertically on each sample and the near edge of the defect was located 5/8" from the edge of the sample horizontally. It was determined that the robot could be programmed to raise and lower its gripper in a strictly vertical path, moving in single steps of 9/16". Five ink marks with 9/16" separation between them were made on each test sample such that the center mark was in the center of the sample, and thus marked the center of the defect. The test sample was then positioned on the styrofoam sheet so that the antennas were in line with the uppermost ink mark (Position #1). After two upward increments of the robot arm, the antennas were transmitting directly through the defect (Position #3). The arm would increment twice more to the lowermost ink mark (Position #5) and then lower back down to the home position.

For each type of material tested in the second experimental stage, a graduated cylinder and eyedropper were used to precisely measure out four different volumes of water in milliliters at room temperature. In this stage, the robot arm remained stationary during testing and was used simply as a fixture to support the styrofoam sheet. Using the eyedropper to distribute the water evenly over the surface of the material sample under test, four different samples of the same material with the same square surface area were used, one for each different water volume. Each sample was allowed two minutes for complete absorption before it was affixed to the styrofoam for testing. One identical sample of each material was also tested dry.

RESULTS

Statistically, the form of the experimental procedure had been assumed to be that of a single factor blocked design. The null hypothesis is that there is no significant difference between the average value for each treatment level. We wish to reject this hypothesis. The alternate hypothesis is that <u>at least one</u> average treatment value differs significantly from the others. In the case of the first experimental stage, the treatment factor was the position of the modeled defect with respect to the antennas. In the second stage, the treatment was the level of moisture absorption in the sample. The blocking factor in both cases was the variation of frequency. An analysis of variance (ANOVA) was performed upon all sets of data obtained. Considering that both experimental stages of this study consisted of five different treatment levels and 201 blocking factor levels (frequency points), the number of degrees of freedom was the same for all tests. A comparison of all F_0 values obtained from ANOVA is provided in Table II.

Table II

F_0 values obtained from Analysis of Variance in each experiment (Arranged in descending order)
SMA = sleeve monopole antennas used, CMA = conical monopole antennas used
PT = Power Transmission (dB), IPD = Insertion Phase Delay (Degrees)

Variable = Defect position with respect to antennas	F_0	Variable = Moisture level absorbed by the sample	F_0
SMA IPD through Hexcel Honeycomb (1/16" cell)	2750.504	SMA PT through 100% cotton jean fabric	7565.332
SMA IPD through Hexcel Honeycomb (1/8" cell)	1860.742	SMA IPD through 100% cotton jean fabric	4854.649
SMA IPD through Plascore Sandwich	1159.896	SMA PT through Contex™ C120 fabric	3666.903
SMA PT through Hexcel Honeycomb (1/8" cell)	1033.506	SMA PT through Hexcel Honeycomb (1/16" cell)	2546.989
SMA PT through Unidirectional Fiberglas	1031.258	SMA PT through 50% polyester/50% cotton fabric	2431.791
SMA PT through Hexcel Honeycomb (1/16" cell)	613.224	SMA PT through Hexcel Honeycomb (1/8" cell)	1868.755
SMA PT through SL22 microwave absorber	420.594	SMA PT through Contex™ C160 fabric	1166.229
SMA PT through Polyimide Glass	352.624	SMA IPD through Hexcel Honeycomb (1/8" cell)	1037.219
SMA IPD through SL24 microwave absorber	328.769	SMA IPD through Hexcel Honeycomb (1/16" cell)	876.506
SMA PT through UI80 microwave absorber	214.167	SMA IPD through Contex™ C160 fabric	655.947
SMA PT through Plascore Sandwich	196.011	SMA IPD through 50% polyester/50% cotton fabric	396.830
SMA IPD through Unidirectional Fiberglas	100.709	CMA PT through 100% cotton jean fabric	335.375
CMA PT through UI80 microwave absorber	60.996	SMA IPD through Contex™ C120 fabric	274.053
CMA IPD through Hexcel Honeycomb (1/16" cell)	59.815	CMA PT through Contex™ C120 fabric	238.281
CMA PT through Polyimide Glass	50.751	CMA PT through Contex™ C160 fabric	233.574
CMA PT through Plascore Sandwich	42.905	CMA PT through 50% polyester/50% cotton fabric	232.512
CMA IPD through Hexcel Honeycomb (1/8" cell)	40.669	CMA IPD through Hexcel Honeycomb (1/16" cell)	139.229
CMA PT through Hexcel Honeycomb (1/16" cell)	33.897	CMA PT through Hexcel Honeycomb (1/16" cell)	111.471
SMA IPD through Polyimide Glass	31.528	CMA IPD through 100% cotton jean fabric	104.737
SMA PT through SL24 microwave absorber	29.569	CMA IPD through Contex™ C120 fabric	103.843
CMA PT through Hexcel Honeycomb (1/8" cell)	28.759	CMA IPD through Hexcel Honeycomb (1/8" cell)	46.632
CMA PT through Unidirectional Fiberglas	27.613	CMA PT through Hexcel Honeycomb (1/8" cell)	42.692
CMA IPD through Plascore Sandwich	25.867	CMA IPD through Contex™ C160 fabric	35.729
SMA IPD through UI80 microwave absorber	20.354	CMA IPD through 50% polyester/50% cotton fabric	8.618
SMA IPD through SL24 microwave absorber	10.589		
CMA PT through SL24 microwave absorber	9.661		
CMA PT through SL22 microwave absorber	8.373		
CMA IPD through SL22 microwave absorber	6.425		
CMA IPD through Polyimide Glass	5.824		
CMA IPD through SL24 microwave absorber	5.459		
CMA IPD through UI80 microwave absorber	5.330		
CMA IPD through Unidirectional Fiberglas	1.660		

The magnitude of the F_0 value indicates the significance of the difference of at least one treatment average from the other treatment averages. To determine whether to reject the null hypothesis, F_0 is compared to the F-value from a statistical table for a specific confidence level using the given number of degrees of freedom in the experiment. If F_0 is greater than the F-value from the statistical table, the null hypothesis can be rejected. The F-value for rejection of the null hypothesis with a 95% confidence level, given 4 treatment degrees of freedom and 800 error degrees of freedom, is 2.383. Thus, it can be seen from Table II that only one set of measurements (insertion phase delay through the unidirectional fiberglas for the different positions of the defect w.r.t the CMA antennas) does not show a significant difference between the average values for the different treatments. What can also be seen is that the measurements taken using the pair of sleeve monopole antennas show the largest deviation of at least one treatment average from the other averages. From this point on, we chose the two sets of measurements with the highest F_0 values for further study.

Once the ANOVA has established a statistically significant difference of at least one treatment average from the others in the experiment, Duncan's Multiple Range Test can be used to perform a relative comparison of each treatment average to the others. This is done by arranging the treatment averages in ascending order and determining the range (difference between the averages) for all possible pairwise comparisons. Each of these ranges is then compared with the critical range obtained by multiplying the standard error value by the significant studentized range (critical value) corresponding to the number of experimental degrees of freedom. The results of Duncan's Test on the two sets of measurements are shown in Tables III and IV.

Table III				Table IV			
Duncan's Test on Insertion Phase Delay through 1/16" cell Hexcel Honeycomb with Modeled Defect (using SMA antennas)				Duncan's Test on Power Transmission through 100% cotton jean fabric for various moisture levels (using SMA antennas)			
Standard Error (S) =	0.60733			Standard Error (S) =	0.03310		
Treatment Averages		Critical Ranges		Treatment Averages		Critical Ranges	
Position 3	-55.051	$R5 = S * r_{.05,5,800}$	1.88	3.0 ml	-4.9031	$R5 = S * r_{.05,5,800}$	0.102
Position 2	-35.575	$R4 = S * r_{.05,4,800}$	1.83	2.0 ml	-4.8836	$R4 = S * r_{.05,4,800}$	0.100
Position 1	-26.742	$R3 = S * r_{.05,3,800}$	1.77	1.0 ml	-1.7387	$R3 = S * r_{.05,3,800}$	0.097
Position 5	-26.408	$R2 = S * r_{.05,2,800}$	1.68	0.5 ml	-0.9718	$R2 = S * r_{.05,2,800}$	0.092
Position 4	-23.833			Dry	-0.2697		
#4 - #3	31.2181	> R5		Dry - 3.0	4.63334	> R5	
#4 - #2	11.742	> R4		Dry - 2.0	4.61382	> R4	
#4 - #1	2.90947	> R3		Dry - 1.0	1.46898	> R3	
#4 - #5	2.57528	> R2		Dry - 0.5	0.70201	> R2	
#5 - #3	28.6428	> R4		0.5 - 3.0	3.93134	> R4	
#5 - #2	9.16676	> R3		0.5 - 2.0	3.91181	> R3	
#5 - #1	0.33419	< R2		0.5 - 1.0	0.76698	> R2	
#1 - #3	28.3087	> R3		1.0 - 3.0	3.16436	> R3	
#1 - #2	8.83257	> R2		1.0 - 2.0	3.14484	> R2	
#2 - #3	19.4761	> R2		2.0 - 3.0	0.01952	< R2	
Conclusion: Only the average values for Position #1 and #5 are not significantly different from each other.				Conclusion: Only the average value for 2.0 ml and 3.0 ml are not significantly different from each other.			

The corresponding graphs of power transmission and insertion phase delay versus frequency for the two sets of measurements are shown in Figures 1 and 2.

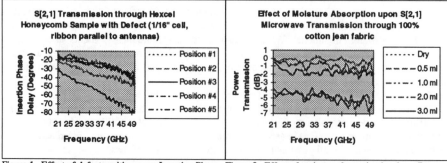

Figure 1. Effect of defect position upon Insertion Phase   Figure 2. Effect of moisture absorption levels on Power
Delay through 1/16" cell Hexcel Honeycomb   Transmission through 100% cotton jean fabric

Information on statistical techniques for analysis of experimental design was obtained from
Montgomery.[9]

CONCLUSIONS

The results of this study show that VNA measurements of microwave transmission through a
material are capable of distinguishing between different levels of moisture in the material and also
of indicating the presence of some surface defect in the materials. Statistical analysis of variance
and Duncan's Multiple Range Test are applied to the resultant data. Suggestions for further
research include the testing of other types and sizes of modeled defects to determine resolution of
the method and the investigation of other statistical procedures which may be more suited toward
this type of experiment.

ACKNOWLEDGMENTS

The authors would like to gratefully acknowledge the aid of Hans Bank, David Allabastro,
Kevin Beasley and Dr. Marek Szary for their assistance in antenna fabrication, as well as Dr. Don
Purinton of Texas Instruments for the provision of material samples. Special thanks to Southern
Illinois University College of Engineering and to the Lac Vieux Desert Band of Lake Superior
Chippewa Indians for providing financial support.

REFERENCES

1. R. Schneiderman, Microwaves & RF 33, 33-34+ (1994).

2. N. Gopalsami, S. Bakhtiari, S.L. Dieckman, A.C. Raptis, and M.J. Lepper, Mater. Eval. 52,
    412-415 (1994).

3. J.C. Bolomey and N. Joachimowicz in Dielectric Metrology via Microwave Tomography:
    Present and Future, edited by M.F. Iskander, R.J. Lauf, and W.H. Sutton (Mater. Res. Soc.
    Proc. 347, Pittsburgh, PA, 1994) pp. 259-268.

4. R. Zoughi, S.D. Gray, and P.S. Nowak, ACI Mater. Journal 92, pp. 64-70 (1995).

5. W.L. James, Y.H. Yen, and R.J. King, <u>A Microwave Method for Measuring Moisture Content, Density, and Grain Angle of Wood</u>, Res. Note FPL-0250 (U.S. Department of Agriculture, Forest Service, Forest Products Laboratory, Madison, 1985).

6. J.V. Candy and C. Pichot, IEEE Trans. on Antennas and Prop. **39** (3), 285-290 (1991).

7. E.E. Funk and S.E. Saddow, <u>Low Cost Conical Ultra-Wideband Antenna</u>, Report. No. ARL-TR-302 (U.S. Army Research Laboratory, Adelphi, 1994).

8. A.W. Rudge, et al, ed., <u>The Handbook of Antenna Design</u>, (Peregrinus on behalf of the Institution of Electrical Engineers, Stevenage, 1986), pp. 278-279.

9. D.C. Montgomery, <u>Design and Analysis of Experiments</u>, (Wiley and Sons, New York, 1976).

MAPPING THE ε" OF CONDUCTING SOLID FILMS *IN SITU*

M.J. WERNER, R.J. KING
KDC Technology Corp., 2011 Research Dr., Livermore CA 94550, mike@aimnet.com

ABSTRACT

Two nondestructive microwave methods for sensing and mapping the local loss factor ε" of conducting solid films are described. Example results of both methods are compared to the low frequency sheet resistance as measured by two- and four-point probe methods. Specimens ranged from 26 to 10,000 ohms per square. Depending on the test frequency, spatial resolution of about 1 cm is achievable, as demonstrated by maps of ε".

INTRODUCTION

The microwave technology described here forms part of a portable sensor system for on-line nondestructive measurement and mapping of the dielectric properties of the conducting films which are often encountered in aerospace equipment and elsewhere: electromagnetic shields, coated canopies, static-suppressive paint, moisture in radomes, radar-signature-modifying structural materials and the like.

The first method uses a one-port reflection-mode resonant sensor which can be scanned over the film surface, either in contact with the film or in close proximity, to generate a map of ε". This sensor is an open resonator in microstrip with one side exposed so that the near field of the resonator fringes into the film; see Fig. 1. Thereby variations in the ε" of the material dramatically affect the input resistance (R_0) at resonance of the sensor (R_0 is proportional to $1/\varepsilon$"). To some extent the resonant frequency f_r is also affected. Details of the resonant sensor's operating characteristics are described elsewhere [1].

The second method involves measuring the local insertion loss, due to the film, of a plane propagating wave at oblique incidence. This noncontacting "modulated scatterer" [2] method requires a transmitting horn on one side and a small nonlinear field probe on the opposite side; see Fig. 2. The test material is conveyed past these to generate a map of ε".

EXPERIMENT

Both types of sensor were tested for sensitivity and dynamic range using thin conducting films deposited on plastic. Tests of the resonant sensor's dynamic range continued using microwave-absorbing conductive cloth. Microwave sensor data from the film and the cloth were compared with DC sheet resistance $R_\square$ readings taken elsewhere. The resonant sensor's spatial resolution was tested by mapping voids in microwave-absorbing honeycomb.

Thin Conductive Films on Plastic

A canopy manufacturer provided a sheet of plastic covered with an electrically conducting transparent film, together with 4 point probe DC data. In use, this film is laminated after deposition, rendering conventional 4-point probe measurements of the distribution of sheet resistance of the finished product impossible. The manufacturer did not identify the material composing the film, but it resembled vapor-deposited gold. In one corner of the plastic sheet ($R_\square \sim 26$ ohms/square) the film was thick enough to be plainly visible, tapering off toward the opposite edge ($R_\square \sim 130$ to 150 ohms/square). Fig 3 is a three-dimensional map of the film's sheet resistance.

Mat. Res. Soc. Symp. Proc. Vol. 430 © 1996 Materials Research Society

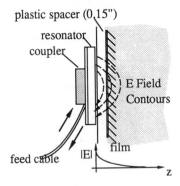

plastic spacer (0.15")

resonator
coupler

E Field
Contours

feed cable |E| film

z

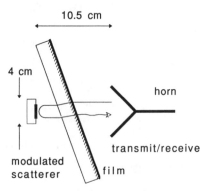

10.5 cm

4 cm

horn

transmit/receive

modulated
scatterer   film

Fig. 1. Schematic drawing of the resonant
sensor's profile. The electric field is the
near field of a 1 GHz standing wave
on the resonator. The field decays
roughly exponentially into the test
material, with penetration depth of
several millimeters depending on the
f_r, resonator shape and test material.

Fig. 2. Block diagram of 4.5 GHz non-
contact 2-way transmission system.
The modulated scatterer is a non-
linear field probe which amplitude-
modulates the transmitted signal
at 455 kHz. The back-scattered
signal is coherently detected with a
homodyne detection system.

Using the first method, we mapped the distribution of ε'' with a 1 GHz resonant sensor.
A 3.5 inch long linear dipole sensor was scanned over the film, taking data automatically every
5 cm over a 4 x 5 grid. Fig. 4 is a map of R_0 (in ohms, proportional to $1/\varepsilon''$); chosen rather than
ε'' for ease of comparison to $R_\square$ in Fig. 3. The resemblance is evident. Using sensors of
average sensitivity and penetration depth, it was necessary to stand the sensor off from the film
by 0.15 inch, because the electric field would have been shorted out entirely by direct contact
with the most conductive corner (26 ohms/square).

Fig. 5 plots the microwave against the DC data. It shows that the input resistance R_0 is
a fairly linear function of the logarithm of $R_\square$, but with considerable variance about the linear

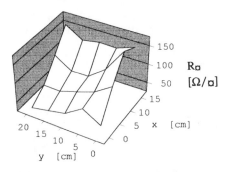

150
100   $R_\square$
50    $[\Omega/\square]$
15
10
20
15   5   x   [cm]
10   5
y   [cm]   0   0

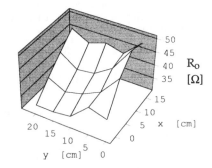

50
45
40   R_0
35    $[\Omega]$
15
10
20   15   5   x   [cm]
10   5
y   [cm]   0   0

Fig. 3. Map of the sheet resistance $R_\square$ of
a thin conducting film deposited on
plastic, as measured by the supplier.

Fig. 4. Map of input resistance R_0 of the
resonant sensor of Fig 1, in contact
with the film of Fig. 3.

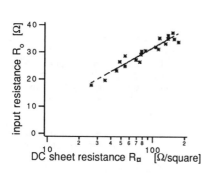

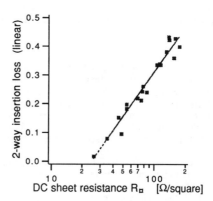

Fig. 5. 1 GHz microstrip resonator sensor on thin conductive film on plastic. R_0 appears to be linear in $\log(R_\square)$.

Fig. 6. Modulated scatterer's insertion loss through thin conductive film at 4.5 GHz. Insertion loss also appears to track $\log(R_\square)$

fit. The variance may be attributed to poor spatial resolution caused by the large footprint of this particular sensor, which dates to an early stage in the development of microwave sensors.

The curious dependence on $\log(R_\square)$ is at present unexplained. From elementary electro-magnetic theory we would have expected R_0 to be proportional to $\sqrt{R_\square}$. We do not have a model of the microstrip resonator sensor because, owing to the laminar geometry, it is difficult to simulate the field distributions of these sensors in closed form. In order to corroborate Fig. 5 to some degree, we resolved to determine whether other types of microwave sensor, confronted with this and other types of sheet conductor, also responded with the same functional relationship to $\log(R_\square)$.

We mapped the distribution of ε'' again with the "modulated scatterer" method at 4.5 GHz, the frequency of a convenient system on hand. As depicted in Fig. 2, a cw carrier is transmitted through the film. A small nonlinear field probe is used to modulate this transmitted signal at 455 kHz. Note that the measured back-scattered signal is from the field probe, and not the film itself. The footprint of the "modulated scatterer" is approximately the 4 cm diameter. Besides simplicity and the need for only one transmit/receive antenna, a major advantage of this method is that the modulated scatterer only senses the field over a small spot, of the order of half a wavelength, without need for a focusing lens, and without contacting the specimen. The two-way insertion loss is the same regardless of the direction of propagation because of reciprocity.

Fig. 6 shows that the two-way insertion loss (linear, not in dB) is also a fairly linear function of the logarithm of $R_\square$. The variance from linearity is smaller than that of the resonant sensor, which can be attributed to the smaller footprint of the modulated scatterer.

With the present unoptimized system, the signal at $R_\square = 26$ ohms/square is approaching the noise floor. It is likely that the dynamic range of the system could be improved by using a one-way transmission system, using both a transmit and a receive horn, but at the expense of less sensitivity and greater complexity.

Comparing Figs. 5 and 6 it appears that the dynamic range of the resonant sensor exceeds that of the modulated scatterer, albeit with less sensitivity. Certainly from the standpoint of convenience the resonant sensor is a superior means of mapping ε'' for most purposes. Therefore, further tests were confined to various types of the resonant sensor.

Conductive Cloth

A greater sample population of $R_\square$ was required to adequately test the dynamic range of the resonant sensor. Rather than trying to extend the range of $R_\square$ with thin metal films, we acquired fifteen 8 x 11 x 0.06 inch sheets of conductive cloth from a textile manufacturer. The sheets are coated with doped polypyrrole, a conducting polymer. The nominal DC sheet resistivity as measured by the manufacturer using a 2-point resistivity meter ranged well over three orders of magnitude, from 27 to 10,000 ohms/square. The cloths are useful to the military as a type of camouflage sometimes known as "space cloth".

Tests were performed by taking a single data point of R_0 for each sheet of cloth, using a 1/8 inch foam spacer between the microstrip resonant sensor and the cloth to avoid damping out the sensor's electromagnetic wave completely. The use of foam rather than plastic for the spacer material greatly increased the sensitivity of R_0 to variation in $R_\square$.

In Fig. 7 it is seen that R_0 is again linear in $\log(R_\square)$, but over a much wider dynamic range. A different resonator shape than a linear dipole was used--a ring with a small gap, resembling a piston ring--so the sensitivity is not the same as in Fig. 5. The data fits a straight line better than Fig. 5, probably because the sensor footprint is much smaller than any variation in $R_\square$. Errors may be attributed to the manufacturer's sampling techniques for estimating $R_\square$.

Coaxial Resonator Sensor

It occurred to us that the dependence on $\log(R_\square)$ might be due to the geometry of the microstrip resonator. The microstrip resonator is inductively coupled to the feed cable in such a way that the feed cable can "see" the test material directly. In order to remove this effect, we fabricated a coaxial resonator sensor; a section of coaxial transmission line which is one quarter-wavelength long and open circuited on one end, see Fig 8. The coaxial resonator differs from the microstrip resonator in that the standing wave propagates back and forth in air, in a direction normal to the plane of the test material, rather than in that plane. In consequence, the coaxial sensor's coupling mechanism is not directly affected by the test material, because the feed cable is on one end of the sensor, and the test material is on the other end.

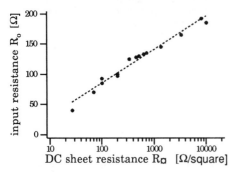

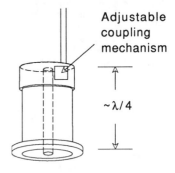

Fig. 7. 1 GHz microstrip "ring-with-gap" resonator sensor on conductive cloth. R_0 appears to track $\log(R_\square)$ linearly over more than three decades.

Fig. 8. Schematic drawing of the coaxial resonator sensor.

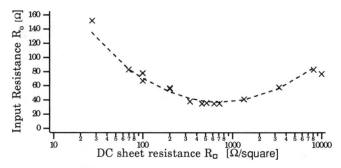

Fig. 9. R_0 data for conductive cloth using coaxial resonator sensor. The dashed line is a parabola fitted to all but the two outliers at either end.

Incidentally, the quality factor Q_0 is generally larger, and the sensitivity lower than that of the microstrip resonator; since test material affects the admittance of the line only at the end, rather than all along the line via the distributed admittance as it would for a microstrip sensor.

Upon testing the conductive cloth of various sheet resistivities with the coaxial resonator, the results of Fig. 9 were obtained. R_0 is now a nearly parabolic function of the logarithm of the cloth's sheet resistivity, falling to a broad minimum at about 600 ohms/square. The breadth of the minimum is somewhat unfortunate, because the effect could be used to advantage in the testing of "space cloth" if the minimum were sharp enough.

The dependence on $\log(R_\square)$ is modified but not escaped entirely by the change from microstrip to coaxial geometry. The parabolic dependence is unexplained at present. However, the existence of a minimum in R_0 does accord with physical intuition. The ratio of E/H (the electric field intensity divided by the magnetic field intensity) for the TEM wave inside a coaxial cable is 377 ohms, the intrinsic impedance (η_0) of free space. If the wave travelling down the coaxial transmission line encounters an object with about the same impedance as the intrinsic impedance of the wave, then the reflection from the object would be nearly zero, the dissipation in the transmission line would be a maximum, therefore R_0 should be a minimum.

The finite minimum is centered at a sheet resistance of about 600, rather than 377 ohms/square, presumably because the spacer layer increases the effective surface impedance of the cloth and prevents complete extinction of the reflected wave.

Conductive honeycomb panels

We mapped the ε'' of a microwave-absorbing honeycomb panel similar to those used in certain airframes, in order to test the capability of the microstrip resonant sensor in a realistic application. The panel had been deliberately damaged by scorching the honeycomb in two places before attaching the skin, creating two voids under an otherwise featureless composite skin material. Something like this can happen in an actual airplane, progressively damaging the structure, if hot exhaust gases from a jet engine are not properly ducted. The effect is not visible to the naked eye and cannot be detected without special equipment. The dielectric properties of the panels were not known to us in advance.

We used the same sensor as the one used to generate Fig. 7, rastering it over the panel and taking data every 4 mm. Fig. 10 is the resulting map of ε''. The raw R_0 data has been

71

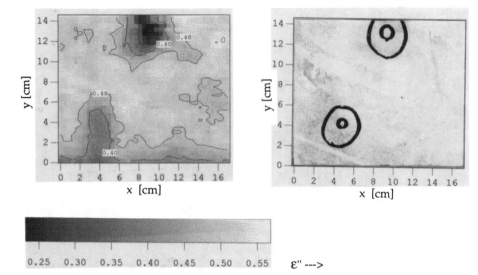

Fig. 10. Sensor map of damaged honeycomb panel. Pixel size is 4 x 4 mm

Fig. 11. Outline of damaged areas obtained by visual inspection before assembly of thepanel of Fig. 10. To scale.

converted to the equivalent ε'' of an infinite half-space using the RLC model of the sensor [1].

The two voids show up as areas of reduced ε''. Their position agrees with Fig. 11 ; a representation, to scale of the felt-pen-on-clear-plastic template of the voids made before the skin was attached. From Fig. 10, the spatial resolution of the sensor appears to be on the order of 1 cm.

CONCLUSION

We have examined two microwave methods for sensing and mapping ε'' of solid films. The advantage of the resonant sensor is its larger dynamic range, demonstrated here, and the simplicity of having a single compact probe. However, the resonant sensor needs to contact the specimen under test. For very thin films the modulated scatterer system appears to have greater sensitivity and linearity, and it need not contact the test material. The resonant sensor in microstrip and the modulated scatterer system are both linear in the logarithm of the DC sheet resistivity, while the coaxial resonant sensor is parabolic in $\log(R_\square)$.

ACKNOWLEDGMENTS

Metal films deposited on plastic by Pilkington Aerospace, conductive cloth manufactured by Milliken Research Corporation. Laboratories of these organizations measured the $R_\square$ data.

REFERENCES

1. Ray J. King, U.S. Patent No. 5 334 941 (2 August 1994).

2. R.J. King and Y.H.Yen, IEEE Trans. on Microwave Theory and Techniques MTT-29 (11), 1225 - 1231,(November 1981).

MATERIALS CHARACTERIZATION AND DIAGNOSIS USING VARIABLE FREQUENCY MICROWAVES

J. Billy Wei*, Zak Fathi*, Denise A. Tucker*, Michael L. Hampton*, Richard S. Garard*, and Robert J. Lauf **
* Lambda Technologies, Inc., 8600 Jersey Court, Suite C, Raleigh, NC
** ORNL, Oak Ridge, TN

ABSTRACT

Product quality control is a crucial part of manufacturing and usually involves materials characterization and diagnosis. Though various microwave assisted nondestructive evaluation (MA-NDE) systems have been fabricated for materials inspection, none of the systems can be applied to materials within a mold or reactor. A broadband variable frequency microwave based, resonant mode MA-NDE was studied as an alternative for characterization of materials within a cavity. The main advantage of the resonant mode MA-NDE are non-intrusive and volumetric diagnosis of the material inside a mold. The principles and possible applications of the resonant mode MA-NDE are discussed. Resonant mode MA-NDE was fully demonstrated by using Vari-Wave to trace material status during microwave curing of Diglycidyl Ether of Bisphenol A (DGEBA) /Diaminodiphenylsulphone (DDS) epoxy samples.

BACKGROUND

Microwave assisted nondestructive evaluation (MA-NDE) techniques have been used in many areas under various names, such as microwave thermography [1], microwave imaging technique [2], and microwave sensors [3]. All these techniques share many common principles, theories, and measurement tools. All MA-NDE systems can be catalogued into two types: passive and active systems. Steinberg provided a good illustration on the difference between the two [4]. Passive systems, such as radio astronomy and microwave thermometry, utilize the microwave energy or noise emitted from the body to extract the property information, such as temperature. Sources, such as $f_1(t)$ and $f_2(t)$, provide independent signal and the radiation field at the sensing antenna array, $e(x,t)$, varies spatially and temporally. Active systems, on the other hand, require transmitters to shine microwaves on the object and extract property information from received reflected or scattered energy. The transmitter launches a sequence of T spaced pulse waves, $p(t)\exp(j\omega t)$, at targets. The target re-radiates reflected (or transmitted, or scattered) waves in the direction of the sensing antenna array, such as $a_1 p(t-\tau_1)\exp(j\omega(t-\tau_1))$ and $a_2 p(t-\tau_2)\exp(j\omega(t-\tau_2))$. The radiation field at the sensing antenna array, $e(x)$, only varies spatially.

Passive MA-NDE Systems

The principle for passive systems such as microwave thermography is that the intensity of radiation $I(v)$ emitted by a body at temperature T is given by the well-known Planck Radiation Law given by

$$I(v) = e(v) \frac{2 h v^3}{c^2} [e^{hv/kT} - 1]^{-1}$$

where v is the frequency of the emitted radiation, c is the velocity of propagation, h is Planck's constant, k is Boltzmann's constant, and e(v) is the material emissivity which varies between 0 and 1. The typical value of the emissivity at microwave frequency is 0.5 [1]. I(v) is then related to the body temperature at various depth according to the penetration depth of the frequency. Also, Rayleigh-Jeans approximation can be applied for microwave radiation because hv<< kT.

$$I(v) = e(v) \frac{2 h v^3}{c^2}$$

Clearly, microwave thermography works in a similar way as infrared thermography. The main difference between the two is that infrared is mainly for surface monitoring due to its short wavelength while microwaves can be used for either surface or subsurface monitoring depending on the frequency. Main applications of passive MA-NDE systems are in medical (such as for breast cancer detection [1], subsurface temperature measurement (such as for subsurface temperature measurement during asphalt pavement [5] and radio astronomy. Bocquet et al. provided a good summary of what is involved in microwave thermography [5]. In an essence, the low noise power, which is directly related to the body temperature via Nyquist's law and is about 10^{-11} watts for a temperature of 300K and a bandwidth of 1 Ghz, is detected through a receiver and amplified. Then the noise power is compared to a reference signal from a known temperature through a lock-in detector to convert to temperature readings.

Active MA-NED Systems - Propagating Mode

All active MA-NDE systems are based on the same principle: a continuous or pulsed microwave beam, either single frequency or variable frequency, shines on the object and one or multiple receivers collect the transmitted, reflected, double transmitted, or scattered signal for extraction of desired property information. The detected signal is determined by the interaction between microwaves and the object. The amplitude and the phase of the received signal depends not only on the magnitude of the local change of the materials dielectric properties to be detected, but also on its shapes, dimensions, and the position of the change in the object. The received signal is then converted into a distribution of dielectric properties. Materials dielectric properties can in turn be related to material viscosity, humidity, temperature, composition, etc. Usually, different applications require different frequency for optimum results depending on the nature of materials. The best frequencies are those frequencies at which a small variation in the measured property could lead to a big change in the material's dielectric properties. Typical optimum frequencies are between 1 and 20 Ghz, which can be determined via experiments or numerical modelling. The active MA-NDE principle has been used to construct instruments for civil structure inspection [6], for medical inspections [7], for defect detection in dielectric composites and thermographic inspection [8], for in-situ quality control and optimization of silicon film [9], for microwave imaging in materials nondestructive evaluation [4,7,10,11], and for remote sensing of sea ice [12].

Usually, different algorithms are used in different applications to convert the received microwave signal into desired property information. The microwave assisted imaging technique deserves special attention as an alternative for nondestructive inspection of industrial materials due to its recent technological advancement, especially for conveyed products and possibly interior properties of thick materials [10,11]. As any other imaging instrument, microwave imaging is also based on the concept of point response. The image is based on the back-calculated dielectric properties of the materials by using theoretical transmission or reflection coefficients or by comparison to the response of standard materials. Commonly used methods are: direct monitoring, back-projection in free space, diffraction tomography, and inverse scattering technique. Direct monitoring and back-projection are based on superposition of plane waves (both propagating and evanescent waves) with a maximum spatial resolution up to $1/2\lambda$. This method is best suited for inspection of "slab' materials conveyed at speeds up to a few meters per second. Diffraction tomography takes into account wave scattering (but neglects multiple scattering) inside the materials and is best suited for "looking" inside the materials with a maximum spatial resolution up to $1/2\lambda$. The inverse scattering technique is the ultimate goal of quantitative dielectric tomography for materials inspection and the technique is not as advanced as others. The unknown discretized dielectric constant distribution calculated from the inverse scattering technique requires a significant amount of independent data, which is usually derived from multi-view or multi-frequency scanning. Though the spatial resolution is not limited by the wavelength, it is limited by the sensitivity of the measurement (or the signal over noise ratio) and the computing time [10].

All MA-NDE techniques mentioned so far are based on samples in free space with property information extracted from signal generated via interaction of propagating microwaves and an object. For this reason, this type of MA-NDE is classified as propagating mode MA-NDE. None of the propagating mode MA-NDE techniques can be adequately used to evaluate products within a reactor or mold. However, materials manufacturing is generally confined within some kind of mold or reactor. Currently, various probes are used to detect properties at local points within the mold and no methods are available yet to sense the properties through out the entire product. This paper presents a novel microwave assisted nondestructive evaluation (MA-NDE) system for rapid, on-line, non-intrusive, non-destructive, and volumetric monitoring of product properties within a metallic reactor or mold based on interactions between resonant microwaves and materials--resonant mode MA-NDE.

Principles of the Resonant Mode MA-NDE

The resonant mode MA-NDE technique is based on a variable frequency microwave concept and interactions between microwaves and materials. When a microwave signal of a given frequency is launched into a metallic cavity which is fully or partially filled with materials, the microwaves will reflect back and forth between cavity walls and travel through the material many times before establishing a final standing wave condition. As a result of all these wave reflections and attenuations, the microwave signal may be: 1) totally confined within the cavity, 2) totally reflected back to the launcher, or 3) partially confined and partially reflected. The final condition depends on the microwave frequency, cavity dimensions, and material properties. The materials properties include the physical geometry, dielectric properties, internal or surface defects, and chemical and physical properties. The ratio between the reflected signal and the input signal can be monitored and plotted as a function of the frequency. This signal ratio versus frequency curve is called the microwave reflective spectrum.

For a given frequency range and cavity dimension, the dimension of the material, and the location of the material inside the cavity, the spectrum is purely a function of the nature of the material. Therefore this spectrum can be used as a signature curve for the product during processing, for example. In other words, products of high quality will share a common characteristics curve which is different from that of lesser quality products.

The signal ratio at a given frequency is a summation of all interactions between materials and microwaves at that frequency. However, the contribution to the final condition in this summation is not equal for all the interactions. The signal ratio is mainly determined by the interactions at locations of high electric fields. In other words, dielectric properties at those high field locations can be back calculated from the final conditions of related frequencies and use of the proper software. When sufficiently broad, variable frequency microwaves are launched into the cavity, every point inside the mold could have high fields under certain frequencies as demonstrated in heating experiments. Therefore, with a microwave reflective spectrum over a proper frequency range, it is possible to calculate the distribution of the dielectric properties inside the mold. Once developed, this technology can be applied to polymers, composites, ceramics, chemicals, pharmaceutical products, foods, and other products during or after processing.

There are two levels of applications for the resonant mode MA-NDE. Level one is to identify the status of a product by comparing the microwave reflective spectrum of the product to standard spectra. By comparing the in-situ microwave reflective spectrum during manufacturing to the standard spectra, a computer control system will be able to identify the status of the product and to output commands to other modules for proper adjustment of process variables. It also can be used for the end product to check the product quality. Successful application of this technique depends on the database gathered, and can greatly reduce or eliminate defected parts.

The level two application of the MA-NDE system is to map the distribution of dielectric properties inside the mold based on the reflective spectrum using dielectric modeling software. The dielectric properties can then be used to interpret the extent of cure, degree of impurity, temperature, humidity, fiber content, and voids inside the mold etc. This level of application is the ultimate goal of the resonant mode MA-NDE. A fully developed MA-NDE system can be used for both process development (level two application), on-line process control (level one and two applications) and QC (level one application) in the actual manufacturing process. The features of the resonant mode MA-NDE are summarized as follows:

1. Rapid monitoring -- follows changes in a mold that occurs in seconds.
2. Quantitative monitoring - has a capability to quantitatively follow changes of the materials chemistry inside a mold.
3. Non-contact, Non-intrusive and Non-destructive -- monitoring will not affect product properties.
4. On-line capability -- for measurement during processing.
5. Volumetric monitoring -- obtain properties of the entire product rather than local areas.

EXPERIMENTS AND RESULTS

To demonstrate the resonant mode MA-NDE concept, a Vari-Wave (Model 1500, Lambda Technologies, Inc, Raleigh, NC) was used to obtain microwave reflective spectra during microwave curing of a DGEBA/DDS epoxy sample. The Vari-Wave system was used for both curing and resonant mode MA-NDE. Basic features of the model 1500 Vari-Wave are: output frequency ranging from 6 to 18 Ghz, output power adjustable from 0 to 120 W, and a cavity size of 10" x 10" x 8". A stoichiometric DGEBA/DDS epoxy sample of 14.73 g was microwave cured at 145°C isothermally for 65 min. using the following variable frequency microwaves:

frequency range of 9.25 to 10.75 Ghz, sweep rate of 1 sec., and input power of 70 W. The sample was contained in a 100 ml Pyrex beaker and was placed in the center of the cavity on the top of a Teflon Block (5" in diameter and 2.75" in height). A fiberoptic temperature sensor was used to monitor the temperature during cure. The initial sample temperature was 45°C and the sample reached 145°C after being microwave heated for 6 min. A closed-loop temperature control mechanism with an on/off microwave power switch was used to maintain the sample temperature at 145°C. Microwave reflective spectra were taken at the following moments during cure to monitor the cure status of the epoxy: A) before cure with sample at 45°C, B) when sample reached 145°C, C) after sample isothermally cured for 30 min., D) after sample isothermally cured for 60 min., and E) after sample isothermally cured for 65 min. All spectra were taken in a frequency range from 15.75 Ghz to 16 Ghz as shown in Figure 1, with 6% off set in y-axis between each spectrum.

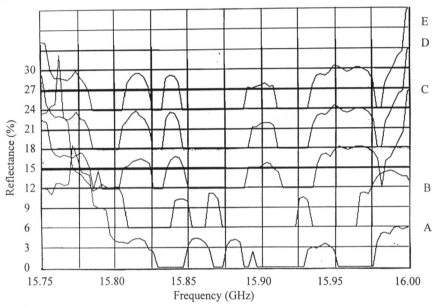

Figure 1. Microwave Reflective Spectra during Isothermal Epoxy Curing at 145°C

In general, the spectra shift to the left during curing and the changes in peak location and shape are directly related to the change of the dielectric properties. Comparing spectrum A to B, the peaks not only shift to the left but also change in shape dramatically. Also, the peak at 15.894 Ghz in spectrum A has totally disappeared in spectrum B. Comparing spectrum B and C, again the peaks shift dramatically. However, there is only a small change between spectrum C and D, and there is essentially no change between spectrum D and E. In other words, the epoxy almost reached ultimate extent of cure after isothermal curing at 145°C for 30 min. and essentially no further reaction occurred after 60 min. Clearly, MA-NDE is an excellent tool for in-situ status monitoring during epoxy curing inside the mold. This concept can also be applied to most processes inside a mold, such as reaction injection molding and autoclaving.

CONCLUSIONS

The resonant mode MA-NDE is based on interactions between microwaves and materials at the molecular level. Just as there is a mitation in the propagating mode MA-NDE, which is only for samples in the free space, the resonant mode MA-NDE is only for samples inside a cavity. There are two levels of applications for the resonant mode MA-NDE. Level one is to identify the status of a product by comparing the microwave reflective spectrum of the product to standard spectra. The level two application of the MA-NDE system is to map the distribution of dielectric properties inside the mold based on the reflective spectrum using dielectric modeling software. The level one application of the resonant mode MA-NDE was demonstrated in this paper. However, resolution and limitation of the level one application are yet to be defined. Currently, little work has been done towards the level two application and this will be the focus for the future research.

REFERENCES

1. Alan H. Barrett and Philip C. Myers, "Basic Principles and Applications of Microwave Thermography", in Medical Applications of Microwave Imaging, Lawrence E. Larsen and John H. Jacobi eds., p41, IEEE Press, (1986).
2. Bernard D. Steinberg and Harish M. Subbaram, Microwave Imaging Techniques, John Wiley & Sons, Inc. (1991).
3. E. Nyfors and P. Vainikainen, Industrial Microwave Sensors, Artech House, (1989).
4. Bernard D. Steinberg and Harish M. Subbaram, Microwave Imaging Techniques, p3, John Wiley & Sons, Inc. (1991).
5. B. Bocquet, Van De Velde J.C, Mamouni A., and Leroy Y., "Non Destructive Thermometry by Means of Microwave Radiometry", Mat. Res. Soc. Symp. Proc. Vol. 347, p 143, (1994).
6. C.W. Sohns and D.W. Bible, "Microwave Based Civil Structure Inspection Device", Mat. Res. Soc. Symp. Proc. Vol. 347, p189, (1994).
7. Lawrence E. Larsen and John H. Jacobi, Medical Applications of Microwave Imaging, IEEE Press, (1986).
8. G. d'Ambrosio, R. Massa, M.D. Migliore, A. Ciliberto, and C. Sabatino, "Microwave Based and Microwave Aided Non Destructive Test Methods", Mat. Res. Soc. Symp. Proc. Vol. 269, p497, (1992).
9. C. Swiatkowski, A. Sanders, M. Kunst, G. Seidelmann, C. Haffer, and K. Emmelmann, "In-situ Quality Control of the Production of Semiconductoe Devices by Microwave Photoconductivity Measurements", Mat. Res. Soc. Symp. Proc. Vol. 269, p491, (1992).
10. J. Ch. Bolomey and Ch. Pichot, "Microwave Imaging Techniques for Non-Destructive Testing of Materials", Mat. Res. Soc. Symp. Proc. Vol. 269, p279, (1992).
11. J.CH. Bolomey, G. Cottard and B.J. Cown, "On-line Transverse Control of Materials by Means of Microwave Imaging Techniques", Mat. Res. Soc. Symp. Proc. Vol.189, p49, (1991).
12. Frank D. Carsey, Microwave Remote Sensing of Sea Ice, American Geophysical Union, (1992).

MICROWAVE SEPARATION MEASUREMENTS OF BULK LIFETIME AND SURFACE RECOMBINATION VELOCITIES IN Si WAFERS WITH VARIOUS SURFACE PROPERTIES

Y. OGITA*, M. MINEGISHI*, H. HIGUMA*, M. SHIGETO**,
K. YAKUSHIJI**
*Dept. of Electrical & Electronic Engineering, Kanagawa Institute of Technology, Atsugi, Kanagawa, 243-02 Japan
**Showa Denko K.K., Chichibu, Saitama, 369-18, Japan

ABSTRACT

The separation measurements of a bulk lifetime τ_b, front and back surface recombination velocities S_0, S_w have been investigated for both 620 μm and 1.08 mm thick Si wafers with various surface treatments. The separation was made by applying bisurface photocoductivity decay method to the photoconductivity decay curves measured by 500 MHz microwave reflection under the bias light illumination. The bulk lifetime and the surface recombination velocities have been determined, respectively, to be 624-659 μs and 18.4-66.1 cm/s for the oxidized sliced-surface and 6285 cm/s for the sliced surface. And they have been also determined, respectively, to be 412-422 μs and <1 cm/s for the oxidized mirror polished-surface or the oxidized etched surface, 565-626 μs and 1042-1112 cm/s for the oxidized sandblasted-surface. The noncontact microwave BSPCD method with bias lights makes it possible to determine separately the bulk lifetime and surface recombination velocities in the front surface and backsurface in Si wafers with the wafer thickness for the practical use.

INTRODUCTION

Reduction of a device size and increase of number of the device process demand strict evaluation of Si wafer quality. The minority recombination carrier lifetime is very sensitive to the defects, heavy metal contaminations and surface properties in a Si wafer [1-2]. The noncontact lifetime measurement is one of the most powerful technique to evaluate the crystal quality in Si wafers and the wafer processes for the device fabrication. The most popular method for measuring the lifetime is the photoconductivity decay technique. Ogita has proposed a separation measurement method of the bulk lifetime and surface recombination velocities in the wafer with different surface called as Bisurface Photoconductivity Decay (BSPCD) method [3]. The method has successfully determined the bulk lifetime and the surface recombination velocities for the sliced wafers with 0.7-2.0 mm thick [4] and for the poly-silicon extrinsic gettering Si wafers [5] . However, for the wafer with a thickness less than 0.7 mm, the bulk lifetime has been determined to be smaller than that of thicker wafers. In oxidized wafers, the bulk lifetime has not been determined exactly, because of the round shape decay as unexpected theoretically.

In this article, we have examined the separation using the BSPCD method between the bulk and surface recombination velocities from the photoconductivity decay curves measured under the bias light illuminations for thinner 620 μm thick and thicker 1.08 mm thick Si wafers. We have also

Mat. Res. Soc. Symp. Proc. Vol. 430 © 1996 Materials Research Society

examined the separation for the wafers having different surface-properties between the front and back surfaces, as the oxidized sliced-surface and the sliced-surface, the oxidized sliced-surface and oxidized-sliced surface, the oxidized mirror polished-surface and oxidized etched-surface, the oxidized mirror polished-surface and oxidized sandblasted-surface.

MEASUREMENTS AND SAMPLES

The separation was made using BSPCD method [3]. Two photoconductivity decay curves were measured by photoexciting the respective surface. The respective intercept was determined by the extrapolation to the t=0 of the respective asymptote of the tail decay. S_0 and S_w were determined by solving the simultaneous equation about the intercepts, and τ_b was extracted from the S_0, S_w and the apparent (or effective) lifetime τ_a obtained directly from the gradient in the tail decay.

The experimental setup to measure the PCD curves is shown in Fig. 1. The PCD curves was measured without contact by the reflection of 500 MHz electromagnetic wave. The wafer sample was placed apart by 1 mm from the coplanar strip line antenna to couple with 500 MHz wave supplied through the stub tuner and the circulator from the UHF oscillator with a output power of 100 mW. The 500 MHz employed here is lower than the 10 GHz which has been used extensively. The reason is to prevent the influence on the PCD curve due to the skin effect, to get more exact PCD curves. The sample was photoexcited from both sides of the sample by the laser diode with a optical power of several watts, a wavelength of 904 nm, and a pulse width of 60 ns. In order to kill the effect of the trapping center, the sample is also illuminated by the lamp (Manabeam, Matsushita Densi K.K.) with a power of 60 watts and with a wave length>700 nm as the bias light. The reflected wave modulated by the photoconductivity decay signal was received by the same coplanar antenna, detected by the diode and amplified by 10-100 times by the amplifier and A/D converted, and eventually displaced on the computer monitor.

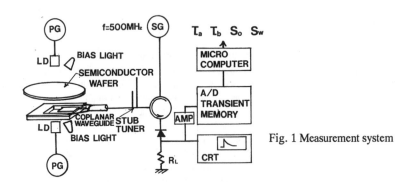

Fig. 1 Measurement system

Two different thickness wafers were used to examine the separation for the thinner wafer, as shown in Figs. 2 and 3 . The wafer samples were 5 inches p type Si CZ wafers with 10 Ωcm and (100) plane, which were sliced from the same ingot. The various surfaces of the wafer were prepared to examine the separation. The sample labeled as sample A is the wafer with a thickness of 1.08 mm and with the S_0 surface having the oxidized sliced-surface and the S_w surface whose half is the oxidized sliced-surface and the residual surface is the sliced-surface obtained by removing the oxidized layer by the wet etching. The oxidation was made by the wet oxidation at 1000 °C to be an oxide thickness of 800 Å. The another sample labeled as sample B is the wafer with a thickness of 620 μm and with the S_0 surface having the oxidized mirror surface and the S_w surface whose half is the oxidized etched-surface and the residual half is the oxidized sandblasted-surface. The oxidation was made by the wet oxidation at 1100 °C to be an oxide thickness of 1000 Å.

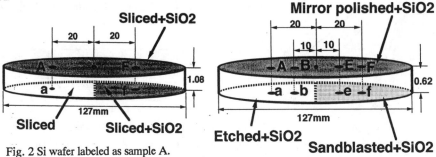

Fig. 2 Si wafer labeled as sample A.

Fig. 3 Si wafer labeled as sample B.

EXPERIMENTAL RESULTS AND DISCUSSIONS

PCD curves measured at the points Aa and Ff in sample A are shown in Figs. 4 and 5, respectively. The notation written near the curves indicates the photoexcited surface ; e.g., the curve notated as "Sliced +SiO₂" was measured when the "oxidized sliced-surface" was photoexcited. However, only in Fig. 4, the photoexcited surface is denoted by the point F or f. Fig. 6 shows S_0 dependent PCD curves theoretically calculated when a S_0 surface is photoexcited [3]. Figs. 7 and 8 show two typical examples of carrier profiles calculated for $S_0 > S_w$ and $S_0 < S_w$, respectively [3] .

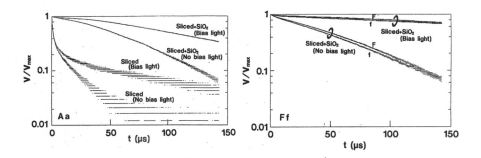

Fig. 4 PCD curves measured for Aa in sample A.

Fig. 5 PCD curves measured for Ff in sample A.

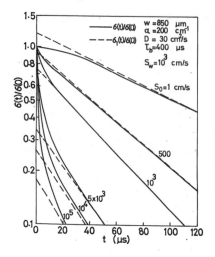

Fig. 6 S_0 dependent PCD curves calculated.

	INT1	INT2	τ_{a0} (μs)	τ_{aw} (μs)	τ_a (μs)	τ_b (μs)	S_0 (cm/s)	S_w (cm/s)
Aa	1.256	0.151	111	115	113	624	18.4	6285
Ff	1.003	0.958	420	424	422	659	30.7	66.08

Table 1 τ_b, S_0, and S_w determined by BSPCD method for sample A (τ_{a0} and τ_{aw} obtained for photoexcitation of S_0 surface and S_w surface, respectively)

	INT1	INT2	τ_{a0} (μs)	τ_{aw} (μs)	τ_a (μs)	τ_b (μs)	S_0 (cm/s)	S_w (cm/s)
Aa	1.040	1.048	411	413	412	412	<1	<1
Bb	1.041	1.055	422	422	422	422	<1	<1
Ee	1.271	0.635	83.2	83.5	83.3	565	<1	1042
Ff	1.258	0.619	81.0	81.3	81.2	626	<1	1112

Table 2 τ_b, S_0, and S_w determined by BSPCD method for sample B (τ_{a0} and τ_{aw} obtained for photoexcitation of S_0 surface and S_w surface, respectively)

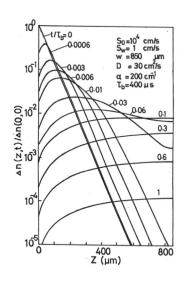

Fig. 7 Carrier profiles calculated.

The "Sliced+SiO$_2$" curves in Fig. 4 seem to correspond to the curve in the S_0=1 cm/s in Fig. 6. The decay curve does not decrease initially. This can be understood from that confined carriers initially near the S_0 surface does not decrease near the S_0 surface due to small S_0 as seen in Fig. 6. On the other hand, "Sliced" curves in Fig. 4. seem to correspond to the curves in the large S_0 in FIg. 6.

The carrier near the S_0 surface decreases abruptly as seen in Fig. 7. Thus, the curves decay abruptly due to the large S_0. We notice that the bias light illumination drastically raises the curves as seen in Figs. 4 and 5. In Fig. 5, the respective curve for photoexcited F or f decays with the same gradient since the surface of F is the same as the surface of f. The parameters S_0, S_w, τ_b were determined as shown in Table1 by applying the BSPCD method to the PCD curves measured under the bias light illumination as shown in Figs. 4 and 5. The bulk lifetime of 624 μs at Aa closes to 659 μs at Ff, in spite of the points having different surfaces. Fig. 9 shows the wafer thickness dependent bulk lifetimes determined by BSPCD method for the bare wafers sliced from the same ingot and also plots the bulk lifetimes determined here. The bulk lifetimes denoted as "Sliced+SiO₂, Sliced+SiO₂" and "Sliced", "Sliced+SiO₂" in the top inset in the figure nearly equal to that measured for the bare wafers. Thus, we can said that the separation was made successfully.

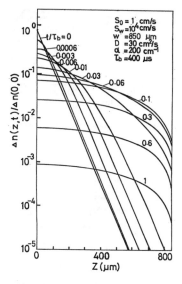

Fig, 8 Carrier profiles calculated.

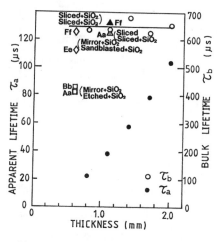

Fig. 9 Wafer thickness dependent bulk lifetimes measured for the bare wafers and the bulk lifetimes measured for the sample A and B

Figs. 10 and 11 show the PCD curves measured at Aa and Ff, respectively, in sample B. The PCD curves notated as "Mirror polished+SiO₂" at Aa nearly equal to the curve of "Etched+SiO₂" as seen in Fig. 10. This implies that the surface recombination velocity in oxidized mirror polished-surface nearly equals to that in oxidized etched-surface. S_0, S_w and τ_b determined at these 4 points are shown in Table 2. τ_b is also shown in inset of Fig. 9. As seen in Table 2, the surface recombination velocities at Aa and Ff are small to be <1 cm/s. As seen in Fig. 11, the curve at F decays slowly due to slow decrease of carrier in Fig. 8, on the contrary, the curve at f decays quickly. This implies that the surface recombination velocity at F is smaller than that at f. The difference of the surface recombination velocity as seen in the data at Aa, and Ee, Ff in Table 2 will explain the difference of surface properties between the oxidized mirror polished-surface and the oxidized sandblasted-surface. The bulk lifetime for only the wafer with the oxidized mirror polished-surface or the oxidized etched-surface is smaller slightly compared to that for the wafer with the oxidized sliced-surface, the sliced surface, or the oxidized sandblasted-surface as known in Tables 1 and 2 or in Fig. 9. The bulk lifetime will be reduced by the introduction of the contaminants in the oxidation process. Thus, the bulk lifetimes in "Mirror+SiO₂, Etched+SiO₂" would be reduced as seen in Fig. 9. As seen in Fig. 9, however, the bulk lifetime in "Sliced+SiO₂, Sliced+SiO2", "Sliced, Sliced+SiO₂", and "Mirror+SiO₂, Sandblasted+SiO₂" will recover to larger value by the gettering effect in the oxidation of the sliced surface or the sandblasted surface. Thus, the behavior of the bulk lifetimes will be able to explain.by the introduction of contaminants and the gettering effect in the oxidation process.

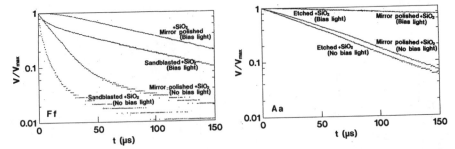

Fig. 10 PCD curves measured for sample B.   Fig. 11 PCD curves measured for sample B.

CONCLUSIONS

The 500 MHz reflection under the bias light illumination has detected the true PCD curves for the p type Si wafer sample with the surfaces as sliced, oxidized sliced, oxidized mirror, oxidized etched, oxidized sandblasted surface. The bias light illumination has been very effective to get true PCD curves. BSPCD method applying to the PCD curves has extracted reasonable τ_b and surface recombination velocities. τ_b has been extracted to be 624 μs for the sample with the sliced surface and the oxidized sliced-surface, 659 μs for the sample with both oxidized sliced-surfaces, 413-422 μs for the sample with the oxidized mirror polished-surface and the oxidized etched-surface, and 565-626 μs for the sample with the oxidized mirror polished-surface and the oxidized sandblasted-surface. The difference of τ_b determined has been explained by the gettering effect of the surfaces in the oxidation process. The surface recombination velocity has been extracted to be 6285 cm/s for the sliced surface, 18.4-66.1 cm/s for the oxidized sliced-surfaces, <1 cm/s for the oxidized mirror polished-surface or the oxidized etched-surfaces, and 1042-1112 cm/s for the oxidized sandblasted-surface. τ_b, S_0 and S_w has been extracted successfully for the thinner 620 μm wafer for the practical use and the thicker 1.08 mm wafer.

REFERENCES

1. T. Abe, T. Itoh, Y. Hayamizu, K. Sunagawa, S. Yokota, and H. Yamagishi in Defect Control in Semiconductors, edited by K. Sumino (NORTH-HOLLAND, Vol. I, Amsterdam, 1990), p. 297
2. M. Hourai, T. Naridomi, Y. Oka, K. Murakami, S. Sumita, N. Fujino, and T. Shiraiwa, Jpn. J. Appl. Phys., 27, L2361 (1988
3. Y. Ogita, J. Appl. Phys., 79, 9 (May 1, 1996) in print
4. Y. Ogita in Extended Abstracts of The 177th Society Meeting, (The Electrochemical Society, 90-1, Pennington, NJ, 1990), p. 702
5. H. Daio, Y. Uematsu, and Y. Ogita in Extended Abstracts of the 42th Spring Meeting, The Japan Society of Applied Physics and Related Society, (The Japan Society of Applied Physics, No. 2, Tokyo, 1995), p. 834 (31a-H-7)

A NEW 915 MHz LOW POWER GENERATOR FOR MATERIALS TESTING UTILIZING AN EVANESCENT MODE CAVITY

CHARLES R. BUFFLER*, PER O. RISMAN**
*Microwave Research Center, 126 Water Street, Marlborough, NH 03455, cbuff@top.monad.net
** Microtrans AB, Box 7, S-43821 Landvetter, Sweden, risman@por.se

ABSTRACT

The economy of utilizing 915 MHz vs. 2450 MHz microwave equipment has long been acknowledged for many industrial processes. A severe roadblock to advancing 915 MHz processing has been the lack of proper equipment to run test trials. At 2450 MHz low power equipment is readily available and inexpensive, even resorting to the use of the consumer microwave ovens. At 915 MHz, however, prototyping systems have only been available using very expensive 5 kW magnetrons. Experimenters have been forced to use 2450 MHz low power equipment to explore scale-up potential for high power 915 MHz systems. Because of wavelength and dielectric property differences, this procedure may give highly erroneous results, sometimes leading to the purchase of production equipment which does not work.

This paper describes an inexpensive, low power, microwave system for prototype testing at 915 MHz. A 600 watt generator is coupled to a specially designed evanescent mode applicator/ cavity for the trial testing prior to making a scale-up decision. A description of the generator and evanescent cavity design are presented.

INTRODUCTION

In general the size of a microwave generating tube such as a magnetron depends upon the wavelength associated with the frequency of operation. This fact comes about because the internal resonating structures of the tube are dependent upon the wavelength. As frequencies become higher, the internal structure of magnetrons decreases in size. As the tube structure becomes smaller, it is more difficult to extract and dissipate the heat generated during the operation of the tube, and other types of microwave generating tubes must be utilized if high average power is required. For this reason, technology to date has been able to fabricate only 2450 MHz magnetrons with output powers as high as 15 kW. 915 MHz tubes, which are typically 3.6 times as large, are being developed to deliver 100 kW from an individual tube. Because of the large difference in power available from a single magnetron, it is much more economical to generate power at 915 MHz vs. 2450 MHz. 915 MHz power generator costs typically are $1.00 per watt. Exceptions are low power 5kW generators costing up to $8.00 per watt. 2450 MHz generators run between $2.00 – $4.00 per watt. Note that the listed costs do not include the actual microwave processing applicator, tunnel conveyor, or cavity. The total cost of the microwave portion of a processing system will depend upon the sophistication of the applicator and/or conveyor system required by the process.

Because of the ease of fabricating high power 915 MHz tubes, there has not been a driving

Mat. Res. Soc. Symp. Proc. Vol. 430 © 1996 Materials Research Society

economic motive to fabricate a low-power 915 MHz magnetron. This situation has unfortunately left the microwave process developer without prototyping systems. In the past, magnetrons which were used in the GE Versatronic microwave oven were utilized. This magnetron was taken out of production as a replacement tube in the late 1970s. The last inventory of 915 GE magnetrons disappeared around 1987.

To rectify the situation, the Microwave Research Center has been involved in the acquisition of a unique low power 915 MHz magnetron over the past several years. This magnetron has been developed at a European magnetron manufacturing facility and has recently been introduced into the United States by the Microwave Research Center. Table 1 lists typical 915 and 2450 MHz power sources available in the US.

Table I – Available Power from Laboratory and Industrial Microwave Generators (US manufacturers)

Manufacturer	Power (kW) at: 915 MHz	2450 MHz
Cober Electronics, Inc.[3]	1.2/1.9/3.0/6/10/15	5/30/60/75
Continental Electronics [4]		1MW
Ferrite Components, Inc. [6]	0-75	
Microdry, Inc. [9]	5/30/60/75/100	1.2/6/10/15/20
Microwave Research Center [10]	0.6	1.4
Amana, a Raytheon Company [1]	5 – 75	1.0

DIFFICULTIES OF SCALING TO INDUSTRIAL-SIZE SYSTEMS

Because of the lack of availability of a low power 915 MHz power source, experimenters and researchers have to rely on low power, 2450 MHz systems which are readily available. Typically consumer or commercial ovens are used for this purpose. Using 2450 MHz as a base to scale up to a 915 MHz industrial power source can be replete with problems.

In review, it is known that the microwave power absorbed per unit volume, P_r, is given by:

$$P_r = C^{ste} f \varepsilon'' |E|^2 \tag{1}$$

where C^{ste} is a constant depending upon the units chosen; f, the frequency of operation, ε'', the dielectric loss factor and E the electric field *within the sample*. The equation illustrates the problem of using 2450 MHz to scale up for 915 MHz. First the frequency differs by a factor of 2.7. Second ε'' may bear no resemblance whatsoever at 915 MHz to its value at 2450 MHz.

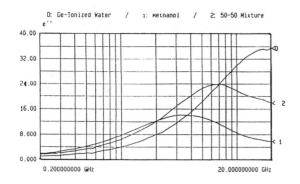

0: De-Ionized Water   /   1: Methanol   /   2: 50-50 Mixture

Figure 1. Loss Factor of Water-Methanol Mixture [5]

Figure 1 illustrates the difference in ε'' for methanol as measured by a Hewlett Packard open-ended probe [8]. At 2450 MHz $\varepsilon'' = 13$, while at 915 MHz $\varepsilon'' = 8$. Note that this difference in value is exacerbated by the fact that in industrial processing ε'' varies with temperature, and this temperature variation can differ at the two frequencies. An even more difficult problem arises from the difference between the magnitudes of electric fields at 915 MHz and 2450 MHz. Unfortunately, the electric field E inside the material is a function of its geometry, the dielectric constant ε', and the loss factor ε''. ε', ε'', and λ_0, the free space wavelength, combine to give an instructive parameter, d_p, defined as the penetration depth, given approximately by:

$$d_p = \frac{\lambda_0 \sqrt{\varepsilon'}}{2\pi\varepsilon''} \qquad (2)$$

The penetration depth is a measure of how far the microwave energy penetrates before being reduced to $1/e = 37\%$ of its original value, as measured from the inside surface of the material being processed [2]. This concept assumes a plane wave impinging on a planar surface. Because ε' and ε'' may be strong functions of temperature, which can be different at the two frequencies, major differences in the heating behavior at 2450 MHz and 915 MHz most probably will occur. Table 2 gives some illustrative values of ε', ε'' and d_p for the two frequencies for a number of different materials.

Examples

Unfortunately not nearly as many measurements on materials have been made at 915 MHz and 2450 MHz as have been made for food products. A vivid example of the difference in heating performance between 915 MHz and 2450 MHz is seen in the defrosting of pails of frozen product, typically food with a high water content. One might expect that 2450 MHz would be the perfect frequency for the thawing of the surface layer of a frozen food product in its container so that the product could be easily dumped into a conventional heating vat. This expectation comes from the fact that the penetration depth is 2/3 as shallow at 2450 MHz. What is not taken into consideration is that a phenomenon called "focusing" can occur under certain conditions of penetration depth and geometry.[2, 12]. In this particular example, it turns out that at 2450 MHz, focusing occurs and a pocket of boiling water is produced inside the frozen product before the surface thaws sufficiently to release the product from the container. At 915 MHz however, the heating is considerably more uniform and the product is released from the pail with little effort. Note that the microwave behavior in this example depends very much on the salt content of the product.

A contrary example is the tempering of large blocks of frozen meats to raise their temperature to just below the freezing point. At 2450 MHz, edge heating effects, boil water off the corners and edges of the frozen block, leaving the center of the block frozen solid. The application of 915 MHz microwave energy ameliorates the situation by raising the temperature of the frozen blocks considerably more uniformly. Thus, with these two examples, it can be seen that data taken at 2450 MHz can lead to erroneous conclusions when applied to heating performance at 915 MHz. One could easily expect similar errors for other types of materials used in industrial materials processing.

Table II – Dielectric Properties and Penetration Depth of Selected Materials (ca. 25° C)

Material	915 MHz			2450 MHz		
	ε'	ε''	$d_p(cm)$	ε'	ε''	$d_p(cm)$
Barium titanate [7]	1350	270	0.71	400	120	0.33
Blood [14]	56.8	17.8	2.23	57.5	17.1	3.87
Fused quartz [2]	4.0	0.002	5200	4	0.002	1950
Ice (-12°C)	3.25	0.005	1880	3.2	0.003	1162
Manganese oxide [11]	68	6.5	6.6	62	10.7	1.4
Mica[11]	8.9	.11	140	837	0.09	64
Oats (dry) [14]	2	0.16	46	1.97	0.14	20
Rubber (red) [14]	2.63	0.045	188	2.60	0.47	67
SiC (powder; 60 grit)	–	–	–	10.4	0.9	7.0
Soil (clay; 20% H_2O) [14]	15	7	3	11	3	2.2
Teflon® [2]	2.1	.0006	12,600	2.1	0.0006	4700
Wood (40% H_2O) [14]	6.7	1.21	11.3	5.1	1.12	4.0
Water [13]	78.4	3.47	13.3	77.4	9.2	1.87
Water (2% NaC1)	77	22	2.1	75	35	0.5

® A registered Trademark of Dupont

MICROWAVE GENERATOR AND POWER SUPPLY

The novel, unique and inexpensive low power microwave generator prototyping test system described here is based on the utilization of a newly-developed 600 watt magnetron. This magnetron is a permanent magnet structure with air cooling, very similar in size and weight and external appearance to the standard microwave cooking magnetrons. The magnetron operates

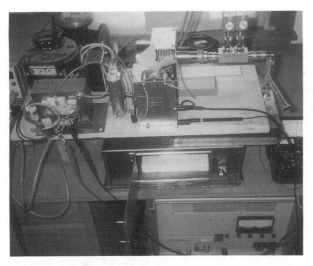

Figure 2. Breadboard layout of magnetron, power supply and the cavity

using standard, inexpensive power supply components of either the half wave doubler or full wave doubler configuration. At present we are utilizing the full wave doubler system because of some early instabilities observed in the magnetron with the half wave doubler supply. These instabilities are attributed to the high peak anode current required for the half wave operation in order to obtain the specified 600 watt microwave output.

In Figure 2, an open breadboard configuration for the breadboard prototyping system, including the magnetron, power supply and applicator cavity are shown. Power control is effected by changing the anode voltage to the magnetron by means of the variable autotransformer shown. For industrial applications, the magnetron and power supply are housed in a small benchtop, 12" x 14" x 9" cabinet. Power can be controlled by an external variable autotransformer. The generator output is by means of a 7/16 standard coaxial connector allowing custom configuration to the experimental needs. The generator output connector orientation can be configured in either a horizontal or a vertical direction. These two configurations allow the experimenter flexibility in putting together their own microwave test system. Typical operating and mechanical characteristics of the magnetron are given in Table 3.

Table III – Preliminary 915 MHz Microwave Prototyping system Specifications

Frequency: 915 MHz, nominal
Power Output: 600 watts, nominal 800 watts maximum (matched load)
Power Input: 120 volts A.C., single phase, 60 Hz, 10 amps (approximate)
        Output Connector: 7/16 coaxial (IEC169–4, DIN 47223)
        Connector Orientation: H (horizontal); V (vertical)
Dimensions: 12" x 14" x 9"
Weight: 27 lb.

Accessories:
Variable power auto-transformer for power control; 10 amps
        Open case
        Bench top in case with line cord
7/16 right angle adapter
Directional Coupler: forward, 60 dB, reverse, 50 dB

Applicator Cavity

In the past, experimenters have used 2450 MHz multi-mode cavities, usually in the form of a consumer or commercial microwave oven. At lower frequencies, for example, at 915 MHz, the number of modes in a multi-mode cavity are drastically reduced, and in some cases, eliminated. At 915 MHz, a 12" x 12" x 9" cavity could support only one TE_{10} type mode. If one wishes to keep the cavity volume and size similar to the benchtop ovens used at 2450 MHz, the dimensions would have to be increased to 20 inches to support the next higher TE/TM_{10} type modes.

In order to obtain a practical heating pattern, a hybrid TE/TM_{21} is suitable. The heating pattern for such a single-mode cavity is not completely uniform over the cavity cross section, but is more effective in the cavity center (Figure 3.)

The occurrence of heating is due primarily to currents induced by the magnetic field, not the electric field, as is usually the explanation for microwave heating. The product is heated by the displacement currents set up by the magnetic fields and thus may be coined "microwave induction" or "microwave displacement" heating. The conventional explanation analogy is $P = E*E/R$. Here we employ $P = I*IR$. This hybrid mode is preferred, because of its ability to greatly reduce edge overheating effects in low profile samples. The hybrid mode also provides a low impedance match between the microwave energy in the cavity and the dielectric properties of typical loads being heated. A good match assures high coupling and efficiency of the system. The heating pattern of the hybrid mode is similar to the standard conventional TM_{21} mode and thus cannot be sustained in a benchtop-sized cavity. The dimensions of a TE/TM_{21} hybrid mode cavity would typically need to be on the order of 20 inches. It was thought that this size might be a bit large for some practical applications. It was thus decided to explore utilizing a zero mode cavity/applicator as an evanescent mode structure with a small, convenient benchtop dimension. At first glance, it seems incongruous to utilize a non-propagating mode for the coupling of microwave energy into a load for heating purposes. It, however, should be remembered that an evanescent mode does not decay instantaneously, but over a distance given by its decay depth where the energy is reduced to 1/e of the original energy entering the structure. If an applicator/cavity is designed so that the load is placed within one decay depth of the exciting port, it is possible to couple energy into it from the structure. The cavity con-

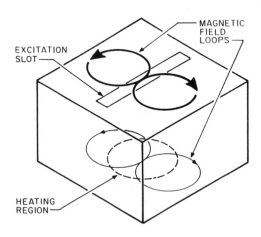

Figure 3. TM_{21} Mode Configuration

MAGNETIC
FIELD
LOOPS

EXCITATION
SLOT

HEATING
REGION

structed was utilizing an evanescent TM/TE_{21} hybrid mode with cavity dimensions of 14-3/4 inches width and 10-1/2 inches depth. The height of 5-1/2 inches was chosen so that the load would be within one decay depth of the exciting port. The cavity is also shown in Figure 2.

A major advantage of this type of mode structure is that the cavity volume mode electric field propagating in the y direction is nearly negligible Thus choke structures for doors are considerably more easily designed and fabricated than with conventional multi-mode structures. A disadvantage to such an evanescent mode applicator is the difficulty of exciting the evanescent mode with standard techniques. Loop coupling is difficult because the loop size must be so large that it no longer has its conventionally needed inductive properties. Similarly E field probes have the same problem with their capacitive hats being so constructed that they do not have intrinsic capacitive properties. In order to circumvent this problem, a matching structure was designed coupling to the magnetic field lines of the evanescent mode. This matching structure consists of a coaxial to wave guide transition, a short, low impedance wave guide section, and then a field matching structure using a low impedance slot between the waveguide and the applicator cavity. The matching structure with its slot located at the midpoint of the applicator is also shown in Figure 1.

CONCLUSIONS

Since the low-power 915 MHz generator system described is coaxial in nature, it is easily modified into any custom configuration desired by the experimenter. Modular components such as a variable power auto transformer for power regulation, directional couplers, power monitors, tuners, matched dummy loads, etc. may all be configured into experimental systems incorporating the cavities or applicators of the researchers choice. With such flexibility, the experimenter has available a low cost means of prototype testing at the desired frequency of operation, and need not encounter costly mistakes resulting from the use of 2450 MHz microwave oven systems.

REFERENCES

1. Amana (A Raytheon Corporation) (1996) Amana, IA 52204

2. Buffler, C. (1994) *Microwave Cooking and Processing: Engineering Fundamentals for the Food Scientist.* Van Nostrand Reinhold, New York

3. Cober (1996). Cober Electronics, Inc. 151 Woodward Ave., Norwalk, CT 06854 (Web site: www.connix.com/~myone/cober/cober.htm)

4. Continental Electronics Corporation, 5215 S. Buckner Blvd., Dallas, TX 75227

5. Engelder, D. and Buffler, C. (1991). Measurement of dielectric properties of food products at microwave frequencies. Microwave World 12(2):6-15.

6. Ferrite Components, Inc. (1996) 24 Flagstone Dr., Hudson, NH 03051

7.   Harvy, A. (1963). *Microwave Engineering*, Academic Press, New York.

8.   Hewlett Packard (1991) HP 85070A Dielectric Probe Kit, Data Sheet 5952-2382. HP Material Measurement Software, Data Sheet 5952-2382. Hewlett Packard Co., Santa Rosa, CA

9.   Microdry, 7450 Highway 329, Crestwood, KY 40014

10. Microwave Research Center. 126 Water Street. Marlborough, NH 03455 (Web site: www.rubbright.com)

11. Nelson, S, Lindroth, D. and Blake, R. (1989) Dielectric properties of selected and purified minerals at 1 to 22 GHz. J. Microwave Power 24(4):213–220

12. Ohlsson, T. and Risman, P. (1978) Temperature distribution of microwave – spheres and cylinders. J. Microwave Power 13(4):303–310

13. Risman, P. (1988) Microwave properties of water in the temperature range +3to 140° C. Electromagnetic Energy Reviews Volume 1.

14. Tinga, W. and Nelson, S. (1973) Dielectric properties of materials for industrial processing – tabulated. J. Microwave Power 8(1):23–65

MICROWAVE SKIN EFFECT IN
NONCONTACT PHOTOCONDUCTIVITY DECAY MEASUREMENTS IN Si WAFERS

Y. OGITA, S. TAKAHASHI
Dept. of Electrical & Electronic Engineering, Kanagawa Institute of Technology, Atsugi, Kanagawa, 243-02 Japan

ABSTRACT

A formula to calculate the photoconductivity decay curve influenced by the microwave distribution within a silicon wafer has been derived analytically. The difference of photoconductivity decay curves measured at 500 MHz and 10 GHz has been explained by the skin effect and by the comparing the experimental curves to those estimated by the analytical formula. The wafer resistivity and wafer thickness dependent photoconductivity decay curves calculated based on the skin effect have been demonstrated. The results have led to the conclusion that 10 GHz microwave probe scarcely influences on the photoconductivity decay for the silicon wafer with the resistivity >10 Ωcm and a wafer thickness <0.7 mm. For the wafer with a resistivity<10 Ωcm and a wafer thickness>0.7 mm, the photoconductivity decay curve is influenced by the skin effect at 10 GHz. However, the apparent lifetime (or effective lifetime) obtained from the gradient of the tail decay is out of the influence by the microwave skin effect.

INTRODUCTION

The reduction in device dimension in ULSI demands extremely high quality of a silicon wafer. The minority carrier lifetime is very sensitive to defects, metal contaminants, surface properties and damages [1-6]. Thus, noncontact carrier lifetime measurements have been used extensively to evaluate or to characterize the silicon wafer quality. It is very powerful technique due to nondestructive, noncontacting, speedy and easy ways to measure. The most popular technique is a photoconductivity decay (PCD) method. The PCD can be measured without contact using a probe of an electromagnetic wave as a microwave. To measure exactly the PCD curves, the microwave has to penetrate throughout the whole volume involving all photocarriers so as to detect the photoconductivity induced by photocarriers. However, generally, the microwave does not necessarily penetrate throughout the wafer due to the skin effect. The penetration depends on the resistivity, the probe frequency and the wafer thickness. Thus, in a low resistivity silicon wafer and/or in a thick wafer, the microwave can not penetrate throughout the wafer. In such cases, we can not exactly measure the PCD. That is, the bulk lifetime, the apparent (or effective) lifetime and surface recombination velocity extracted from the decay may involve some error. However, the influence of the skin effect on the PCD measurements has not been considered except in the reference [7]. However, in this reference, the PCD curve was calculated numerically. There has not been any reports about the analytical solution and the experimental study about the influence of the skin effect on the PCD. Nevertheless, the apparent lifetime has been determined from the 1/e decay or the gradient of the photococnductivity decay measured under the impulse photoexcitation.

In this article, the influence of the skin effect on the PCD measurements is considered theoretically and experimentally. The formula for calculating the photoconductivity decay is

Mat. Res. Soc. Symp. Proc. Vol. 430 © 1996 Materials Research Society

derived analytically. The validity of the formula is confirmed from the coincidence between the theoretical PCD curves and the experimental ones. The PCD measurements were made for thin (501 μm) and thick (2.01 mm) wafers with an identical resistivity (10 Ωcm) using high (10 GHz) and low (500 MHz) frequency probe. Because, we will be able to observe the changes of the PCD curves caused by the difference of the thickness or the frequency based on the skin effect. And we can compare the changes to the theoretical ones calculated by the formula derived here. The PCD curve is drastically influenced by a probe frequency, a wafer resistivity and a wafer thickness based on the skin effect. So, the wafer resistivity and thickness dependent PCD curves calculated from the formula are demonstrated. In order to prevent the influence, some relations between the wafer resistivity , the wafer thickness and the probe frequency are discussed and resulted.

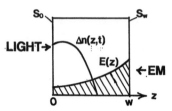

Fig. 1(a) Calculation model. The case of S_0 surface photoexcitation and S_w surface electromagnetic wave irradiation.

Fig. 1(b) Calculation model. The case of S_w surface photoexcitation and S_w surface electromagnetic wave irradiation.

MODEL AND THEORY

   The wafer model for calculating the photoconductivity decay is shown in Fig. 1. τ_b, S_0, S_w, and w is a bulk lifetime, a surface recombination velocity, a backsurface recombination velocity and a wafer thickness, respectively. The light for the photoexcitation irradiates on the S_0 or S_w surface of the wafer. The electromagnetic wave as the detection probe also irradiates S_0 or S_w surface of the wafer. Thus, four combinations in the irradiation exist. Fig. 1 also schematically shows the photoexcited excess minority carrier profiles $\Delta n(z,t)$ and the electric field component profiles $E(z)$ of the electromagnetic wave within the wafer, as two examples of the combinations. The photoexcitation light and the electric field are assumed to be single exponential decay within a wafer. And the internal multireflection is assumed to be negligible. The photoexcitation is assumed to be impulse. The effect of the internal static-electric field is neglected to be small due to low level injection condition. The excess minority carrier profiles are obtained by solving the continuity equation. The analytical solution without considering the skin effect has been derived in reference [8]. The signal output V(t) of the probe (here, under the linear detection) for the photoconductivity change is proportional to the product of the excess carrier concentration and the electric field intensity . Thus, V(t) normalized by V(0) can be written as in Eq. 1. Where q and μ are a magnitude of an electric charge and a mobility of a carrier, respectively.

$$\frac{V(t)}{V(0)} = \frac{q\mu \int_0^w \Delta n(z,t) E(z)\,dz}{q\mu \int_0^w \Delta n(z,0) E(z)\,dz} = \frac{q\mu \int_0^w \sum_{i=1}^{\infty} \Gamma_i F_i(t)\Phi_i(z) E(z)\,dz}{q\mu \int_0^w \Delta n(z,0) E(z)\,dz} \qquad E(z) = E_0 \exp\frac{-(w-z)}{\delta} \qquad (2)$$

$$= \frac{q\mu \sum_{i=1}^{\infty} \Gamma_i F_i(t) \int_0^w \Phi_i(z) E(z)\,dz}{q\mu \int_0^w \Delta n(z,0) E(z)\,dz} \qquad (1)$$

$$E(z) = E_0 \exp\left(\frac{-z}{\delta}\right) \qquad (3)$$

Γ_i, $F_i(t)$ and $\Phi_i(z)$ are a numerical coefficient, time dependent terms and spatial terms, respectively. These explicit functions are given by reference [8]. The integral terms can be easily calculated analytically. The electric field intensity $E(z)$ is given by Eq. (2) for S_0 surface photoexcitation (S_0-L) and S_w surface electromagnetic irradiation (S_w-EM), and by Eq. (3) for S_w surface photoexcitation (S_w-L) and S_w surface electromagnetic irradiation (S_w-EM).

EXPERIMENTS

To confirm the validity of Eq. (1), two different thick wafers as 2.01 mm and 0.501 mm , and two different probe frequency as 10 GHz and 500 MHz were used in the noncontact PCD measurement. The former is to observe the skin effect due to the thickness, and the latter is to observe the skin effect due to the probe frequency. The samples are p type Si wafer with resistivity 10 Ωcm, as sliced. The measurements of the photoconductivity decay was made by detecting the reflection wave of 500 MHz or 10 GHz probe wave from the wafer sample [4]. The photoexcitation was made by the LD with the pulse width of 40-80 ns.

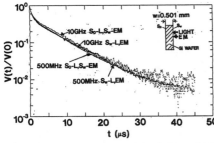

Fig. 2 shows the photoconductivity decay curves measured, for various combinations of the photoexcited surface and the electromagnetic wave irradiated surface, and for two probe frequencies. The measured decay curves seems to be the same curve. This shows that the skin effect does not influence on the decays for both 10 GHz with 2.1 mm skin depth and 500 MHz with 7.2 mm skin depth, because of thin wafer. Figs. 3 and 4 show the PCD curves measured using 500 MHz for 2.01 mm thick wafer. Fig. 3 shows it for S_0 surface photoexcitation and S_w surface electromagnetic wave irradiation. Fig. 4 shows it for S_w surface photoexcitation and S_w surface electromagnetic wave irradiation. Both curves seem to be same. Because they are not influenced by the skin effect due to deep skin depth due to the low frequency. The theoretical curves in the figures show the ones calculated from Eq. (1) by using parameters S_0=9000 cm/s, S_w=9000 cm/s and τ_b=600 μs which were determined by BSPCD method using both curves measured here [8].

Fig. 2 PCD curves measured for 0.5 mm thick wafer i various irradiations of the photoexcitation light and electromagnetic prove wave.

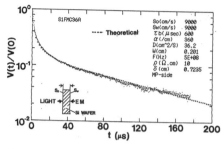

Fig. 3 PCD curve measured for 2.01 mm thick wafer using 500 MHz probe in S_0 surface photoexcitation and S_w surface electromagnetic wave irradiation.

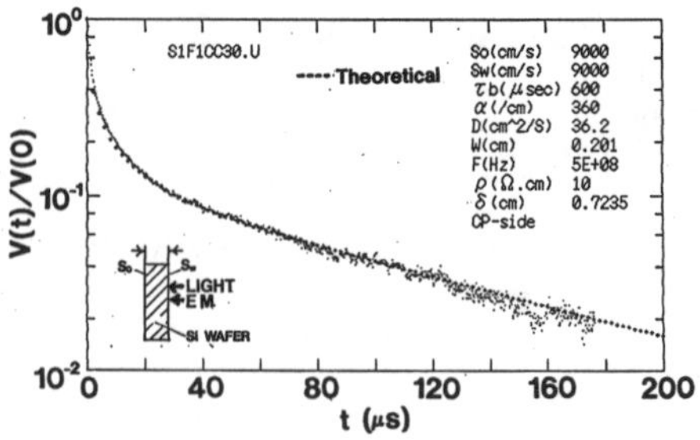

Fig. 4 PCD curve measured for 2.01 mm thick wafer using 500 MHz probe in S_w surface photoexcitation and S_w surface electromagnetic wave irradiation.

We can see that the experimental data well fall on the theoretical curves. On the other hand, Figs. 5 and 6 show the decay curves measured using 10 GHz for the same sample, for the same photoexcitation and irradiation surfaces as in Figs. 3 and 4, respectively. . The measurement data is noisy because of very thick wafer sample. The curve for S_0 surface photoexcitation decays quickly compared to the curve for S_w surface photoexcitation.

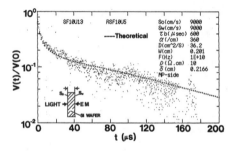

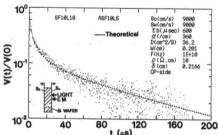

Fig. 5 PCD curve measured for 2.01 mm thick wafer using 10 GHz probe in S_0 surface photoexcitaion and S_w surface electromagnetic wave irradiation.

Fig. 6 PCD curve measured for 2.01 mm thick using 10 GHz probe in S_0 surface photoexcitaion and S_w surface electromagnetic wave irradiation.

This can be explained by that the extraction of the photoexcited carrier from the electromagnetic region as shown in Fig. 1(b) is observed as if the carrier decreases as seen in Fig. 6, on the other hand, the entrance of the carrier into the electromagnetic wave region as shown in Fig. 1(a) is observed as if the carrier increases as seen in Fig. 5. The theoretical curve calculated using the same values of the parameters as above falls on the respective data. Thus, we could confirm that the formula of Eq. (1) derived analytically well estimates the PCD curves influenced by the skin effect.

RESISTIVITY AND THICKNESS DEPENDENT DECAYS

Figs. 7 and 8 show wafer resistivity dependent PCD curves calculated from Eq.(1) for 2.01 mm thick wafer irradiated by 10 GHz microwave. In the S_0 surface photoexcitation and the S_w

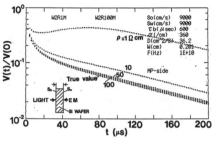
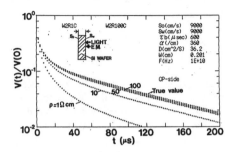

Fig. 7 Wafer resistivity dependent PCD curves in S_0 surface photoexcitation and S_w surface electromagnetic wave irradiation.

Fig. 8 Wafer resistivity dependent PCD curves in S_w surface photoexcitation and S_w surface electromagnetic wave irradiation.

microwave irradiation as seen in Fig. 7, the PCD goes up with decreasing of the resistivity. When the resistivity is 1 Ωcm, the curve becomes to be swollen. As the resistivity decreases, the skin depth also decreases so that the carrier enter late into the microwave region. Thus, the signal is observed as if the carrier genetrates late. On the other hand, in the S_w surface photoexcitation and the S_w surface microwave irradiation as seen in Fig. 8, the PCD curve goes down quickly with decreasing of the resistivity. This is due to the extraction of the carrier from the microwave region. However, in both figures, we should notice that the gradient of sufficiently time-elapsed tail decay is the same. This can be explained as followings. In a large time region, the PCD curve decays with exponentially. This means that the carrier decreases keeping the same spatial carrier profile, i.e., it decreases with the same time rate, i.e., it is the steady state. Thus, the carrier change can be detected not depended on the microwave profile within the wafer. Because the carrier in anywhere within the wafer decreases with the same time rate. Therefore, we conclude that the apparent (or effective) lifetime obtained from the tail decay curve is not influenced by the skin effect due to the wafer resistivity.

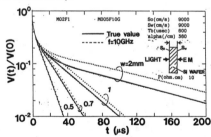
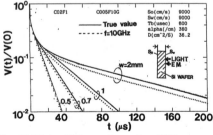

Fig. 9 Wafer thickness dependent PCD curves in S_0 surface photoexcitation and S_w surface electromagnetic wave irradiation.

Fig. 10 Wafer thickness dependent PCD curves calculated in S_w surface photoexcitation and S_w surface electromagnetic wave irradiation.

Figs. 9 and 10 show wafer thickness dependent PCD curves calculated for a 10 GHz probe. The decay deviates upward with increasing wafer thickness when the photoexcited and irradiated surfaces are contrary, as seen in Fig. 9. When the potoexcited and irradiated surface are identical, it deviates downward as seen in Fig. 10. However, those gradients in the tail decay curves are same as the ones of the true curves. Thus, we can say that the apparent (or effective) lifetime is not influenced by the skin effect .

CONCLUSIONS

The formula to calculate the PCD influenced by the skin effect of the electromagnetic wave probe has been derived analytically. The validity of the formula has been confirmed from the agreement between the calculated decay curves and the experimental ones. The wafer resistivity and wafer thickness dependent decay curves calculated using the formula have been demonstrated. It has been shown that for 10 GHz probe and 10 Ωcm Si wafer having a wafer thickness<0.7 mm, the observed PCD curve is scarcely influenced by the skin effect. On the contrary, it has been shown that for a thickness>0.7 mm, the skin effect influences on the decay to increase when the photoexcited surface and the electromagnetic wave irradiated surface are contrary, and to decrease when the photoexcited surface and the electromagnetic wave irradiated surface are identical. It has been shown that the apparent (or effective) lifetime obtained from the gradient of tail decay curve is not influenced by the skin effect.

ACKNOWLEDGEMENTS

Authors would like to thank to Dr. K. Yakushiji of Showa Denko K.K. for the supply of the wafer samples used in this experiment.

REFERENCES

1. A. Y. Liang and C. J. Varker, Lifetime Factors in Silicon, ASTM, STP712, ASTM, Philadelphia, PA, 1980, p. 73
2. Y. Ogita and H. Takai, Semiconductor World, 1, 10, p.39 (1982)
3. K. Katayama and F. Shimura, Proceedings of Diagnostic Techniques for Semiconductor Materials and Devices, (The Electrochemical Society, Pennington, NJ, 1992), Vol. 92-2, p. 184
4. M. Hourai, T. Naridomi, Y. Oka, K. Murakami, S. Sumita, N. Fujino, and T. Shiraiwa, Jpn. J. Appl. Phys., 27, L2361(1988)
5. Y. Hayamizu, T. Hamagichi, S. Ushio, T. Abe, and F. Shimura, J. Appl. Phys., 69, p. 3077
6. Y. Ogita, K. Yakushiji, and N. Tate in Semiconductor Silicon/1994, edited by H. R. Huff, W. Bergholz, and K. Sumino (The Electrochemical Society, Proc. 94-10, Pennington, NJ, 1994),p. 1083
7. A. Usami, in The 42th Technical Digest of The 145th Committee of The Japan Society for the Promotion Of Science, (The Japan Society for the Promotion of Science, Tokyo, 1988) p.1
8. Y. Ogita, J. Appl. Phys., 79, 9 (May 1, 1996) in print

Part III

Microwave Processing

MICROWAVE PROCESSING OF REDOX CERAMIC-METAL COMPOSITES

R.R. Di Fiore, D.E. Clark
University of Florida, Gainesville, FL 32611, RFIOR@MSE.UFL.EDU

ABSTRACT

Ceramic-metal composites have been produced through the reduction of copper oxide (CuO) and the oxidation of aluminum in a reducing atmosphere. Fine powders of CuO and Al ($<38\mu m$) were mixed with yttria stabilized zirconia, ball milled and uniaxially pressed into disc samples. Microwave hybrid heating was used to process samples using β-SiC and activated carbon powder as susceptors. Pyrolysis of the carbon provided the reducing atmosphere. The resulting $Cu/Al_2O_3/Y$-ZrO_2 composite was analyzed for density, compressive strength and resistivity.

INTRODUCTION

The CuO/Al reactants were chosen based on the Ellingham diagram of free energy (ΔG_0) versus temperature for the oxidation of metals [1]. The reduction of CuO and the simultaneous oxidation of Al results in an exothermic reaction which releases approximately 450 kilojoules of energy per mole of reactants. This energy can be used to aid in heating and densifying the composite. Reduction/oxidation (redox) of CuO/Al forms a Cu/Al_2O_3 composite at high temperatures. This process is known as self-propagating high temperature synthesis (SHS) [2]. When used in conjunction with microwave processing, redox reactions start at the center of the sample and work outward.

Stabilized ZrO_2 was used as a matrix for the Cu/Al_2O_3. This material has desirable dielectric properties, strength and toughness [3]. The ZrO_2 is also stable in a reducing atmosphere when in the presence of Al. Either partially stabilized ZrO_2 (3 wt. % yttria) or fully stabilized ZrO_2 (8 wt. % yttria) was used as the matrix.

One of the advantages to using microwave processing is selective heating. This was achieved by using a combination of microwave hybrid heating (MHH) and microwave absorbent materials within the composite [4]. In MHH, β-SiC and activated carbon powders absorb microwave energy and convert it to heat. The CuO in the composite is a strong susceptor to microwave energy and ZrO_2 is known to suscept well above 600 °C [5].

Mat. Res. Soc. Symp. Proc. Vol. 430 © 1996 Materials Research Society

EXPERIMENTAL PROCEDURE

The CuO (Fisher Scientific, C470-500) and Al (Fisher Scientific, A559-500) powders were mixed in a 3:2 molar ratio. This mixture was added to either 3 wt% Y-ZrO$_2$ (Zirconia Sales (America), Inc., HSY3.0) or 8 wt% Y-ZrO$_2$ (Zircar Products, Inc., zirconia powder type ZYP) in a 1:4 weight ratio of 3CuO/2Al to Y-ZrO$_2$. The mixture was pulverized in a porcelain mortar, then transferred to a polypropylene bottle and mixed for 30 minutes with zirconia media on a ball mill. The powder was removed and sieved through a 75 μm mesh screen to break up agglomerates.

An analysis of behavior during heating was performed using DTA/TGA analysis (Figure 1). Several 200 mg samples were ramped to 1200 °C to determine phase changes and reaction temperatures. The heating schedule was based on the DTA/TGA data.

Samples (0.5 to 1.0 grams) were formed in a cylindrical stainless steel die using a uniaxial press. Samples were first de-aired at a 1000 lb load, then pressed at a 7000 lb load. Bulk green density of the disc compacts was 65% of the calculated theoretical density of the starting materials.

The disc samples were placed within a specially constructed alumina firebrick housing (Thermal Ceramics, K3000) with 20 grams of β-SiC granules (Standard Sand and Silica Co., 16 GRP Carbolon Green SiC) and 2 grams of activated carbon powder (Fisher Scientific, 05-690B) (Figure 2). An inconel shielded, ungrounded tip, K-type thermocouple (Omega Engineering, Inc., XCIB-K-2-3-L) was used to monitor temperature [6].

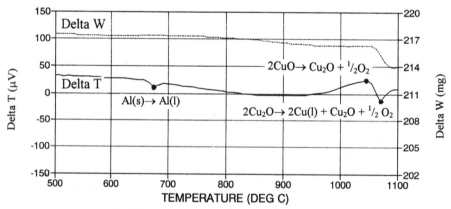

Figure 1. DTA/TGA analysis of the CuO/Al/Y-ZrO$_2$ starting powder showing phase changes in an argon atmosphere.

The sample housing was placed within a modified 1000 watt microwave (Goldstar, MA 1160M) which had been previously mapped for regions of microwave field uniformity (Figure 3). Carbon powder was evenly distributed on an alumina mat, placed in the cavity and subjected to microwaves. The patterns that formed on the mat were used to locate the high energy field areas. Mapping of the hot spots was necessary for optimum sample placement and processing uniformity.

Heating rates ($\approx$100 °C/min.) were carefully monitored to control the redox reaction rate. Samples ramped quickly to 1100 °C (>200 °C/min.) resulted in non-uniform composites. The heat treatment schedule was based on the DTA/TGA analysis. Samples were ramped at 110 °C/min. to 660 °C, then 1100 °C with soak times at each temperature of 5 - 10 minutes. Total processing time within the microwave was 25 - 30 minutes. Samples were quickly cooled to room temperature in flowing air (oxidizing atmosphere) or carbon powder (reducing atmosphere). After processing, samples were visually examined for uniformity and cracking.

Several samples were reheated to 1000 °C and quenched in deionized water (T=23 °C) to evaluate thermal shock resistance. The remaining samples were tested for compressive strength using a Comten Industries compression tester (Model 922MT 20/10).

RESULTS

The microwave used in this study had no mode stirrer, and the turntable originally with the unit was removed. This accounted for the poor distribution of the microwave field. Field pattern mapping of the microwave cavity was necessary to ensure uniform processing of the samples. Samples that were placed within a hot

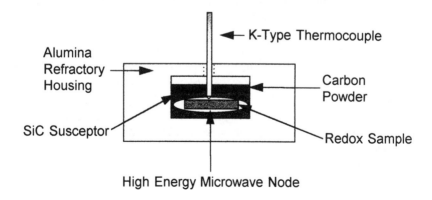

Figure 2. Microwave refractory housing and sample placement.

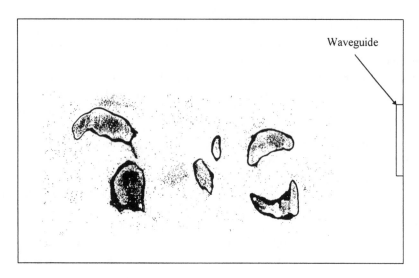

Figure 3. Hot spot patterns within the microwave cavity .

spot showed no signs of preferential shrinkage due to non-uniform densification. Samples processed before node mapping showed signs of incomplete reaction as well as density gradients. There was a clear pattern typical of microwave processed SHS specimens [2,7]. The redox reaction began at the center of the sample and gradually worked its way outward. Completely reacted samples were black in color, while partially reacted samples remained gray at the outer edges.

Bulk density measurements on fired samples were 85% of the calculated theoretical density (ρ_B = 4.21 g/cc). Additional firing (to 1200 °C for 20 min.) of selected samples increased bulk density by 10%.

Samples were tested for resistivity by applying test leads across the diameter on the top or bottom surfaces of the discs. For the air-quenched samples, gradient resistance was observed. The bottom face (which was next to the SiC) had a low resistance, usually less than 200 Ω across the 12.5 mm diameter. The top face (which was next to the carbon powder) had a high resistance, greater than 300 kΩ. Samples quenched in carbon powder were found to have a high resistivity across both surfaces.

Five samples were quenched in deionized water (ΔT = 973 °C). None of the samples cracked. Upon repeated thermal cycling, cracks did develop at the edges of several samples. There was a noticeable change in the physical features of these samples. All samples were black after the initial processing, but changed in hue to dark red after quenching. This may indicate a phase change in the material.

Sectioning of the samples revealed that this was also true of the interior. An increase in the cooling rate of the red samples was also noted. Red samples cooled twice as fast as the black samples under similar conditions (30 seconds versus one minute). Additional characterization is planned.

Compressive strength testing was performed on 0.5, 1 and 2 mm thick samples at a rate of 25.4 mm/sec. In all cases, a 110 kg load did not cause failure. For the 0.5 mm thick samples, the average compressive strengths were greater than 56 MPa. Further strength testing is planned for bar samples in four point flexure.

SUMMARY

Mapping of the hot spots within the microwave cavity aided in processing uniformity. Samples were successfully processed within these hot spots. Microwave processing of $CuO/Al/Y-ZrO_2$ yielded a ceramic-metal composite with unique features. The $Cu/Al_2O_3/Y-ZrO_2$ composites were produced with an average 93.5% theoretical density after firing to 1200 °C. Resistivity of the samples could be altered by modifying the processing parameters. Samples were able to withstand a thermal shock of $\Delta T=973$ °C. Quenched samples exhibited a color change and an increase in the cooling rate of the sample. Compressive strengths of the composites exceeded 56 MPa.

FUTURE EFFORTS

Results of this study have been used to determine other redox systems of interest. These include the $Si/Al_2O_3/ZrO2$, $Mg/Al_2O_3/ZrO_2$ and the $Mg/TiO2$ composite systems. Additional characterization using small angle x-ray scattering and scanning electron microscopy will be used to determine phase changes during processing.

ACKNOWLEDGMENTS

We would like to thank the National Science Foundation (NSF) and the Electric Power Research Institute (EPRI) for their partial financial support.

REFERENCES

1.   Denny A. Jones, Prevention and Protection from Corrosion, Macmillan Publishing Co., New York, (1992).

2.   T. Yiin, M. Barmatz, H. Feng and J. Moore, "Microwave Induced Combustion Synthesis of Ceramic and Ceramic-Metal Composites," *Microwaves:*

Theory and Application in Materials Processing III, Amer. Cer. Soc., Westerville, OH, pp. 541-547 (1995).

3.   M.A. Janney, C.L. Calhoun and H.D. Kimrey, "Microwave Sintering of Zirconia-8 mol% Yttria," *Microwaves: Theory and Application in Materials Processing I*, Amer. Cer. Soc., Westerville, OH, pp. 311-318 (1991).

4.   A.D. Cozzi, D.E. Clark, M.K. Ferber and V.J. Tenney, "Apparatus for the Joining of Ceramics Using Microwave Hybrid Heating," *Microwaves: Theory and Application in Materials Processing III*, Amer. Cer. Soc., Westerville, OH, pp. 389-396 (1995).

5.   R.M. Hutcheon, P. Hayward, B.H. Smith and S.B. Alexander, "High Temperature Measurement - Another Analytical Tool for Ceramic Studies?," *Microwaves: Theory and Application in Materials Processing III*, Amer. Cer. Soc., Westerville, OH, pp. 235-242 (1995).

6.   G. Darby, D.E. Clark, R. Di Fiore, R. Schulz, D. Folz, A. Booyapiwat and D. Roth, "Temperature Measurement During Microwave Processing," *Microwaves: Theory and Application in Materials Processing III*, Amer. Cer. Soc., Westerville, OH, pp. 515-522 (1995).

7.   D.E. Clark, I. Ahmad and R.C. Dalton, "Microwave Ignition and Combustion Synthesis of Composites," Matl. Sci. & Eng., **A144**, pp. 91-97 (1991).

8.   E. Bescher and J.D. Mackenzie, "Microwave Heating of Cermets," *Microwaves: Theory and Application in Materials Processing I*, Amer. Cer. Soc., Westerville, OH, pp. 557-563 (1991).

9.   R. Di Fiore and D.E. Clark, "Microwave Joining of Zinc Sulfide," *Microwaves: Theory and Application in Materials Processing III*, Amer. Cer. Soc., Westerville, OH, pp. 381-388 (1995).

MICROWAVE PROCESSING APPLIED TO CERAMIC REACTIONS

M. GONZALEZ, I. GOMEZ AND J. AGUILAR
Universidad Autónoma de Nuevo León, Facultad de Ingeniería Mecánica y Eléctrica, Apartado Postal 076 "F", Cd. Universitaria, San Nicolás de los Garza, N.L., C.P. 66450, México.

ABSTRACT

The necessary energy for processing ceramics usually comes from gas firing or electricity. With gas firing, ceramics processing is conducted at relative low temperatures. In the case of electric arc furnace, temperatures are high enough to smelt the material and the reaction takes place in a liquid state. In this work we use microwave radiation for conducting reactions between oxides to produce ceramic materials, mainly spinels. The microwave energy was supplied using an 800W magnetron operating at 2.45 GHz. The microstructure and the mineral composition were studied by means of SEM and X-Ray diffractometer respectively.

INTRODUCTION

One of the most important properties of ceramic materials is that they retain their structural properties at high temperatures, they usually present a high melting point and this property makes difficult the process of this kind of materials, because it is necessary to achieve high temperatures. Traditionally, these materials have been produced using gas burners or electric resistance elements. During the last years there has been an upsurge of interest, within the scientific community, on the possible applications of microwave radiation as an energy source for the processing of materials. There are some discussions about the mechanism that governs the absorption of microwave energy, some of these mechanisms are dipolar losses, ion jump relaxation and ohmic loss effects [1], but in many materials the transfer mechanism is still unknown. If the ceramic that is being processed takes enough energy for conducting a process, it would be feasible to carry out processes that, for now, are limited to be smelted in an electric arc furnace.

The objective of this work focuses on the use of microwave energy to produce magnesia-alumina spinels.

Al_2O_3 - MgO System

From thermodynamic considerations MgO and Al_2O_3 should react to form $MgAl_2O_4$, but in practice solids do not usually react together at room temperatures over normal timescales so, for having a reaction, it is necessary to heat a powder mixture of MgO/Al_2O_3 in this case

Mat. Res. Soc. Symp. Proc. Vol. 430 © 1996 Materials Research Society

at higher temperatures, often above 1200 °C in order to have a reaction at an appreciable rate, but in many cases, could be necessary to heat the powder mixture for several days.

If the reaction is taking place in a solid phase, the first stage of reaction would be the formation of $MgAl_2O_4$ nuclei, this nucleation is difficult because of the differences shown in the structure of reactants and products and because the large amount of structural reorganization that is involved in forming the product. $MgAl_2O_4$ has a crystal structure that shows similarities and differences to those of both MgO and Al_2O_3; MgO and spinel have a cubic close packed arrays of oxides, in contrast to Al_2O_3 which has a distorted hexagonal close packed array of oxide ions; on the other hand, the Al^{3+} ions occupy octahedral sites in both Al_2O_3 and spinel, while the Mg^{2+} ions are octahedral in MgO but tetrahedral in $MgAl_2O_4$ [2].

As illustrated in Figure 1 [3], spinel ($MgAl_2O_4$) is the only intermediate compound in the phase diagram of the system MgO-Al_2O_3 and its melting point is 2135 °C. Spinel forms two eutectics, one of them is 45 wt% of magnesia and 55 wt% of alumina with a melting point of 2030 °C; the other eutectic is 97 wt% of alumina and 3 wt% of magnesia with a melting point of 1925 °C.

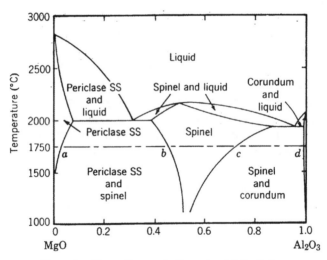

Figure 1. Phase diagram in the system MgO-Al_2O_3

EXPERIMENTAL PROCEDURE

Spinel production tests were carried out using a conventional microwave oven with a magnetron working at 2.45 GHz and 800 Watts. The reagents MgO (1 μm) and Al_2O_3

(50 μm) were mixed thoroughly to get intimate contact between the powders. The ratio used was 1:1 molar (10 gr. Al_2O_3 + 3.95 gr. MgO). Fine grained materials should be used if possible in order to maximize surface areas and therefore reaction rates. The surface area of reacting solids has a great influence on reaction rates since the total area of contact between the grains of the reacting solids depends approximately on the total surface area of the grains.

Some authors had found that MgO and Al_2O_3 are not good absorbents of microwaves at low temperatures [4], but they can absorb microwave radiation if their temperature is increased. In order to preheat the materials to appropriate temperatures, the mixture was placed over a high purity carbon bed (3 gr.) which is a good absorbent of microwaves. Just the bottom part of the reagents layer was in contact with carbon and it heats the sample by means of conduction and then the mixture of reagents could achieve the appropriate temperature for becoming absorbents of microwaves.

The mixture was placed over a carbon bed inside a crucible made of high purity alumina and thermally insulated as shown in Figure 2. This kind of crucible was selected because it can resist temperatures up to 2000°C.

The crucible was placed inside the chamber in a specific place that according to the best heating rate [5]. The samples were heated by microwave energy for times of 20, 40 and 60 minutes and analyzed respectively.

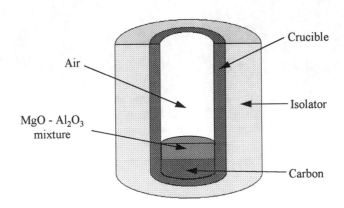

Figure 2. Scheme of the specimen sat in an isolated crucible

RESULTS AND DISCUSSION

Once that the set time was reached, the samples were removed from the crucible and analyzed by SEM and X Rays Diffraction to confirm the spinel formation. The place that was in contact with the carbon bed was partially smelted. The zone close to the carbon bed and above the smelted zone exhibited a shrinkage, due to sintering, while the upon side of the sample was unreacted and no sintering occurred.

The samples exposed for 40 and 60 minutes to microwave energy exhibited similar sintering profile. Spinel formation in the test for the 20 minutes sample was poor. Spinel was mainly produced after 40 minutes of microwave irradiation.

The unsintered material does not show spinel presence (just traces) even when this compound was searched carefully. Figure 3 shows a SEM image of the spinel phase taken to a specimen exposed for 40 minutes to the microwave radiation. This part was chemically analyzed by means of X-Ray Spectrometry and the presence of spinel was confirmed.

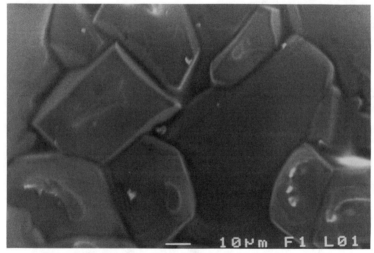

Figure 3. SEM image of the spinel obtained with microwaves

The difference between sintering degrees across the sample could be explained from the difference of temperatures inside the sample, the upper side was uncovered an therefore it was colder than the bottom side of the sample closer to the auxiliary heating source (carbon bed). The side that is in contact with carbon reaches the temperature, at which the materials became absorbent of microwaves, more rapidly than the part that is on the top and once that they absorb microwaves, their temperature is increased because they absorb microwave by themselves. The sample that is on the top would never reach the same temperature than the part of the sample that is on the bottom because the first one is losing heat by radiation.

Taking just the sintered part of the ceramic material, the diffraction pattern shows spinel formation (Figure 4). In this pattern we can see, besides of $MgAl_2O_4$ spinel, the presence of MgO which did not react. We did not find any carbon compound thus we can say that carbon does not take place in the solid reaction.

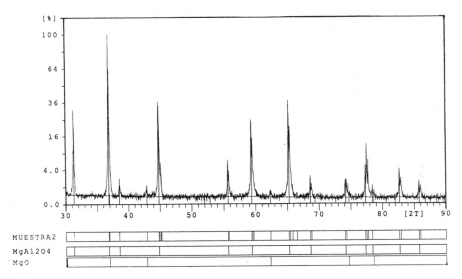

Figure 4. Diffraction pattern of the sintered material. Specimen exposed 40 minutes to microwave radiation.

CONCLUSION

In this work was demonstrated that it is possible to use microwaves as an energy source for the production of ceramic materials such as alumina-magnesia spinel.

Carbon could be used as an auxiliary source of heat for processing ceramic materials that need to be heated to certain temperature, before becoming absorbent of microwave such as magnesia-alumina spinel.

Temperatures around the melting point were achieved in some parts of the samples, the part that was unreacted was on the top and was unprotected from radiative losses, then the temperature was far from the melting point.

ACKNOWLEDGMENTS

We express our gratitude to CONACYT (Mexican Science and Technology Council) for its financial support.

REFERENCES

1. Jon Binner, The potential of microwave processing for ceramics, Materials World, March 1993, pp 152-155.
2. A.R. West. Chemistry of Solid State and its Applications, pp 5-17.
3. R. Dal Maschio, B. Fabbri and C. Fiori, Industrial Applications of Refractories Containing Magnesium Aluminate Spinel, Industrial Ceramics, Volume 8, No. 3, 1988, pp 121-126.
4. S.J. Oda, I.S. Balbaa and B.T. Barber, Proc. Microwave Processing of Materials II (Materials Research Society, San Francisco, CA USA, Apr. 17-20 1990) pp. 391-402.
5. I. Gomez and J. Aguilar, Proc. Dynamic in Confining Porous Systems (Materials Research Society, Boston, MA USA, Nov 27 - Dec 2 1995) pp 391-402.

MICROWAVE INDUCED PLASMA (MIP) BRAZING OF SILICON NITRIDE TO STAINLESS STEEL

M. SAMANDI*, M. BATE*, R. DONNAN*, S. MIYAKE**
*Surface Engineering Research Centre, Department of Materials Engineering, University of Wollongong, Wollongong, NSW 2522, Australia.
**Welding Research Institute, University of Osaka, Osaka, Japan.

ABSTRACT

In an attempt to accelerate the process of joining of metals to ceramics, a new rapid brazing technology has been developed. In this process, referred to as Microwave Induced Plasma (MIP) brazing, a microwave plasma is used to rapidly heat the ceramic and metal to the melting temperature of the reactive braze material. The heating rate obtained by MIP could be many times faster than those achieved by conventional resistive heating in a tube furnace. The fast heating rate has no detrimental effect on the joint quality and in fact results in the formation of a thick interfacial film suggesting significant interdiffusion between the braze and ceramic, possibly stimulated by the microwave radiation. In this paper the experimental arrangement of the MIP system is described. The unique capability of the MIP heating is demonstrated by successful joining of hot pressed silicon nitride to stainless steel using reactive metal brazing. The results of microstructural characterisation of the joints carried out by SEM and EDS will also be presented.

INTRODUCTION

Engineering ceramics have, in general, high temperature strength, good wear resistance, low thermal and electrical conductivity, relatively low density and good corrosion resistance [1]. These properties are attractive for many engineering applications such as electrical insulators, wear resistant parts and high temperature components of heat engines. The disadvantages of engineering ceramics include the difficulty and cost of manufacture, sensitivity to small flaws, brittleness and high material cost. These disadvantages can be, to certain extent, overcome by combining ceramics with metals, with properties that compliment those of ceramics.

The technology of joining ceramics has progressed steadily since its beginning in the early 1930s. Many techniques have been proposed to join ceramics to metals including fusion welding, adhesive bonding, friction welding, etc. However, the most common joining methods for structural applications are metal brazing, glazing and diffusion boding. Reactive metal brazing, in particular, has emerged as a powerful technique for joining of metals to ceramics. In this technique, a small quantity of a reactive element such as titanium or zirconium is added to a conventional braze alloy, eg. Cu-Ag eutectic. At the brazing temperature ($\approx 800°C$) the reactive element reacts with the metalloid in the ceramic (oxygen in oxides and nitrogen in nitride ceramics) and forms an interfacial film which facilitates the wetting of ceramics by the braze. Furthermore, the formation of an interfacial film improves the joint strength by reducing the residual stresses arising from thermal mismatch between ceramic and metal at the faying surface.

In spite of the progress that has been made in the development of suitable braze materials the heating of metals and ceramics is still carried out in conventional vacuum furnaces using very slow heating rates. Microwave radiation has been recently employed for rapid joining of engineering ceramics [2]. However, direct microwave heating cannot be applied for joining of metals to ceramics, because metals are microwave reflectors. In an attempt to accelerate the brazing process it was decided to utilise Microwave Induced Plasma (MIP) for rapid brazing of ceramic to metal.

Pioneering work by Bennett et al [3-4] in the late 60's has already established the possibility of using the plasma generated in a microwave field for rapid heating and densification of oxide ceramics. Bennett et al [3-4] reported higher density, smaller grain size and better room temperature mechanical properties for alumina sintered by MIP at 70 mbar, compared to conventionally densified materials, for the same temperature and time. The suitability of MIP for rapid brazing of alumina to copper was recently demonstrated [5]. It has been realised that MIP enables rapid and direct heating by the localisation of the plasma to the joint area. Further, the MIP system offers reduced processing times and lower energy costs over conventional processing methods. The main thrust of this investigation was to assess the feasibility of rapid brazing of silicon nitride to stainless steel by MIP system.

EXPERIMENTAL PROCEDURES

Polished hot pressed silicon nitride samples (5x5x3 mm^3) were joined to 316 stainless steel testpieces (10x10x3 mm^3). The braze material was Cusil ABA with a composition of Ag63.5/Cu35/Ti1.5.

Fig. 1 shows schematically the set up of the MIP system consisting of a 6 kW, continuous wave microwave power supply and magnetron (operating at 2.45 GHz). The microwave radiation is coupled to the cavity (brazing reactor) via a rectangular waveguide and an axisymmetrical launcher which is separated from the reactor by a quartz window. The plasma is generated and sustained in a tuned multimode stainless steel cylindrical cavity.

The brazing chamber is a double wall, water cooled stainless steel cavity with an inside diameter of approximately 20 cm and a height of 40 cm which is evacuated by a rotary pump to 0.01 torr and backfilled by hydrogen to the working pressure (5-50 Torr). The cavity walls are highly polished in order to reduce total surface area and therefore minimising the absorption of the microwave energy as well as reducing absorption of ambient gas and water vapours on the surface of the cavity walls.

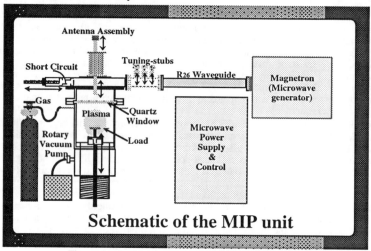

Figure 1. A schematic diagram of the MIP system.

MIP brazing was conducted in a hydrogen plasma. The chamber was evacuated to approximately 10^{-2} Torr and flushed with the process gas several times prior to brazing. A low pressure hydrogen atmosphere was then established in the chamber and partially ionised in the microwave field to form a plasma in the vicinity of the sample. The microwave power was increased incrementally in order to heat the sample until melting of the braze metal was observed. Chamber pressure was also increased in order to confine the size of the plasma. Melting of the braze was achieved in 15-20 minutes. This temperature was held for 15 minutes and the reverse of the heating schedule used as the cooling cycle. On extinguishing the plasma the sample was left to cool in the chamber for a further 20-30 minutes before air admittance and removal of the sample.

Scanning electron microscopy was used to examine the cross section of the joints, in particular, the interfacial layers forming between the silicon nitride and stainless steel with the braze material. Energy dispersive spectroscopy was utilised to perform elemental line scans across the joints.

RESULTS

Excellent joints were formed between silicon nitride and stainless steel even at the highest heating rate (50°C/min). Fig. 2 shows the microstructure of a typical joint. The braze material, approximately 20 μm thick shows characteristic silver-copper eutectic as well as some proeutectic copper rich phase as predicted by the copper-silver binary phase diagram [6]. The interface between braze and silicon nitride indicates the formation of a continuous interfacial layer. No crack or defect was detected at the ceramic interface.

Examination of the interfacial layer between the silicon nitride and braze at higher magnification, Fig. 3., clearly reveals the continuous nature of the interfacial film. The EDS elemental line scans are presented in Fig. 4. The left hand side of the profiles show the interfacial region between the silicon nitride and braze. It can be seen that at this interface there is a sharp increase in nitrogen and titanium contents, suggesting the formation of titanium nitride with a corresponding drop in silicon content. A noticeable increase in titanium concentration in silicon nitride can also be detected which indicates some interdiffusion.

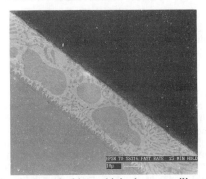

Figure 2. Micrograph showing a typical brazed joint between silicon nitride (dark layer at the top) and stainless steel sample.

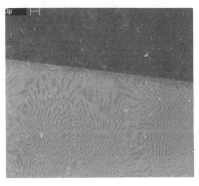

Figure 3. Details of the continuous interfacial layer formed between silicon nitride and braze material.

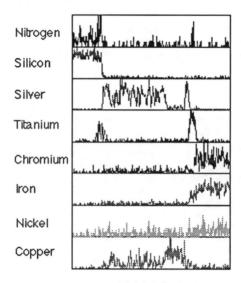

Figure 4. EDS elemental line profiles obtained across the brazed joint between silicon nitride and stainless steel (The ceramic is at the left hand side).

DISCUSSION

When brazing with MIP two facts are apparent. Firstly, much faster heating rates can be achieved than in conventional heating. Secondly, similar microstructures are obtained in MIP heating to those conventionally heated with much longer heating cycles (see ref. 7).

The reason that faster heating rates can be achieved in MIP brazing arises from the fact that MIP is a very direct form of heating. The microwave energy is concentrated in a small plasma which surrounds the sample and directly heats it by ion/atom bombardment. In the conventional heating, eg. by resistive heating elements, the entire system, ie. furnace and the sample, has to be heated to the brazing temperature. Obviously the thermal inertia introduced by the furnace precludes fast heating rates.

The exact reason for the formation of a thick interfacial layer of titanium nitride in MIP brazed joint is not very well understood. As mentioned previously, similar behaviour has been observed for much slower conventional heating rates which allows prolonged diffusion [7]. It is, therefore, not unreasonable to assume that enhanced diffusion, perhaps stimulated by microwave radiation, is responsible for improved interfacial reaction between the active element and the ceramic. In other words as the sample is MIP heated the dielectric loss of the silicon nitride increases too, thus allowing direct coupling with microwave radiation.

Finally it should be emphasised that the formation of titanium nitride between silicon nitride and reactive braze is not unexpected because this phase is thermodynamically favourable under the prevailing conditions in this work [8]. At the brazing temperature, the silicon nitride decomposes according to reactions {1} below. This decomposition occurs by the reaction of silicon nitride with titanium to form titanium nitride. This accounts for the substantial titanium nitride layer which was formed. It is this layer of titanium nitride which leads to the formation of strong bonds as suggested by Xian et al [9].

$$Si_3N_4 + 4Ti \longrightarrow 4TiN + 3Si \qquad \qquad \{1\}$$

CONCLUSIONS

This work has demonstrated, for the first time, the feasibility of using a microwave induced plasma (MIP) for reactive metal brazing of silicon nitride to stainless steel. The direct nature of MIP heating allows short heating up period and total brazing cycle time could be significantly reduced. Furthermore, enhanced diffusion of reactive element in the braze, possibly stimulated by microwave radiation, results in the formation of a thick interfacial film at the faying surfaces which is beneficial in providing strong bonds. The main interfacial layer formed between silicon nitride and braze has been identified as titanium nitride.

ACKNOWLEDGMENTS

The authors wish to thank the Co-operative Research Centre (CRC) for Materials Welding and Joining for providing financial support to carry out this work. Also, the suggestions by Dr C. Montross regarding specimen preparation and characterisation are duly acknowledged.

REFERENCES

1. S. Saito (Ed); Fine Ceramics, Elsevier Applied Science Publishers Ltd, USA, (1988).

2. M. Samandi and M. Doroudian, Plasma Assisted Microwave Sintering and Joining of Ceramics MRS Symposium Proceedings, Volume **347** Materials Research Society, Pittsburgh, Penn. p. 605-615, (1994).

3. C.E.G. Bennett and N.A. McKinnon, Sintering in gas Discharges, Nature, **217**, p.1287-88, (1968).

4. C.E.G. Bennett and N.A. McKinnon, Glow Discharge Sintering of Alumina, in Kinetics of Reactions in Ionic systems; edited by T.J. Gray and V.D. Frechette, Plenum Press, New York, p.408-412, (1969).

5. M. Samandi, Microwave Induced Plasma Joining of Ceramics, Conference Proceedings, Materials in Welding and Joining, Adelaide, Australian Institute of Metals and Materials. p.137-142, (1995).

6. M. Hansen and K. Anderko, Constitution of Binary Alloys (Second Edition), McGraw Hill, New York, (1958).

7. P. O. Santacreu, J. L. Koutny, J. D. Bartout, A Study of The Reactive Brazing Of A Silicon Nitride To Steel, Materials at High Temperature, **12**, pp293-299, (1994).

8. M.G. Nicholas, Active Metal Brazing, in Joining Of Ceramics, Edited by M.G. Nicholas, Chapman & Hall, London, (1990).

9. A. P. Xian and Z. Y. Si., Joining of Si_3N_4 Using Ag57Cu38Ti5 Brazing Filler Metal Journal of Materilals Science, **25**, p.4483-4487 (1990).

CONTINUOUS PROCESSING OF SHEET-LIKE MATERIALS BY MICROWAVE ENERGY

David A. Lewis, Stanley J. Whitehair, Alfred Viehbeck and Jane M. Shaw
IBM T.J. Watson Research Center, P.O. Box 218, Yorktown Hts, NY 10598

ABSTRACT

Initial results for the continuous microwave processing of wide webs using a new microwave applicator are presented. The results show that acceptable uniformity can be obtained using this system and that with the appropriate controls, a very tight process window can be maintained to produce preimpregnated glass cloth (prepreg) for use in circuit board manufacture.

INTRODUCTION

The production of films and sheet and the processing of coatings on films is commercially very important for products such as Mylar, cloths and fabrics, and paper products. A number of techologies are currently being utilized for these processes, including hot air, IR lamps and RF energy. The use of microwave energy for the processing has a number of distinct advantages over the other technologies, however, the lack of a sufficiently uniform microwave applicator and open process control loops have prevented the successful implementation of the technology.

One of the more difficult films to process is preimpregnated woven glass fabric (prepreg) which is utilized in the electronics industry as the precursor to printed wiring boards (PWB). This was a US$30B industry in 1994 with about 25% of the manufacturing based in the US. Prepreg is formed by dipping a woven glass cloth through a bath containing an epoxy-novalac resin diluted with a solvent, typically acetone or methlyethylketone (MEK) to adjust the viscosity and ensure adequate take-up of the resin onto the cloth. The coated cloth is then passed vertically up a tower and dried and the curing advanced to about 25% using hot air or, in some cases, infra-red radiation. While some of the air is recycled, most is exhausted or passed through an incinerator resulting in a thermally inefficient system.

The production of prepreg is a very good test for the microwave sheet processing technology because of the extremely tight tolerances required - higher than for most other processes. If the cure is too high, there will not be sufficient flow during the lamination step which forms the circuit boards, resulting in poor interlaminar adhesion and reliability concerns. If the resin is undercured, too much will "squeeze out" during the lamination step resulting in the dielectric properties of the circuit board, which acts as a transmission line structure, being out of specification, resulting in poor electrical performance. Although the drying step is self limiting for microwave processing (because the dielectric constant and loss factor both decrease sharply as the solvent content decreases), the curing step, which is somewhat concurrent with the drying and is subject to thermal runaway.

Microwave energy has all of the "typical" advantages in this application, including more efficient energy transfer, internal conversion of energy to heat, reduced heating lengths (which would allow towers with a lower height, resulting in lighter cloths and products being run without the threat of web breaks in the towers). However, the primary advantage of microwave processing in this application is the ability to efficiently process water based resins in the future, although there are definite advantages for todays solvent based resins. Current process towers can not handle water based resins.

As residence times are reduced, better process control is required to ensure that voiding and bubbles are eliminated and that the cure is advanced to the correct degree. The very fast response of microwave systems provide an ideal technology for implementing a fast reacting feed-back system.

In this paper, preliminary uniformity and processing results will be discussed.

EXPERIMENTAL

The web handling system is shown in Figure 1. It consists of two towers with 3 drawers each which can either be filled with microwave applicators or air baffles. Typically, the first drawer has been filled with baffles to allow for some air drying of the freshly impregnated cloth before it reaches the first microwave applicator. Initial studies have shown that the solvent content of MEK based resins decreases to about 20 % in 10-20 seconds at room temperature and remains relatively constant after that. By allowing an air drying stage, a large amount of the resin can be removed without the risk of bubble formation in the initial stages when it is most prevalent.

The handling system has tension controls, air gas measurement system to determine the solvent concentration in the exhaust flow and safety features which perform a shut-down if the solvent vapor concentration exceeds 25% of the lower explosion level (LEL) for the solvent being used. It also has capabilities for hot air drying to both provide baseline materials for a direct properties comparison and also allow supplemental heating to prevent solvent or water condensation on the surfaces of the otherwise cold microwave applicator surfaces.

The resin used is a high performance replacement for FR-4 epoxy novalac resin in MEK (approx. 55% solids) in a flame retardant formulation typical of electronics grade materials.

The microwave applicator consists of four zones, each with a dedicated variable power microwave source, operating at 2.45GHz, microwave power sensors and an infra-red pyrometer to monitor the temperature of the web.

Thermal profiles were measured using a Mikron IR Thermal imager, Model TH1100.

The electric field patterns in the microwave applicator were modeled using Maxwell by Ansoft, on an RS/6000, model 590 with 1GB of memory and 18GB of disk drive capacity. Typical models took from 1 to 48 hours, while models were run up to 7 days. The modeling times were primarily due to the large number of elements required to accurately model the very thin web in the large applicator accurately. Design optimization of the applicator was accomplished by first modeling the proposed changes, including launch structures, aperture designs, etc.

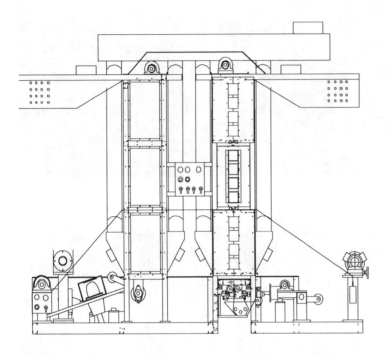

Figure 1. Schematic of lab-scale prototype prepreg tower showing the feed roller on the right and the take-up reel on the left and the two vertical towers approx 8' high in which up to 3 microwave applicators per tower are mounted.

RESULTS

In order to achieve the tight product tolerances, the electric field in the microwave applicator must be extremely uniform across the 12" wide web. Figure 2 shows thermal profiles of a continuously moving web exiting the microwave applicator under different applicator conditions. The applicator is at the bottom of the image and the web is moving vertically upwards. Hence, the web cools after exiting the microwave applicator and the cooling pattern can be clearly seen in Figure 2.

The web used to generate these heating profiles was a previously manufactured prepreg and is the most difficult case to heat since the dielectric constant and loss factor are the smallest they will be in the process. The images were recorded at a web speed of 3.5 m/min and the width of the web filled the x axis of the image at 30 cm (12 inches).

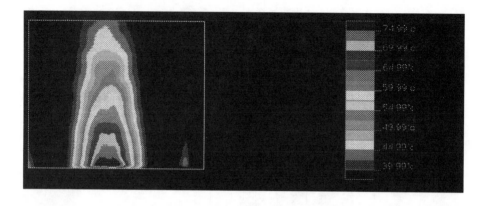

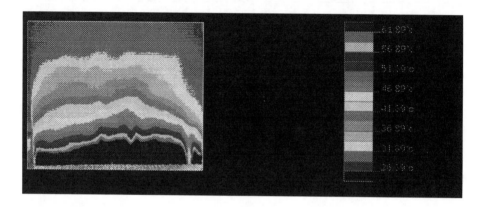

Figure 2. Thermal uniformity of a sample of glass cloth in the microwave applicator formed by holding the cloth stationary in the applicator and removing for a thermal image; (a) when the micrwoave energy is launched in the wrong manner; (b) when the energy is launched in the correct manner.

Figure 2a shows a very strong central heating zone with two small heating patterns on each side of the web. This pattern shows a very strong thermal gradient across the web and would result in a very non-uniform product which would be unacceptable for prepreg (and virtually any other non-self-limiting process). This profile is the result of poor applicator configuration and it is to be avoided. Figure 2(b) shows the thermal profile obtained when the microwave applicator is configured correctly. The relatively uniform profile across the web is acceptable for most applications, although further refinement is necessary, hopefully through optimization through FEA modelling.

It should be noted that the thermal heating patterns obtained experimentally match the electric field patterns obtained by FEA modeling extremely well, with a wide range of dielectric loads. This gives the confidence of the utility of this software and has allowed some design optimization already.

The applicator has four independently powered and controlled zones with an effective space of about 8 cm of relatively low electric fields between zones. As the web passes from zone to zone of the microwave system, the web cools before reaching the next adjacent zone. The rate of temperature decrease (cooling) is dependent on a number of factors, including convective losses (from the air moving across the surface), evaporative cooling (in the drying stage, which does not occur with the material shown in Figure 2) and infra-red losses. Taking a web speed of about 1.0 m/min, therefore, the web temperature can decrease from about 75C to about 48C between zones. At a web speed of 3 m/min, the cooling is only from 75C to 65C. These temperature losses are greater when the temperature is higher.

At the slowest speeds, the web essentially cools to ambient temperature between zones, while at higher speeds the amount of cooling is decreased. Hence, at faster web speeds, since the temperature jump required within adjacent zones is less than at slower speeds (due to less cooling), the microwave power used for each zone actually decreases for modest increases in web speed. This is shown schematically in Figure 3 which shows 3 different temperature profiles of a piece of web as it moves sequentially through each zone. It is desirable for some cooling to occur between zones to prevent thermal runaway and hot spot formation since the hottest regions cool more than surrounding regions, providing a more uniform dielectric material as it enters the next zone. Additional zones provide a tighter process control capability.

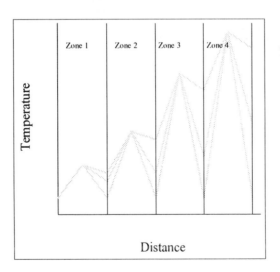

Figure 3. Idealized temperature profile of web as it successively passes through the 4 zones of the microwave applicator as measured by pyrometers placed at the center of each zone and between zones.

The prepregging process is very complicated with both solvent removal and curing taking place sequentially (although the two processes are very difficult to separate). Excessive temperature (due to the heating rate being too high) will result in bubble formation, hence the temperature - time profile must be tightly controlled.

One advantage of microwave processing is that extremely fast, controlled heating rates can be obtained, providing a wide range of temperature- time profiles for the material.

CONCLUSIONS

A microwave technology has been developed which allows the continuous, uniform processing of webs of material. Initial feasibility tests using manual control of applicator properties and microwave power settings have shown that preimpregnated woven glass fiber mats can be successfully produced with microwave energy in a very efficient manner, within the required tolerances. A greater understanding of the physical processes accompanying the processing are expected to lead to better process control and an optimized process. Automated process controls are expected to allow the more consistent fabrication. Scale-up of the applicator design to commercially useful sizes appears to be possible.

ACKNOWLEDGEMENTS

The authors would like to sincerely thank EPRI (Electric Power Research Institute) and NYSERDA (New York State Energy Research and Development Authority) for financial support of this project and the personnel at Cober Electronics for their collaboration in designing the microwave applicator being used in this work.

MICROWAVE MELTING OF ION-CONDUCTING GLASSES

D. J. DUVAL, M. J. E. TERJAK, S. H. RISBUD, B. L. PHILLIPS
CEMS Department, University of California, Davis, CA 95616, djduval@ucdavis.edu

ABSTRACT

Glasses of the system $AgI-Ag_2O-(0.95B_2O_3:0.05SiO_2)$ have been formed by microwave processing using a domestic multi-mode oven operating at 900 watts and 2.45 GHz. Microwave heating resulted in rapid melting times with homogeneity in the quenched glasses equivalent to or better than conventional melting at 730°C. The glass forming region in this pseudo-ternary system is compared with the conventionally melted glass forming region in the system $AgI-Ag_2O-B_2O_3$. A reversible color difference has been observed between glasses conventionally melted and those melted by microwave for all glass compositions in our system.

INTRODUCTION

Inorganic glasses exhibiting very high ionic conductivities, or Fast Ion Conducting glasses (FIC glasses), are being studied for use as solid electrolytes [cf. ref. 1] and chemical sensors [2] because of their ease of fabrication and isotropic conductivity. Mobility of silver ions in some of these glasses can be compared to that of Na^+ in a 5% saline solution [3]. The advantage of processing these glasses by microwave heating include rapid and uniform heating on a macroscopic scale, as well as selective heating on the meso- and microscopic scale [4]. A list of various glass formulations, including FIC and semiconducting glasses, processed by low powered domestic microwave ovens has been compiled in the literature [4]. However, a more detailed study of microwave processing for systems similar to $AgI-Ag_2O-B_2O_3$ had yet to be performed. These glasses are known to exhibit the highest ionic conductivities and highest glass transition temperatures [5,6]. Further, it has been suggested that mixing glass formers (e.g. SiO_2 and B_2O_3) may enhance resulting ionic conductivities [7,8].

This work is intended to present data and observations relating to microwave processing of various glasses in the system $AgI-Ag_2O-(0.95B_2O_3:0.05SiO_2)$. Comparisons will be made with conventionally melted glasses of this system, as well as with published data for conventionally melted glasses of the related system $AgI-Ag_2O-B_2O_3$.

EXPERIMENTAL

Preparation of raw materials was a key factor in creating homogeneous glasses. AgI, B_2O_3, Ag_2O and SiO_2 obtained form Johnson Matthey were ground together in the appropriate ratios in a mortar and pestle with acetone. Powders were then dried and passed through a 200 mesh screen. Ammonium nitrate from Fisher Scientific Co. was then added as an oxidizer such that it constituted 20 wt.% of the total batch, and then ground with the powder in a mortar and pestle with acetone. The resulting mixture was then dried overnight in a vacuum desiccator, passed through a 200 mesh screen, and pressed into pellets with a hand press for microwave melting. Powders to be conventionally melted were not pelletized. Whether in pellet or powder form, batch sizes were 10 g.

Conventionally melted glasses were melted in a box furnace in air at 730°C. The powder was poured as a single charge into preheated alumina crucibles and allowed to melt for 20 min. Pellets for microwave melting were deposited in an alumina crucible at ambient temperature and placed in a fiber refractory-lined 900 watt domestic microwave oven (Sharp Electronics Corp.) and allowed to melt for 2 to 2.5 min. at 100% duty cycle. All glasses were formed by quenching the melt between two copper plates at room temperature.

Specimens for Fourier transform far infrared analysis (FIR) were analyzed on a home-made spectrometer with a bolometer detector and a xenon lamp source in transmission mode. These specimens were prepared by mixing glass powders with polyethylene and pressing pellets. Fourier transform infrared analysis (FTIR) was performed on a Mattson Galaxy FTIR 3000 by mixing glass powder with mineral oil and measuring the spectra in transmission. Pieces of glass

125

Mat. Res. Soc. Symp. Proc. Vol. 430 © 1996 Materials Research Society

with parallel faces measuring ~0.5 mm in thickness were used for ultraviolet-visible spectroscopy (UV-VIS) using a Hitachi U-2000 spectrophotometer in transmission mode. X-ray diffraction data were collected from the glass powder with a Siemens D-500 diffractometer using Cu K_{α} radiation. [109]Ag spectra were obtained for the glass powders at 18.6 MHz with a Chemagnetics CMX-400 solid state NMR spectrometer.

RESULTS AND DISCUSSION

The eutectic composition of $0.95B_2O_3{:}0.05SiO_2$ was chosen as the constant molar ratio between the glass formers. We felt this would enhance melting and homogeneity of the resulting glasses. Molar ratios of other components were chosen for comparison with the published glass forming region in the $AgI-Ag_2O-B_2O_3$ system [5,9]. NH_4NO_3 (bp=210°C) was added as a mild oxidizer to minimize silver reduction during heating [9].

A minimum batch size of 10 g was required to couple with microwave power and melt, and pellets melted more readily than powder. Pelletization was not necessary for conventional melting. No more than 20 wt.% of the batch could be composed of ammonium nitrate, as its boiling caused the pellets to crumble; decreasing the efficiency of the specimen's microwave power absorption. Less ammonium nitrate resulted in silver reduction.

In order to see which of our glasses' components coupled with microwaves, pellets of the individual components were pressed and each placed into alumina crucibles inside the microwave oven. After heating for 4 min. at 100% duty cycle, only the AgI pellet got warm and started to melt. This proved that in our raw powder batches, only AgI crystals coupled with the microwaves and then heated the other powders.

After 15-30 sec. ammonium nitrate was observed to boil. This indicated that power was being absorbed, probably in a manner associated with size-limiting constraints which would decrease the expected heating rate [10]. After this time, a plasma was observed in the specimen. We suggest that the plasma resulted from dielectric breakdown occurring between adjacent AgI particles. Mobility of Ag^+ increases with temperature, thus allowing an induced polarization of AgI crystals. Above a threshold voltage, breakdown can occur, causing fusing or melting in the immediate region. This fusing will increase the characteristic length which limits power absorption, resulting in more efficient microwave heating. Arcing subsided in about 10 sec. in most cases, and rapid heating took place. Power absorbed probably followed the relation [11]:

$$P_{abs} = K'' \, VE^2/8\pi \qquad\qquad (1)$$

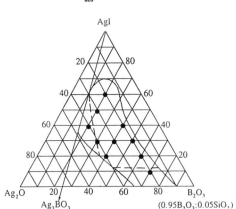

where the power absorbed (P_{abs}) equals the melt's imaginary dielectric constant (K'') times specimen volume (V) times the power of the microwave field ($E^2/8\pi$). Complete melting was assured when the alumina crucible glowed red where in contact with the specimen.

Figure 1. shows the glass forming region for microwave heating of our system in the dotted lines. Solid lines are shown for comparison with published data for the system $AgI-Ag_2O-B_2O_3$. It appears that too little AgI did not allow coupling with microwave energy, resulting in incomplete melting (low AgI limit on glass formation). It also seems that insufficient glass former ($0.95B_2O_3{:}0.05SiO_2$) inhibited the formation of a uniform glass, although significant melting had occurred.

Figure 1. Triaxial with glass-forming regions: a) published data for the system $AgI-Ag_2O-B_2O_3$ in the solid line and b) microwave melting of our system.

Conventional melting of these compositions formed a dark red glass. It is suggested that microwave processing may either enhance devitrification or provide insufficient heat to melt glasses with more than 50 mol% Ag_2O. Glasses from compositions inside the dotted lines were x-ray amorphous.

Most remarkably, all glasses processed by microwave appeared redder than conventionally melted glasses. Figure 2. indicates the UV-VIS absorption spectra for a metaborate and a triborate glass which underwent both microwave and conventional melting. Note red shift in absorption of the microwave-melted glasses. In both compositions, the absorption edge shifted from ca. 400 nm to 486 nm with microwave processing. Further tests indicated that the color change was reversible, i.e. red glasses became yellow after conventional melting, and yellow glasses turned red after microwave melting. Other compositions showed similar color trends, but were too opaque for routine UV-VIS analysis.

FTIR data collected for the diborate glass show a possibility of increased BO_4 concentration for microwave-melted glasses. Bands associated with BO_4 units [6] seem to decrease in relative intensity when compared with conventionally melted glasses (see Fig. 3.). More detailed infrared analyses on more specimens need to be performed before this claim can be substantiated.

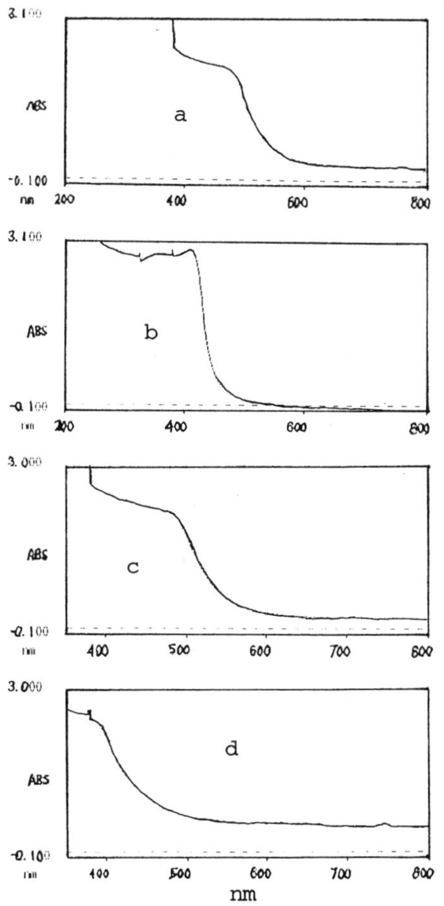

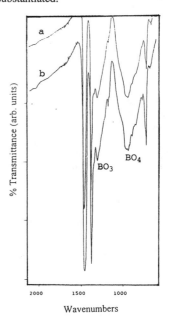

Figure 3. FTIR data for glass 2AgI-Ag_2O-2(0.95B_2O_3:0.05SiO_2). Note difference between BO_4/BO_3 ratios in a) microwave and b) oven melted glass.

Figure 2. UV-VIS spectra for a) microwave and b) oven melted AgI-Ag_2O-3(0.95B_2O_3: 0.05 SiO_2,) and c)microwave and d) oven melted 3AgI-Ag_2O-(0.95B_2O_3:0.05SiO_2)

NMR data showed that Ag^+ exist in similar chemical environments in both microwave and oven melted glasses of the same composition. A chemical shift of 530 ppm was observed for the well-defined peak of both glasses in a high AgI diborate glass (See Fig. 4.). Intensity of the metallic silver peak indicated equal (trace) amounts in both glasses. FIR results on a triborate glass substantiate this observation by showing identical bands at ca. 100 cm^{-1}. Published data [12] assign this band to a convolution of two Ag-O and one Ag-I interaction). NMR data did indicate, however, that residence times at a given site for Ag^+ are different for the two glasses: the microwaved glass exhibited T_1 relaxation times of one half those of the conventionally melted glasses (See Fig. 4.). Temperature studies will be required to elucidate the structural significance of this observation.

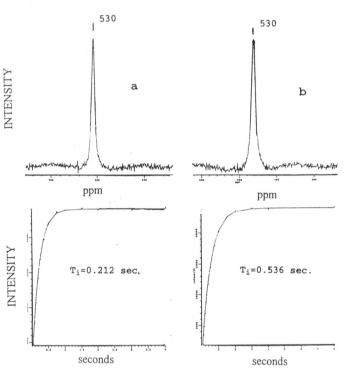

Figure 4. NMR spectra of Ag^+ in glass $2AgI-Ag_2O-2(0.95B_2O_3:SiO_2)$. Upper curves depict the Ag^+ chemical shift in a) microwave and b) oven melted glass. Lower curves show the T_1 relaxation times.

CONCLUSIONS

Our work thus far has shown that a wide range of glasses in the pseudo-ternary system $AgI-Ag_2O-(0.95B_2O_3:0.05SiO_2)$ can be processed by microwave heating in a multi-mode low-powered (domestic) microwave oven without the use of susceptors or supplemental heating. Glasses we melted by microwave heating are consistently redder than the same glasses conventionally melted. We believe the color differences may be due to the way silver ions couple with the microwave electric field, and this may have an effect on the structure of the glass matrix. Furthermore, the color change is reversible upon remelting of the glass frits. Since microwave melting involves significantly different mechanisms than conventional melting, it appears that local motion of the I^- and especially Ag^+ may have an effect on the structure of the glass matrix.

Three stages were observed for microwave melting of these glasses: 1) Local heating of AgI occurred slowly because of limitations on heating rate due to particle size as outlined in references 10 and 11. 2) A hot plasma was eventually generated when, we believe, the Ag^+ obtained sufficient mobility to polarize the AgI crystals in the microwave's electric field. Charge imbalance between neighboring crystals caused dielectric breakdown of the intervening material, generating a plasma which would fuse or melt the material locally. 3) Fusing/melting resulted in a significantly larger path length for Ag^+ ions to move under the microwave's driving electric field.

Size limitations on heating rate are then overcome, and heating rates could increase dramatically. Arcing (plasma generation) ceased once true melting began.

Overall melt uniformity was probably enhanced convection currents caused by large thermal gradients across the batch in the case of microwave melting. Much smaller thermal gradients are expected across conventionally melted glasses, hence requiring their longer melting times to achieve uniformity.

ACKNOWLEDGMENTS

We would like to thank Paul Bruins for providing the FTIR spectra. Scott Johnson and Michael Skolones are thanked for providing the FIR spectra, and Mark Niemeyer for help obtaining the UV-VIS spectra. Thanks are also extended to Peter Klavins for use of the x-ray diffractometer.

REFERENCES

1. A. K. Arof, K. C. Seman, A. N. Hashim, R. Yahya, S. Radhakrishna, Mat. Sci. Eng. B **31** 249-54 (1995)

2. Y. Sadaoka, M. Matsuguchi, Y. Sakai, J. Mat. Sci. Lett. **9** 1028-32 (1990)

3. T. Minami, J. Non-Cryst. Sol. **73** 273-84 (1985)

4. B. Vaidhyanathan, M. Ganguli, K. J. Rao, J. Sol. State Chem. **113** 448-50 (1994)

5. T. Minami, T. Shimizu, M. Tanaka, Sol. State Ion. **9&10** 577-84 (1983)

6. K. M. Shaju, S. Chandra, Phys. Stat. Sol. B **181** 301-11 (1994)

7. R. V. G. K. Sarma, S. Radhakrishna, J. Mat. Sci. Let. **9** (10) 1237-8 (1990)

8. F. Branda, A. Costantini, A. Buri, Phys. Chem. Glass **33** (2) 40-2 (1992)

9. M. C. R. Shastry, K. J. Rao, Sol. State Ion. **37** 17-29 (1989)

10. V. M. Kenkre, M. Kus, J. D. Katz, Phys. Stat. Sol. B **172** 337-47 (1992)

11. V. M. Kenkre, M. Kus, J. D. Katz, Phys. Rev. B **46** (21) 13825-31 (1992)

12. E. I. Kamitsos, J. A. Kapoutsis, G. D. Chryssikos, J. M. Hutchinson, A. J. Pappin, M. D. Ingram, J. A. Duffy, Phys. Chem. Glass **36** (3) 141-9 (1995).

MICROWAVE PROCESSING CaO-Al$_2$O$_3$-SiO$_2$ GLASS USING SOL-GEL TECHNIQUE

Y. Zhou*, O. Van der Biest*, C. Groffils** and P. J. Luypaert**
*Dept. of Metallurgy and Materials Engineering, Katholieke Universiteit Leuven, De-Croylaan 2, B-3001 Heverlee, Belgium.
**Dept. of Electrical Engineering, Katholieke Universiteit Leuven, Kard. Mercierlaan 94, B-3001 Heverlee, Belgium.

ABSTRACT

Gels in the CaO-Al$_2$O$_3$-SiO$_2$ (CAS) system were successfully converted into glass in a single mode, tuneable, cylindrical microwave applicator, operating at 2.45 GHz in the TM$_{012}$ mode. Transparent glasses were formed as a result of the direct microwave heating of homogeneous CAS gels to well above their melting temperature (1170°C). The effect of processing parameters, such as incident power level, sample mass and location, and thermal insulation has been investigated. The excellent coupling of the CAS gel with microwave can be understood from the structural evolution during sol-gel processing.

INTRODUCTION

In high temperature processing of glass and ceramics by microwave heating, a major difficulty encountered is the poor coupling of these materials with microwave energy at low temperature [1]. In this study a new processing route is investigated in which direct microwave heating is combined with sol gel technique to prepare glass.

For many glass and ceramic materials, such as SiO$_2$ and Al$_2$O$_3$, the dielectric loss factor rises very slowly with increasing temperature until the critical point is reached (T_c) at which the loss factor rises steeply with temperature. The low dielectric loss factor at lower temperatures limits glass and ceramic materials from absorbing sufficient energy and hence slow down the heating process.

Using sol-gel synthesis from metal-organic compounds or a mixture of inorganic and metal - organic compounds permits the mixing of constituents at a molecular level [2]. Microwave heating offers uniform thermal processing, which may preserve homogeneity in the starting gels and avoids any segregation of chemical composition due to thermal gradients during heating. Ideally, the chemical species necessarily involved in a sol-gel synthesis will initially couple well with microwaves and heat the gels as well as their solid decomposition products. If these products reach a temperature high enough so that their dielectric loss will cause them to continue to heat up to their critical coupling temperature and further to their melting point, glass formation will result. Thus, this route allows for glass processing to take full advantage of both sol-gel and microwave techniques.

EXPERIMENTAL PROCEDURE

Synthesis of CaO -Al$_2$O$_3$-SiO$_2$ Gels

The composition investigated was the ternary eutectic of 16CaO-9Al$_2$O$_3$-65SiO$_2$ (CAS). The composition was selected because it has the lowest liquid temperature (1170°C) in the CAS ternary system [3]. The precursors used were tetraethoxysilane (Si(OC$_2$H$_5$)$_4$, TEOS), aluminium sec-butoxide (Al(OC$_4$H$_9$)$_3$) and calcium nitrate tetrahydrate (Ca(NO$_3$)$_2$·4H$_2$O). The preparation method consisted of the following steps: (1) partial hydrolysis of TEOS based on

Mat. Res. Soc. Symp. Proc. Vol. 430 © 1996 Materials Research Society

the principle discussed by Yoldas [4]. The solution was prepared by thorough mixing of TEOS and solvent (C_2H_5OH), to which 1/4 stoichiometric molar water was added. HCl was added to catalyse the hydrolysis reaction, (2) adding aluminium alkoxide $Al(OC_4H_9)_3$ to the partially hydrolysed TEOS/ethanol mixture, (3) adding an ethanol dissolved $Ca(NO_3)_2 \cdot 4H_2O$ solution to TEOS/aluminium alkoxide mixture, (4) adding further stoichiometric amount of water to complete hydrolysis of all alkoxides. The resultant mixture was cast into glass dishes, and then gelled and dried in an air oven at 65°C for about 3 days.

Microwave Heating Equipment

Figure 1 is a schematic illustration of the specially designed microwave heating unit. Its main characteristics include: a microwave generator giving a variable power supply between 0 and 1.0 kW operating at 2.45 GHz; a circulator which protects the generator from damage by deviating possible reflected waves towards a matched water-cooled dummy load; a directional coupler for measuring the incident and reflected power; a rectangular wave guide; a transition section containing a 3-stub tuner which realises the coupling of the cavity to the waveguide; a single-mode cylindrical applicator closed with a moving plunger, which allows the control over resonance condition by manually varying the length of the cavity; reflection power meter; an electronic balance (M1212, Precisa); infrared pyrometer for temperature measurement (M90, Mikron); a data logger (DTL 1214) or computer.

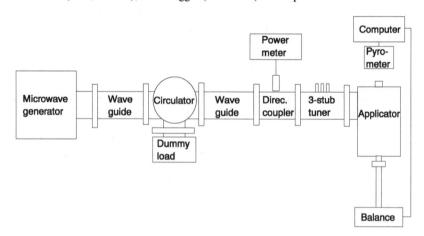

Figure 1 Block diagram of single mode microwave heating set-up.

During microwave heating, the gels were loaded in an Al_2O_3 crucible(99,7 % purity), which was surrounded by porous Al_2O_3 insulation to prevent heat loss. The centre of the specimen coincided with the centre of cavity along the axis, while b/a, i.e., the distance to the top end of cavity (b) relative to the length of cavity (a), is allowed to vary. It is not simple to measure the actual temperature of the sample with good accuracy in a microwave field. Our first attempts with a thermocouple showed that the thermocouple interfered with the microwave field and caused locally an excessive heat up, leading to measurement uncertainties. When using a pyrometer, the emissivity of the sample, which is temperature dependent and varies during the course of heating, could introduce errors. In this work, the emissivity value was calibrated against a thermocouple in a conventional furnace. Despite the limitations encountered, the temperature measured by the infrared pyrometer did provide a basis for some relative information on the thermal behaviour of the samples within this study. This pyrometer only starts to function above 250°C. The temperature reading was recorded using a data logger or computer. The electronic balance was connected to the computer to monitor the change of mass.

RESULTS AND DISCUSSION

The Effect of Input Power Level

Figure 2 shows the effect of microwave input power on the heating of gels, determined at 400 W, 500 W, 600 W and 800 W for a given mass of 10 g. The position of samples relative to the cavity was kept at b/a of 12 cm/24 cm.

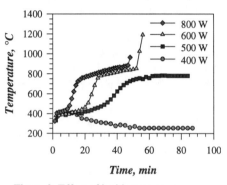

Figure 2 Effect of incident power on microwave heating of CAS gels with sample mass of 10 g.

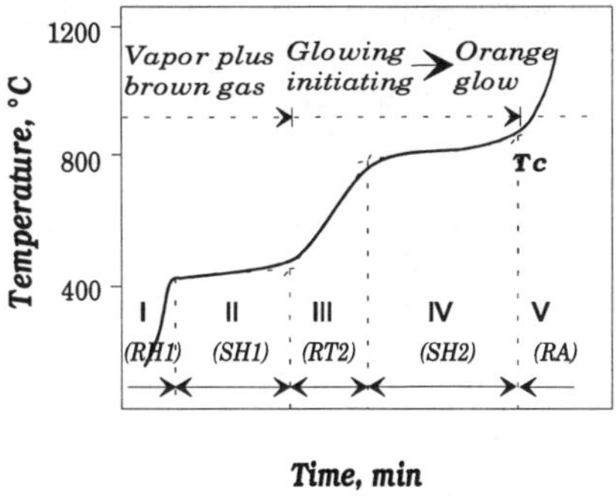

Figure 3 Schematic presentation of microwave heating features for CAS gels at constant power, displaying 5 characteristic heating phases.

The heating features exhibited can be categorised as five heating phases as schematically presented in Figure 3: rapid heating 1 (I), slow heating 1 (II), rapid heating 2 (III), slow heating 2 (IV), and run away (V).

In Phase I, gels showed a very rapid increase to about 400°C within 3 ~ 6 min, and the heating rate was very slightly increased with increase of power level. Some lossy materials are likely to be associated with this type of characteristic heating in Phase I. At this stage of heating, a large amount of water and alcohol was observed escaping from samples, followed by a release of brown gases with a distinct smell.

In Phase II, gels heated slowly to a maximum temperature of about 440°C, at much slower heating rates. The time needed to reach the maximum temperature was dependent on the input power. At 500 W, gels attained this maximum temperature at 30 min, but this time was reduced to 20 min at 600 W and 10 min at 800 W. The slight increase in temperature during this stage is due to a lack of microwave absorber. The dependence of heating rate on power level is because the build up of heat in the gels requires more time at low power inputs after a large amount of coupling materials are removed in the previous heating phase. At 400 W, gels remained fairly constant at 440°C for a few minutes, and then temperature decreases progressively below 250°C.

133

In Phase III, gels attained maximum temperatures of about 730°C ~ 780°C at a fast heating rate which increased with the increase in power. At 500 W, the gel reached this temperature at about 46 min, reduced to 28 min at 600 W and 16 min at 800 W. The fast heating characteristic in Phase III is due to strong interaction of some molecules with microwaves. The absorbing material at this stage of the gel process is considered to be related to calcium nitrate which is probably formed in the gels during the heating [3]. The nitrates are shown to be good coupling materials [5]. The presence of the nitrate coupling material at this stage is crucially responsible for the rapid increase in temperature, providing the coupling at intermediate temperatures.

In Phase IV, the temperature increases slowly, and the heating rate was not a strong function of input power. At 600 W or higher, the temperature eventually rose to the triggering point of approximately 880°C for the thermal run away to occur in CAS gels. At 500 W, a fairly constant temperature remained at 780°C at 30 min exposure. The slow temperature rise within this phase was caused by dielectric loss of CAS gel derived material with absence of liquid and impurity phase(s).

In Phase V, the temperature rose exponentially with increasing time, but no noticeable difference in heating rate was observed as power increased. Gels displayed run-away only at input power of above 600 W. When the heating reached critical point (T_c), further heating to above the melting temperature could be readily achieved. As a result, a dense and transparent glass was obtained.

This set of experiments shows that microwave power does have an effect on the heating behaviour of CAS gels. A power level below 500 W can allow for conversion from wet gel to dry. 600 W-input power is necessary for melting of the CAS gels. In fact a higher power level causes problems due to hot spot formation in the insulation and fracture of the crucible.

Effect of Sample Mass

Figure 4 shows the heating characteristics for CAS gels relative to sample mass of 6 g, 8 g, 10 g, and 12 g The position of the sample was the same as before. A 600 W input power was chosen due to its effectiveness in heating of gels to adequate high temperatures as demonstrated in the previous power level study.

The results showed clearly that CAS gel can only be heated sufficiently to T_c with the mass above 10 g. The heating pattern for 12 g was very similar to that for 10 g, with negligible difference in heating rate through all five phases. For sample mass below 10 g, however, the

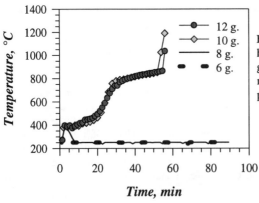

— 12 g.
— 10 g.
— 8 g.
— 6 g.

Figure 4 Effect of sample mass on heating characteristics of CAS gels during exposure to microwaves at 600 W input power.

gels showed a distinctive change in heating behaviour, no significant heating being noted during exposure to microwaves. The initial rise in temperature was related to the presence of good microwave coupling agents, e.g. water and alcohol, attaining a maximum of 400°C for 8 g and 360°C for 6 g in about 5 min. After which the heating ceased, and the temperature decreased quickly to well below 250°C.

Since the mass has such strong an effect on heating characteristics of CAS gel, it is necessary to verify the findings of mass effect at another incident power level of 800 W. The results are consistent with that at 600 W. For mass below 10 g, no significant heating for CAS gel could be induced at 800 W.

The poor heating of gels below the minimum mass is probably not due to significant dissipation of heat to the surroundings, otherwise heating to a certain temperature should occur with increasing time. The influence of sample mass was interpreted as a modification of the coupling of the cavity. In this case, the microwave field was unfavourably influenced due to shift of the cavity resonant condition, due to the change in sample mass which can affect the dielectric properties of a loaded cavity. The position of samples in the cavity was no longer situated at the position of maximum field. Thus, the mass and the volume of the sample relative to the size of the cavity are important factors. It is considered that these factors are material dependant, different from one material to another.

Effect of Sample Location

The tests were performed by microwave heating CAS gels in two different locations in the cavity. In case one, 14 g CAS gels were positioned at 2 cm below the middle of the cavity, i.e. b/a was 14 cm/24 cm. In case two, 12 g CAS gels were located precisely in the middle of the cavity, e.g. b/a of 12 cm/24 cm. A full incident power up to 1.0 kW was applied for both cases in order to observe the full heating process to high temperatures.

The importance of the position of the load in the cavity is demonstrated in Figure 5. The CAS gels positioned at 2 cm below the middle of the cavity could be only heated to a maximum of about 480°C after 30 min at 1.0 kW microwave exposure. When the sample was placed at the exact middle, temperatures reached 1380 °C after 23 min exposure at the same power level and there was no limit to the maximum temperature achieved in this case. This indicates that for effective heating, the location of the gels in the cavity is a significant factor.

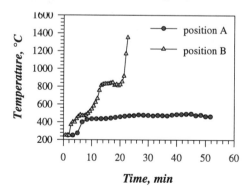

Figure 5 Effect of the position of the sample on microwave heating characteristics for CAS gels at 1.0 kW input power. Position A was for the gels positioned at 2 cm below the middle line of the cavity; Position B was at exact middle.

Thermal Insulation Effect

Figure 6 compares the heating results of gels with a used insulation and that for a new set. The power scheme shown is for the gels with the used insulation, and an incident power of 1.0 kW was applied to gels with the new insulation. Note the samples were loaded at position of b/a of 14 cm/22-23 cm. The gels were melted in a total heating time of about 20 min with the used insulation, whereas a rise in temperature only to a maximum of 500°C (see Figure 6 with triangle marks) during 50 min exposure is shown for the gel with the new set. That indicates that the insulation has a significant effect on the heating behaviour.

A meaningful phenomenon related to the used insulation was the generation of sound with a characteristic frequency, with evidence of hot spots generated in the insulation. Two experiments were designed to find out the cause of the sounds for a used insulation. First, heating without gel samples inside, keeping the other conditions unchanged and second, without sample and crucible. Surprisingly, the sounding and glowing were observed instantly as the microwave power was applied in the first case, but not in the second experiment.

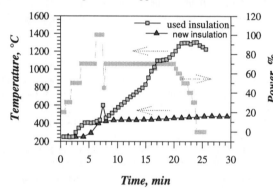

Figure 6 Effect of insulation on microwave heating behaviour of CAS gels. The power scheme shown is for used one and 1.0 kW was used for the new.

The high field established within the gap between crucible and insulation causes electrical breakdown, arcing and excessive heating, which produced a spike in the temperature profile at 7.5 min, as shown in Figure 6. Once the gels themselves started to couple significantly to the microwave field, a normal selective heating of gels by microwaves observed. The temperature was increased progressively till the melting point was reached. No run-away was observed because the position (14 cm from top end) and cavity length (22-23 cm) were not optimised for a maximum field. This limited the maximum temperature that could be reached to 1200 °C.

CONCLUSION

Transparent glass in the CAS system was successfully formed as a result of melting precursor gels by direct microwave heating. The heating features for CAS gels were found to consist of five phases, sequentially including (a) first rapid heating to 400°C in Phase I related to a large quantity of strong absorbers in liquid form, (b) slow heating up to a temperature of 480°C in Phase II associated with a small quantity of residual coupling agents, (c) second fast heating reaching 700°C in Phase III that crucially depends on the chemical nature of the gels (presence of nitrates), (d) second slow heating period progressively to critical point of 880°C in Phase IV due to temperature effect on dielectric loss factor, and (e) drastically increase in temperatures in Phase V resulting from exponential increase in dielectric loss.

For an effective heating, many processing parameters are important, including, the power level, the sample mass, the position of the sample relative to the cavity, the length of the cavity, the condition of the insulation. All the factors could greatly affect microwave heating behaviour of gels. The relationships established and the data and information collected in this part of the study provide insight as to the necessary operating conditions for sol-gel and microwave processing.

REFERENCES:

1. W. H. Sutton in Microwave Processing of Materials III, edited by R. L. Beatty, W. H. Sutton, and M. F. Iskander, Mat. Res. Soc.Symp. Proc. **269**, Pittsburge, PA, 1992, pp. 3-20.
2. C. J. Brinker, K. D. Keefer, D. W. Schaefer and C. S. Ashley, *J. Non-Cryst. Solids*, **48**, 47-64 (1982).
3. F. Pancrazi, J. Phalippou, F. Sorrentino and J. Zarzycki, *J. Non-Cryst. Solids*, **63**, 81-93 (1984).
4. B. E. Yoldas, *J. Mat. Sci.* **12**, 1203-1208 (1977).
5. T. T. Meek, C. E. Holcombe and N. Dykes, *J. Mater. Sci. Lett.* **6**, 1060-1062 (1987).

MICROWAVE SINTERING OF PURE AND DOPED
NANOCRYSTALLINE ALUMINA COMPACTS

R. W. Bruce,* A. W. Fliflet, Plasma Physics Division, D. Lewis, III, R. J. Rayne and B. A. Bender, Material Science and Technology Division, L. K. Kurihara,** G.-M. Chow, and P. E. Schoen, Center for Biomolecular Science and Engineering, Naval Research Laboratory, Washington, DC, 20375 - 5346

ABSTRACT

A single-mode cavity microwave furnace, operating in the TE_{103} mode at 2.45 GHz is being used to investigate sintering of pure and doped nanocrystalline alumina. The purpose of these experiments is to determine the effect of additives on the sintering process in the nanocrystalline regime. Using the sol-gel method, high purity Al_2O_3 nanocrystalline powders were synthesized. These powders were calcined at 700°C and then CIP'ed to 414 MPa, producing 0.4 in. diameter, 0.25 in. high cylindrical compacts. The compacts were heated in the microwave furnace to temperatures between 1100°C to approximately 1800°C and were then brought back to room temperature using a triangular heating profile of about 30 minutes duration. A two-color IR pyrometer was used to monitor the surface temperature of the workpiece. The additives tested in this work lowered the temperature needed for densification but this effect was offset by increased grain growth. Initial grain growth from < 5 nm to ~ 50 nm was closely correlated with the γ to α-alumina phase transition.

INTRODUCTION

The microwave sintering and densification of nanocrystalline alumina compacts is of current research interest[1-7]. The purpose of the present work is to study the effect of the doping of very pure nanocrystalline alumina on the heating, sintering and densification of this material. The addition of dopants (i.e., sintering aids) has been a long standing practice of the ceramics industry for achieving higher densification at lower temperatures than would be achieved with the pure material. A previous NRL study using a 2.45 GHz single-mode microwave furnace had focused exclusively on the very pure material and the results indicated that the final density was limited by the agglomeration of the nanocrystalline grains apparent in the green compacts [1-2]. Another study by Freim, et. al., has emphasized the importance of the γ to α phase transition as a cause of the anomalous grain growth during sintering of nanocrystalline alumina. In this study the effectiveness of using dopants to improve densification and reduce grain growth during sintering of nanocrystalline alumina has been investigated.

EXPERIMENTAL

Sample Preparation

Nanocrystalline powders were prepared using a modified sol-gel technique [8]. The precursor alkoxide was suspended in absolute ethanol while stirring. A solution of distilled de-ionized water (DDW) - ethanol (excess water) was made into an aerosol with nitrogen and sprayed into the alkoxide suspension. The reactants were stirred for an additional 30 minutes after the addition of the water solution and then filtered, washed with water and dried. The

---

* SFA, Inc., 1401 McCormick Drive, Largo, MD 20774-5322
** Department of Biochemisty, Georgetown University, Washington, DC 20057

Mat. Res. Soc. Symp. Proc. Vol. 430 © 1996 Materials Research Society

precursor powders were calcined in air at 700°C for 2 hours to remove the organic moieties. The dopants were either added during the sol-gel synthesis (in solution) or after calcination (as powder). The x-ray diffraction patterns for each of the calcined powders are given in Figure 1. These patterns indicate that the calcined powders were nanocrystalline (< 5 nm) with the additives appearing as microcrystalline peaks only for the nanocrystalline alumina with yttria.

These powders were then uniaxially pressed to 35 MPa and cold isostatically pressed (CIP'ed) to 414 MPa. Initial green densities are given in Table I. The samples were placed in a box oven at 200°C until the sintering operation to minimize the absorption of water vapor prior to sintering.

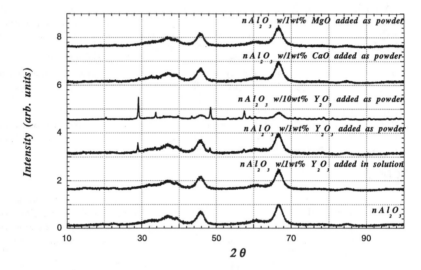

Figure 1. X-ray diffraction plots of the calcined powders.

Table I
Designation of Powders

Sample Designation	Dopant	Initial Rel. Density	Initial Grain Size
Al	Pure	41.6%	3.8 nm
AlYS1	1 wt% Y_2O_3 added in solution	41.0%	4.0 nm
AlYP1	1 wt% Y_2O_3 added as powder	42.2%	3.9 nm
AlYP10	10 wt% Y_2O_3 added as powder	42.1%	3.8 nm
AlCP1	1 wt% CaO added as powder	43.3%	3.4 nm
AlMP1	1 wt% MgO added as powder	43.4%	3.2 nm

Sintering Experiments

Each sample was placed in a casket made from fiber insulation and/or a susceptor material [9]. The casket assembly was then placed near the center maxima of the TE_{103} microwave furnace. Runs with various heating profiles were then made to determine the final densities that could be obtained at various peak temperatures (typically 1100°C through 1800°C) and dwell times (typically 0 to 30 minutes). The microwave power was increased during the run to bring the sample up to temperature in approximately 15 minutes. The cooling time of the samples was approximately 15 minutes, also. All sintering runs were done in air at one atmosphere of pressure. Surface temperature was monitored through a hole in the casket using a two-color pyrometer. Pyrometer temperature measurements were calibrated to a thermocouple using a tube furnace and nanocrystalline alumina compacts as the standard material.

RESULTS AND DISCUSSION

Densification

The green and sintered densities were calculated from the weight and dimensions of the compacts. The relative density of the sintered compacts as a function of peak surface temperature is plotted in Figure 2(A) for pure alumina and alumina with 10 wt% yttria. The data of Freim et al. for sintering of commercially available gas condensation synthesized nanocrystalline γ-alumina is shown for comparison [7]. Sintered densities ranged from little change for the lowest processing temperatures (~1100°C) to over 90% relative density for temperatures >1700°C. Densification data for nanocrystalline alumina with other dopants is shown in Figure 3. The scatter in these data is attributed mainly to problems of non uniform heating and thermal runaway in the somewhat lossy casket material. These effects could result in the interior temperature of the sample being substantially different from the monitored surface temperature. Severe cases of thermal runaway led to melting of the casket material and shifting of the sample. Improvements in casket design were implemented while these data were being obtained [9].

As expected, the NRL pure nanocrystalline material shows the least amount of densification at a given temperature. Among the doped materials, the samples containing Y_2O_3 densified the most with the 10 wt% samples achieving the highest values [Figure 2(A)]. Comparison of our data with the data of Ref. 7 for pure nanocrystalline alumina shows significantly less densification for our compacts particularly at the lower temperatures. This is attributed to the lower starting densities (42% versus 55% relative density) of our compacts even though they were cold-pressed at higher pressure than the compacts prepared in Ref. 7 (414 vs. 350 MPa). This difference in initial density may be caused by an increased agglomeration or polydispersion of our powders.

Particle Size

Subsequent to each sintering run, the average particle size of the densified compact was determined from an x-ray diffraction measurement (Scherrer technique). Data for the pure nanocrystalline alumina, the nanocrystalline alumina with 10 wt% yttria, and the data from Ref. 7 (both XRD and BET results) are shown in Figure 2(B). Grain size is shown to increase rapidly with densification—10% densification of the green compact causes the grain size to increase from a few nanometers to > 50 nm. At above 70% TD the grain size is ≥ 1 μm which is too large to be measured with the Scherrer method. Thus our grain size data points indicating 1 μm should be viewed as a lower bound on the actual size.

The data for the yttria-doped alumina indicates that the increased densification with temperature is offset by increased grain growth. Consistent with the results of Ref. 7, our data shows that the initial increase in grain size is closely correlated with the γ to α-alumina phase transition. This transition occurs at about 100°C higher temperature in the pure alumina than in the yttria-doped alumina. Compared to the other dopants, the yttria containing samples had a greater amount of grain growth at lower temperatures than did the other additives except for the calcia-doped samples which showed 1 μm sized grains as low as 1200°C. The magnesia doped samples showed the least amount of grain growth but also had the lowest sintered densities.

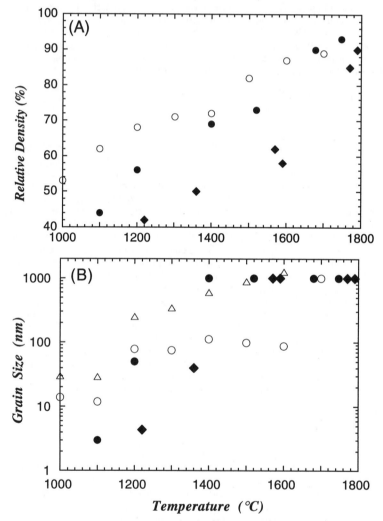

Figure 2. (A) Relative Density versus Temperature; (B) Grain Size versus Temperature. {Legend: ♦: Pure nanocrystalline alumina; •: nanocrystalline alumina with 10 wt% yttria; ○: X-Ray Data of Freim et al. [7]; △: BET Data of Freim et al. [7]}

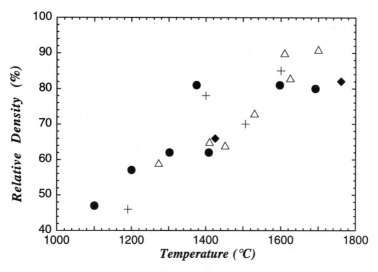

**Figure 3: Density versus temperature for nanocrystalline alumina with additives. Legend:
+: 1 wt% yttria added in solution; Δ: 1 wt% yttria added as powder; ●: 1 wt% calcia added
as powder; ♦: 1 wt% magnesia added as powder.**

CONCLUSIONS AND SUMMARY

From the analysis of the data, the following points can be made:

• The use of additives leads to densification at lower temperatures, but this densification is achieved at the cost of increased grain growth.

• The rapid grain growth that occurs during the initial phases of sintering seems to correlate with the γ to α phase change. This agrees with the data of Freim et al. [7]

• The onset of the γ to α phase transition occurred approximately 100°C higher in the pure nanocrystalline alumina compared to those with additives. Freim et al. observed a similar difference between the phase transition temperature of their pressed and un-pressed powder. They attribute the lower temperature transition in the pressed powder to a greater number of nucleation sites. The presence of dopants may also increase the number of such nucleation sites.

One goal of this study is to further explore the effect of calcination temperature upon agglomeration: lower calcination temperatures should yield less agglomeration and is the desired result. This should allow for higher green densities. Achieving the γ to α-alumina phase change without significant grain growth through an appropriate calcination method and/or microwave treatment will also be looked at. Finally, we intend to modify the cavity furnace to allow for an oxygen or vacuum environment to overcome the inhibiting effects upon densification due to the presence of nitrogen in the air atmosphere currently used [10].

ACKNOWLEDGMENT

This work is supported by the Office of Naval Research.

REFERENCES

1. R. W. Bruce, A. W. Fliflet, L. K. Kurihara, D. Lewis, III, R. Rayne, G.-M. Chow, P. E. Schoen, B. A. Bender and A. K. Kinkead in *Microwaves: Theory and Applications in Materials Processing III*, edited by D. E. Clark, D. C. Folz, S. J. Oda, and R. Silberglitt (*Amer. Cer. Soc.. Trans.* **59**, Westerville, OH, 1995) pp. 407-414.

2. A. W. Fliflet, R. W. Bruce, D. Lewis, III, R. Rayne, B. Bender, L. K. Kurihara, G.-M. Chow, P. E. Schoen and A. K. Kinkead, "Design and Initial Operation of a 6 kW 2.45 GHz Single-Mode Microwave Cavity Furnace," NRL Memorandum Report NRL/MR/6793--95-7745, 1995.

3. W. H. Sutton, *MRS Bulletin* **23** (11), (1993).

4. M. Aliquat, L. Mazo and G. Desgardin in *Microwave Processing of Materials II,* edited by W. B. Snyder, Jr., W. H. Sutton, M. F. Iskander and D. L. Johnson (*Mater. Res. Soc. Proc.* **189**, Pittsburgh, PA, 1991) pp. 229-235.

5. J. A. Eastman, K. E. Sickafus, J. D. Katz, S. G. Boeke, R. D. Blake, C. R. Evans, R. B. Schwarz and Y. X. Liao in *Microwave Processing of Materials II*, edited by W. B. Snyder, Jr., W. H. Sutton, M. F. Iskander and D. L. Johnson (*Mater. Res. Soc. Proc.* **189**, Pittsburgh, PA, 1991) pp. 273-278.

6. D. Vollath, R. Varma and K. E. Sickafus in *Microwave Processing of Materials III*, edited by R. L. Beatty, W. H. Sutton and M. F. Iskander (*Mater. Res. Soc. Proc.* **269**, Pittsburgh, PA, 1992) pp. 379-384.

7. J. Freim, J. McKittrick, J. Katz and K. Sickafus in *Microwave Processing of Materials IV,* edited by M. F. Iskander, R. J. Lauf and W. H. Sutton (*Mater. Res. Soc. Proc.* **347**, Pittsburgh, PA, 1994) pp. 525-530.

8. L. K. Kurihara, G.-M. Chow and P. E. Schoen, "Low Temperature Processing of Nanoscale Ceramic Nitride Particles Using Molecular Precursors," Patent Disclosure, US Navy Case 82,737 (1995), application pending.

9. R. W. Bruce, A. W. Fliflet and A. K. Kinkead, presented at the 1996 MRS Spring Meeting, San Francisco, CA, 1996 (unpublished).

10. R. L. Coble, *J. Amer. Cer. Soc.* **45**, 123 (1962)

PREPARATION OF YBCO PLATELETS AND FILMS BY MOD AND MICROWAVE HEATING

C. Manfredotti[a], M .Castiglioni[b], P. Polesello[a], N. Rizzi[a], P. Volpe[b]

a Experimental Physics Dept. University of Turin, Torino (ITALY)

b General and Organic Applied Chemistry, University of Turin, Torino (ITALY)

Abstract

Metal Organic Deposition (MOD) technique, which represents a flexible method for obtaining both platelets and films of YBCO, is based on baking, pyrolyzing and annealing stages of neodecanoates solutions. The annealing step is generally very long. The whole annealing stage can be simply substituted by a heat treatment (about 1 h) in a commercial microwave (MW) oven, followed by a 15 min. annealing at 900°C in a resistive furnace without oxygenation. With this process, YBCO platelets, films and wires were obtained starting from powders prepared by MO (Metal-Oxide). XRD patterns are indicative of a strong texturing by the evidence of (003) and (005) lines. By the same technique, YBCO films were prepared by spray deposition on Al_2O_3 for a total thickness of about 5 μm, each spray step, of 1 μm, was followed by 5 min pyrolysis at 500 °C and 1 h MW annealing.

The MW technique was also used for preparation of quartz and ceramic fibers covered by YBCO.

Introduction

The $YBa_2Cu_3O_{7-x}$ (YBCO) perowskitic phase with $0<x<0.2$ is a well known and studied ceramic superconductor, though other superconductor ceramics with higher critical temperature have been synthesized. The main advantage is the simple and easily reproducible, though time consuming, preparation.

In our laboratories the MOD technique has been used for the preparation of YBCO pellets [1], and for the deposition of thin YBCO films on flat surfaces and on metal or ceramic wires. The MOD method requires three steps: the deposition of the MO (Metal-Oxide) compounds on the supports, the pyrolysis of the MO and the long annealing under oxygen atmosphere. MW heating of the reacting mixtures for the preparation of YBCO, proposed by Mingos et al.[2], has been used by us for the superconductor synthesizing also from the MO starting materials and has been demonstrated to be safe and time saving.

Mat. Res. Soc. Symp. Proc. Vol. 430 © 1996 Materials Research Society

Materials and methods

Organic salts (metalorganic) as precursors are usually neodecanoates. Their preparation has been performed following the reactions:

$$NH_3 + HOOC_{10}H_{19} \rightarrow NH_4OOC_{10}H_{19}$$
$$nNH_4OOC_{10}H_{19} + M(C_2H_3O_2)_x \rightarrow M(OOC_{10}H_{19})_n + CH_3COONH_4$$

where M is yttrium, barium or copper. The salt mixtures, in stoichiometric proportions to obtain the desired superconductor -YBCO in this case- are directly annealed to obtain the powders to be pressed in pellets. For comparison some pellets were prepared starting from the mixtures of the inorganic compounds, namely Y_2O_3, CuO, using as a Ba source one of the followings: BaO, BaO_2, $BaCO_3$ or $Ba(NO_3)_2$ [3, 4].

For film deposition the salts are first dissolved in 1:1 pyridine-xylene mixture and subsequently sprayed with a commercial aerograph onto the substrates and pyrolized at 650 °C; the two last steps are repeated many times. The wires are covered with a similar procedure, but substituting the spray with a simple dipping.

The conventional sample annealing is performed in a tubular oven (Carbolite Co) under oxygen flux for very long times (about 3 days for pellets, 1 day for films and wires); alternatively the annealing was performed in a commercial microwave 2.45 GHz oven (SMEG Co., 600 W) for about 1 h, sometimes followed by only a 10-15 min annealing at 900°C in a resistive furnace without further oxygenation.

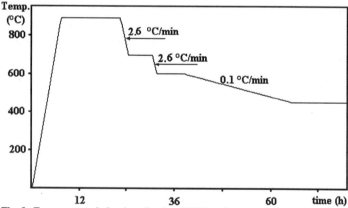

Fig. 1- Temperature behaviour for the YBCO pellets annealing in a resistive oven.

Sample preparation and characterization

A)Platelets. Many pellets have been prepared using different methodologies with the aim both to compare the obtained materials and to understand the mechanisms involved.

1) Pellets from neodecanoate solution and thermally annealed.

The solution of neodecanoates was dried under stirring for 3 h, then ground and annealed at 900 °C for 3 h, ground again, cold pressed at 10 ton/cm^2 and finally annealed following the temperature program showed in Fig.1

2) Pellets from salts and oxides mixture annealed in MW oven..

A mixture of the Y_2O_3, BaO and CuO powders appears unreacted after annealing in MW (600 watt) oven for 60 min. (two 30 min. steps); this fact is attributable to the not very high temperature reached by the barium oxide with respect to its melting point (m.p. 1918 °C).

Substituting the oxide with the barium peroxide (m.p. 450 °C) in the above stoichiometric mixture, the YBCO is obtained as $YBa_2Cu_3O_{6.5}$ by a MW heating for 30 min.

Barium nitrate also decomposes at relatively low temperature (about 600 °C), losing NO_2. Then to prevent a damage to the oven, it should be purged with an air flow to remove the gases. Generally the YBCO is formed after 2 or 3 annealing cycles of about 30 min, but with low oxygen content. Higher oxygenation to obtain $YBa_2Cu_3O_7$-d with d<0.2 can be obtained by a further heating step in a resistive oven for 24 h at 450°C under O_2 [5].

If Ba carbonate is used instead of the nitrate, ventilation of the oven will facilitate the loss of CO_2 by the carbonate.

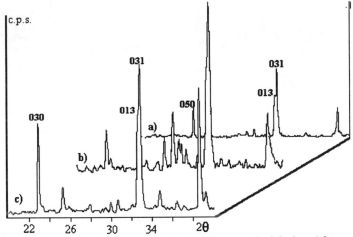

Fig. 2 - XRD spectra of platelets prepared with different methodologies: a) from neodecanoates and annealed in the resistive oven; b) from inorganic oxides(BaO_2) and annealed in MW oven; c) from neodecanoates and annealed in MW oven

3) Pellets from neodecanoates annealed in the MW oven.

The powders resulting from the pyrolysis of the salts solution are ground, then MW annealed for 15 min. and ground again. The powders are then pelletized by pressing at 10 ton/cm^2. The YBCO phase is then obtained by a further annealing (1 h) in MW oven. If the MW oven is modified to eliminate the residual CO_2 no further annealing under O_2 in the resistive oven is needed (see pellets from Ba carbonate).

Fig. 2 shows XRD spectra of pellets prepared by procedure 1), 2) (barium peroxide) and 3) for comparison.

MW annealing shows the advantage (at least for YBCO) of a very short reaction times in comparison with the conventional annealing. Moreover it shows an exclusive property: it can be used to recrystallize and to improve the crystal texture of the YBCO platelets . In fact it is known that repeated annealing can produce an increase of the microcrystal dimensions, but during the slow increase of the temperature, as in a resistively heated oven, the "green phase"(Y_2BaCuO_5) builds up; the very fast heating of the particles caused by MW absorption and the following very fast cooling as the MW are turned off prevents the formation of the green phase. Fig. 3 and 4 show an YBCO platelets before and after a reannealing by microwaves.

Fig. 3 MOD prepared YBCO platelet   Fig. 4 Same as Fig. 3, reannealed by MW.

The scale is 2 μm for 3 mm   The scale is 2 μm for 3 mm

B) Films. The films deposition was done by spraying the MOD solution onto different substrates, the best results were obtained using silver as a substrate . The conventional annealing , performed in O_2 atmosphere, is time consuming mainly because of the slow cooling step, which requires at least 500 min. The samples prepared for the microwave annealing were sprayed onto Al_2O_3 because it is transparent to microwaves. The complete annealing step in MW oven takes only 1 h.

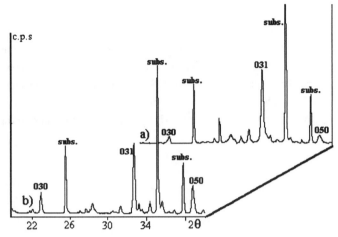

Fig. 5 - XRD analysis of films prepared by MOD and sprayed on Al₂O₃
a) annealed in a resistive oven; b) annealed in a microwave oven

If the oven is modified to purge the carbon dioxide released from the $BaCO_3$, the sample does not need further treatments. Fig. 5 shows XRD comparative analysis of films obtained by the two (conventional and MW) processes.

Wires. Wires covered with MO salts mixture were obtained by dipping 0.125 - 0.250 mm diameter of Ag, Cu or other metals wires for the conventional annealing, which do not differs from the one performed for films. The samples annealed in MW oven were obtained by covering quartz or ceramic fibers; the annealing was 1h long for all the samples which result relatively smooth in surface, as shown by the comparison of SEM images reported in Figs. 6-7

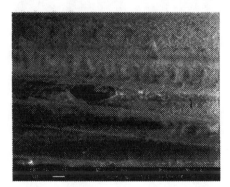

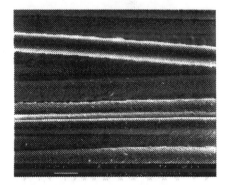

Fig. 6: YBCO covered ceramic fiber annealed in a resistive oven

Fig. 7: Same as Fig. 6, annealed in a MW oven

Conclusions

MW 2.45 GHz heating both speeds up the annealing step for YBCO platelets and improves their quality, in particular way the texturing. The advantages for film annealing could be more widely investigated by using better substrates than MgO ($SrTiO_3$, for instance). As regards wires, metals should be avoided and the process is more complicated by the lack of suitable substrates. Nevertheless, the results are promising, particularly towards a faster annealing process for YBCO.

Acknowledgements

The authors acknowledge Prof. G. Chiari for diffraction measurements.

References

[1] C. Manfredotti, F. Fizzotti, M. Boero, P. Rellecati, L. Mauro, P. Volpe, D. Andreone, Mat. Res. Soc. Symp. Proc., **275**, 257 (1992)
[2] Mingos et al., Nature, **332**, 311,(1988).
[3] Laishevtseva et al., Russ.J.Inorg.Chem., **37**, 855 (1992).
[4] Zakharon et al., Russ.J.Inorg.Chem., **35**, 163, (1990).
[5] Chandrasekhar Rao et al., Superc.Sci.Tech., **7**, 713, (1994).

NOVEL MICROSTRUCTURES IN MICROWAVE SINTERED SILICON NITRIDE

M. E. Brito#, K. Hirao, M. Toriyama and M. Hirota*
National Industrial Research Institute of Nagoya, Nagoya 462, JAPAN
*Industrial Technology Department, New Energy and Industrial Technology Development Organization (NEDO), Tokyo 170, JAPAN
#e-mail: brito@nirin.go.jp

ABSTRACT

Preliminary results on microwave sintering of seeded silicon nitride show that a well defined bi-modal grain size distribution is attainable in Si_3N_4-Y_2O_3-Al_2O_3-MgO sintered bodies by microwave sintering at 28 GHz of materials seeded with ß-Si_3N_4 particles (2 vol. %). A positive effect on the mechanical performance is anticipated for these microstructurally controlled silicon nitride ceramics

INTRODUCTION

Microwave sintering of ceramics materials has earned the attention of researchers pursuing the production of ceramics with unique microstructures and properties currently unattainable by conventional processing.

As for oxide ceramics (e.g. alumina and zirconia), it has been recognized that microwave heating affects sintering and grain growth rates[1,2]. Tiegs et al. [3] demonstrated a similar trend for silicon nitride ceramics sintered with Y_2O_3 and Al_2O_3 as additives, being this a material that need to be densified via a liquid phase. Considering the low microwave absorption factor of Si_3N_4, heating in a microwave field should be associated then to the preferential heating of the additives, and heating of the liquid phase generated by eutectic reactions of the additives with the silica present at the surface of Si_3N_4 raw powder, during the process at high temperatures. Plucknett and Wilkinson[4] reported on unusually grain size distribution in microwave sintered silicon nitrides. Microwave sintering at 2.45 GHz resulted in a very fine grain size with a distribution narrower than the generally observed in conventional sintered materials. Among the possible explanations to this grain refinement, the authors invoked a higher nucleation rate which in turns is the result of mass transport rate enhancement. The supersaturation of dissolved Si_3N_4 in the intergranular liquid becomes then a cause for the creation of numerous nucleation sites.

On the other hand, to improve mechanical properties of silicon nitride ceramics, Hirao et al. have proved the efficiency of seeding this material with ß-Si_3N_4 crystals of regulated morphology[5]. The seeding allows the development of bi-modal microstructures where elongated grains are well distributed in a fine grained matrix. Figure 1 schematically shows the distribution in grain size obtained by seeding. Further improvements in mechanical performance can be anticipated if the microstructure of silicon nitride can be controlled in such a way that a narrower grain size distribution is attained in both the matrix and in the elongated and thick grains. This work presents preliminary results on the microstructural control of silicon nitride by combining microwave sintering and seeding techniques.

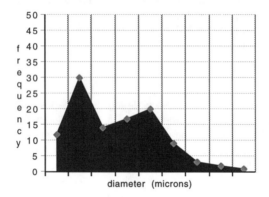

Fig. 1 Schematic of grain size distribution based on experimental data[5] of seeded and non-seeded silicon nitride sintered by conventional methods.

EXPERIMENTAL

Microwave sintered samples were prepared by mixing $\alpha-Si_3N_4$ powder (E10 grade, UBE Industries Ltd., Japan) with 5 wt. % Y_2O_3, 2 wt. % Al_2O_3 and 1 wt. % MgO sintering aids. Seeds consisted of $\beta-Si_3N_4$ particles with a mean diameter of 1.4 μm and a mean length of 5.6 μm. Samples with 2 vol. % seeds addition and without seeds addition were microwave sintered in a multimode cavity operating at 28 GHz. Experiments were conducted under flowing nitrogen atmosphere at temperatures varying between 1600 and 1750 °C and for sintering times of 2 and 6 hr. For the sake of comparison, a similar set of experiments were conducted using conventional heating techniques. For all the experiments, heating and cooling rates (20 °C /min and 10 °C/min respectively) were fixed to match those attainable using conventional heating.

The densities of all samples were determined via immersion in distilled water. The microstructure was evaluated from plasma-etched, polished surfaces using scanning electron microscopy (SEM) and the identification of crystalline phases was obtained using X-ray diffraction (XRD).

RESULTS AND DISCUSSION

The densities of micro-wave sintered samples were slightly higher than that of conventional sintered. However, all the samples tested could be densified up to

97.4% of their theoretical density. XRD patterns showed a complete transformation from the original α-Si$_3$N$_4$ phase into the β-phase for all the samples, independently of the used heating technique.

A comparison of the microstructures developed during conventional and micro-wave sintering of non-seeded silicon nitride is found in Fig. 2. The samples were sintered at 1700 °C for 2 hr. The effects of microwave sintering on the microstructure homogenization and on grain growth are evident from the micrograph of Fig. 2b.

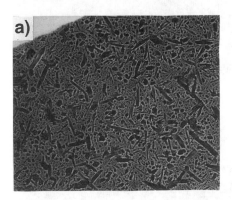 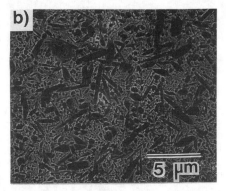

Fig. 2 SEM micrographs of polished and plasma etched cross sections of conventionally (a), and microwave (b) sintered samples.
Sintering conditions: 2 hr @ 1700 °C.

A similar trend is found for seeded samples. The low magnification SEM micrographs of Fig. 3 reveal the tendency to develop bi-modal microstructures in both, conventional (Fig. 3a) and microwave sintered (Fig. 3b) materials. Besides the enhancement in grain growth observed, it seems to be differences regarding population of elongated and thick grains, being it higher in the microwave sintered sample.

More detailed characteristic of these microstructures can be observed at the higher magnification micrographs of Fig. 4. The characteristic core-rim contrast observed in bigger grains is due to slight changes in composition between highly pure Si$_3$N$_4$ seeds and dilute β-sialons (Al and O incorporated into the lattice) that grow from seeds.

Although, as expected, bigger grains grow from seeds in both cases, microwave sintered microstructures presenta distinguishable group of uniform thick and elongated grains than the conventionally sintered ones. This fact suggest that mass transport conditions nearby seeds under microwave radiation is much more uniform. The broadening in grain size distribution for both elongated and matrix grains in turn can be attributed to non-uniform heating conditions within the samples sintered by conventional methods.

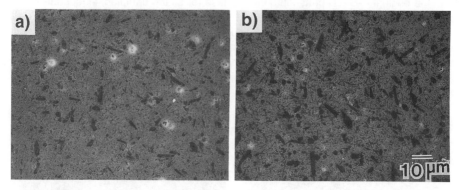

Fig. 3 SEM micrographs of polished and plasma etched cross sections of conventionally (a) and microwave (b) sintered and seeded samples. Sintering conditions: 2 hr @ 1700 °C.

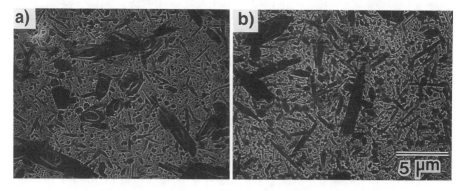

Fig. 4 Details at higher magnification of polished and plasma etched cross sections of conventionally (a) and microwave (b) sintered and seeded samples. Sintering conditions: 2 hr @ 1700 °C . Notice the charactersitic core-rim contrast in the larger grains.

The effectiveness of combining seeding techniques to microwave sintering to develop ideal bi-modal microstructures in silicon nitride is demonstrated in Fig. 5, which compares microstructures of non-seeded (Fig. 5a) and seeded (Fig. 5b) materials microwave sintered at 1750 °C for 2 hr. Needless to say the microstructures seen in Fig. 5b with well distributed thick and elongated grains within a fine grained matrix and with a narrow grain size distribution approaches the ideal bi-modal grain size distribution schematized in Fig. 6. The figure shows two well defined groups

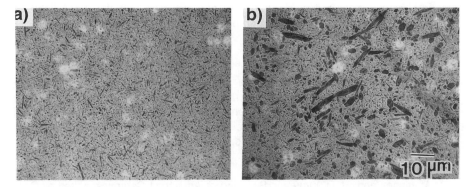

Fig.5 SEM micrographs of polished and plasma-etched cross sections of microwave sintered samples without seed addition (a) and with seed addition (b). Sintering conditions: 2 hr @ 1750 °C.

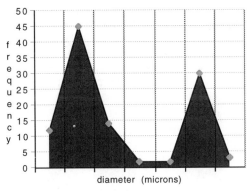

Fig.6 Schematic of the ideal grain size distribution for bi-modal microstructures that should be reached by seeding and microwave sintering silicon nitride. The figure does not shown experimental data but idealized grain size distribution.

corresponding to matrix grains and grains grown from seeds, presenting both narrow grain size distributions.

A quantitative analysis using image analyzer and other related techniques is indeed required to reach a conclusion. Nevertheless, the uniformity in grain growth achieved using microwave sintering and/or annealing[6] is remarkable and indicate us that an accurate control of microstructures in silicon nitride is possible and it will depend heavily on the original grain size distribution of seeds.

Speculating on the probable mechanical properties displayed by this type of microstructures, it can be said that as narrower the grain size distribution of the two groups of grain, further improvements can be attained in strength, fracture toughness and reliability (Weibull modulus) of silicon nitride ceramics.

ACKNOWLEDGMENTS

Research is partially sponsored by the New Energy and Industrial Technology Development Organization (NEDO), Japan. The authors gratefully acknowledge fruitful discussions with Mr. T. Saji (Fuji Dempa Kogyo KK). Experimental support and important suggestion by our colleague Mr. Y. Shigegaki (Synergy Ceramics Laboratory, Nagoya) are really appreciated.

REFERENCES

1. M.A. Janney and H. D. Kimrey in Sintering of Advanced Ceramics, edited by C. A. Handwerker, J. E. Blendell and, W. A. Kaysser (Ceramic Transactions 7, Am. Ceram. Soc., Westerville, OH 1990), p. 328.

2. M.A. Janney and H. D. Kimrey in Microwave Processing of Materials -II, edited by W. B. Snyder, W. H. Sutton, D. L. Johnson and, M. F. Iskander (Mater. Res. Soc. Proc. 189, Pittsburgh, PA 1991).

3. T. N. Tiegs, J. O. Kiggans and H. D. Kimrey, Ceram. Eng. Soc. Sci. Proc. 12, 1981 (1991).

4. K. P. Plucknett and D. S. Wilkinson, J. Mater. Res. 10, 1387 (1995).

5. K. Hirao, T. Nagaoka, M. E. Brito and S. Kanzaki, J. Am. Soc. 77, 1857 (1994).

6. M. Hirota, M.E. Brito, K. Hirao, K. Watari, M. Toriyama and T. Nagaoka, this volume.

A COMPARISON OF MM-WAVE SINTERING AND FAST CONVENTIONAL SINTERING OF NANOCRYSTALLINE Al_2O_3

G. Link[1], V. Ivanov[4], S. Paranin[4], V. Khrustov[4]
R. Böhme[2], G. Müller[5], G. Schumacher[2], M. Thumm[1,3], A.Weisenburger[2]
Forschungszentrum Karlsruhe; [1]ITP, [2]INR; P.O.Box 3640, 76021 Karlsruhe, Germany
[3]University Karlsruhe, IHE, Kaiserstr. 12, 76128 Karlsruhe, Germany
[4]Institute of Electrophysics, RAS, Komsomolskay 34, Ekaterinburg 620219, Russia
[5]Institut für Plasmaforschung, Pfaffenwaldring 31, 70569 Stuttgart, Germany

ABSTRACT

The phase transformation and densification behavior under high power millimeter-wave (mm-wave) radiation of a 30 GHz gyrotron and during fast conventional sintering of nanocrystalline γ-Al_2O_3 powder have been investigated and compared. The powder used for compacts was synthesized from aluminum metal by application of the exploding wire technique in an oxidizing atmosphere. The particle size distribution of this powder has a maximum at about 20 nm. Magnetic pulse technique was applied for the compression of samples up to 80% of the theoretical density (TD). Both mm-wave sintering and fast firing in a conventional electrical resistance furnace enable the densification and a complete phase transformation into α-Al_2O_3 already at a temperature of approximately 1150 °C. The average grain size of the sintered ceramic is in the range of 50 to 100 nm. With mm-waves densification starts at about 50 °C lower temperatures compared to conventional techniques and higher final densities were obtained already at 150°C lower temperatures.

INTRODUCTION

The demand of structural ceramics is growing steadily. Many contemporary ceramic materials are superior to metals in hardness, strength, chemical resistivity, wear resistance, high melting temperature or low density. However the use of ceramics is seriously hampered by their brittleness, their high energy consumption for manufacturing and the high cost for net-shape machining.

Advantages of ultrafine microstructures for synthesis, processing and property modifications of advanced ceramic materials have long been suspected. Properties of nanostructured materials are directly related to their unusual microstructure that features extremely small grains and a large fraction of highly disordered interfaces [1]. Research is not only motivated by basic science aspects but also by the expectation that novel properties will lead to new technological applications. For example sintering and creep are altered drastically by the reduction of the particle size to the nanometer range. The equations for densification rate $d\rho/dt$ and strain rate $d\varepsilon/dt$ show a strong dependence on grain size d, i.e. $(1/d)^q$, with q in the range of 2 to 4. Thus it becomes possible to tailor the mechanical properties by controlled grain growth.

In order to achieve improved material properties with nanocrystalline ceramics, one has to overcome the problem of grain growth during the sintering process. The diffusion controlled processes of grain growth and densification which are strongly dependent on temperature. It is apparent that the kinetic processes which accompany conventional sintering schedules favor grain growth over densification and the enhanced sintering behavior of nanocrystalline ceramic powder cannot be exploited.

Nowadays, two methods of conventional sintering exist which allow to some extent to overcome the problem of the dramatic grain growth, i.e. hot pressing and hot isostatic pressing [2]. But both of these methods have their own disadvantages such as very high cost, capability of operating with a limited assortment of materials and with workpieces of the simplest shape. Therefore, an adequate procedure for densification of nano-size ceramic/composite powders is still lacking.

Great interest was concurrently aroused in the new approach to high temperature sintering of ceramics based on using microwave energy for their heating. It has been demonstrated that this is an effective technique for sintering of nanostructured TiO_2 and ZrO_2 to high densities without the occurrence of significant grain growth [3;4]. But microwave sintering of nanostructured γ-Al_2O_3 produced by the gas condensation method was not very successful up to now. Strong grain growth has been observed during the polymorphic phase transformation from γ-Al_2O_3 to α-Al_2O_3 which adversely affects the sintering behavior, leading to low final densities and large crystallite sizes [5]. The sintering behavior of nanostructured γ-Al_2O_3 synthesized by the exploding wire technique and compacted by a magnetic pulse technique which allows green densities of up to 80% of the theoretical density has been investigated. The influence of green density and different sintering techniques on densification will be discussed.

EXPERIMENTS

Conventional Heating Technique

Nanostructured alumina compacts have been sintered in a conventional electro-resistive furnace at the Institute of Electrophysics in Ekaterinburg, Russia, with program control at temperatures up to 1400 °C. Several modes of heating the samples were applied: linear heating with rates of 2, 5, and 10 °C/min. and fast firing, where samples were inserted into the preheated furnace. Temperatures were measured by means of a thermocouple close to the sample for linear heating and on furnace walls for fast firing experiments.

The dynamics of sintering have been studied by means of an optical cathedometer. This allows the measurement of linear shrinkage perpendicular to the specimen's compaction axis.

MM-Wave Sintering Technique

For mm-wave sintering a compact Gyrotron Oscillator Technological System (GOTS) at the Forschungszentrum Karlsruhe, Germany was used. It is a special design of the Institute of Applied Physics in Nizhny Novgorod, Russia, for high temperature processing of materials with mm-wave radiation. The gyrotron is operated in the continuos wave regime with smoothly controlled output power up to 10 kW at a frequency of 30 GHz and an electronic efficiency of about 25 % [6].

The mm-wave radiation generated by the gyrotron passes a Vlasov type mode converter and a quasi-optical transmission line to a highly oversized untuned microwave furnace. By using a quasi-optical transmission line it is possible to diminish the influence of the radiation reflected back from the furnace upon the operation of the gyrotron to an admissible level. The high ratio of furnace dimension to wavelength ($L/\lambda \sim 50$) and operation of a mode stirrer are essential prerequisite for homogeneous heating of materials (Fig. 1).

For thermal isolation the samples were places into a box built up of alumina ceramic fiber boards and dense alumina ceramic plates (Fig. 2). Temperatures were measured on the sample surface by means of a thermocouple with an accuracy not worse than ± 5 °C. The sintering process is temperature controlled via a feedback loop realized with a personal computer.

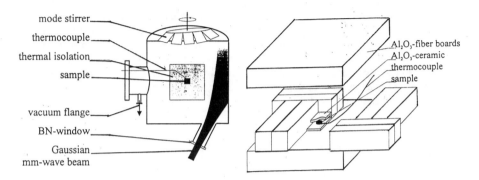

Fig. 1: Arrangement for mm-wave sintering experiments. Fig. 2: Thermal isolation, made of Al_2O_3 ceramic fiber material.

Sample Preparation

For the experiments γ-Al_2O_3 powder was utilized produced by the exploding wire technique [7]. This method uses a high current pulse which disperses an aluminum wire into ultrafine particles in an oxidizing atmosphere. The specific surface area, determined by BET measurements, was about 60 m^2/g, which corresponds to an average particle size of about 30 nm. The size distribution is rather wide, from 2 nm to more than 100 nm. Disk shaped specimens with 15 mm in diameter and 1.0 to 1.5 mm thick were compacted by the magnetic pulsed method at a pressure amplitude up to 2.2 GPa. The pressure pulses are generated as a result of the interaction of a pulsed magnetic field of an inductor with the conducting surface of a concentrator which actuates the piston [8], since the Lorentz force pushes any conductor out of the pulsed magnetic field region. The densities of the green specimens were in the range from 2.5 to 2.9 g/cm^3 (68% to 80% TD of γ-Al_2O_3). The densities of green and sintered specimen have been determined by Archimedes' method. The phase composition was monitored by X-ray phase analysis.

RESULTS AND DISCUSSION

Figure 3 shows typical kinetic curves of densification during sintering experiments with linear heating up to 1300 °C. During conventional sintering the linear shrinkage was optically measured. It has been used, together with green density, to deduce the density as a function of temperature. This data is shown in Figure 3 together with the densities of several mm-wave sintered samples of similar green density dependent on final sintering temperatures. This explains the more pronounced scattering of mm-wave data than of conventional sintering data.

The curves can be split into three characteristic parts with different intensity of shrinking. The initial low temperature part with very slow and small shrinking. The "active" shrinking starts when some critical temperature is reached. Some time later the "active" stage is followed by a stage of somewhat slower shrinking. The transition temperatures for the different shrinking stages depend slightly on the heating rate and the green density. An increase of the heating rate and a reduction of the green density shift these transition temperatures to higher values. It was shown by X-ray phase analysis that for mm-wave as for conventional sintering there is a phase transition to δ– and θ-phase during the active stage. The beginning of the final stage of slower shrinking correlates with the end of a polymorphic γ-to-α transition. Therefore the increase of the second transition temperature with an increase of the heating rate can be explained by assuming that the polymorphic transition is connected with a definite volume change and demands a definite time. The acceleration of the phase transition processes with the increase of the green density is related to an increase of the number of intergrain contacts and of the concentration of α-alumina crystallization centers [9].

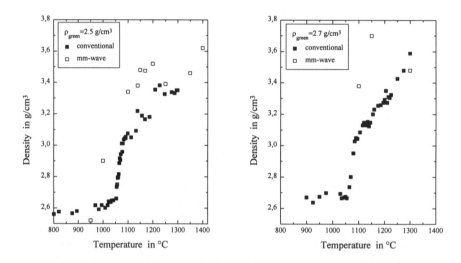

Fig. 3: Comparison of linear conventional heating and mm-wave heating for samples with green densities of 2.5 and 2.7 g/cm^3, respectively.

The densities of the sintered samples are listed in Table 1. It can be seen that their density strongly depends on the green density. Raising of the green density increases the final density in cases of fast firing as well as in cases of sintering by linear or mm-wave heating. In particular, the largest density (3.76 g/cm^3) of sintered ceramics corresponds to the densest green compacts, and the low-density green compacts lead to the lowest densities after the sintering (3.22 g/cm^3). The influence of the heating rate is not significant. Taking the final density as sole measure fast firing does not seem to be advantageous for compacts made of nano-sized Al$_2$O$_3$ powder. As can be seen from Table 1 the fast fired samples are less dense than those sintered by linear heating. It should be noted that the fast fired samples have been held at the sintering temperature for more then 100 minutes. Kinetic curves show that the densification finishes within 10 to 15 minutes. This duration is commensurate with the time span in which the sample acquires the temperature of the furnace.

Table 1: Resulting densities after sintering with different techniques.

sintering parameters		green densities in g/cm^3 (% TD)		
temperature °C	heating rates °C/min.	2.5 (68)	2.7 (74)	2.9 (79)
1300	2	3.32 (83)	3.16 (79)	3.66 (92)
1300	5	3.22 (81)	3.62 (91)	3.76 (94)
1300	10	3.29 (82)	3.60 (90)	3.73 (93)
1300	fast firing	-	3.56 (89)	3.67 (92)
1150	30 (with mm-waves)	3.57 (89)	3.70 (93)	-

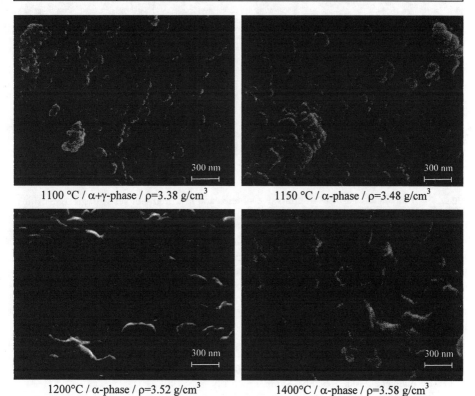

1100 °C / α+γ-phase / ρ=3.38 g/cm^3

1150 °C / α-phase / ρ=3.48 g/cm^3

1200°C / α-phase / ρ=3.52 g/cm^3

1400°C / α-phase / ρ=3.58 g/cm^3

Fig. 4: SEM of fracture surfaces of mm-wave sintered samples with green density of 2.5 g/cm^3

With mm-wave heating higher densification can be achieved already at temperatures 150 °C lower than with conventional sintering techniques, as can be seen from Table 1. mm-Wave sintering of samples with green densities of 2.5 g/cm^3 leads to a final density of 3.57 g/cm^3 already at 1150 °C, while the maximum density with conventional sintering techniques to 1300°C was 3.32 g/cm^3. For samples with green densities of 2.7 g/cm^3 the obtained final densities were 3.70 g/cm^3 and 3.62 g/cm^3 for mm-wave and conventional sintering, respectively.

Sintering under mm-wave irradiation leads to an enhancement of the densification process, as can be seen from Figure 3. With mm-waves the "active" stage of sintering starts at temperatures approximately 50°C lower than for conventional sintering. An increase of the sintering temperature >1200°C gives no essential increase of final densities but a distinct increase of grain size (cf. Figure 4) to average values of about 200 nm at 1400 °C. In the initial stage of sintering at 1100 °C small grain growth to an average size of about 50 nm and the beginning of the phase transformation to α-Al_2O_3 can be observed. At 1150 °C the phase transformation is completed. It is accompanied by a further grain growth to values close to 100 nm and a neck linking of the grains.

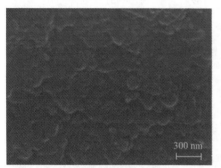

mm-wave sintered: 1150°C / 30°C/min.
ρ=3.70 g/cm^3

fast fired: 1245°C
ρ=3.46 g/cm^3

<u>Fig. 5:</u> SEM of fracture surfaces of mm-wave sintered and fast fired samples with green densities of 2.7 g/cm^3

If one compares the microstructure of conventionally fast fired samples with mm-wave sintered samples (cf. figure 5), no specific difference can be seen. Coarsening of grains is slightly more pronounced in the fast fired sample compared with the mm-wave sintered one. But there is a clear difference in the final density which is about 7% higher for the mm-wave sintered sample already at 100°C lower temperatures.

REFERENCES

1. J. Karch, R. Birringer, H. Gleiter, Nature, **330** (10), 556 (1987).
2. H. Hahn, NanoStructured Materials, **2**, 251 (1993).
3. J.A. Eastman et al.; Mat. Res. Soc. Symp. Proc., **189**, 273 (1991).
4. Zh. Jinsong et al.; Mat. Res. Soc. Symp. Proc., **347**, 591 (1994).
5. J. Freim et al., NanoStructured Materials, **4** (4), 371 (1994).
6. Yu.V. Bykov. et al., MIOP '95, Conf. Proc. of 8th Exhibition and Conference on High Frequency and Engineering, Sindelfingen, Germany, 321 (May 30.-June 1. 1995).
7. I.V. Beketov et al., Conf. Proc. 4th Euro Ceramics, Riccione; **1**, 77 (October 2.-6. 1995).
8. V.V. Ivanov et al., Conf. Proc. 4th Euro Ceramics, Riccione; **2.**, 169 (October 2.-6. 1995).
9. B.G. Adamenko, P.O. Pashkov, Poroshcovaya Metallurgia, **190**, 93 (1978) (in Russian).

SPANISH ACTIVITIES (RESEARCH AND INDUSTRIAL APPLICATIONS) IN THE FIELD OF MICROWAVE MATERIAL TREATMENT

José Manuel Catalá Civera, Elías de los Reyes Davó.
Departamento de Comunicaciones, Universidad Politécnica de Valencia, Camino de vera s/n, C.P. 46022 Valencia. Spain, jmcatala@dcom.upv.es

ABSTRACT

The GCM (Microwave Heating Group) within the Communications Department at the Technical University of Valencia is dedicated to the study of microwaves and their use in the current industrial processes in the Valencian Community and in Spain. To this end, a microwave heating laboratory has been developed and the benefits of incorporating microwave technologies into current industrial processes have been demonstrated. In this paper some of the industrial applications which are being investigated are presented.

INTRODUCTION

The GCM is an engineering group within the Communications Department at the Technical University of Valencia in Spain. This group has been dedicated to microwave research during the last five years and its main activity consists of studying the feasibility of changing the current and old industrial processes to another new ones where microwaves are the main technology, in order to improve the overall performance.

Valencian Community is a region placed in the East of Spain in Europe. Within the economical sectors of Valencian Community and Spain some of the most important are: ceramics, food products, footwear industry, furniture and wood products.

Most studies and projects developed by the GCM have been focused on such sectors having studied the following processes, namely within ceramic sector: slip-casting drying and ceramic firing; within footwear industry: rubber vulcanization and textile drying; within furniture and wood sectors: wood drying and coating adjustment heating and finally within food products sector: fruits tempering, thawing and vegetable sterilization. In the next sections some of these studies that are being performed within the GCM laboratory in Valencia are presented.

MICROWAVE STUDIES CARRIED OUT IN MAIN SPANISH INDUSTRIAL SECTORS

CERAMIC SECTOR

The application of microwaves to ceramic materials involves the slip-casting drying process, the ceramic firing and also the drying of ceramic materials for isolators in electrical transformers.

Slip Casting process

The drying of artistic ceramic by means of microwaves is a very important application due to the relevance of this economical sector in Spain. The GCM has developed projects for the Lladró company, one of the world leader companies in artistic ceramic.

Mat. Res. Soc. Symp. Proc. Vol. 430 © 1996 Materials Research Society

The conventional slip casting process used in traditional ceramic companies consists of pouring a mixture of clay and water "slip" into plaster moulds and allowing the slip to set during a fixed time (5-30 minutes for casting-up time) until a fine cast layer more compact (greenware) is settle over the mould. The driving force for the extraction of liquid during slip casting is the suction pressure created by the porous structure of the mould. Once the cast layer is formed a moisture gradient is created and the process becomes one of diffusion. Since diffusion of the liquid through the cast is much slower than through the mould the former becomes the rate controlling step. Thus casting rate is determined to a large extent by permeability of the cast. In order to the mould absorbs water from the cast or greenware that is being set, the mould must be completely dried or with a very low moisture percentage. The casting process can be found modeled in [2,3,4].

Once the greenware is formed, the excess slip is drained off and the mould/casting is allowed for sit during an additional period of time until the greenware can be safely removed from the mould. Later on and before the greenware is glazed or fired, it needs a final drying process of an hour approximately. The conventional drying processes usually have the following disadvantages: inefficient use of energy, long drying periods of time, large ovens and large floor spaces for the drying process. The ceramic slip casting process is shown in figure 1.

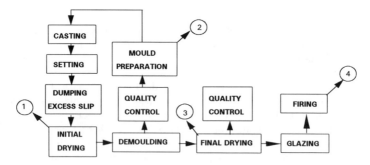

Figure 1. Ceramic Slip-Casting Process

The conclusion of studying this process was that the microwaves can be used with benefits within four stages of the slip casting process [2,4], namely during the casting period, after draining off the excess slip, to dry de greenware after demoulding and to dry the moulds ready for re-use.

In order to study the drying process, a prototype has been built consisting of a cylindrical cavity of $1m^3$ volume, 2 kW of microwave power and with the possibility of working with low pressures making use of a controllable vacuum pump.

The necessity of working with vacuum appears as a consequence of accelerating the process whose advantages are: the evaporation rate improvement, the input energy minimization and the modification the temperature of the mould and the greenware, guaranteeing the integrity of the moulds and increasing the product life as they do not reach temperatures above 50-55 °C.

The introduction of microwaves to the drying process reduces the drying from several hours to 10-15 minutes increasing the quality and consistency of the ceramic piece [4]. Nevertheless, the main advantage is the drying of the ceramic mould in order to be re-used.

Ceramic moulds play a fundamental role in the ceramic industry, thus under normal conditions its use is limited to one cast per mould per day, because they need this time to recover the moisture conditions they need for being used again.

This fact implies that a very high mould stocks are needed in order to reach high production rates. As a consequence, more personal, warehousing and reduced flexibility for changes is necessary, which limits the production very much.

Starting with conventional moulds of plaster and after small modifications for being used with microwave energy, the drying time are considerably reduced and mould life is widely increased. The use of microwave energy and vacuum processing in the drying of ceramic moulds reduces the re-using time from 12-24 hours to around 15-20 minutes, allowing for better qualities in the mould features, thus reducing the total number of moulds, lowering the fabrication expenses and adding new substantial reductions of extra costs [4].

Ceramic Firing

Once the greenware is glazed, it must be heated again in order to give consistence to the painting and getting its final aspect. This kind of process is carried out at very high temperatures and it is necessary a chemical process among the materials that integrated the greenware: porcelain, glycerin, etc. Currently this process is performed at high temperature gas or electric ovens.

The ceramic firing process does not have very suitable features for using the microwaves directly because the materials have low levels of moisture content and very high temperatures must be reached. In order to solve this problem with microwaves combined solutions must be found. The solution is reached making use of *dissipate ceramics* [5] .

Dissipate ceramics are ceramic materials that has been previously mixed and they have been completely dried. They are able to absorb microwave energy in order to reach high temperatures and then they re-radiate this power in the infra-red band heating the ambient and allowing the ceramic to be fired. This kind of materials are developed in the ENSEEIHT (National Technical Institute of Toulouse, France) [5].

Using this hybrid procedure, we get high temperature ovens with very good energetic efficiency rate. The temperature limitations are given by the maximum temperature that the isolated material can reach without losing its properties. Nowadays, it is possible to obtain absorbent materials that can reach up to 2000 °C without degradation.

The GCM, under co-operation with the ENSEEIHT has implemented an oven with the following features: a volume of 16 liters (20×20×40 cm.), microwave power of 4 kW. (four 1 kW. magnetrons) and an insulating material of 15 cm. over the whole contour.

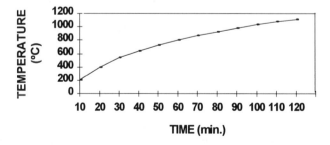

Figure 2. Temperature reached versus time needed in the high temperature oven

In order to get the final temperature, after the thermal equilibrium is reached, the speed of the process depends on the delivered power, the cavity volume to be heated and the depth of the thermal isolating layer. In figure 2, the graphic temperature versus time for the implemented oven is presented. The thermal equilibrium is produced around 1300 °C.

The use of this heating technique for the ceramic firing present the following advantages: the total cleanliness of the ambient during the firing (gas ovens produce impurities that can stain the painting), low losses (complete isolation of the cavity), simple control of the oven temperature (desired curves of temperature versus time can be easily obtained) and high speed when reaching the maximum temperature.

FOOD PRODUCTS INDUSTRIAL SECTOR

Microwaves are being used to some extent by the GCM in a certain number of industrial few food product applications like tempering, thawing, sterilization and pasteurization.

Microwave Enzymatic Inactivation

Once harvested, the mushroom suffers a gradual darkening called enzymatic browning, due to the formation of pigments, known as melamines, which result from the oxidation of some substances such as molecular oxygen, a reaction catalyzed by the enzyme polyphenyloxidase (PPO)[6]. This problem shortens the shelf life of the mushroom, making its marketing as a fresh product difficult and increasing the cost of its conservation and cooled transportation, controlled environment, etc. At the industrial level drastic treatments are employed such as blanching which inactivates the PPO, but at the same time causes weight losses, nutritional deterioration, loss of fragrance, etc. in the mushrooms.

Microwave application allows temperatures of up to 75 to 85°C to be reached with energy penetration, and considerably reduces the process time, and consequently improves the appearance, nutritional value and structure of the mushrooms, with lower losses of weight.

Continuing with food products, another area in which the group is involved is the tempering and thawing of frozen fruits which currently present some problems for industry, these are namely: apples, peaches, kiwis, blackcurrants, apricots and strawberries.

Microwave Tempering and Thawing.

Preservation of food by freezing, with the emphasis on fruit products, is an excellent method of ensuring the long term retention of the original characteristics, almost unchanged, especially of perishable materials.

Freezing destroys cell integrity and compartmentation thereby increasing the probability of undesired chemical and biochemical reactions. The consequence is that during frozen storage, there is a gradual cumulative and irreversible loss of quality with time, as represented by color , flavor and texture deterioration.

It has been claimed that factors such as cultivation, selection and methods of harvesting, cryostabilization pretreatments, freezing techniques, frozen storage, shipping, and thawing, affect the final quality of frozen fruits. The main goals of this project will be concerned both with a better understanding of the mechanism responsible for the changes of the quality parameters at a molecular level and with the improvement of cryostabilization and thawing technologies.

Microwave thawing suffers from three fundamental difficulties: the low dielectric loss factor of frozen products, the relatively large amount of energy required to produce the change of state and the superficial thermal runaway effect[7].

To resolve these problems, several tests are being carried out at the GCM laboratory making use of vacuum processing and surface cooling and results are indicating that microwave thawing may be a very successful application to improve the final frozen fruits quality. The main advantages of microwave tempering and thawing are speed and better quality of frozen products.

The results obtained in the food products sector and the quantity of industrial processes that can incorporate microwaves to some extent in its procedures, means that the food products sector is one of the most successful industrial sectors where the microwaves can be used.

FOOTWEAR INDUSTRIAL SECTOR

The successful application of microwave rubber vulcanization in the automobile sector is well established [8,9,10]. The footwear sector presents enough similarities in the materials involved to suggest that the current vulcanization processes, which is discontinuous and very inefficient energetically and temporally could be carried out with microwave energy.

Nowadays, footwear vulcanization processes are performed using laminated rubber sheets and heating technology employing conduction through hot oil and in batches.

A project of the group, together with footwear materials technicians in Spain, is investigating the vulcanization process of these rubber laminated sheets with microwave technology. The main objective is to change the current processes for continuous processes with higher energy efficiency, taking up less floor space and increasing the speed and quality of the final product.

FURNITURE & WOOD INDUSTRY SECTORS

Within the furniture industry, the wood drying and the heating of sheet materials for coating adjustment and fixing are the main applications in which GCM has been involved.

The process of drying wood can be considered as one of the most complicated and time consuming processes in present day industry. The feasibility studies which have been carried out indicate that microwaves could be an important factor in accelerating this process.

In order to study these sheet materials, several microwave prototypes have been built in the GCM laboratory which showed a considerable improvement in process time compared to currently used processes.

CURRENT LINES OF RESEARCH

The research lines which the GCM is following has lead on to the study of new microwave heating processes, the design of microwave applicators for industrial use, and the study of methods for monitoring these processes.

One of the advantages of the use of microwaves in heating processes is that the amount of energy required decreases as the process evolves. In drying process for example, the amount of energy needed decreases as the moisture content is reduced. There is a clear need , therefore, to implement mechanisms to control the progression of the microwave process and act upon the process or the power supplied by the microwave generator.

The GCM is studying this control through passive devices. By taking measurements with electrical and magnetic probes at different positions in the microwave applicator, and after a period materials characterization with known responses, we can make use of inverse scattering techniques allowing us to follow the evolution of the heating process to be monitored.

CONCLUSIONS

As it has been frequently demonstrated, microwaves offer in most industrial heating and drying applications the best solution in terms of speed and energy efficiency. They also have many other advantages such as: better product quality , reduced floor space, improved working environment and improved process control [1,10]. The GCM is involved in the study of these applications within the main Valencian and Spanish industry with a view to modernizing and improving their processes.

REFERENCES

1. A.C. Metaxas and R.J. Meredith, in Industrial Microwave Heating, (Peter Peregrinus Ltd., London, UK, 1993), pp. 313-319.
2. A.C. Metaxas and J.G.P. Binner, in Microwave Processing of Ceramics, (Advanced Ceramic Processing and Technology, Vol. 1, chapter 8, 1990) , pp. 285-370.
3. S.J. Oda and I.S. Balbaa. in Microwave Processing at Ontario Hydro Research Division, edited by W.H. Sutton, M.H. Brooks and I.J. Chavinsky. (Mater. Res. Soc. Proc. Vol. 124., Pittsburgh, USA, 1988), pp. 303-309.
4. J.M. Catalá. in Secado de Ceramica por Microondas. (GCM private communication in the Communications Department of the Polytechnic University of Valencia).
5. A. Bouirdene, S. Lefeuvre, M.Audhuy in Les fours thermiques micro-ondes, edited by Comité Français de L'Electricité, (Micro Ondes et Hautes Fréquences. International Congress Proc. Nice 8-10 Octobre 1991), pp. 71-82.
6. J. Thuery, in Microwaves: Industrial, Scientific and Medical Applications (Artech House Inc., London, 1992), pp. 339-357.
7. P. Jones, A. Millns, in Defrosting and tempering of frozen foods by radio-frequency heating, (I.M.P.I. 26th Microwave Power Symposium Proc., August 5-7, 1991), pp 17-18.
8. B. Krieger, in Industrial Vulcanization of rubber, edited by Comité Français de L'Electricité, (Micro Ondes et Hautes Fréquences. International Congress Proc. Nice 8-10 Octobre 1991), pp.708-737
9. B. Krieger and C.R.Buffler in New Curing Technology Enhances Rubber Processing,. (Proceedings of the American Chemical Society, Rubber Division, Cleveland, Ohio, Oct.1, 1985).
10. J. Irving and J. Chabinsky, in Applications of microwave energy. Past, present and future, Brave new worlds, edited by W.H. Sutton, M.H. Brooks and I.J. Chavinsky (Mater. Res. Soc. Proc. Vol. 124. Pittsburgh, USA, 1988), pp. 17-32.

PROCESSING AND CHARACTERIZATION OF MICROWAVE AND CONVENTIONALLY SINTERED BULK YBCO HIGH-T_C SUPERCONDUCTORS

I.A.H AL-DAWERY[†], J.G.P. BINNER[†], T.E. CROSS[‡]
† Department of Materials Engineering and Materials Design
‡ Department of Electrical and Electronic Engineering
The University of Nottingham, University Park, Nottingham, NG7 2RD, UK

ABSTRACT

The use of microwave energy for the sintering and annealing of high-T_C YBCO superconductors has been investigated with a view to taking advantage of the opportunities presented by this heating technique. It has been found to offer the possibility of sintering 'from the inside out' due to the nature of the temperature profile developed and as a result bulk YBCO bodies measuring 35 mm in diameter by 5 mm thick have been produced with completely uniform and high oxygen content ($x = 7$ in $YBa_2Cu_3O_x$) and densities of $\approx 98\%$ of theoretical. These properties were achieved using processing times approximately one sixth of those required conventionally. A model for the microwave heating process is proposed.

INTRODUCTION

Current bulk polycrystalline yttrium barium copper oxide (YBCO) high-T_C superconductors suffer from the disadvantage that it is extremely difficult to obtain the correct oxygen stoichiometry throughout the bulk of the material [1]. This occurs because during conventional sintering the interior of large samples heats by thermal conduction from the hotter surface. This results in the surface densifying before the interior, cutting off the latter from the oxygen-rich atmosphere required during sintering. This effectively creates a shell of superconducting phases surrounding a central core which is non-superconducting [2]. The current solution to the problem is to utilize very slow heating rates and long sintering cycles. The former reduces the temperature gradients across the sample whilst the latter allows time for oxygen to diffuse throughout the body. However these processing conditions do not always produce the desired microstructure and properties in the final component [3].

As has been well documented microwave heating can result in inverse temperature gradients, ie the centre of the body is hotter than the surface [4]. The idea behind the work presented in this paper was to investigate whether the inverse temperature profile could be controlled sufficiently to produce very high density bodies composed entirely of the relevant superconducting phases and hence possessing superior properties to those currently available. Previous attempts to microwave sinter YBCO have focused on the potential for lower sintering temperatures due to the so-called microwave effect and hence reduced impurity segregation and microcracks [5] or simply demonstrated that microwaves could be used [6].

EXPERIMENTAL

Disks of SS-ACS grade YBCO powder (Superconductive Components Inc, Ohio, USA) measuring 35 mm in diameter and 5 mm thick were single action die pressed at 200 MPa. This resulted in green bodies with densities of 45-50% of theoretical. Sintering and annealing studies were performed using a controlled atmosphere, 5 kW microwave sintering furnace operating at 2.45 GHz and a conventional horizontal tube furnace. Temperature was monitored in the microwave furnace using a black body tipped optical fibre thermometer (Accufibre Inc,

169

Oregan, USA) whilst a conventional Pt/Pt13%Rh-type thermocouple was used for the conventional furnace. All sintering and annealing heat treatments were performed using a flowing oxygen atmosphere at a rate of 100 cc min^{-1}. The arrangement inside the microwave furnace is currently confidential however heating was by microwaves only. In contrast to Kim et al [5] the samples microwave heated readily from cold. In Table 1 the sample number indicates the processing conditions used, viz: MP750-1 indicates microwave processed at 750°C for 1 hour. Similarly CP950-24 indicates conventionally processed at 950°C for 24 hours. For all the MP samples the heating rate to the sintering temperature was 600°C/hr and the cooling rate was 200°C/hr. An annealing hold was inserted during cooling at 505°C for 4 hours. The equivalent conditions for the CP samples was 150°C/hr heating rate; 60°C/hr cooling rate and annealing at 700°C for 24 hours.

The resulting samples were characterized with respect to critical transition temperature T_c, oxygen content, density, phases present and microstructure. T_c was determined via electrical resistivity measurements from room temperature to 20 K using an AC four-probe technique. The electrical contacts were made with fine platinum wire and silver paste; the specimen ends being polished mechanically using 10 μm grade diamond emery paper and subsequently washed in acetone. For the measurements a constant current of 4 mA was used and the voltage difference was measured to ± 0.1 mV. The oxygen content in the samples, x in $YBa_2Cu_3O_x$, was determined from the lattice parameters a, b and c. These were in turn calculated using data obtained from x-ray diffraction performed on the surface and centre of each sample. A small number of samples had their oxygen content measured across the whole thickness by repeatedly grinding away material. Electron beam diffraction patterns obtained using the TEM were used to confirm the lattice parameter values. The oxygen contents were determined from correlations of lattice parameter to oxygen content [3]. Density was measured by the Archimedes method using mercury; theoretical density was assumed to be 6.46 g cm^{-3}. Phase analysis involved XRD, whilst SEM and TEM were used for microstructure characterisation.

RESULTS AND DISCUSSION

Oxygen content, critical transition temperature and density measurements of the samples are shown in Table 1. The first observation to be made is an apparent reduction in the sintering temperatures and time required with microwave heating, illustrated in Figure 1. Over recent years many investigators have reported lower sintering temperatures/times in a wide variety of ceramic systems as a result of using microwave energy. In the YBCO system for example, Kim et al [5] found that equivalent densities could be obtained at temperatures approximately 40°C lower with microwave sintering than with conventional sintering. However temperature measurement was only at the surface and, in addition, required a correction factor to allow for a 3 mm gap between specimen and thermocouple to prevent arcing and chemical reaction.

The current results are not believed to provide evidence for or against a microwave effect for two reasons. Firstly, a different temperature measurement technique was used in the two sets of experiments although to reduce the effect of this the Accufibre probe was checked against the thermocouple within the tube furnace. The experiments showed that a maximum of 10°C difference in temperature measurement resulted. Secondly, it should be remembered that in both the conventional and microwave processing experiments it was the surface temperature which was measured. For the microwave heating case this will have been lower than the internal temperature, which was the whole point of the research.

There still remains the much faster processing times arising from using microwaves, however care must also be taken when interpreting these results. Firstly there is the fact that the precise internal temperature within the microwave heated body, known to be higher than

Table 1: Physical property measurements of the samples

Sample number	Surface oxygen content	Interior oxygen content	Critical transition temperature, T_c K	Relative bulk density %
MP750-1	6.74	6.74	78	80
MP800-1	6.88	6.88	85	87
MP850-1	6.94	6.94	88	89
MP900-1	6.90	6.90	88	93
MP925-1	7.00	7.00	90	95
MP925-2	7.00	7.00	91	95
MP925-6	7.00	7.00	91	96
MP950-1	7.00	7.00	90	97
MP950-2	7.00	7.00	90	98
MP950-6	7.00	7.00	90	98
CP950-24	6.88	6.78	85	84
CP960-24	6.90	6.78	86	85
CP970-24	6.88	6.78	85	87
CP980-12	6.92	6.80	87	89
CP980-24	6.90	6.78	86	90
CP980-48	6.78	6.70	80	93

the surface temperature measured, is unknown although, as explained below, it is not believed to be high enough to result in densification times as short as one or two hours. Secondly the heating process is believed to be quite different and capable of allowing the faster processing.

A very significant result is the uniformity of the oxygen content throughout all the MP samples. This is indicative of the inverse temperature profile developed allowing the centre to densify whilst still connected via open porosity to the oxygen atmosphere. None of the CP samples yielded a uniform oxygen content. Figure 2 illustrates a typical variation across the thickness for an MP and CP sample. It is also informative to consider the variation in density across the sample, Figure 3. This clearly demonstrates that the density in the CP sample decreased slightly towards the centre of the body. For the MP sample however, no variation in density was found within the accuracy of the technique used, $\pm \frac{1}{2}\%$. This is believed to illuminate the process occurring during microwave sintering. Prior to describing this it should be noted that, although not reported in this paper, the dielectric properties of both green and sintered YBCO have been measured as a function of temperature [7]. From these the penetration depths of the microwaves into the ceramic [4] have been calculated, see Figure 4.

The proposed model for the sintering process is as follows. Initially an inverse temperature profile develops as for most materials heated by microwaves. However Figure 4 shows that by a temperature of 800°C the depth of penetration has fallen from about 600 mm in the green body to about 0.5 mm; this decreases even further as sintering occurs. Thus as the body heats and densifies the microwave field is effectively excluded from the centre of the sample and the hot zone, ie the region which is sintering, moves towards the surface. This ensures that the body densifies completely uniformly and is believed to be the explanation for the faster processing times observed with microwave sintering rather than a 'microwave effect'. The values for penetration depth also indicate that bodies substantially bigger than those used during the present work should be able to be sintered uniformly using microwave energy.

The sintering process just described also allowed much faster heating and cooling rates to be utilised without damaging the specimen. With microwave processing heating rates of 600°C/hr and cooling rates of 200°C/hr could be used; conventionally the rates were limited to 150°C/hr and 60°C/hr respectively, approximately a factor of four lower. Combined with the shorter time at temperature required for both sintering and annealing, the result is a substantially shorter total sintering cycle. Typically about 12 hours versus at least 72 hours.

With respect to the microstructure, the temperature gradients between the surface and interior with conventional processing resulted in grain sizes up to 30 μm diameter at the surface and about 15 μm in the interior regions. When samples more than 5 mm thick were sintered the interior region was often found to display a poor degree of sintering with minimal connection between the neighbouring grains. Although as expected the microstructures of the microwave processed samples were found to depend on the exact heating arrangements used, for the optimal conditions developed a uniform grain size of 10-15 μm was observed. X-ray diffraction results indicated that secondary phases such as Y_2BaCuO_5 (211) often existed in the surface layers of the CP samples. In contrast this phase was rarely found in the MP samples. When it did occur it was in very small quantities and located in the centre of the body. Since it is not normally formed at temperatures below 980-1000°C [8] it is possible that local hot spots developed. The quantities of the phase were too small to be located by either SEM or TEM investigation so the local microstructure around the regions could not be analysed.

Results for the TEM analysis indicated the presence of microcracks and substantial secondary phases in the CP samples; these included the 211 phase already mentioned and also phases such as CuO and $BaCuO_2$. The latter are well known as agents which reduce the critical transition temperature whilst microcracks can also reduce critical current density. In contrast, for the MP samples the grain boundaries were clean of secondary phases and the grains

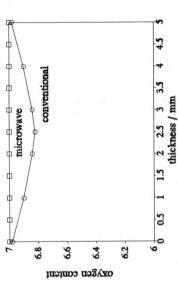

Figure 1: Densification of bulk YBCO superconductor as a function of surface sintering temperature

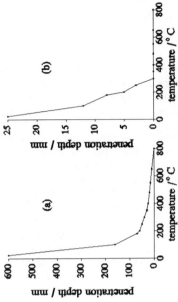

Figure 2: Oxygen content versus thickness for samples MP950-2 (microwave) and CP980-12 (conventional)

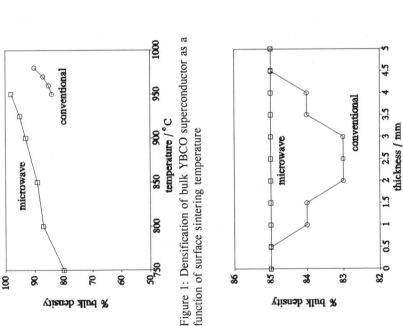

Figure 3: Relative bulk density versus thickness for samples MP800-0.5 (microwave) and CP950-24 (conventional)

Figure 4: Penetration depth-temperature relationship of a) a 45 % dense green and b) a 94 % dense sintered bulk YBCO sample

173

themselves were full of twins. The presence of twins is believed to be beneficial to current carrying characteristics since they act as strong pinning centres preventing flux motion [9].

CONCLUSIONS

Conventional and microwave processing of bulk high-T_c YBCO superconductors has been investigated. The former resulted in low and non-uniform oxygen contents and the presence of a range of secondary phases. These combined to limit the values of T_c; the highest value, 87 K, was only achieved when the bulk density was 89%. When the maximum bulk density of 93% was achieved, T_c fell to 80 K. Combined sintering/annealing cycles also took at least 72 hours for the best properties achieved. When even longer sintering/annealing times were tried undesirable repercussions on grain growth resulted.

In contrast, microwave heating offers the ability to achieve perfectly uniform and high oxygen content ($x=7$) and density (98%) throughout bulk samples. The T_c values were also higher, typically 90-91 K, and grain size was uniform and finer than in the conventionally processed samples. In addition total processing time was approximately a factor of six faster. In part this was achieved by the ability to use shorter hold times at temperature for both the sintering and annealing part of the schedule. However in addition microwave processing also allowed heating and cooling rates approximately a factor of four faster than with conventional processing without the sample suffering from thermal shock.

The reason for the above improvements in sample characteristics is believed to be due to the nature of the temperature distribution throughout the body during processing. Initially an inverse temperature profile is developed as usual, however the rapid increase in dielectric loss with increasing temperature and densification results in a severe reduction in penetration depth of the microwaves. This in turn results in the microwave field being excluded from the centre of the sample and the hot, or sintering, zone sweeping outwards towards the surface. This movement of the hot zone is believed to explain the faster processing times achievable rather than the more usually offered explanation of a 'microwave effect'. The size of the penetration depth at low temperatures also suggests that much larger samples than those investigated here should be amenable to microwave processing.

REFERENCES

1. J.W. Ekin, Adv. Ceram. Mat. 2 (3B) p.586 (1987).
2. M. Kogachi, K. Nakanishi, H. Sasakura and A. Yanase, Jpn. J. Appl. Phys. 28 (4) p.L609 (1989).
3. K. Salama, V. Selvamanickam, L. Gao and K. San, Appl. Phys. Lett. 54 (23) p.2353 (1989).
4. A.C. Metaxas and J.G.P. Binner, in Advanced Ceramic Processing Technology Vol 1, edited by J.G.P. Binner (Noyes Publications, New Jersey, 1990) p.285.
5. H.L. Kim, H.D. Kimrey and D.J. Kim, J. Mat. Sci. Lett. 10 p.742 (1991).
6. I. Ahmad, G.T. Chandler and D.E. Clark, in Microwave Processing of Materials, edited by W.H. Sutton, M.H. Brooks, I.J. Chabinsky (Mater. Res. Soc. Proc. 124, Pittsburgh, PA, 1989) p.239.
7. M. Arai, J.G.P. Binner and T.E. Cross, J. Am. Ceram. Sco. 78 (7) p.1974 (1995).
8. K. Salama and V. Selvamanickam, Supercond. Sci. & Tech., 5 (S1) p.S85 (1992).
9. K-Chen, S.W. Hsu, T.L. Chen and P.T. Wu, Appl. Phys. Lett., 56 (26) p.2675 (1990).

MICROWAVE REACTION SINTERING OF
TUNGSTEN CARBIDE COBALT HARDMETALS

T. GERDES*, M. WILLERT-PORADA*, K. RÖDIGER**, K. DREYER**

*   University of Dortmund, Dept. Chem. Eng., Dortmund, FRG
**   WIDIA GmbH, Essen, FRG

ABSTRACT

Microwave reaction sintering of tungsten- carbon-, cobalt mixtures is described as a method that combines the liquid phase reaction sintering of tungsten carbide cobalt hardmetals with the enhanced densification behavior of hardmetals in the microwave field. Starting with green parts composed of W-, Co- powder, soot as carbon source and with varying amount of additives a dense and extremely fine grained hardmetal can be obtained in one pressureless microwave reaction sintering step.
By this method hardmetals with improved mechanical properties can be obtained in a drastically simplified processing cycle, without time consuming steps such as carburizing and milling.

INTRODUCTION

In the past application of a very high temperature during synthesis of refractory materials such as WC was not only required for the carbide synthesis reaction, but also for purification of the carbide powder [1]. Despite the improved purity of current raw materials, the commercial process to obtain WC for powdermetallurgical products, such as cemented carbides, still requires temperatures of 1700-2300 °C during carburization.
Mainly because of the high synthesis temperature of numerous carbides alternative methods of powder production, such as self propagating high temperature synthesis, SHS, were developed during the last 30 years [2,3] with the main goal of using „adiabatic" conditions to facilitate the synthesis reaction and sintering. However, SHS based processes usually do not yield dense products, either due to volatile by-products or the absence of a liquid phase which is necessary for pressureless densification. On the way to refractory materials with new properties, reaction sintering processes are still of great interest, because of the high potential to create new microstructures due to the extreme processing conditions of a combustion heated synthesis, and with the aim to reduce the number of processing steps by combining the synthesis and the sintering process.
The increasing demand for durable, low cost, high quality cutting tools requires the development of new processing methods that will not only reduce the production costs but improve the microstructures of the refractory material. Surprisingly, reaction sintering of commercially very successful materials like WC-Co cutting tools can not be find among the systems investigated in SHS and related processes.
The implementation of a reaction sintering process into large scale production is regarded to be very difficult, because of the complicated control of the ignition and propagation of the reaction front. In this context, microwave heating has been suggested as a feasible method for homogeneous ignition of the reaction within the volume of a sample. However, the only systems reported to be reaction sintered by microwaves are dispersion ceramics like $TiC-Al_2O_3$ [4-7]. For materials with a high electronic conductivity, such as hardmetals, microwave reaction sintering is not expected to provide volumetric ignition within a sample. Nevertheless,

175

microstructural refinement is observed upon microwave sintering of hardmetals [8]. Combined with the ability of microwave heating to increase the heating rate and follow the reaction exotherm independently of heat conduction, and with the possibility to heat a batch volumetrically [9] microwave reaction sintering is an attractive method for producing cemented carbides without a separate powder synthesis step.

EXPERIMENTAL

Green parts with approximately 12 g weight / tool bit, composed of commercial tungsten powder, soot (6 wt%) and cobalt powder (6 wt%) as well as 2 wt.% of a paraffin lubricant were obtained by axial pressing at 300 MPa.
A typical microstructure of the reaction mixture green part is shown in Figure 1.

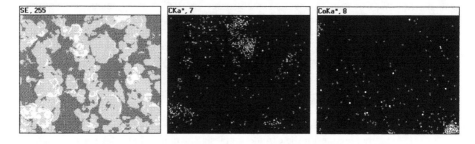

Figure 1: Microstructure and elemental distribution of C (middle) and Co (right) within W, C, Co-reaction mixture green parts

In addition graded powders with 5 wt.% of TaC/NbC and/or ZrC, TiC, TiN were employed.
Sintering is performed on single cutting tools as well as batches of samples in a microwave sintering apparatus described in [10] and, in comparison with this, in a vacuum furnace equipped with graphite heating elements.

RESULTS AND DISCUSSION

All sintering experiments are subdivided into a binder-burnout, reaction and sintering step. As shown in Figure 2 the most time consuming step of a microwave reaction sintering process is the lubricant removal at temperatures of 20-600°C. The following exothermic carburization reaction W + C → WC (-38 kJ/mol) is a very fast process that is completed within < 10s for typical batch sizes. As shown by the heating profile in Figure 2 adjustment of the forwarded microwave power enables the take up of a part of the reaction exotherm, resulting in a heating rate of approximately 2500°C/h from the reaction temperature to sintering temperature. Microwave sintering of the reacted mixture is completed after a 10 min. hold at 1200°C, yielding a > 92% dense cutting tool. To reach > 99% density sintering temperatures of > 1400°C are necessary, as shown in Figure 3.

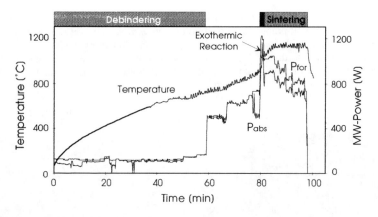

Figure 2.   Temperature and power profile of a MW-reaction Sintering cycle

The comparison between microwave and conventional heating of reactive W, C, Co- mixtures containing different additives indicates an accelerated densification upon microwave reaction sintering regardless the composition of the reaction mixture (Figure 3). Full density is achieved by microwave reaction sintering at approximately 100°C lower temperature.

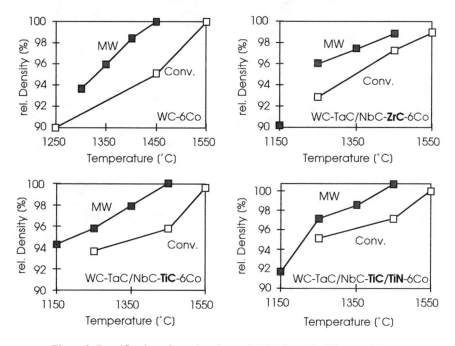

Figure 3. Densification of reaction sintered WC-6Co with different additives

The accelerated densification of *microwave sintered reaction mixtures* results in a lower grain size of the fully densified product as compared to *conventional reaction sintering*, as shown in Figure 4. Compared with *microwave sintering* of WC-Co-cutting tools, obtained from a carburization reaction starting with the same W-powder as in the case of the reaction mixture, grain growth is significantly suppressed due to microwave reaction sintering, e.g. from approximately 3-4 μm WC to 0,8 - 1,2 μm.

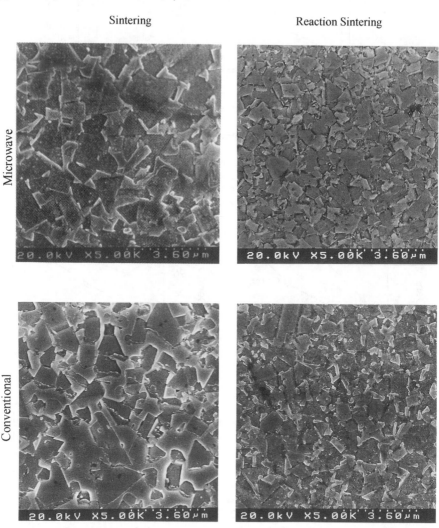

Figure 4. Comparison of the WC-6Co microstructure for different sintering methods

The reduction of WC-grain size in the microwave reaction sintered WC-Co-cutting tools is important for the increase of hardness without changing the amount of the metallic binder. As visible in Figure 5 fully densified microwave reaction sintered cutting tools of WC-6Co display an increase in hardness of 10% in comparison with commercial product.

Densification of a reaction sintered material without excessive grain growth is only possible at a low sintering temperature. The prerequisite of such a sintering process is the appearance of a reactive liquid phase. At low metallic binder contents a homogeneous distribution of the liquid forming element, e.g. Co in WC-Co is difficult to achieve, as visible in Figure 1. Therefore, the accelerated densification upon microwave reaction sintering is probably connected with an accelerated distribution of a liquid phase formed immediately after the reaction between W and C. In case of cemented carbides containing WC and Co this liquid phase is formed by dissolution of WC in Co. The distribution of the eutectic W-C-Co liquid phase throughout the porous WC-Co compact will govern the homogeneity of the densification process. Insofar the accelerated densification upon microwave reaction sintering could be traced back to an enhanced distribution of the liquid phase by the action of a changing electromagnetic field.

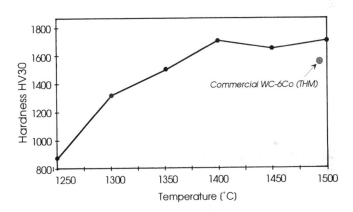

Figure 5. Hardness of MW-reaction sintered WC-6Co hardmetals

In summary microwave reaction sintering of W, C, Co offers a significant reduction of the number of processing steps, lower energy costs and, most important, optimized mechanical properties due to a microstructure improvement. The order of magnitude of technological advantages is shown in Figure 6, by comparing the number and temperature-time requirements of a conventional hardmetal production process with the microwave reaction sintering of these materials.

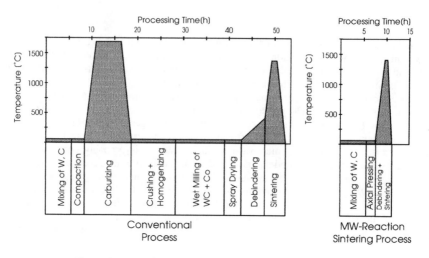

Figure 6. Manufacturing steps in the hardmetal production

REFERENCES

1. **H. Kolaska**, <u>Pulvermetallurgie der Hartmetalle</u>, Fachverband Pulvermetallurgie, Hagen, 1992

2. **R. Pampuch, J. Lis, L. Stroierski**, Self Propagating High-Temperature Synthesis of Ceramic Powders, Sci. of Ceramic **14**, 1989, p. 14-26

3. **A.G. Merzhanov,** Self Propagating High-Temperature Synthesis of Refractory Compounds, Moscow Vestnik Akademii Nauk SSSR, **10**, p.76-84, 1976

4. **S. Adachi, T. Wada, T. Mihara**,High pressure self-combustion sintering of alumina-titanium carbide ceramic composition, J. Am. Ceram. Soc., 73, **5**, p. 1442-1451, 1990

5. **J.F. Crider**, Synthesis and densification of oxide-carbide composites, Ceram. Eng. Sci. Proc., **3** [9-10] p. 519-526, 1982

6. **R.A. Cutler, A.V. Virkar, J.B. Holt**, Synthesis and densification of oxide-carbide composites, Ceram. Eng. Sci. Proc., **6** [7-8], p. 715-728, 1985

7. **M. Willert-Porada, B. Fischer, T. Gerdes**, Application of Microwave Heating to Combustion Synthesis and Sintering of Al_2O_3-TiC Ceramics, Ceram. Trans. 36, p. 365-375, 1993

8. **T. Gerdes, M. Willert-Porada**, Microwave Sintering of Metal-Ceramic and Ceramic-Ceramic Composites, MRS Symp. Proc. Vol. 347, p. 531-537, 1994

9. **T. Gerdes, M. Willert-Porada, H. Kolaska, K. Rödiger**, Guidelines for Large Scale MW-Processing of Hardmetals, Ceram. Trans. 59, p. 423-431, 1995

10. **T. Gerdes, M. Willert-Porada, K. Rödiger**, Microwave Sintering of Tungsten Carbide Cobalt Hardmetals, MRS Proceedings, 1996

CAVITY EFFECTS AND HOT SPOT FORMULATION IN MICROWAVE HEATED CERAMIC FIBERS

G. A. Kriegsmann, New Jersey Institute of Technology, Department of Mathematics and the Center for Applied Mathematics and Statistics, Newark, NJ. 07102

ABSTRACT.

Recently the heating of a thin ceramic cylinder in a single mode applicator was modeled and analyzed assuming a small Biot number and a known uniform electric field through out the sample. The resulting simplified mathematical equations explained the mechanism for the generation and growth of localized regions of high temperature. The results predicted that a hot-spot, once formed, will grow until it consumes the entire sample. Most experimental observations show that the hot-spot stabilizes and moves no further.

A new model is proposed which incorporates the effect of the cavity and the nonuniform character of the electric field along the axis of the sample. The resulting simplified mathematical equations indicate that these effects stabilize the growth of hot-spots.

INTRODUCTION.

When a thin cylindrical ceramic sample is heated in a single mode waveguide applicator interesting thermal phenomena often occur. A localized hot spot forms along the axis of the sample and begins to propagate outward elevating the temperature of the ceramic sample [1-3]. In most instances the spot eventually becomes stationary. These interesting phenomena occur even though the electric field intensity is essentially constant along the axis of the cylinder.

A simple mathematical model, based upon a small Biot number limit, has recently been presented [4]. Cavity effects were neglected so that the electric field was assumed uniform and known throughout the sample. This model identified the mechanism which triggers the formation of these hot-spots. However, it predicted that upon generation the spots propagate outward and engulf the entire sample. Thus, it could not describe the situation in which the spot stabilizes and becomes stationary.

The purpose of this paper is to incorporate cavity effects into this model in an effort to explain the stabilization process. The new model is still simpler and easier to analyze than the one used in [1] because the variations of the temperature in the cross-section have been averaged out and this yields a diffusion equation in only one spatial variable. Moreover it takes into account the effect by which the heated sample detunes an initially high Q cavity.

THE MODEL.

A thin ceramic cylinder of length H and radius R is positioned at the center of a single mode TE_{103} waveguide applicator as shown in Figure 1. The ceramic sample is held in place by two thermally insulated and microwave transparent push-rods. Unlike previous models which took the electric field to be known and constant along the axis of the fiber, the model presented here incorporates cavity effects into the heating process. The only physical assumptions made are that the Biot number, hR/K is small and that the fineness ratio, $\epsilon \equiv R/H$ is much smaller than the Biot number. This is the case for ceramic fibers.

Mat. Res. Soc. Symp. Proc. Vol. 430 © 1996 Materials Research Society

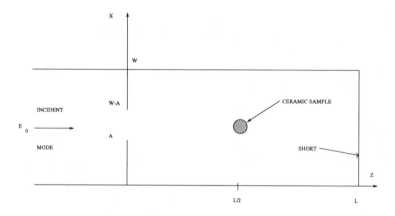

Figure 1: Waveguide Geometry.

Combining the analyses used in References 4 and 5, the temperature, T, can be shown to satisfy

$$\rho C_P \frac{\partial T}{\partial t} = K \frac{\partial^2 T}{\partial Y^2} + \frac{\sigma}{2} ||\mathbf{E}||^2(Y) \frac{|1 + r_1|^2}{|1 - r_1 \Gamma_1(T)|^2} - 2[\frac{h}{R}(T - T_A) + \frac{se}{R}(T^4 - T_A^4)], \quad 0 < Y < H, \tag{1a}$$

where ρ is the density of the ceramic, C_P is its specific heat, K is its thermal conductivity, $\sigma(T)$ is the effective electrical conductivity, h is the effective heat transfer coefficient, s is the radiation heat constant, e is the emissivity of the surface, and T_A is the ambient temperature of the surrounding medium. All of these parameters, except σ, are taken to be independent of T so that the interplay between cavity detuning and the electrical conductivity can be examined. Finally, the thermal boundary and initial conditions are taken as

$$\frac{\partial T}{\partial Y} = 0, \qquad Y = 0, H; \quad T(X, Y, Z, 0) = T_A. \tag{1b}$$

The parameter r_1 in (1a) is the reflection coefficient for an inductive iris in a homogeneous and infinite waveguide; an excellent approximations for a small aperture is given by [6]

$$r_1 = -1 + i\beta_1 \pi \delta^2, \quad \delta \equiv [\frac{1}{2} - \frac{A}{W}] \tag{2}$$

where A is the iris height and β_1 is the lowest modal wave number. The reflection coefficient Γ_1 takes into account the field reflected by the fiber and the short at $Z = L$, without the iris. The term involving the electric field in (1a) is defined by

$$||\mathbf{E}||^2(Y) = \frac{1}{\pi R^2} \iint_C |\mathbf{E}(X, Y, Z)|^2 \, dX \, dZ \tag{3}$$

where C denotes the cross-section of the fiber. The effect of the cavity is contained in the term

$$G = \frac{1 + r_1}{1 - r_1 \Gamma_1}. \tag{4}$$

The determination of $\mathbf{E}$ and Γ_1 is difficult in the present problem. The asymptotic and variational formulae available in the literature assume that there is no variation of electrical properties along the axis of the fiber. This is clearly not the case when a hot spot forms; the electrical conductivity in this region will be substantially higher than in the cooler parts of the fiber. Asymptotic formulae that take these variations into account have recently been developed [7]. They are:

$$||\mathbf{E}||^2(Y) = 4E_0^2[1 + O(Y, \epsilon^2)] \tag{5a}$$

$$\Gamma_1(T) = [-e^{2i\beta_1 L} + \frac{\pi\epsilon^2 k^2(1 - N^2)}{\beta_1}\frac{H}{W}] - i\frac{k^2\pi\epsilon^2}{\omega\epsilon_0}\frac{1}{W}\int_0^H \sigma(T)\,dY \tag{5b}$$

where k is the wave number of free space, $N^2 = \epsilon_1/\epsilon_0$, and $\epsilon_{1(0)}$ is the permittivity of the fiber (free space). The bracketed term in (5b) is due to the short and to the fiber, when losses are neglected. If it is denoted by γ_1, then $|\gamma_1| \simeq 1$ as required by the conservation of power. The remaining term in (5b) takes into account the loss in the fiber. Finally, the second term in (5a) represents the variation of the electric field in Y.

In the analysis that follows the second term in (5a) will be ignored, i.e., the Y variations of $\mathbf{E}$, because they are $O(\epsilon^2)$. However, the $O(\epsilon^2)$ terms in Γ_1 will be retained, because they are important in a high Q cavity. This is seen by inserting (5b) and (2) into (4) to obtain

$$G = \frac{i\beta_1\pi\delta^2}{(1 + \Gamma_1) - i\beta_1\pi\Gamma_1\delta^2}. \tag{6a}$$

By choosing the approximate resonant length $L = \pi(3 + \frac{1}{2}\delta^2)/\beta_1$, so that $\gamma_1 \simeq -1 - i\beta_1\pi\delta^2 + \beta_1^2\pi^2\delta^4/2$, (6a) reduces to

$$G = \frac{-2i\beta_1\pi\delta^2}{\beta_1^2\pi^2\delta^4 + 2i\dfrac{k^2\pi\epsilon^2}{\omega\epsilon_0}\dfrac{1}{W}\int_0^H \sigma(T)\,dY}. \tag{6b}$$

Now in the absence of losses $\sigma = 0$ and (6b) is $O(1/\delta^2) \gg 1$ which shows the resonance effect of the cavity. For the fiber losses to have a significant detuning effect on the cavity the two terms in the denominator must be of the same size. This gives the scaling law $\epsilon \sim \delta^2$ which relates the aperture opening to the fineness ratio of the fiber. Setting $\chi = \frac{\epsilon}{\delta^2}$ and inserting (5a) and (6b) into (1a), the heat equation finally reduces to

$$\rho C_P \frac{\partial T}{\partial t} = K\frac{\partial^2 T}{\partial Y^2} + \frac{8\chi^2 E_0^2}{\beta_1^2\pi^2\epsilon^2}P(T) - 2[\frac{h}{R}(T - T_A) + \frac{se}{R}(T^4 - T_A^4)], \quad 0 < Y < H, \tag{7a}$$

$$P(T) = \frac{\sigma(T)}{1 + 4\alpha^2(\dfrac{k^2\pi}{\omega\epsilon_0}\dfrac{1}{W}\int_0^H \sigma(T)\,dY)^2} \tag{7b}$$

where $\alpha = \chi^2/(\beta_1\pi)^2$.

DISCUSSION.

Equation (7) is very similar to the equation derived in [4] and used in [2]. The fundamental difference is the power term; in the previous works this was proportional to $|\mathbf{E}|^2\sigma(T)$ where $\mathbf{E}$ was assumed constant. The presence of the average value of σ, in the denominator of (7b) is caused by the detuning effect of the heated fiber. Its profound effect on the heating process will now be demonstrated.

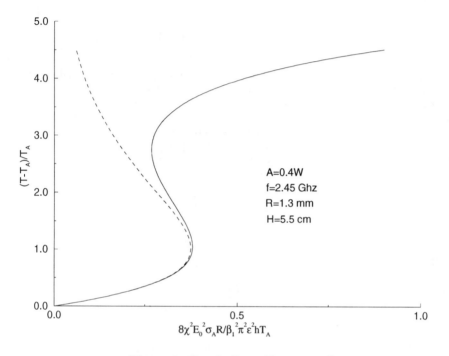

Figure 2: Steady State Response Curves.

In the discussion that follows published data for alumina will be used [8]. First, observe that equation (7) and its boundary and initial conditions (1b) support a spatially homogeneous solution which satisfies the ordinary differential equation

$$\rho C_P \frac{dT}{dt} = \frac{8\chi^2 E_0^2}{\beta_1^2 \pi^2 \epsilon^2} \frac{\sigma(T)}{1 + 4\alpha^2(\frac{k^2\pi}{\omega\epsilon_0}\frac{H}{W}\sigma(T))^2} - 2[\frac{h}{R}(T - T_A) + \frac{se}{R}(T^4 - T_A^4)]. \tag{8}$$

The steady state solutions of this equation lay on the s-shaped response curve shown as the solid line in Figure 2. The upper branch is caused by the presence of σ in the denominator of (8). If it is removed, then the steady state solutions lie on the dashed curve in Figure 2 which has no upper branch. It is this upper branch that allows hot-spots to form [4].

Next choose a value of E_0 which gives a dimensionless power slightly less than the critical value, 0.38 shown in Figure 2. This corresponds to a temperature $\sim 300^\circ C$. A small positive perturbation, which is symmetric about $H/2$, is added to this temperature and the sum is used as an initial condition for (7). The evolution of the temperature is very interesting: At first diffusion dominates, the temperature smooths out, and its average increases. Then after a period of time, the remnants of the initial perturbation begin to grow back into a highly localized structure, as shown by the solid curve in Figure 3. This hot-spot has a sharp maximum of about $2400^{\circ}C$ and cooler regions of about $30^{\circ}C$. A portion of the fiber is melted. Decreasing the radius of the fiber makes the localization even sharper, but does not significantly raise the maximum. If the radius of the fiber is increased to $1.76mm$, the dynamics are similar, but now the peak is somewhat lower ($\sim 1800^{\circ}C$) and the spot more diffuse. This is shown as the dashed curve in Figure 3. Increasing the radius further continues this trend until a threshold a point is reached where a hot-spot is no longer possible to sustain. For the present example this occurs at a radius of $2.4mm$. Choosing a radius greater than this value makes the response curve monotonically increasing, i.e., the s-shaped character is gone. According to the present theory, this indicates that hot-spots will no longer be generated. This feature is in qualitative agreement with the experiments on heating tows of ceramic whiskers.

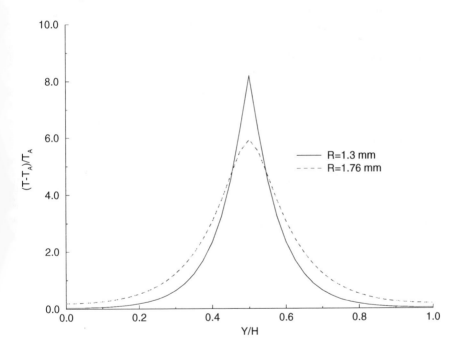

Figure 3: Hot Spot Formation.

CONCLUSIONS.

A simple mathematical model, which describes the heating of a ceramic fiber in a single mode microwave cavity, has been presented. By incorporating the dependence of the reflection coefficient on temperature and its concomitant detuning effect, the model explains the mechanism for the generation of localized regions of high temperature. The model also explains the dynamics of these hot spots and indicates that the smaller the radius of the fiber, the more localized and stable the spot. The effects of temperature dependent thermal parameters and phase changes have been neglected in the present model to focus upon the effects of cavity detuning. Their inclusion is the subject of ongoing research.

ACKNOWLEDGEMENTS.

This work was supported by the Air Force Office of Scientific Research Under Grant No. F49620-94-1-0338, and the Department of Energy Under Grant No. DE-FG02-94ER25196.

Bibliography.

1. Y. L. Tian, Microwaves: Theory and Practice in Microwave Processing, Ceramic Transactions, **21**, 283 (1991).

2. J. R. Thomas, W. P. Unruh, and G. J. Vogt, Microwave Processing of Materials IV, MRS Symposium Proceedings **347**, 363 (1994).

3. Johnson, D. L., Private Communications, 1993-94.

4. G. A. Kriegsmann and S. I. Varatharajah, Microwaves: Theory and Applications in Microwave Processing II, Ceramic Transactions **36**, 221 (1993).

5. G. A. Kriegsmann, Microwaves: Theory and Practice in Microwave Processing III, Ceramic Transactions, **59**, 269 (1995).

6. L. Lewin, Theory of Waveguides, (John Wiley, 1975).

7. G. A. Kriegsmann, Center for Applied Mathematics and Statistics, Report #9608, (Department of Mathematics, New Jersey Institute of Technology, 1996).

8. D. G. Watters, "An Advanced Study of Microwave Sintering", Ph.D. Dissertation, Northwestern University (1988).

Microwave Heating of Ceramic Laminates

J.A. Pelesko and G.A. Kriegsmann
Department of Mathematics
Center for Applied Mathematics and Statistics
New Jersey Institute of Technology
University Heights
Newark, NJ 07102

ABSTRACT

The microwave heating of a ceramic laminate composed of three layers is modeled and analyzed. Two materials with widely disparate effective electrical conductivity's comprise the laminate. The ratio of the these conductivities is exploited as a small parameter in the development of an asymptotic theory. Two physically distinct situations are considered. In the first, a low-loss ceramic is surrounded by lossy material. In the second situation, a lossy material is surrounded by a low-loss material. The results are physically interpreted to attain insight into the dynamics and parameter dependence of the microwave heating of ceramic laminates.

INTRODUCTION

Investigators in the field of microwave processing of ceramics are beset with numerous difficulties. Some materials of commercial interest, such as Alumina, are essentially transparent to microwaves. These microwave transparent, or low-loss, materials require large amounts of power to heat to desired processing temperatures. Further, due to the exponential dependence of their electrical conductivity's, these materials often suffer from uncontrollable thermal runaway. Other materials of commercial interest, such as Silicon Carbide, absorb microwaves readily. These microwave absorbent, or lossy, materials heat well from within but still lose heat readily at their surface. This leads to unavoidable and undesirable thermal gradients during the heating process. In recent years, numerous investigators have explored a variety of hybrid heating techniques in an effort to overcome these difficulties [1],[2].

In this paper, we construct a simple mathematical model of two idealized hybrid heating schemes, which yields insight into the success of hybrid heating in the control of power requirements, thermal runaway, and thermal gradients. First, we consider a low-loss semi-infinite ceramic slab, bounded by two lossy semi-infinite ceramic slabs. The outer lossy slabs are commonly referred to as susceptors. An asymptotic theory, based upon the exploitation of the ratio of the two materials room temperature effective electrical conductivity's is developed. Second, we extend the analysis to the case of a single lossy semi-infinite ceramic slab, bounded by two low-loss semi-infinite ceramic slabs. In this scenario, the inner lossy slab is referred to as the target, while the outer slabs are typically called insulation. The similarities in the physical scenarios are reflected in the mathematical models. This allows the asymptotic analysis to proceed in a nearly identical manner.

FORMULATION

We begin by considering a ceramic laminate comprised of three thin isotropic ceramic slabs as shown in figure 1. The outer two slabs are considered to be identical materials of equal thickness. Further, the composite is irradiated by identical microwaves from both sides. We chose this symmetric radiation to simplify the analysis to follow; the system of governing equations is symmetric about the centerline of the laminate.

We further assume that the incident electromagnetic wave and the electric field within the materials are time harmonic, while the temperature distribution in each material is time dependent. While the governing equations do not admit such a solution it has been shown in [3] that the equations we consider are the leading order equations of an asymptotic theory. This theory is based on the assumption that the time required for heat to diffuse an electromagnetic wavelength is much greater than the period of a microwave.

With these assumptions in mind, we formulate the equations governing the temperature distributions in each material for the susceptor ceramic configuration. Again, recalling our symmetry assumption, we only need two heat equations and a symmetry boundary condition. First, we have:

$$\rho_1 c_1 \frac{\partial T_1}{\partial t'} = K_1 \frac{\partial^2 T_1}{\partial x'^2} + \frac{|E_1|^2}{2} \varsigma_1(T_1) \quad -a < x' < 0 \tag{1}$$

$$\rho_2 c_2 \frac{\partial T_2}{\partial t'} = K_2 \frac{\partial^2 T_2}{\partial x'^2} + \frac{|E_2|^2}{2} \varsigma_2(T_2) \quad 0 < x' < b \tag{2}$$

Here ρ denotes density, c specific heat, K thermal conductivity, $|E|^2$ the electric field intensity, and ς_j denotes the effective electrical conductivity of the jth material which is a known function of the temperature. The variation with temperature of $|E|^2$ and ς_j makes this a non-linear problem.

The susceptor and the ceramic are assumed to be in perfect thermal contact at $x' = 0$ and hence we impose the following boundary conditions at this interface

$$K_1 \frac{\partial T_1}{\partial x'} = K_2 \frac{\partial T_2}{\partial x'} \quad at \quad x' = 0 \tag{3}$$

$$T_1(0, t') = T_2(0, t') \tag{4}$$

At the left hand boundary of the susceptor, $x' = -a$, we assume that heat is lost through both convection and radiation and hence impose

$$K_1 \frac{\partial T_1}{\partial x'} = h(T_1 - T_A) + s\epsilon(T_1^4 - T_A^4) \quad at \quad x' = -a \tag{5}$$

where h is a convective heat transfer coefficient, s is the Stefan-Boltzmann constant, ϵ is the emissivity, and T_A is the ambient temperature of the surrounding environment. At the axis of symmetry, that is at $x' = b$ we impose the condition

$$\frac{\partial T_2}{\partial x'} = 0 \quad at \quad x' = b \tag{6}$$

We assume that both slabs are initially at the ambient temperature of the environment. That is

$$T_1(x', 0) = T_2(x', 0) = T_A \tag{7}$$

Finally, as mentioned in the introduction, the formulation for the susceptor-ceramic case is mathematically identical to that for the insulated target case. Hence, we will only give the equations for the susceptor-ceramic configuration.

THE SIMPLIFIED THEORY

We choose dimensionless temperature and length scales and rewrite the conductivity's as their value at the ambient temperature multiplied by a dimensionless function of the scaled temperatures. Further, we scale time with respect to the diffusive time of the ceramic. This yields the new variables

$$v = \frac{T_1 - T_A}{T_A} \quad u = \frac{T_2 - T_A}{T_A} \quad \varsigma_1 = \sigma_1 g(v) \quad \varsigma_2 = \sigma_2 f(u) \quad t = \frac{K_2 t'}{\rho_2 c_2 b^2} \quad x = x'/b \quad (8)$$

Introducing these into (1)-(7) the following dimensionless parameters naturally arise

$$\mu = \frac{\rho_1 c_1}{\rho_2 c_2} \quad \gamma = \frac{K_2}{K_1} \quad p = \frac{b\sigma_2 E_0^2}{2hT_A} \quad \delta = \frac{\sigma_2}{\sigma_1} \quad B = \frac{hb}{K_2} \quad d = a/b \quad R = (s\epsilon T_A^3)/h \quad (9)$$

The parameters μ and γ are dimensionless ratios of material properties. The parameter p is a dimensionless ratio of power absorbed from the electric field to power lost at the boundaries by convection. The Biot number, B, measures the relative effects of convection and conduction, R measures the relative effects of convection and radiation, and δ is a dimensionless ratio of the two ceramic's effective electrical conductivities at the ambient temperature.

The parameter δ is exploited as a small parameter in the development of an asymptotic theory. We omit the details of the analysis and simply note that in this limit, the scaled susceptor temperature, v, and the scaled ceramic temperature u are given by

$$v(x,t) = v_0(t)(1 + O(\delta^{1/2}))$$
$$u(x,t) = u_0(x,t)(1 + O(\delta^{1/2}))$$

Further, since to leading order, the susceptor temperature is only a function of time, we may replace the susceptor with an effective boudary condition on the leading order problem for the ceramic temperature, $u_0(x,t)$. This reduced system is:

$$\frac{\partial u_0}{\partial t} = \frac{\partial^2 u_0}{\partial x^2} \quad for \quad 0 < x < 1 \tag{10}$$

$$u_0(x,0) = 0 \tag{11}$$

$$\frac{\partial u_0}{\partial x} = 0 \quad at \quad x = 1 \tag{12}$$

$$\mu d \frac{\partial u_0}{\partial t} = \frac{\partial u_0}{\partial x} - BL(u_0) + BPg(u_0) \, ||V_0||^2 \quad at \quad x = 0 \tag{13}$$

where

$$L(u) = u + R((u+1)^4 - 1) \tag{14}$$

189

Here, V_0 is the leading order electric field in the susceptor and the double bars indicate that it is to be integrated over the susceptor. This field can be computed explicitly from Maxwell's equations, because the conductivity is spatially homogeneous to leading order. The leading order susceptor temperature is easily recovered as $u_0(0,t) = v_0(t)$.

This reduced system is amenable to further analysis. First, setting all time derivatives to zero, we find that the steady-state temperature distribution is a constant and given by:

$$P = \frac{L(u_0^*)}{g(u_0^*)\,||V_0||^2} \tag{15}$$

Next, using classical Laplace transform techniques as first developed in [6] we find the dynamics of the temperature distribution are given by the non-linear Volterra integral equation:

$$u_0(x,t) = \int_0^t F(u_0(0,\tau))H(x, t - \tau)d\tau \tag{16}$$

where

$$F(u_0^*) = BPg(u_0^*)\,||V_0||^2 - BL(u_0^*) \tag{17}$$

and

$$H(x, t - \tau) = \frac{1}{1 + \mu d} + \sum_{n=0}^{\infty} \frac{2e^{-\alpha_n^2(t-\tau)}\cos(\alpha_n(1 - x))}{\cos \alpha_n(1 + \mu d + (\mu d\alpha_n))^2} \tag{18}$$

Here the $\alpha_n's$ are the solutions of

$$\tan(\alpha) = -\mu d\alpha \tag{19}$$

Equations of this form have been studied extensively. Below, we will discuss the physical implications of the analysis of such equations. Finally, we note that in the insulated target scenario, the final mathematical step varies. In particular, the dynamics are given by a pair of coupled non-linear Volterra integral equations rather that a single equation. Again, we refer the interested reader to [4].

DISCUSSION

As noted above, in the susceptor-ceramic configuration, the leading order steady-state temperature distribution within the ceramic is a constant and is given by (15). In figure 2, we plot several steady-state power-response curves for various values of the parameter d. We note that they have the now familiar S-shape previously discussed in [5]. Further, we note that the lower and upper branches represent stable steady-state solutions, while the middle branch is unstable. Analysis of the non-linear Volterra integral equation (16) allows us to understand the approach to steady-state. In particular, we find that for a given power, P, and a given initial condition, solutions monotonically approach the nearest stable branch. Examination of figure 2 allows us to draw several physical conclusions. First, as d is increased, the curve moves towards the left. This implies that as susceptor thickness increases, power requirements decrease. In the limit as d approaches zero, our model approaches that of Kriegsmann [5] for the heating of a solitary low-loss ceramic target. Comparing our model to that of Kriegsmann, we find that the use of a susceptor smoothes thermal gradients. This is due to the fact that in the steady-state, we found that thermal gradients are $O(\delta^{1/2})$, in [5] it was found that

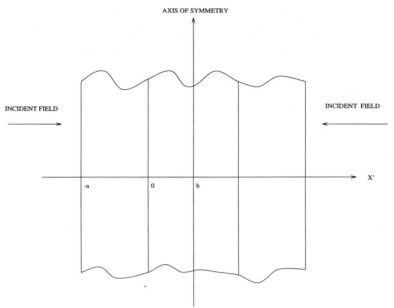

AXIS OF SYMMETRY

INCIDENT FIELD

INCIDENT FIELD

-a   0   b   X'

Figure 1 - Sketch of the model geometry

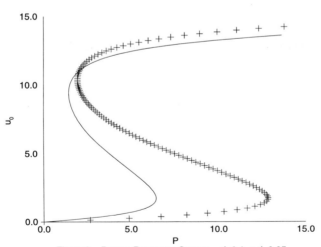

Figure 2 – Power–Repsonse Curves, – d=0.1, + d=0.05

without a susceptor, thermal gradients are $O(B)$. For real materials $\delta^{1/2} \ll B$. Next, we note that the upper branch of the S-shaped curve is lower for larger d. This may have implications for thermal runaway. If initially the upper branch is above the desired processing temperature the addition of a thicker susceptor may bring the branch down to desired processing temperatures. This may allow thermal runaway to be avoided, however the height of the upper branch also depends on material properties and these must be taken into account in any experimental situation.

Finally, we note that similar conclusions may be drawn from an analysis of the reduced system for the insulated target case. The interested reader is referred to [4].

ACKNOWLEDGEMENTS

This work was supported by the Air Force Office of Scientific Research under Grant number AFOSR F49620-94-1-0338 and the Department of Energy under Grant number DE-FG02-94ER25196.

References

[1] G.J. Vogt and W.P. Unruh, "Microwave Hybrid Heating of Alumina Filaments", Microwaves:Theory and Applications in Materials Processing II, The American Ceramic Society, 1993.

[2] M.A. Janney, H.D. Kimrey and J.O. Kiggans, "Microwave Processing of Ceramics: Guidelines Used at the Oak Ridge National Laboratory", Microwave Processing of Materials III, Materials Research Society Symposium Proceedings, 1992.

[3] G.A. Kriegsmann, "Microwave Heating of Dispersive Media", SIAM Journal of Applied Mathematics, vol 53, p. 655, 1993.

[4] J.A. Pelesko and G.A. Kriegsmann, "Microwave Heating of Ceramic Laminates", Journal of Engineering Mathematics,submitted Jan 1996.

[5] G.A. Kriegsmann, "Thermal Runaway in Microwave Heated Ceramics: A one-dimensional model", Journal of Applied Physics, v. 71, 1992.

[6] W.R. Mann and F. Wolf, "Heat Transfer Between Solids and Gasses Under Nonlinear Boundary Conditions", Quart. Appl. Math, v. 11, p. 163, 1951.

ADHESIVE BONDING VIA EXPOSURE TO MICROWAVE RADIATION AND RESULTING MECHANICAL EVALUATION

FELIX L. PAULAUSKAS*, THOMAS T. MEEK**, C. DAVID WARREN*
*Oak Ridge National Laboratory, Engineering Technology Division
Oak Ridge, TN 37831-8048
**University of Tennessee, Materials Science and Engineering Department
Knoxville, TN 37996-2200

ABSTRACT

Adhesive bonding/joining through microwave radiation curing has been evaluated as an alternative processing technology. This technique significantly reduces the required curing time for the adhesive while maintaining equivalent physical characteristics as the adhesive material is polymerized (crosslinked). This results in an improvement in the economics of the process. Testing of samples cured via microwave radiation for evaluation of mechanical properties indicated that the obtained values from the single lap-shear test are in the range of the conventionally cured samples. In general, the ultimate tensile strength, σ_B, for the microwave processed samples subjected to this single lap-shear test was slightly higher than for conventionally cured samples. This technology shows promise for being applicable to a wide range of high volume, consumer goods industries, where plastics and polymer composites will be processed.

KEY WORDS: Composite Bonding, Adhesives, Microwave Processing, Single Lap-Shear Test.

INTRODUCTION

The intent of this work was to produce high quality bonds with a substantial reduction in the required cure time. The bonds were to have mechanical and physical properties equivalent to conventionally cured samples. The experimental results of this work indicate that the cure time is substantially reduced by the application of microwave radiation when compared to thermal curing. Substrate materials discussed in this work are glass and a urethane-based composite with glass fiber reinforcement (SRIM-part). Glass substrates (annealed soda lime silicate slides) were bonded via microwave radiation curing of epoxy to eliminate the variable represented by the absorption of microwave energy by different substrates. This permits, as a first step, the study of the coupling characteristic of the pure neat adhesive resin independent of any "lossy" effect of the substrates. Lossy is a material characteristic or property indicated by a large tangent delta which means electrically very resistive and consequently couples very efficiently to microwave radiation.

In spite of the original intention to study only pure neat resin (BFGoodrich EXP 582E adhesive), it was decided to extend the study to include resin additives. These would enhance the coupling efficiency to microwave radiation and possibly improve the kinetic reaction rate during the curing process of the adhesive. Adhesives used were 100% Goodrich 582E (epoxy based), 582E plus 10 wt% ZrO_2, 582E plus 20 wt% ZrO_2, 582E plus 25 wt% ZrO_2. Further compositions consisted of 582E plus 10 wt% Tosoh-Soda (Tosoh-Soda: 80 wt% ZrO_2 and 20 wt% Al_2O_3; herewith referred to as TZ), 582E plus 25 wt% TZ, 582E plus 0.1 wt% carbon black powder, 582E

193

plus 0.3 wt% carbon black powder, 582E plus 1.0 wt% carbon black powder, and 582E plus 10 wt% carbon black powder.

The effects of additives were studied only to achieve further reductions in the required cure time when compared to microwave processed samples without additives through the improved lossy characteristics of the doped adhesive. The effects of these additives on thermally cured adhesives was outside the scope of this work, therefore, property changes resulting from chemical changes as a result of the additive are not addressed.

The data presented in this paper with respect to adhesive additives will be limited to carbon black powder. The basis for this is that previously published results [1] indicated that zirconia or Tosoh-Soda (TZ) will not improve the microwave energy deposition into the adhesive during the curing process.

SAMPLE PROCESSING AND RESULTS

For sample preparation and subsequent microwave processing of these samples, the various types of substrates were coated using the adhesive compositions noted above. In order to maintain uniform bond line thickness a few glass beads of 30 mil diameter were embedded in the adhesive. Subsequently, all samples were exposed to varying power levels of microwave radiation. At selected time intervals during the processing, substrates were visually inspected to determine acceptance for degree of polymerization (crosslinking). Figure 1 shows typical exposure times versus input power for glass and urethane substrates and the adhesive without additives. Indicated also in this figure is the effect of a high dosage of microwave energy with the subsequent generation of bubbles in the adhesive region. The result of this art of microwave processing is a substantial reduction in the required curing time of the samples. Figure 2 indicates the dependence of the required adhesive curing time on the input microwave power in a 2.45 GHz environment using the identical adhesive condition as in Figure 1. Also noted is the first order power law for this specific material. Substrate type, adhesive composition, exposure time, forward power, and reflected power were recorded for each trial in this experimental program.

The microwave system used was a Cober SF6 power supply which provided up to 5.5 kw of 2.45 GHz radiation into the 61 x 61 x 61cm multimode cavity. Depending on the requirements of the specific process or the part to be processed, any applicator may be used to accomplish a suitable joint. However, depending on part geometry and the joint area geometry, an applicator can be properly designed to accomplish a suitable power density in the required area which optimizes the energy incident on the required area. An analysis of the data reveals that in all cases, epoxy cure time approximately followed theory in that as electric field intensity increased, cure time decreased [2]. The experimental data empirically indicates that power (P) varies as the inverse square root of the curing time of the adhesive. The interrelationship between adhesive cure time and incident microwave power was discussed in [1]. Summarizing the results of that work, it may be stated that two regions in the experimental data are recognizable. In the high power region the experimental data indicates that the power varies as the inverse square root of the cure time, while for the low power data, the input power is inversely proportional to the exposure or cure time. Through analyzing the experimental data presented in Figure 2, urethane substrate/adhesive without additive, an appropriate empirical function was determined which describes these data to a high

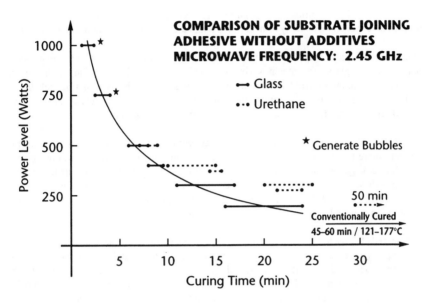

Figure 1. Curing time vs power level.

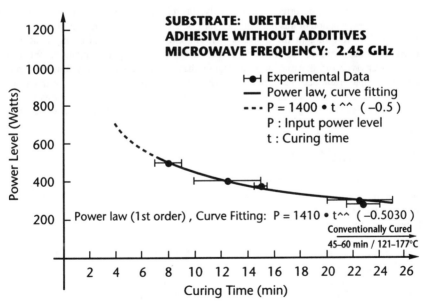

Figure 2. Curing time vs power level. Power law approximations.

degree of accuracy. The empirical relationship which describe the observed data is a power law function with an exponent essentially of negative 0.5 and a constant term of approximately 1400.

Figures 1 and 2 show data obtained for the various systems investigated. The first system investigated was the glass/substrate/epoxy-based adhesive without additives. Cure times ranged from less than one minute to over twenty minutes. Samples with urethane-glass fiber substrates, exhibited cure times which ranged from nine minutes to twenty-five minutes. This is very attractive when compared to a conventional cure time of forty-five minutes.

The selection of an acceptable curing time is dependent upon the industry and manufacturing environment where this technology will be utilized. Because there is a maximum power level above which bubble generation will take place, a minimum permissible curing time to achieve full crosslinking exists. For example, the data depicted in Figure 1 suggests a reasonable and acceptable curing time of between 5 and 15 minutes for many high rate manufacturing environments, as apposed to 45 - 60 minutes for conventional cure.

As expected, for each kind of substrate/adhesive system studied, the required curing time decreased with an increase in the level of the microwave input power, and with an increase in the concentration of the active additive in the adhesive system that responded favorably to the microwave energy deposition. Figure 3 clearly indicates this effect. Our current knowledge indicates that microwave curing will occur within a third to a quarter of conventional curing time. Results, obtained to date with the 582E adhesive indicate an approximate cure time of fifteen minutes or less through exposure to microwave radiation.

During the experimental work, it was noticed that a decrease in the reflected microwave power as a function of curing time occurred; and eventually, this reflected power leveled off and became steady-state. An explanation of this phenomenon may be the deposition of the microwave energy into the adhesive, resulting in the molecular crosslinking process. This changes the lossy characteristic of the adhesive, resulting in a change in the degree of polymerization. When the reflecting power levels off, this may be an indication that crosslinking is largely finished and the curing of the sample is approaching completion. Also, during the processing of the specimens only a marginal increase in the urethane-glass fiber composite substrate temperature was noticed. Further studies are needed in this area to better understand the mechanisms of microwave energy deposition as a function of the molecular crosslinking process.

Characterization of Joint Mechanical Strength

The characterization of the bond strength in the processed samples was determined using single lap-shear samples. Figure 4 shows a schematic of this test. A standard Instron Tensile testing machine, Model 1125, was used to perform this testing. In this evaluation, only urethane-glass substrates were studied.

From the previous study [1] of the cure time together with the subsequent evaluation of the joint mechanical strength, it was noted that when a physically (visually) acceptable joint was obtained, failure occurred as a fiber tear "peel" of the urethane substrate. This condition may be observed in Figures 5 and 6. From these results it was determined that discrete regions of the prior power density versus cure time curve (Fig. 1) needed to be evaluated and related to joint

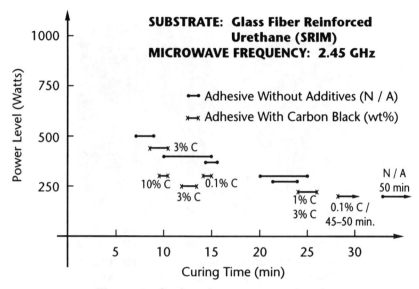

Figure 3. Curing time vs power level.

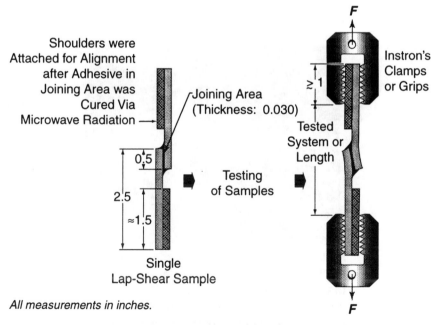

Figure. 4. System Deformation / Displacement In The Single-Lap-Shear Test

Figure 5. A satisfactorily microwave processed sample, when submitted to a single lap-shear test, demonstrated a fiber-"peel"-type fracture directly at the substrate surface.

Figure 6. Microwave processed samples showing localized bubble formation in the adhesive joint area. Under test conditions these samples demonstrated partial peel-type fracture with good values of mechanical strength.

mechanical strength. For this reason four discrete regions on the power density vs. cure time curve were selected for further study. This data is summarized in of Tables I and II.

As seen in Table I, relatively little variation in σ_B (ultimate tensile strength) was observed in the samples with processing times (curing times) from 10 to 40 minutes. In general, the average σ_B ranged from a value of 2600 psi to 3000 psi. For certain specific cases σ_B was as high as 3150 psi. A satisfactorily microwave processed sample when submitted to the single lap-shear test demonstrated a fiber tear "peel" type fracture directly on the urethane substrate as indicated in Figure 5. For the case of no additives in the adhesive and for short curing time with high input power a low average of 1918 psi was observed for σ_B. Upon the initial visual observation of these tested parts, it was observed that a nonuniform formation of bubbles occurred in the adhesive in the joint area, as seen, for example, in Samples A10 and A18 as shown in Figure 6.

In some samples, in spite of the bubble formation directly in the adhesive located in the joint area, a satisfactory value of the joint strength was obtained. For example, in samples A18 and A10 (Figure 6) it is clearly observed that at least half of the joint area is covered by bubbles, a partial "peel" fracture in the substrate is observed in these samples when they were submitted to the single lap-shear test. The "peel" region in the substrate occurred in the area which lacked bubbles. Surprisingly good values of joint mechanical strength were observed even though a high density of bubble formation occurred in some sections of the joint region.

A comparison of samples containing 0.1 wt% carbon black powder and 1 wt% carbon black powder indicated, in general, a good overlapping of σ_B values. The "**short**" cure time samples upon visual observation showed a lower density of bubbles in the interfacial region which could account for the increased σ_B when compared with the above non-additive samples processed at equal conditions (short time and high power). It appears that the presence of carbon black in the adhesive partially inhibits the formation of bubbles in the interface region. For samples not containing voids, fracture occurred by near surface fiber tear of the composite. Since the failures were cohesively through the adherend, any improvements in adhesive mechanical properties due to the incorporation of carbon would not be measurable in this study.

As observed in Table I, there may exist a maximum of σ_B with respect to cure time. This maximum may be observed within the range of moderate cure time (14-17 and 20-25 minutes), for all the microwave processed specimens. For the short sample cure time, a plausible explanation for this observed phenomena may be based on the high electric-field intensity and the subsequent high energy deposition which results in the generation of a high density of bubbles. For the long sample cure time, the observed results may be explained based on the extended/extensive exposure to microwave radiation resulting in possibly over-processing of the polymer (e.g. molecular degradation, increasing the adhesive brittleness, etc.).

The next analysis of mechanical strength data is shown in Table II. Shown in this table are the total crosshead displacement (δ) of the Instron tensile testing machine versus microwave sample processing characteristics. The complexity of the test geometry of single lap-shear samples (e.g. two nonaligned substrates bending in the joint area during testing, as indicated in Figure 4, stresses in the joining region which are difficult to define, and an adhesive with unknown material characteristics, etc.) makes the term "strain" for this specific case not applicable as known academically. The test length of the specimen may be described as a multiple component system

Table I. Single Lap-Shear Test Data of Microwave Process Samples Urethane / Glass Composites Substrates. Adhesive (epoxy based): 582E
Ultimate Tensile Strength (σ_B) vs Microwave Cure Time (CT)

Curing characteristic	Adhesive without Additives			Adhesive with 0.1 wt% carbon			Adhesive with 1.0 wt% of Carbon		
	Cure time range (min)	σ_B, psi average	σ_B, psi range	Cure time range (min)	σ_B, psi average	σ_B, psi range	Cure time range (min)	σ_B, psi average	σ_B, psi range
Short Cure Time	10 – 13	—[1]	3130 ↔ 1450	10 – 11	**2874**	2950 ↔ 2800	—	—	—
Low-Moderate Cure Time	14 – 17	**2875**	2950 ↔ 2800	14 – 16	**2733**	3000 ↔ 2600	12.5 – 14	**2772**	2930 ↔ 2620
Moderate Cure Time	20 – 25	**2950**	3150 ↔ 2840	19 – 23	**2885**	3100 ↔ 2600			
Long Cure Time	≈ 40	**2745**	2900 ↔ 2600	≈ 30	**2662**	2800 ↔ 2550	25 – 30	**2720**	2900 ↔ 2600
average:	**2845**			average:	**2802**		average:	**2755**	

[1] Majority of samples in this series were unacceptable (bubbles in joining area).

Table II. Single Lap-Shear Test Data of Microwave Process Samples
Urethane / Glass Composites Substrate. Adhesive (epoxy based): 582E
Total Cross-head Displacement (δ) vs Microwave Cure Time (CT)

Curing characteristic	Adhesive without Additives			Adhesive with 0.1 wt% carbon			Adhesive with 1.0 wt% of Carbon		
		δ, inch, average			δ, inch, average			δ, inch, average	
	Cure time range (min)	at σ_B	at 1000 lbs load	Cure time range (min)	at σ_B	at 1000 lbs load	Cure time range (min)	at σ_B	at 1000 lbs load
Short Cure Time	10 – 13	—¹	—¹	10 – 11	**0.0859**	0.0495	—	—	—
Low-Moderate Cure Time	14 – 17	**0.0646**	0.0354	14 – 16	**0.0806**	0.0472	12.5 – 14	**0.0800**	0.0473
Moderate Cure Time	20 – 25	**0.0721**	0.0371	19 – 23	**0.0755**	0.0401	25 – 30	**0.0850**	0.0533
Long Cure Time	≈ 40	**0.0606**	0.0378	≈ 30	**0.0731**	0.0438			
	average:²	**0.0652**	0.0368	average:²	**0.0790**	0.0457	average:²	**0.0819**	0.0495

¹ Majority of samples in this series were unacceptable (bubbles in joining area).
² Average of all acceptable samples in all series.

in series. For this reason, it is important to describe it only in terms of the total crosshead deformation/displacement (TCHD) of the sample system in the single lap-shear test.

For comparison reasons the total crosshead displacement (δ) was evaluated for all the samples/series with and without adhesive additives at σ_B, and at a load of 1000 lbs. The TCHD at σ_B will indicate the maximum tolerable deformation that the system can withstand at maximum load. The TCHD at a load of 1000 lbs. was selected as a reference value which corresponds to a stress of approximately 2000 psi which is well below the ultimate strength (material is still physically sound) and is located at the lower limit of the range of observed mechanical strength data.

As seen in Table II, the addition of an adhesive additive such as carbon increases δ both at σ_B and at 1000 lb. load. The variation seen in TCHD between 0.1 wt % and 1 wt % carbon data is found to be negligible for σ_B and at a 1000 lb. load. No clear trend of the TCHD values was observed as a function of the curing characteristic (cure time) for all systems evaluated.

Table III presents the consolidated data obtained on conventionally cured and microwave cured samples in this experimental work. Data for ultimate tensile strength, total TCHD at ultimate strength, and TCHD at 1000 lbs. load is summarized in this table. In general, the data indicates nearly equivalent results for the conventional and microwave processed samples when comparing σ_B. Table III also demonstrates that higher maximum ultimate tensile strength (σ_B) values were consistently obtained for the microwave processed samples compared with the conventionally cured samples. This was observed with and without additives to the adhesive. The data indicates that the microwave processed samples possess ultimate strengths slightly higher or equivalent values than the conventionally cured specimens. As measured, the TCHD for the conventionally processed samples showed a significantly lower value than those obtained for microwave processed samples at corresponding equal load values. The greatest difference in TCHD was observed to occur when comparing conventionally cured and microwave processed samples (especially those specimens with high carbon additive in the adhesive). This indicates that the conventionally processed samples show higher rigidity/stiffness when submitted to the single lap-shear test. A complete, fundamental comparison between mechanical properties of conventionally processed and microwave processed samples is not possible within the scope of this work. For example, in the single lap-shear test, the system under study, (despite equivalent predrying processing of the substrates) may be assumed to be a multiple component system with each component acting in series. When this system is exposed to different processing methodologies (microwave vs. conventional) the affect on each system component (two segments of substrate transmitting forces at each end of a nonaligned adhesive joint region) can be significantly different and a function of what these segments undergo in each process methodology. Another indication of different sample characteristics caused by process methodologies may be found in the samples which experienced total peel failure. Visually, the new surface (cohesive composite failure) of the microwave processed samples shows more resinous characteristic than the conventionally processed samples.

Figures 7-9 represent data obtained in a single lap-shear test for conventionally cured samples and microwave processed samples cured over different levels of input power and for different lengths of time (short, moderate, and long cure times). Data expressed in Figures 7-9 represents load not stress. Associated with each curve is the corresponding ultimate tensile stress (σ_B). When σ_B is evaluated, it must include the joining area which varies very slightly from sample to sample.

**Table III. Comparison Of Single Lap-Shear Test Results
Conventionally Cured Samples vs Microwave Processed Samples**

	Conventionally cured samples (45 min at 150° C)		Microwave Processed Samples					
			Adhesive without additives		Adhesive with 0.1 wt % carbon		Adhesive with 1.0 wt % carbon	
	avg.	(range)	avg.	(range)*	avg.	(range)*	avg.	(range)*
Ultimate Tensile Strength (σ_B), psi	2736	2840 ↕ 2406	2845	3100 ↕ 2600	2802	3100 ↕ 2600	2755	2930 ↕ 2600
Total Cross-Head Displacement, at Ultimate Strength (σ_B), in.	0.0546		0.0652		0.0790		0.0819	
Total Cross-Head Displacement at 1000 lbs load, in.	0.0333		0.0368		0.0457		0.0495	

*Approximate range.

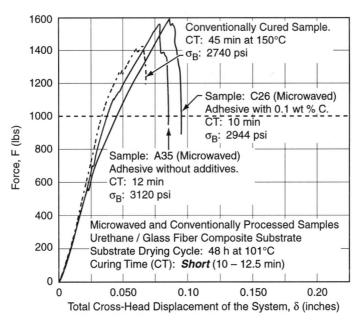

Figure 7. Single Lap-Shear Test Diagram

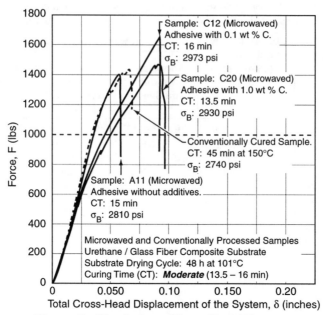

Figure 8. Single Lap-Shear Test Diagram

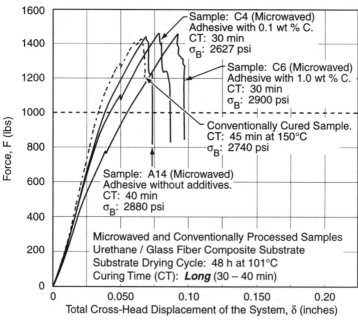

Figure 9. Single Lap-Shear Test Diagram

It is clear from the data on the diagrams (Fig. 7-9) that conventionally cured samples demonstrate a higher rigidity than all the microwave processed samples. From these diagrams it is clear that the higher the concentration of carbon in the adhesive, the larger is the deformation that the tested system can tolerate. Within each figure, data can be compared or analyzed to determine trends since all specimens were subjected to similar processing conditions (power level and cure time). In Figure 8 (**moderate** cure time) four specimens with varying adhesive formulation and processing are compared. Clearly recognizable is an increase in the TCHD as a function of carbon content in the adhesive. Since the samples were subjected to very nearly the same processing conditions, the different results may be attributed directly to the adhesive carbon content. It is evident that the carbon content affects the tolerable state of deformation. Similar conclusions may be drawn from Figures 7 and 9 for the short and long curing times.

From Figures 7-9, in spite of the similarity of the general shape of load versus deformation curves for the conventional and microwave cured samples, it is clear that the conventionally cured samples exhibit a more elastic behavior than do the microwave processed samples, which exhibit more plastic behavior. It is not clear what the reasons are for this difference and it is not possible within the scope of this work to draw any conclusions. This may be attributable to the complex stress state present in the test samples due to the test configuration of the specimens.

For a comparison of microwave vs. conventionally processed samples, a possible explanation for obtained results is discussed below. The higher degree of deformation under load for the microwave processed samples is based on the ability for the long wavelength microwave radiation (possessing faster depth of penetration through the substrate sample than conventionally curing processes) to interact directly at the interface throughout the complete curing process. Due to the known clear chemical affinity between epoxy and isocyanate, this affinity may be increased or enhanced in the presence of microwave radiation and may be a strong function of interface temperature during processing. The relatively fast energy deposition in the joint area is enhanced by the intrinsic polar characteristic of the adhesive at early stages of the curing process. Under these conditions, it is expected that a better chemical bond may be made between the epoxy and urethane directly at the interface boundary. The above results are directly analogous to those observed during the process of glass-ceramic joining. In this process, conventionally made glass-ceramic joints experience mechanical bonding and not diffusional bonding; however, exposure to microwave radiation results in complete diffusion bonding and in many cases the joint is not visible in a SEM.

Basic microscopy of the substrate/adhesive interface indicates a nonhomogeneous distribution of the glass fiber in the substrate. We thus assume that a nonhomogeneous fiber distribution exists through the substrate. From the microscopy it is observed that fiber concentration exists in the form of bundles surrounded by resin. This nonuniform distribution of glass fibers throughout the substrate will severely distort the microwave electric field present in the substrate and the substrate joint region. This nonuniform electric field will almost certainly result in localized high electric field intensity (hot spots) in regions of low composite dielectric constants. Regions which have a high concentration of glass fiber (bundles) should be heated less than regions with higher concentration of resin. This condition may be extremely beneficial for microwave energy deposition in the joint area.

Vast improvements may be possible in the optimization of this process including adhesive reformulation to increase the lossy nature of the adhesive and speed the reaction kinetics of the microwave process. No reformulation of the adhesive was undertaken in this study beyond the inclusion of particulate additives known to be good microwave energy enhancers. Any changes in the chemical constituency of the adhesive to reduce the thermal cure time are expected to also decrease the required microwave cure time due to the excellent coupling characteristic of the epoxy adhesive to a broad spectrum of microwave energy.

CONCLUSION

•   The application of microwave technology for joining of substrates using epoxy based adhesives will significantly reduce the curing time to only a third to a quarter of the conventional cure time. This is accomplished while maintaining equal or slightly higher values of the ultimate tensile strength obtained through the single lap-shear test.

•   Microwave processed samples, when tested as single lap-shear specimens exhibit less rigidity but more plasticity than do conventionally processed samples.

•   Coupling of the Goodrich EXP 582E epoxy based adhesive to the 2.45 GHz microwave radiation is extremely efficient. This coupling characteristic is enhanced by an additive such as carbon black.

•   Total crosshead displacement (TCHD) for conventionally processed samples exhibit a significantly lower value than those obtained for microwave processed samples at correspondingly equal load values. Clearly recognizable is also an increase in the TCHD as a function of carbon content in the adhesive for the microwave processed samples.

•   Microwave processed samples with an additive such as carbon black powder exhibit more plasticity than those without an additive. As additive concentration in the adhesive increases so does plasticity. In all cases increased plasticity was observed with equal or greater ultimate tensile strength (σ_B).

•   This technology may be extended to multiple-layered panels or components.

ACKNOWLEDGMENT

Research was performed at Oak Ridge National Laboratory and sponsored by the Office of Transportation Material, U.S. Department of Energy, under contract No. DE-AC05-96OR22464 with Lockheed Martin Energy Research Corp.

REFERENCES

1.   F. L. Paulauskas, T. T. Meek, C. D. Warren, SAMPE International Technical Conference Proceedings, 27, 114 (1995).

2.   T. T. Meek, J. of Materials Science Letters, 6, 638 (1987).

Part IV

Microwave System Design

APPLICATION OF RIDGED WAVEGUIDE IN MICROWAVE SINTERING

Zhou Jian, Cheng Jiping, Liu Xianjun, Tang Yuling and Chen Lei, Advanced Material Research Institute, Wuhan University of Technology, Wuhan, 430070, P. R. China.

ABSTRACT

In this paper, a ridge rectangular waveguide is designed, and its cutoff frequency, impedance and electric field intensity are given by formulas or curves. A few ceramic samples are sintered in it by microwave energy. It is concluded that the device can be as a satisfactory microwave sintering cavity.

INTRODUCTION

As far as we are aware, the published information of design on ridged waveguide can be traced back to a paper by S. B. Cohn [1]. Ridged waveguide is used extensively because of its wider single—mode bandwidth, the lower cutoff frequency and lower impedance than a common rectangular waveguide of the same width and heigth. Recently, we have applied it to microwave sintering and obtained a series of results.

Microwave sintering is new technology for materials sintered. A lot of experiments have stated that microwave sintering makes ceramic materials. That not only have more excellent properties than made with conventional sintering, but also can be made in less time and can save energy.

This paper has been written to extend the design data of ridged waveguide. Tungsten carbid—cobalt(WC—Co)hard—metals samples were sintered with microwave energy in the ridged waveguide. The results indicate that the hard metal samples can be sintered, and its properties are excellent in the ridged waveguide.

THEORY ANALYSIS

The cross—section of ridged waveguide is shown in Fig. 1. Fig. 1(a) and (b) show the single— and double—ridged cross—section, their equivalent circuit is shown in Fig. 1 (c).

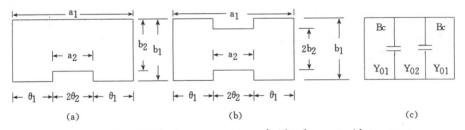

Fig. 1 The cross—section of ridged waveguide.
(a). single—ridge (b). double—ridge (c). equivalent circuit.

The symbols in the design equations is shown in Fig. 1. Where a_1, a_2, b_1 and b_2 are inside dimension in centimeters, θ_1 and θ_2 are the electrical angles in terms of the cutoff wavelength $\lambda c'$ in free space at the ridged—guide, i. e.

Abuot this article: The research was supported by 863 plan (NAMC)

Mat. Res. Soc. Symp. Proc. Vol. 430 © 1996 Materials Research Society

$$\theta_2 = 360 \times a_2/(2\lambda c') \qquad (1)$$

The cutoff of the TE_{10} mode occurs when the lowest root of the following equation is satified

$$b_1/b_2 = (\cos\theta_1 - Bc/Y_{01})/\tan\theta_2 \qquad (2)$$

Where Bc is the equivalent susceptance introduced by the discontinuities in the cross section, as explained in reference [2].

Equation (2) is accurate if proximity effects are taken fully into account in calculating Bc. In the curves of this paper, proximity effects are neglected, but the results are highly accurate so long as $(a_1 - a_2)/2 > b_1$

In terms of θ_1 and θ_2, $\lambda c'$ is given by

$$\lambda c' = (90°/(\theta_1 + \theta_2))\lambda c \qquad (3)$$

Where $\lambda c = 2a_1$, which is the cutoff wavelength of the guide without the ridge. θ_1 and θ_2 are values satifying (2).

In deriving the impedance equation for ridged waveguide, it will be assumed that b_1/a_1 is small, so that the discontinuity susceptance at the edges of the ridge may be neglected.

If the TE_{10} mode alone is set up in the waveguide, the E fieldvalue distridution is the same at all frequencies, including $f = fc$ and $f = \infty$. The E fieldvalue can be calculated easily at the cutoff frequency by the approach used in deriving the cutoff equation. At $f = \infty$, the wave impedance is that of free space. Hence, if the E fieldvalue is known, the H fieldvalue is given by $H = E/(120\pi)$. Both E and H are completely transverse at $f = \infty$, and the current on the top or bottom of the waveguide is completely longitudinal. The current per unit width is equal to the H field intensity at the surface of the conductor.

The guide impedance at infinite frequency will now be defined as the ratio of voltage across the center of the guide to the total longitudinal current on the top face.

$$Z_{0\infty} = \frac{V_0 b_2 E_0}{2\int_0^{a_1/2} E dx} = \frac{120\pi b E}{2\int_0^{a_1/2} E dx} \qquad (4)$$

The impedance at any other frequency is related to $Z_{0\infty}$ by the expression

$$Z_0 = \frac{Z_{0\infty}}{\sqrt{1 - (fc'/f)^2}} \qquad (5)$$

where fc' is the cutoff frequency of the ridged waveguide.

Since the guide has been assumed thin, the voltage across the step will be continuous. The voltage distribution in right half of the cross—section will therefore be the same as that along the shorted composite transmission line, shown in Fig. 2.

Since the input impedance is infinite at open end, the voltage across the guide is a maximum at that point. Transmission—line theory shows that the voltage distribution over the θ_2 range is given by

$$v = v_0 \cos\theta \qquad\qquad \text{form } \theta = 0 \text{ to } \theta = \theta_2.$$

Over the θ_1 range it is given by
$$v = v_1 \sin(\theta_1 + \theta_2 - \theta)/\sin\theta_1 = v_0 \cos\theta_2 \sin(\theta_1 + \theta_2 - \theta)/\sin\theta_1 \qquad \theta_2 \leqslant \theta \leqslant \theta_1 + \theta_2$$

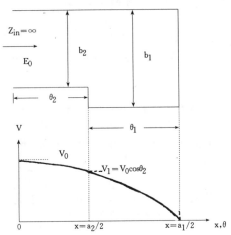

Fig. 2. Appoximate voltage distribution across half of the cross—section.

The E field value is equal to v/b. Thereforce
$$E = E_0\cos\theta, \qquad 0 \leqslant \theta \leqslant \theta_2$$
$$E = b_2/b_1 E_0\cos\theta\sin(\theta_1 + \theta_2 - \theta)/\sin\theta_1 \qquad \theta_2 \leqslant \theta \leqslant \theta_1 + \theta_2 \quad (6)$$
The integral in (4) may be evaluated as follows
$$2\int_0^{a_1/2} E\,dx = \lambda c'/(2\pi) \int_0^{\theta_1 + \theta_2} E\,d\theta$$
$$= E_0\lambda c'/(2\pi)\{\sin\theta_2 + b_2/b_1\cos\theta_2\tan\theta_1/2\}$$

Substiluting this relation in (4) gives

$$Z_{0\infty} = 120\pi^2 b_2/\{\lambda c'[\sin\theta_2 + b_2/b_1\cos\theta_2\tan\theta_1/2]\} \qquad (7)$$

The new cutoff wavelengths for the ridged waveguides are sketched in Fig. 3 (a) and (b) [3].

It is foud to be very closed form to the value calculated and curves plotted by the foregoing method.

EXPERIMENT AND RESULTS

Using the foregoing method, a single ridged waveguide is designed. Its ridge is of transverse sizes that $a_2 \times b_2 = 50 \times 25\,mm^2$ and is assembled in a BJ$-$22 rectangular waveguide, i. e. $a_1 \times b_1 = 109.\,2 \times 54.\,6\,mm^2$, longitudinal length is 500mm.

The ridged waveguide was used to construct a travelling $-$ wave cavity resonator. The electric field intensity is calculated. It shows that the E field value in ridged waveguide is 1.6 times as much as the E field value in rectangular waveguide of the same width and height with the same power.

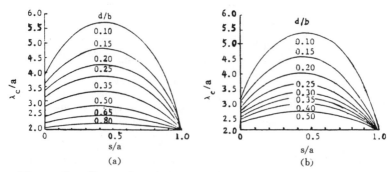

Fig. 3. Cutoff wavelength λc of the lowest order, modified TE_{10} mode of
(a). single—ridged waveguide and (b). double—ridged waveguide.

Put 12 sheets green tungsten carbid—cobalt(WC—Co) hard—metals into the cavity, and equidistant arrangement along the longitudinal direction. The geometry and size of the samples are 18 mm in diameter and 5 mm in height, cylinder, The end of cavity connected with a water load, After finishing above process. the samples are sintered with travelling—wave energy from a 2450MHz/5KW microwave source only in five minutes. The results indicate as follows:

1: The ridged waveguide cavity can be as a satisfactory microwave sintering cavity.

2: It can be increased to the electric field intensity and the homogeneity of the longitudinal direction in the ridged waveguide comparing with a plain rectangular waveguide of the same width and height.

3: The hard metals have excellent material properties with the microwavw sintering in the ridged waveguide. Listed in Table 1.

Table 1. Fine—grained WC—Co hard—metals

sintering method	microwave sintering	conventional sintering
sintering temperature	1200°C	1465°C
sintering time	1. 0min(in vaccum)	60min(in vaccum)
average grain size	0. 8μ	1. 2μ
bending strength	211MPa	150~160MPa
hardness(HRA)	90. 5	89. 3

ACKNOWLEDGMENT

We thank Prof. Jin Pan and Youde Huang in the understanding of microwave waveguide.

REFERENCES

1. S. B. Cohn, Properties of Ridge Wave Guide, Proc. I. R. E, August, 783(1947).
2. J. R. Whinnery and H. W. Jamieson, Epuivalent Circuits for Discontinuities in Transmission Lines, Proc. I. R. E. 32(2), 98—116(1944).
3. Om P. Gandhi, Microwave Engineering and Applications, (Pergamon Press Inc. New York, 1981)P. 90.

THE USE OF MICROWAVE-ASSISTED PROCESSING FOR THE PRODUCTION OF HIGH T_c SUPERCONDUCTORS

A.T. ROWLEY, F.C.R. WROE
EA Technology, Capenhurst, Chester, CH1 6ES, UK., anr@eatl.co.uk

ABSTRACT

High critical temperature superconducting (HT_cS) bulk ceramics with the high values of critical current densities (J_c) necessary for practical applications, are only achieved through careful control of the material microstructure. Consequently, many stages in the production of HT_cSs are extremely time consuming; often taking a number of days to proceed to completion, and usually requiring precise control of sample temperature.

The volumetric nature of microwave heating will clearly have an impact on the processing of large quantities of these materials, particularly when it is balanced with the simultaneous application of radiant heating. In this way, the production of thermal gradients can be avoided, even when relatively fast heating rates (for these materials) are used.

Perhaps more interesting, is the possibility of non-thermal microwave effects (due to the presence of an electric field) enhancing the processing times and even the material properties of HT_cSs.

This work assesses the impact of microwave-assisted processing on a number of stages involved in the production of bulk pieces of Bi-2212 and Y-123 based superconductors, and demonstrates quite clearly that non-thermal as well as thermal microwave effects are relevant to these materials.

INTRODUCTION

At present, it is difficult to manufacture superconducting metal oxide ceramics in bulk forms which have good superconducting properties (e.g. a high critical current density in a large magnetic field). Several processing stages are usually necessary, many of which involve heating or cooling over long periods of time (sometimes weeks) and require very precise control of the temperature of the material.

Nevertheless, laboratory results have been impressive - small pieces (i.e. a few cm) of $YBa_2Cu_3O_{7-\delta}$ based material (Y-123) have been produced with critical current densities in excess of $5 \times 10^4 Acm^{-2}$ at 77K in a field of 1T [1]. Similarly, quite complex shaped pieces (e.g. cylinders) of $Bi_2Sr_2CaCu_2O_{8+\delta}$ (Bi-2212) have been produced with current densities of up to $3000 Acm^{-2}$ in zero applied field at 77K [2].

However, scaling up these production routes using conventional processing will lead to problems. The duration of many of the processes will have to be increased still further to avoid the production of excessive temperature gradients within the samples. In addition, it will become increasingly difficult to control the processes to the precision required, and materials with non-homogeneous superconducting parameters (e.g. T_c, J_c, and critical field, B_c) will be produced.

It is already well known that, when balanced with appropriate conventional surface heating, the volumetric nature of microwave dielectric heating can produce extremely constant temperatures throughout the volume of a ceramic piece. This leads to the ability of heating large products extremely quickly without the production of thermal gradients within the material [3,4].

Moreover, microwave-assisted sintering of other ceramic materials has shown that the diffusion processes which cause sintering to occur are substantially enhanced in the presence of a

Mat. Res. Soc. Symp. Proc. Vol. 430 © 1996 Materials Research Society

high frequency electric field [5-7]. This gives the potential for increased sintering rates or sintering at a lower temperature. Since sintering and other diffusion based processes are employed in the production of high temperature superconductors, it is likely that similar benefits may be seen in these materials.

At present, there are several processing stages involved in the production of high temperature superconductors. A typical production route includes: calcination to produce precursor HT_cS powders; mixing of precursors powders with plastisizers and the forming of the bulk shape; binder and plastisizer removal; melt (or partial melt) processing to give required microstructure; oxygenation or oxygen removal to obtain optimum superconducting parameters. Alternatively, the precursor powders may be pressed into the bulk shape, the green body sintered to achieve the required strength and density, followed by melt-processing of the fired sample and subsequent oxygenation or annealing process.

POTENTIAL FOR MICROWAVE-ENHANCED PROCESSING

Calcination: Microwave-assisted processing should permit large quantities of powder in bulk form to be calcined in much the same way as small quantities or thin layers are presently processed. This will only be possible if a flow of air or inert gas can be maintained through the powder to allow any evolved gases to escape.

Binder removal: Volumetric heating should allow binders and plastisizers to be removed more quickly; providing that any evolved binder can be removed through the surrounding material at the rate that it is produced.

Sintering: The volumetric microwave heating should ensure that large pieces of material can be processed at least as fast, and in the same way, as small pieces (i.e. very much faster than the conventional sintering of large pieces). Moreover, the non-thermal enhancement of diffusion in the presence of a high frequency microwave field should lead to the possibility of even faster firing at the same temperature, or a similar sintering rate at lower temperatures. The latter of these could lead to micro-structural advantages.

Melt processing: This form of processing usually employs fairly rapid heating of the sample to somewhere just above its melting point, followed by slow controlled cooling back through the melting point at which point the highly textured and aligned microstructure develops. Microwave heating should permit much larger pieces of material to be processed in this way, and may lead to a reduction in processing time.

Oxygenation/Annealing: As in the sintering process, oxygenation (and oxygen removal) is a diffusion process which may be enhanced in the presence of an electric field. Therefore the rate of diffusion may be increased, and the process time reduced. Furthermore, the volumetric nature of microwave heating will lead to large pieces of material being produced with homogeneous oxygen content, leading to very uniform superconducting materials throughout the material.

EXPERIMENTS

The furnace used for these experiments was specially designed to allow microwave and conventional heating sources to operate simultaneously [8]. The system, shown schematically in figure 1, is essentially a conventional furnace built within a multi-mode 2.45GHz microwave cavity powered by a 1.2kW industrial-standard magnetron. Care must be taken in the design and construction of the radiant heating elements to avoid any unwanted interaction between the two heating sources. For example, the radiant heating elements can severely affect the distribution of the high frequency fields within the cavity.

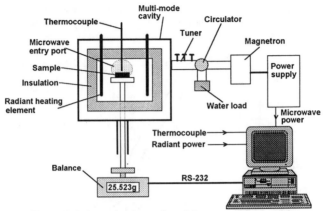

Figure 1: Schematic of experimental arrangement used for microwave-assisted processing of high T_c superconductors.

The thermocouples used in the furnace are encased in a metal sheath, and are positioned to minimise their effect on the microwave field distribution. For the bulk of these experiments a thermocouple is used to control the power to the radiant heating elements; the amount of microwave power is set manually.

For some experiments such as melt processing and oxygenation (as shown in figure 1) the sample weight is monitored during processing; in others, such as sintering, a dilatometer is used in place of the balance to measure the length of the sample.

RESULTS

Calcination

A bulk quantity (i.e. instead of a thin layer) of a stoichometric mixture of spray dried yttrium, barium and copper nitrate powders (~10cm³) was calcined at 770°C for 35 hours to form the Y-123 precursor powder. Two thermocouples, one at the centre of the powder and one at the surface were used to monitor the temperature. When microwave-assisted processing was used, the two thermocouples could be maintained at the same temperature throughout the whole calcination stage, and the powder heated to 770°C at a faster rate. Conventional processing resulted in up to 30°C temperature difference between the two thermocouples. If larger quantities were processed conventionally in this way, this temperature difference would be even greater, and it would be difficult to guarantee the production of highly uniform precursor powders.

Binder removal

In one material processing route Bi-2212 precursor powders are mixed with binders and plastisizers to form a material which can be formed into the shape required. Prior to further processing, this binder is removed in a slow heating stage at 10°C/hour to 500°C. The effect of microwave-assisted processing on the binder removal stage for a small (~5g) sample is shown in figure 2. Clearly, the binder is removed faster in the presence of microwaves. This is probably due to a more uniform temperature distribution throughout the whole of the sample, and this effect will become more pronounced as larger and larger pieces are processed.

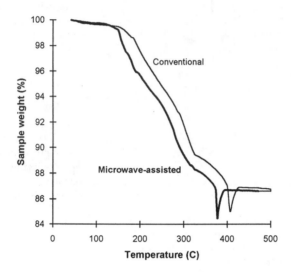

Figure 2: Graph of sample weight versus temperature for conventional and microwave-assisted binder removal from Bi-2212 material.

Sintering

In one production route of bulk Y-123 pieces, the precursor powders are cold die pressed and sintered prior to melt processing. Figure 3 shows the difference between conventional and microwave-assisted sintering of small (~2.5g) pieces of Y-123. In this particular case, the green body is heated to ~950°C at 5°C/min and held at this temperature for 3 hours prior to cooling at 5°C/min back to room temperature.

Two effects are observed - firstly the microwave-assisted sintering produces higher final densities, and, secondly, the microwave-assisted sintering begins at a lower temperature. this should allow the Y-123 to be sintered at a lower temperature, producing a dense body with smaller grains, and will lead to retained benefits when this material is melt processed.

As larger samples are processed, the thermal benefits of microwave-assisted heating will be observed, allowing much larger pieces to be sintered in the same way as small pieces are processed conventionally.

Oxygenation/annealing

During melt processing, oxygen is lost from the Y-123 samples, and has to be replaced to achieve the high values of Jc required. Usually this is done by heating the sample to a set temperature and then cooling it extremely slowly. As the sample cools, oxygen diffuses back into the material structure. To test the effect of microwave-assisted processing on oxygenation, a small (~2cm diameter) piece of melt processed Y-123 is heated to 500°C and held there for 20 hours. Figure 4 shows a comparison of conventional and microwave-assisted processing. In the

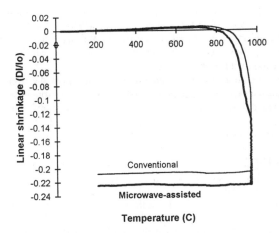

Figure 3: Graph of linear shrinkage versus temperature for conventional and microwave-assisted sintering of Y-123 material

microwave case, the sample gains weight much more quickly, showing more rapid absorption of oxygen in the presence of a microwave field. Calculations on this data show that the effective diffusion coefficient is increased by about 20% by this process.

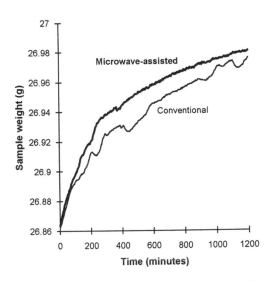

Figure 4: Graph of sample weight versus time for conventional and microwave-assisted oxygenation of Y-123 melt processed material.

CONCLUSIONS

There are many stages in the production of high T_c bulk superconductors which could benefit from the application of microwave-assisted processing. As well as thermal benefits, due to the volumetric nature of microwave heating, there are non-thermal effects which give further advantages over conventional processing.

When large pieces of material are manufactured, microwave-assisted processing allows the materials to be heated in a very controlled way leading to uniform and optimised material properties together with a reduction in processing times. Furthermore, diffusion based processing steps such as sintering and oxygenation are improved due to enhanced diffusion in the presence of a high frequency electric field.

ACKNOWLEDGEMENTS

This work has been undertaken by EA Technology through its Core Research programme on behalf of its members. The authors would like to thank the Directors of EA Technology for permission to publish it here. The authors would like also to thank the Interdisciplinary Research Centre in Superconductivity, Cambridge University for provision of samples.

REFERENCES

1. M. Murakami, Supercond. Sci. Technol. **5**, p. 185 (1992).
2. R. Watson, M. Chen, and J.E. Evetts, Supercond. Sci. Technol. **8**, p. 311-316 (1995).
3. W.H. Sutton, Ceramic Bulletin, **68**(2), p. 376-386 (1989).
4. J. Wilson and S.M. Kunz, J. Am. Ceram. Soc. **71**(1), p. 40-41 (1988).
5. R. Wroe and A.T. Rowley, accepted for publication J. Mat. Sci. (1996).
6. M.A . Janney and H.D. Kimrey, M.A. Schmidt and J.O. Kiggans, J. Am. Ceram. Soc. **74**(7), p. 9-18 (1991)
7. K.I. Rybakov and V.E. Semenov, Phys. Rev. B. **49**(1), p. 64-68 (1994).
8. EA Technology Ltd., International patent application, No. PCT/GB94/01730 (1990).

Overview of Component and Device Characterization

by Doug Rytting, Research and Technology Manager
Santa Rosa Systems Division, Hewlett Packard
1400 Fountain Grove Parkway, Santa Rosa, CA 95403

Abstract

The electronics market is constantly pushing the state of the art in design to reduce cost, size, weight, and power consumption. New designs are emerging causing rapid changes in technology driving high design turnover. This drives the need for component and material measurements that will reduced design cycles and time to market. In the design and measurement of linear devices, error corrected S-parameters are traditionally measured with a network analyzer. the network analyzer combines magnitude with phase measurements for improved accuracy. Time domain techniques are used to get a better physical understanding of the device characteristics. Error correction procedures have been improved to provide high accuracy and ease of use. New methods to measure impedance results in high precision capacitor and inductor models necessary for both surface mount and integrated circuit applications. Note: In the following paper the relevant text follows each slide.

Agenda

✔ S-parameters

✔ Time Domain

✔ Error Correction

✔ Impedance

This paper is a brief overview of network analysis techniques that can be applied to component and materials measurements. S-parameters are the primary measurement technique at high frequencies. Conversion to time domain allow a physical view vs. distance of a distributed network. Error correction techniques are continuing to advance which provides better accuracy and ease of use. Impedance measurement techniques have been improved to provide the accuracy required for modern components.

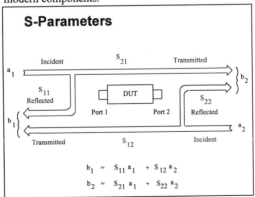

S-Parameters

$$b_1 = S_{11}\, a_1 + S_{12}\, a_2$$
$$b_2 = S_{21}\, a_1 + S_{22}\, a_2$$

The most complete method for describing high frequency devices uses the S-Parameters. By applying a stimulus to a 2-port device in both the forward and reverse directions, and measuring both reflected and transmitted signals. A set of equations combine the resulting input and output signals to calculate S-parameters [1], [2].

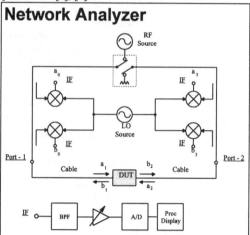

Magnitude and phase (vector measurements) for linear and non-linear devices can be measured using a Vector Network Analyzer. This tuned receiver, combined with a synthesized sweeper, is most often configured with an S-parameter test set enabling full 2-port transmission/reflection device measurements. Vector error correction is used to provide accurate characterization of both coaxial and non-coaxial devices. The receiver's tuned IF filter provides immunity from harmonics and spurious responses to enhance measurement accuracy.

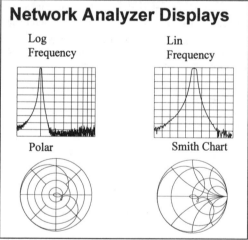

Modern Vector Network Analyzers can convey measurement information in a variety of display formats. Magnitude (linear, dB or SWR), phase or group delay characteristics may be displayed in log and linear frequency display formats. Magnitude and phase may also be displayed in a polar format or, for impedance, in the Smith Chart format.

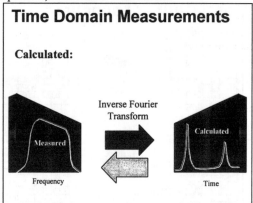

With the data measured in the frequency domain using the network analyzer, the inverse Fourier transform can then be used to view the data in the time domain [3], [4]. The data must be properly prepared by "windowing" the frequency domain data so that the ringing (Gibbs phenomenon) is reduced when viewed in the time domain. The time domain is very useful in looking at reflections that are spaced out in time so that their unique location can be observed. The time resolution, or ability to resolve between closely spaced reflections, is improved as the frequency bandwidth is increased. The range, or how far out in time you can view before the data starts to repeat, is determined by the frequency step size. The smaller the step size the further out the alias free range will be.

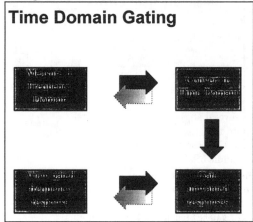

After the data is converted to the time domain, any unwanted reflections can be gated out (filtering in the time domain). This leaves just the responses that are desired. The data can then be transformed back to the frequency domain to observe the frequency response with the unwanted reflections removed.

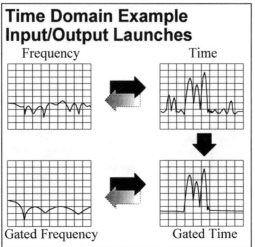

This example is a microstrip passive component in a fixture with coax to microstrip launches at both ports. The data in the frequency domain shows a ripple response but not much insight into what is happening. When transformed to the time domain the effects of the package launches can be easily observed at the start and end of the data. The transitions are then gated out so that only the microstrip device is observed in the time domain. Then transforming back to the frequency domain gives the real data of the microstrip device without the effects of the transitions.

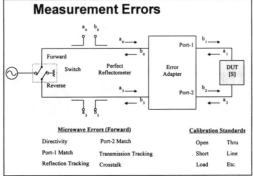

Systematic (repeatable) measurement errors are caused by imperfections in the measurement hardware. There are 6 types:

Error	Primary cause
Tracking (2)	Frequency response of the system
Match (2)	Port mismatch at the test ports
Directivity (1)	Reflection leakage before device
Crosstalk (1)	Leakage around device

In a 1-port reflection measurement, there are three systematic errors: reflection tracking, port-1 match, and directivity. For a 2-port device, measuring both reflection and transmission, there are

12 total systematic errors (6 in the forward and 6 in the reverse direction). Calibration standards include well defined opens, shorts, loads, "thru" connections, and others. Generally, correcting more errors requires more standards. Random errors (such as noise and connector repeatability) and drift errors (due to temperature, etc.) can never be removed with calibration. Their effects can only be minimized with careful measurement practice [5], [6].

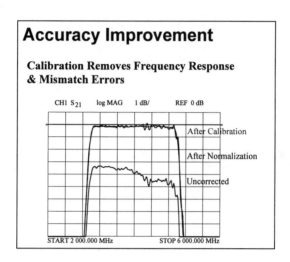

To better understand the true characteristics of the filter under test, it is necessary to isolate the characteristics of the filter from the characteristics of the test system. Performing a normalization prior to the measurement of the filter will remove the frequency response of the system from the measurement, allowing us to more accurately measure the filter's true frequency response. Notice that the normalized filter's frequency response contains ripple that causes its insertion loss to periodically appear greater than 0 dB. What exactly does this mean? How can a passive device appear to have insertion gain? It turns out that the ripple on the normalized measurement is not a characteristic of the filter under test, it is instead a response due to an interaction between the system's port-1 and port-2 match characteristics. This ripple is most prevalent on low insertion loss devices that offer very little attenuation between the port-1 and port-2 mismatch signals. Just as normalization is used to remove the frequency response of the system from a measurement, vector error correction can be used to mathematically remove the effects of all systematic errors from the measurement of a device under test. These errors are port-1 and port-2 match, transmission and reflection frequency response, directivity and crosstalk. The measurement shown after error correction is the most accurate of the three measurements shown.

Error Correction Methods

	Thru (T) with known S-parameters [4 conditions]	3 known Reflects (OSL) on port-1 [3 conditions]	3 known Reflects (OSL) on port-2 [3 conditions]
TOSL	Thru (T) with known S-parameters [4 conditions]	3 known Reflects (OSL) on port-1 [3 conditions]	3 known Reflects (OSL) on port-2 [3 conditions]
TRL & LRL	Thru (T) or Line (L) with known S-parameters [4 conditions]	Unknown equal Reflect (R) on port-1 and port-2 [1 condition]	Line (L) with known S_{11} and S_{22} [2 conditions]
TRM & LRM	Thru (T) or Line (L) with known S-parameters [4 conditions]	Unknown equal Reflect (R) on port-1 and port-2 [1 condition]	Known Match (M) on port-1 and port-2 [2 conditions]

A Known Reference Impedance and a Port-1 to Port-2 Connection are Required
A Thru Measurement with Terminations on Port-1 and Port-2 is also Required for Leakage Calibration

There are numerous error correction methods used in the industry today [7], [8], [9], [10], [11]. The first method (TOSL) developed back in the 1960s was based on the 12 term model. There were 6 errors in the forward direction and 6 in the reverse direction. In the forward direction these errors are the directivity, port-1 match, reflection tracking, transmission tracking, port-2 match, and crosstalk. The first 3 errors can be determined by measuring 3 know calibration standards on port-1 and solving the 3 resultant simultaneous equations. Typically these standards are an open, short, and load (fixed or sliding). Then with the test ports connected, the port-2 match and transmission tracking can be determined. The crosstalk is measured by placing two load on the test ports and measuring the isolation. This whole process is then repeated for the reverse direction. In the 1970s the calibration problem was looked at much closer and it was determined that the 12 term model could really be simplified to 10 terms with a model that did not change as the measurement system switched from measuring in the forward direction to the reverse direction. The two crosstalk terms could be easily measured with terminations on the ports which left only 8 terms to solve for. Since S-parameters are ratio measurements the error terms could all be referenced to one error term reducing the number of unknown to 7. So the real calibration problem is to come up with 7 known conditions to solve for the 7 error terms of the measurement system. The first technique to take advantage of the 7 term error model was the TRL technique developed by Glen Engen and Cletus Hoer at NBS for the six-port network analyzer. This technique used a thru connection, a delay line (ideally 1/4 wavelength long at the center of the frequency band) with known impedance (S_{11} and S_{22} known), and an equal but unknown reflection connected to each test port. These provided enough conditions to solve for the 7 error terms, and the process used also determined the remaining characteristics of the line (S_{21} and S_{12}) and the reflection coefficient of the unknown load. Multiple lines are typically used to cover a broad frequency range. The real advantage of the method is that the requirement of knowing the parameters of the open and short are removed and the errors caused by the imperfect knowledge of these standards is eliminated. The only critical standards are the impedance of the line and the requirement that the reflect be equal at each port. Measurement accuracy was greatly improved in coax and waveguide measurements. Later this approach was augmented to use another line as the thru connection whose characteristics are known or match those of the line (LRL method). The difference in the length of the two lines provided the necessary delay. The TRL and LRL methods are the most traceable techniques for measuring on wafer and are presently being formalized by NIST as the method to support the wafer probing industry. It was also determined that the line could be replaced by a known match and this led to the TRM and LRM techniques. This method is inherently broad band and is very easy to use.

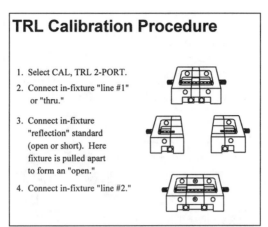

TRL Calibration Procedure

1. Select CAL, TRL 2-PORT.
2. Connect in-fixture "line #1" or "thru."
3. Connect in-fixture "reflection" standard (open or short). Here fixture is pulled apart to form an "open."
4. Connect in-fixture "line #2."

The TRL method is used to calibrate a fixture by first connecting the thru, then the fixture is separated and the resulting opens are used as the high reflect standards. Finally a line is connected to complete the calibration. Many times fixtures can be calibrated by putting in calibration substrates for the thru, reflect and line and don't need to be physically separated as shown in this example.

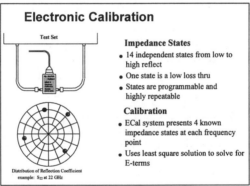

Electronic Calibration

Impedance States
- 14 independent states from low to high reflect
- One state is a low loss thru
- States are programmable and highly repeatable

Calibration
- ECal system presents 4 known impedance states at each frequency point
- Uses least square solution to solve for E-terms

Recently, a very convenient calibration method has been developed by ATN Microwave and HP [12]. Instead of using separate opens, shorts, and loads connected to the test port, an electronic calibration module can be used. The electronic calibration (ECal) method uses 4 known programmable impedance states at each frequency and a programmable thru state as the calibration standards. There are 14 independent impedance states at each frequency, and the 4 states that cover the largest area of the Smith Chart are selected. Since only 3 states are required, the software uses a "least square" solution to solve for the 3 error terms. Stability and repeatability are the key factors for a good electronic calibration system. The ECal module is temperature controlled and precision slotless connectors are used to enhance the repeatability. The ECal method is actually a transfer standard. The ECal module is calibrated using a high quality factory calibration and the calibration data is stored on a EPROM inside the module. As a quality check one of the states not used during calibration can be measured and compared to the factory data. There is software available to recalibrate the ECal module either back at the factory or in the customers facility.

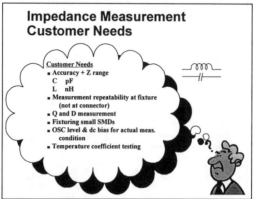

Most wireless products are getting smaller, thinner, lighter, and shorter; which means that, components in wireless products are requested to do the same thing. Recent wireless products are using 1nH inductors and 1pF capacitors. These need to be measured more accurately than ever for good matching with other circuits. Especially, they want to reduce the number of adjustments by using accurate components. In addition, lower power consumption components are used for these products since most of these are operated by batteries. Also, these components need to be evaluated at their operating condition since components characteristics changed if frequency and test level are changed.

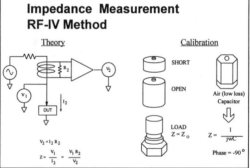

The RF-IV method was invented to reply to these customers' request for impedance measurement [13]. Recent SMDs (1nH and 1pF) can be measured with reasonable accuracy since impedance is calculated directly from voltage and current drop across devices. The I-V, or probe technique, measures two voltages one of which is proportional to the vector current through the device. V_1 is measured across the probe and V_2 across a low-loss balun transformer. This technique is bandwidth-limited because of the pick-up transformer. However, a new balun has been developed to extend the measurement range to 1.8 Ghz. Also, low loss devices can be measured accurately since the RF-IV method can compensate for phase by doing an air-capacitor calibration added to the regular open/short/load calibration.

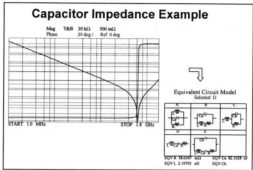

Capacitor Impedance Example

For frequency ranges up to several GHz, circuit D is an equivalent circuit. By knowing parasitic values of a capacitor, a circuit designer can predict the real response of the circuit. Some Impedance Analyzers have the ability to calculate these values from frequency characteristic by pressing one button. It is very important to know there is no ideal capacitor.

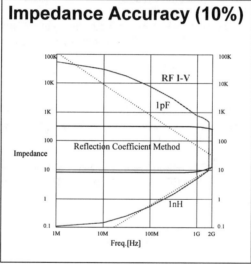

Impedance Accuracy (10%)

This graph shows the comparison of 10% accuracy area of the RF-IV method and the standard Reflection method. At low frequencies the RF-IV method is most accurate for low and high impedance measurements.

Conclusion

With the advent of new calibration techniques and better high impedance RF-IV measurement technology, the difficult component and materials measurements are improved. With this improved accuracy, combined with faster and easier to used methodologies, the design time can be reduced.

References

[1] L. Weinberg, "Fundamentals of Scattering Matrices," Electro-Technology, pp 55 - 72, July 1967

[2] R.B. Marks and D.F. Williams "A General Waveguide Circuit Theory," J. Res. Natl. Inst. Stand. Technol. 97,533, Sept. 1992

[3] M. E. Hines, H. E. Stinehelfer, "Time Domain Oscillographic Microwave Network Analysis Using Frequency Domain Data", IEEE Trans. on Microwave Theory and Techniques, MTT-22-3, pp 276 - 282, Mar. 1974

[4] D. K. Rytting, "Let Time Domain Response Provide Additional Insight Into Network Behavior", Hewlett Packard RF & Microwave Symposium, March 1984

[5] K. H. Wong, "Using Precision Coaxial Air Dielectric Transmission Lines as Calibration and Verification Standards", Microwave Journal, pp 83 - 92, December 1988

[6] S. B. Donecker, "Determining the Measurement Accuracy of the HP8510 Microwave Network Analyzer", Hewlett-Packard RF & Microwave Symposium, March 1985

[7] J. Fitzpatrick, "Error Models for Systems Measurement", Microwave Journal, pp 63 - 66, May 1978

[8] G. F. Engen, C. A. Hoer, "Thru-Reflect-Line: An Improved Technique for Calibrating the Dual 6-Port Automatic Network Analyzer", IEEE Trans. on Microwave Theory and Techniques, MTT-27-12, pp 987 - 993, Dec. 1979

[9] H. J. Eul, B. Schiek, "A Generalized Theory and New Calibration Procedures for Network Analyzer Self-Calibration", IEEE Trans. on Microwave Theory & Techniques, vol. 39, pp. 724-731, April 1991

[10] H. J. Eul, B. Scheik, "Reducing the Number of Calibration Standards for Network Analyzer Calibration", IEEE Trans. Instrumentation Measurement, vol 40, pp. 732-735, August 1991

[11] Kimmo J. Silvonen, "A General Approach to Network Analyzer Calibration", IEEE Trans. on Microwave Theory & Techniques, vol 40, April 1992

[12] V. Adamian, "A Novel Procedure for Network Analyzer Calibration and Verification," ARFTG Digest, Spring 1993

[13] "Electronic Materials Measurement Seminar," Hewlett Packard Seminar Publication

Part V

Dielectric Properties Measurements and Analysis

MEASUREMENTS OF DIELECTRIC PROPERTIES
FOR INTENSE HEATING APPLICATIONS

J. M. BORREGO, K. A. CONNOR, J. BRAUNSTEIN
ECSE Department, Rensselaer Polytechnic Institute, Troy, NY 12180-3590

ABSTRACT

Dielectric measurements for frequencies in the tens of gigahertz are discussed. Interpretation of measurements obtained for mixtures are presented based on some simple analytical models and finite-element analysis of the wave structure in the measurement apparatus.

INTRODUCTION

Radio frequency electromagnetic waves can deliver heat to materials according to the expression

$$P_\mu = \pi \ \epsilon \ \tan\delta \ f \ E^2 \qquad (1)$$

where P_μ is the power per unit volume, ϵ is the dielectric constant, f is the frequency, and E is the intensity of the electric field of the wave. Note that the higher frequencies are more effective as is high source power as expressed in the electric field intensity. High power sources were first developed at lower frequencies, but recently continuously available power of the order of a megawatt is now being provided by microwave tube manufacturers at higher frequencies such as 60 - 140 GHz. There are, thus, two choices we can make to deliver very high microwave power densities to some material -- high frequency or high source power. The choice depends on the particular material processing approach being pursued.

There is a rich and varied literature available on the subject of dielectric measurements. The data that are available in the literature show a good deal of scatter and apply almost exclusively at lower frequencies (below 10-20 GHz), room temperature and low electric fields. Also, one of the most important issues associated with high power heating involves the properties of mixtures, since most applications involve combinations of materials. A variety of methods have been proposed to predict the properties of mixtures without having to measure them all directly. Unfortunately, the formulas do not generally give the same results. In the work reported on here, we decided to look at measurements in the range of 25 - 40 GHz and see if we could use finite-element analysis of experimental conditions to identify the model of dielectric mixtures that seems to work the best.

EXPERIMENT

A great deal of high power microwave work has involved making ceramics, vitrification of soils, processing of minerals, etc. We, therefore, decided to focus on the properties of some common minerals. We chose sand for its obvious abundance and then looked for a mineral with very different properties and for which there are some recent publications -- rutile. The dielectric constant ϵ_r and the dissipation factor $\tan\delta$ of powder samples were determined in two frequency ranges: near 1 MHZ and from 26.5 to 40 GHZ. These frequencies were chosen because they bracket available published data and we had the necessary network analyzers. Minerals come in many different forms. Researchers using a variety of valid techniques have found the electrical conductivity of rutile to vary over a range of 0.03-10 (ohm-cm)$^{-1}$.[1-3] Here we have ignored results from researchers who lack understanding of RF engineering basics and thus can get almost any result. Because materials can vary so much, we performed the 1 MHZ measurements as a benchmark.

The dielectric properties of the powder samples were determined using appropriate jigs to contain them. Working with powdered materials means that measurements apply to a mixture of air and dielectric material and thus must be corrected for packing density, which varied from about 0.5 to 0.6 in our measurements. There is apparently no generally accepted, simple approach to applying this correction at all frequencies.

The dielectric properties at 1 MHZ were measured using a parallel plate capacitor jig that was fabricated to contain the powder samples. The capacitance of the jig was measured empty and filled with the two powders. From this the dielectric constant and loss factor were determined.

TABLE 1 Measurements at 1 MHZ

Sample	ϵ_{eff}	ϵ_r(VH)	ϵ_r(L)	$\tan\delta_{eff}$	$\tan\delta$
Rutile	8.1	33	19.3	0.008	0.01
Sand	2.2	4.1	3.6	0.0035	0.009

where ϵ_{eff} and $\tan\delta_{eff}$ are the effective dielectric constant and loss tangent of the mixture and ϵ_r and $\tan\delta$ are the corrected values accounting for the packing fraction. At this frequency, Von Hippel's relationship for dielectric constant gives generally reasonable results [3].

$$\log\epsilon_{eff} = f_1\log\epsilon_1 + f_2\log\epsilon_2 \qquad (2)$$

where f_i is the packing density of the material with ϵ. This expression is derived by combining simple capacitors. When one of the materials is air, as in our case, this expression becomes very simple

$$\epsilon_r = \epsilon_{eff}^{1/f} \qquad (3)$$

The above values from (3) check relatively well with the value of dielectric constants given in the literature for fused quartz. (von Hippel lists $\epsilon_r = 3.78$ and $\tan\delta = 2\times10^{-4}$. Many natural quartz minerals show similar values.[3]) Looyenga derived an alternative formula [4]

$$\epsilon = (((\epsilon_{eff})^{\frac{1}{3}} + f - 1) / f)^3 \qquad (4)$$

which is in agreement with published quartz data, but does not work as well with rutile (or Ilmenite which should be close). This seems to support the use of von Hippel's simple formula.

For the loss tangent, we derived a simple expression based on the definition of quality factor, which is inversely proportional to the loss tangent.

$$\tan\delta = \tan\delta_{eff} (f+(1-f)\epsilon_r) \qquad (5)$$

At microwave frequencies, the material constants were determined from propagation and attenuation measurements in a piece of Ka-band waveguide of known length filled with the powdered material[5, 6]. The waveguide was shorted at one end and the material was held in place by a thin window. The phase angle of the reflection coefficient was determined, referenced to the speed of light in air (empty waveguide). The value of c found for air was about 2% off from the expected value because of inaccuracies in the network analyzer calibration and in the measurement of the waveguide length. Measurements of both reflection angle and attenuation were noisier for rutile than for sand.

TABLE 2 Measurements at 25 - 40 GHz

Sample	Freq. (GHz)	c (10^8m/s)	ϵ_{eff}	Return Loss (dB)	$\tan\delta_{eff}$ (10^{-4})
Sand	37-39	2.2	1.9	0.7	3.6
Rutile	34-39	1.0	7.9.	3.0	35.

Again the actual values of the dielectric constant differ according to the formula used: ϵ_r is 3.2 and 31.3 from von Hippel, 2.9 and 18.7 for Looyenga. Now the latter number seem to be closer to published data.

It is the measurement of the insertion loss that is most significant in assessing the application of high frequency power to a material. Since this measurement involves propagation in a single direction along the waveguide, the cross-sectional area of the material in powder form and fully packed into solid form can be taken as the same. The packing fraction then scales with the length of the material. Another way of looking at this is to realize that the wave has to propagate through the same material, regardless of its density. Thus, the product of the effective material length L_{eff} and the effective attenuation constant α_{eff} should equal the product of the actual L and α, from which we obtain that

$$\frac{\sqrt{\epsilon_{eff}}\tan\delta_{eff}}{\sqrt{1 - \frac{\omega_{c_{eff}}^2}{\omega^2}}} = \frac{f\sqrt{\epsilon}\tan\delta}{\sqrt{1 - \frac{\omega_c^2}{\omega^2}}} \qquad (6)$$

The square root terms in the denominators can be neglected without producing much error, as long as the frequency ω is much larger than the cutoff frequency ω_c. This expression is in good qualitative agreement with published data on materials with similar properties to those considered here. Note that comparison's can only be made to measurements made with the same technique. The loss tangent for the two materials should thus be multiplied by 1.4 and .85 for sand and rutile using the von Hippel numbers for ϵ_r, and by 1.5 and 1.1 for the Looyenga numbers, respectively. The values for rutile are lower than one would expect from the literature. The clever measurements of Kudra and his colleagues [7] and Parkhomenko's compilation of rock data[2] list quite a bit larger (3 to 10 times) values for both ϵ_r and $\tan\delta$. Since rutile has such distinctively large values for these parameters and since mineral composition can vary a great deal, we conclude that the particular sample of rutile we analyzed is on the low end for these parameters.

NUMERICAL MODEL OF MEASUREMENT

The two formulas used for finding the dielectric constant imply an overly simple picture of the measurement process. Several papers have been written on how to incorporate more of the actual details of the measurement geometry [6,8]. However, these expressions are sufficiently simple that it would be helpful if we could use them with some confidence. In order to test their approximate validity, we simulated the measurement using a finite element code (with absorbing boundary conditions) that we had originally written to characterize open radiators such as Vlasov launchers [9]. The geometry we used was a shorted parallel plate transmission line with a section loaded with dielectric near the end. The mode used in the measurements was the TE_{10} mode so the parallel plate TE_1 mode has the same properties. The geometry is the same as that shown in Figures 1 and 2 of [6]. The finite element mesh for this problem was chosen to have a density with at least 20 elements per wavelength.

First we considered an empty waveguide and then a waveguide with its end completely filled with dielectric to be sure that we had a high enough density mesh and that the wave solution looked reasonable. Then, to model the mixture with a 50% packing fraction of dielectric in air, we only put dielectric material in every other triangular region defined by the mesh. The results are shown in Figures 1 and 2 for two different frequencies 33 and 40 GHz. The material properties were assumed to be roughly those of an organic material we have measured, $\epsilon = 2 - j0.06$, to make the features of the results easier to recognize.

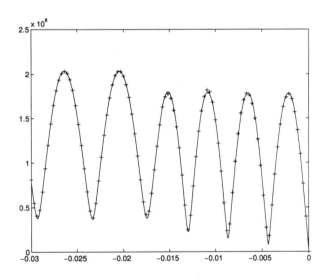

FIGURE 1 Numerical Results for 33 GHz
Standing Wave Pattern In Air (Left Two Peaks) and Dielectric Mixture (Right Four Peaks)

The crosses on the plots are data points from the finite element solution. To interpret the data, we fit the points to the characteristic standing waves that should obtain in the waveguide:

$$f(x) = A_d[e^{-\alpha x}e^{-jk_d x} - e^{\alpha x}e^{jk_d x}] \qquad (7)$$

234

in the dielectric and

$$f(x) = A_s[e^{-jk_sx} + \Gamma_x e^{jk_sx+\phi_s}] \qquad (8)$$

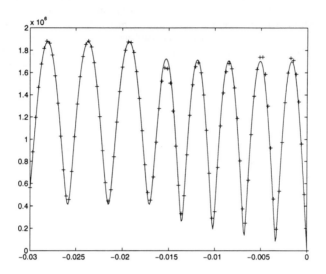

FIGURE 2 Numerical Results for 40 GHz
Standing Wave Pattern in Air (Left Two Peaks) and Dielectric Mixture (Right Four Peaks)

in free space. The parameters such that the curves fit the data are the wavenumber in the open region $k_s = 530$ and 715, the wavenumber in the dielectric $k_d = 726$ and 924, and the decay constant in the dielectric $\alpha = 9.5$ and 12, at 33 and 40 GHz, respectively. Note that the excellent fit of the ideal analytic form for the field to the numerical results shows that the mesh size was appropriately small. From ω_c, k_d, and ω we get the effective dielectric constant of the half filled dielectric region of $\epsilon_{eff} = 1.5$. From

$$\alpha = \frac{k^2 \tan\delta}{2k_d} \qquad (9)$$

we get $\tan\delta_{eff} = 0.018$.

Applying equations (3) and (4) to determine which comes closer to the actual conditions of the simulated measurement, we find from (3) that $\epsilon_{eff} = 2.25$ and from (4) that $\epsilon_{eff} = 2.14$. It appears then that, while neither expression is correct, Looyenga's expression gets us closer to the actual answer. Applying equation (6) we find that a measurement of this attenuation would give us $\tan\delta = 0.03$, as expected. We conclude, then that the combinations of equations (3), (4), and (6) give qualitatively correct results, when (3) and (4) are in rough agreement. However, when we have

235

large differences between their predictions, we must apply the measured numbers conservatively when making predictions of how a microwave processing system will work.

SUMMARY

We have measured the electrical properties of some typical powdered minerals that might be encountered in microwave processing applications. The measurement technique is standard and should be reliable if the material conditions of the process match the conditions in the measurement apparatus. However, to check the results, it is always useful to have some simple formulas that permit comparison with published data on similar materials. We checked two of the more popular models and found that they tend to give larger values of ϵ_r and thus smaller values of $\tan\delta$ for a simulated measurement. Thus, if one uses published data to predict the heating of a material mixture, power deposition per unit volume will be somewhat over estimated.

REFERENCES

1. S. Nelson, D. Lindroth, R. Blake, Dielectric properties of selected and purified minerals at 1 to 22 GHz., J. Microwave Power and Electromagnetic Energy 24 (1989) 213-220

2. E.I. Parkhomenko, Electrical Properties of Rocks, Plenum Press (1967)

3. A. R. von Hippel, ed., Dielectric Materials and Applications, John Wiley and Sons, New York (1954).

4. H. Looyenga, Dielectric constants of heterogeneous mixtures, Physica 31 (1965) 401-406

5. S. Roberts and A. von Hippel, A New Method for Measuring Dielectric Constant and Loss in the Range of Centimeter Waves, J. Appl. Phys. 17 (1946) 610-616.

6. S. O. Nelsen, L. E. Stetson, and C. W. Schlaophoff, A General Computer Program for Precise Calculation of Dielectric Properties from Short-Circuited-Waveguide Measurements, IEEE Trans. on Instr. and Meas. IM-23, (1974) 455-459

7. T. Kudra, G.S.V. Raghavan, F.R. van de Voort, Microwave heating of rutile, J. Applied Physics 73 (1993) 4534-4540

8. J. Baker-Jarvis, E. J. Vanzura, W. A. Kissick, Improved Technique for Determining Complex Permittivity with the Transmission/Reflection Method, IEEE Trans. on Microwave Theory and Techniques 38 (1990) 1096-1102

9. K. Connor, P. Shin, S. Salon, The hybrid finite element-boundary element solutions of waveguide problems, IEEE Trans. on MAG 25 (1989) 3943-3945 and L. Nicolas, K. A. Connor, S. J. Salon, B. G. Ruth, L. F. Libelo, 2 Dimensional Finite Element Analysis of Open Microwave Devices, Journal de Physique III 2, (1992) 2101-2107.

FDTD SIMULATIONS AND ANALYSIS OF THIN SAMPLE DIELECTRIC PROPERTIES MEASUREMENTS USING COAXIAL PROBES

S. BRINGHURST, M. F. ISKANDER, and M. J WHITE
Electrical Engineering Department, University of Utah, Salt Lake City, UT 84112

ABSTRACT

A metallized ceramic probe has been designed for high temperature broadband dielectric properties measurements. The probe was fabricated out of an alumina tube and rod as the outer and inner conductors respectively. The alumina was metallized with a 3 mil layer of moly-manganese and then covered with a 0.5 mil protective layer of nickel plating. The probe has been used to make complex dielectric properties measurements over the complete frequency band from 500 MHz to 3 GHz, and for temperatures as high as 1000 °C.

A 3D Finite-Difference Time-Domain (FDTD) code was used to help investigate the feasibility of this probe to measure the complex permittivity of thin samples. It is shown that by backing the material under test with a standard material of known dielectric constant, the complex permittivity of thin samples can be measured accurately using the developed FDTD algorithm. This FDTD procedure for making thin sample dielectric properties measurements will be described.

An uncertainty analysis to quantify the errors due to differential thermal expansion of the inner and outer conductors of the coaxial probe was conducted using a 2D cylindrical FDTD code. The FDTD results were compared with error-analysis data, based on analytical solutions for the special case when an air gap exists between the probe and the material under test [1], and good agreement was observed. It is shown that for a differential thermal expansion between the inner and outer conductor of 0.1 mm an error as large as 170% may result in the measurement of the magnitude of the input impedance. These as well as other error-analysis results and experimental data on thin-sample dielectric measurements will be presented and discussed.

INTRODUCTION

Open-ended coaxial probes have been used for dielectric properties measurements for several years [2,3]. The attractiveness of the open-ended coaxial probe measurement procedure has been in the relative ease of the measurement procedure and in the broadband measurement capabilities of the probe. Probes are available that can make measurements over a band of 10 GHz. In the last few years the use of probes has been extended to the measurement of thin samples [4,5]. This has not been done in the past due to the problem of satisfying the infinite sample criteria often required in the calculation procedure.

Furthermore, none of the available probes were suitable for making measurements at temperatures as high as 1000 °C. This is due to the problems with the differential thermal expansion between the inner and outer conductors of metal probes. Even the use of metals of small thermal expansion coefficients, such as kovar, has resulted in limited success and provided reasonable results for temperatures limited to 800 °C at which the thermal-expansion coefficient of kovar increased significantly and hence affected the

Mat. Res. Soc. Symp. Proc. Vol. 430 © 1996 Materials Research Society

accuracy of the measurements [6]. We have designed a metallized ceramic probe that does not have the differential thermal expansion problems inherent in metal probes [7]. Furthermore, with the utilization of an FDTD calculation procedure, it was possible to extend measurements to thin samples. With these new additional measurement capabilities, it was important to conduct a detailed error analysis to help guide this rather delicate measurement procedure, particularly when thin samples were used. In the following sections the thin sample dielectric properties measurements procedure and results from the uncertainty analysis will be presented.

THIN SAMPLE DIELECTRIC PROPERTIES MEASUREMENTS

A 3D Finite-Difference Time-Domain (FDTD) model is constructed around the probe. The computational domain is typically 44 x 44 x 120 cells, with each cell being 1.0 mm^3. First-order absorbing boundaries are placed around the probe to limit the computational domain. The use of the first-order radiation boundaries is justified due to the limited radiation from the open-ended probe. The excitation plane is place about three-fourths of a wavelength from the open end of the probe. Then a forward moving (towards the measurement end) TEM wave is launched at the excitation plane of the coaxial probe. Once steady-state has been reached the potential difference between the coaxial conductors and current on the center conductor were calculated from the resulting field distribution. The input impedance is then calculated and used to calculate the dielectric properties of the material under test.

To make thin sample measurements, thin samples of various thicknesses and complex permittivities are modeled with FDTD simulations. The calculated input impedances are then graphed versus complex permittivity for different thicknesses and frequencies. To help standardize the measurement procedure all samples are backed with a standard material such as a highly conducting (metal) plate. The impedance of the desired sample (backed with the standard material) is then measured with the open-ended probe using a network analyzer. The measured impedances are then compared with the FDTD graphs and the value of the dielectric constant, as a function of frequency, is obtained through interpolation. Figure 1 shows the results of two different samples measured using this method versus cavity methods.

A 2.5 mm thick sample of ZrO_2 + 8 mol% Y_2O_3 and a 5.0 mm thick sample of Teflon were measured over the frequency range from 500 MHz to 3 GHz using the metallized ceramic probe. The results for the ZrO_2 are compared with the value ε_r' = 17.9 which was obtained using a cavity perturbation technique [8]. The results for Teflon are compared with the value found in literature of ε_r' = 2.1 [9]. It is shown in Figure 1 that the FDTD obtained results agree quite well with available data. The worst case error for the above results is 8%.

ERROR ANALYSIS

The metallized ceramic probe was designed to eliminate the differential thermal expansion which is inherent in metal probes. The differential thermal expansion in metal probes is caused by the outer conductor heating faster during the heating cycle, and cooling faster during the cooling cycle. During the heating cycle the outer conductor heats faster, and therefore expands faster than the inner conductor, which causes an air gap between the inner conductor and the material being measured. During the cooling cycle the outer conductor cools faster and therefore shrinks faster than the inner conductor, which causes an air gap between the outer conductor and the material under test. It is of interest to quantify the errors in dielectric properties measurements as a result of the differential thermal expansion in metal probes.

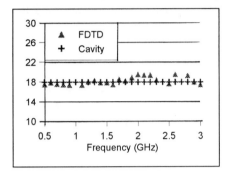 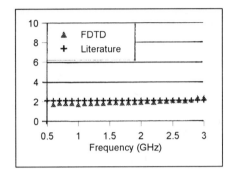

(a)   (b)

Figure 1: The dielectric constants ε_r' for two different samples are shown versus frequency over the frequency range from 500 MHz to 3 GHz. (a) ZrO_2 + 8 mol% Y_2O_3, and (b) Teflon. The results from the used FDTD procedure are compared with data measured using cavity perturbation methods or available in literature.

The differential thermal expansion between the inner and outer conductor is generally on the order of a fraction of a millimeter. To model numerically an air gap on the order of 0.1 mm, a typical 3D FDTD model would require a computation domain on the order of 46 million cells. A model that is large and far exceeds available computational resources in most places in both memory and time constraints. For this purpose a cylindrical 2D FDTD code was developed which makes use of the phi-symmetry in coaxial probes. With the cylindrical 2D FDTD code, using a cell size of 0.1 mm^2, the same size model, that was much too large using a 3D code, is expected to be around 180,000 cells.

The input impedance of the material under test is again calculated as described earlier and used to calculate the dielectric properties of the material under test. To verify the accuracy of the developed cylindrical 2D FDTD code the impedance calculations were compared with analytical results available in literature [1]. The model consisted of a probe with inner conductor of radius 1.9 mm and outer conductor of radius 7 mm. The probe was filled with a filler material of dielectric constant equal to 2.54, and the sample under test had complex permittivity value of $\varepsilon_r^* = 10 - j1$. A uniform air gap between the probe and the sample was modeled from zero to 1.0 mm in increments of 0.1 mm. The results of the reflection coefficient as calculated using the FDTD model versus the analytical results are shown in Figure 2. The magnitude of the reflection coefficient is shown in Figure 2(a), and the phase is shown in Figure 2(b). It is shown in Figure 2 that the calculated results are accurate as compared to available analytical results.

An uncertainty analysis was then conducted using the FDTD code to quantify the errors caused by the differential thermal expansion which is inherent in metal probes. It should be noted that the air gaps due to differential thermal expansion are different than the analytical results. The analytical results are for a uniform air gap between both the inner and outer conductors, and the surface of the sample under test, rather than an air gap between either the inner conductor or the outer conductor, and the surface of the sample under test. The probe was first modeled flush against the material under test, and then an air gap of 0.1 mm to 1.0 mm (in increments of 0.1 mm) between the inner conductor and the material under test was modeled. The same procedure was repeated, but with an air gap between the outer conductor and the material under test.

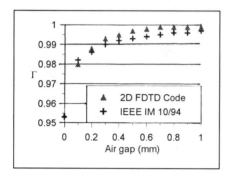
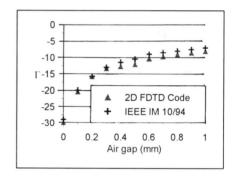

(a)   (b)

Figure 2: Results of cylindrical 2D FDTD code versus analytical results for calculating the reflection coefficient (Γ) in the coaxial probe when placed in contact with a material of complex permittivity value $\varepsilon_r{}^* = 10 - j1$. Air gaps were created between the probe and the surface of a dielectric sample. (a) magnitude of Γ and (b) phase of Γ.

The magnitude of the errors due to differential thermal expansion depend on the material under test. The error analysis results are shown for both a low loss material (Al_2O_3) and for high loss (SiC) materials. Figure 3 shows the errors in the measured input impedance due to the differential thermal expansion during the heating cycle which results in an air gap between the inner conductor and the material under test.

Figure 4 shows the errors in the measured input impedance due to the differential thermal expansion during the cooling cycle in which an air gap between the outer conductor and the material under test is expected.

Another analysis was completed to examine the effects of differential transverse thermal expansion of the center and outer conductors which causes changes in the characteristic impedance (Z_0) of the probe and hence causes errors in the dielectric measurements. The ratio between the diameters of the inner and outer conductors was changed by ±5% and ±10% from its original value. The new characteristic impedance of the probe was then calculated and used to find the new complex permittivity values. The new complex permittivity values were then compared to the values obtained for a probe with a characteristic impedance of 50 Ω, and the difference is characterized as the resulting error. Table I gives the results of these comparisons.

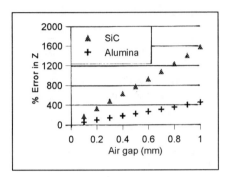
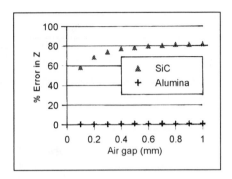

(a)   (b)

Figure 3: The percent error in the measured input impedance (a) magnitude and (b) phase, due to an air gap between the inner conductor and the material under test. Air gap is due to different thermal expansion during the heating cycle in which the expansion of the outer conductor is larger than the inner one.

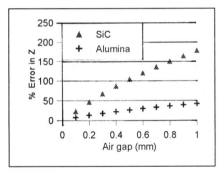
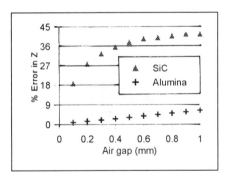

(a)   (b)

Figure 4: The percent error in the measured input impedance (a) magnitude and (b) phase, due to an air gap between the outer conductor and the material under test. Air gap is due to different thermal expansion during the cooling cycle in which the expansion of the inner conductor is larger than the outer one.

Percent difference in:	Change in characteristic impedance by:			
	+ 5%	- 5%	+ 10%	- 10%
ε_r'	1.67%	2.48%	4.5%	2.76%
ε_r''	45.0%	123.0%	73.9%	157.0%

Table I: Comparison of complex permittivity results when characteristic impedance values are changed due to differential transverse thermal expansion between the inner and outer conductors of the probe.

From Table I it may be observed that while changes or uncertainty in knowing the characteristic impedance of the coaxial probe may result in a small error in calculating ε_r', this error will result in significantly larger errors in the measured values of ε_r''.

CONCLUSIONS

A method of measuring the dielectric constant of thin samples through the use of FDTD has been developed. The thin sample results for ZrO_2 and Teflon were shown to be accurate over the frequency range from 500 MHz to 3 GHz with worst case error of only 8 %. A cylindrical 2D FDTD code has been developed to compliment our existing 3D FDTD codes in calculating the input impedance of a material under test. The FDTD code has been validated by comparison with analytical results available in literature.

An uncertainty analysis was performed to quantify the errors caused by differential thermal expansion. The analysis results suggest that an air gap as small as 0.1 mm due to differential thermal expansion between the inner and outer conductors may result in serious (as high as 170%) measurement errors in the complex-permittivity values. The analysis results also suggest that high loss materials are more susceptible to errors from air gaps than low loss materials. Because of the delicate nature of the dielectric properties measurements using the open-ended coaxial probe, it is suggested that a probe with minimal differential thermal expansion like the metallized ceramic probe be used for high-temperature measurements.

REFERENCES

[1]   J. Baker-Jarvis, M. D. Janezic, P. D. Domich, and R. G. Geyer, "Analysis of an open-ended coaxial probe with lift-off for nondestructive testing," *IEEE Trans. Instrum. Meas.*, Vol. 43, pp. 711-718, 1994.
[2]   E. C. Burdette, F. L. Cain, and J. Seals, "*In vivo* probe measurement technique for determining dielectric properties at VHF through microwave frequencies," *IEEE Trans.*, MTT-28, pp. 414-427, 1980.
[3]   O. M. Andrade, M. F. Iskander, and S. Bringhurst, "High temperature broadband dielectric properties measurement techniques," *Microwave Processing of Materials III*, MRS Vol. 269, pp. 527-539, 1992.

[4]   C. L. Li and K. M. Chen, "Determination of electromagnetic properties of materials using flanged open-ended coaxial probe - full-wave analysis," *IEEE Trans. Instrum. Meas.*, Vol. 44, pp. 19-27, 1995.

[5]   P. De Langhe, K. Blomme, L. Martens, and D. De Zutter, "Measurement of low-permittivity materials based on a spectral-domain analysis for the open-ended coaxial probe," *IEEE Trans. Instrum. Meas.*, Vol. 42, pp. 879-886, 1993.

[6]   M. F. Iskander and J. B. Dubow, "Time- and frequency-domain techniques for measuring the dielectric properties of rocks: a review," *Journal of Microwave Power*, 18(1), pp. 55-74, 1983.

[7]   S. Bringhurst and M. F. Iskander, "Open-ended metallized ceramic coaxial probe for high temperature dielectric properties measurements," *IEEE Trans. Micro. Theory. Tech.*, accepted for publication.

[8]   S. Bringhurst, O. M. Andrade, and M. F. Iskander, "High-temperature dielectric properties measurements of ceramics," *Microwave Processing of Materials III*, MRS Vol. 269, pp. 561-568, 1992.

[9]   C. A. Balanis, <u>Advanced Engineering Electromagnetics</u>, p. 79, (John Wiley & Sons, Inc., 1989)

DIELECTRIC MEASUREMENT OF MULTI-LAYERED MEDIUM USING AN OPEN-ENDED WAVEGUIDE

O. TANTOT, M. CHATARD-MOULIN, P. GUILLON
I.R.C.O.M. 123 av. A. Thomas 87060 Limoges Cedex FRANCE, E-mail: tantot@ircom.unilim.fr

ABSTRACT

The use of a circular waveguide radiating into a multi-layered media allows the characterization of heterogeneous and fluid subtances. Many microwave measurement devices, based on reflection coefficient measurements, are subjected to air gap problems that introduce some inaccuracy in the determination of the unknown complex permittivity of the materials. Our purpose is to try and take the air gap into account in these measurements.

INTRODUCTION

Measurement methods based on reflection measurement allow the determination of dieletric or magnetic properties of materials, one parameter being known. Studies on circular waveguide radiating into two media have been realized. The analysis using a greater number of media layers is a significant improvement and permits the isolation of the material under test. Air as a first layer between the flanged waveguide and the sample allows for a non contact characterization and minimizes the air gap problems of flanged waveguides.

THE METHOD

A multi-layered dielectric sample is placed between the flange at the open waveguide end and a known last medium (air, metal or dielectric medium) assumed to be infinitely thick (Figure 1).

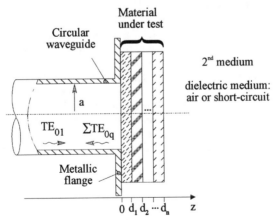

Figure 1: Circular waveguide radiating into a mutli-layered medium

Relations between the reflection coefficient of the incident mode (TE_{11} or TE_{01}) on which the

Mat. Res. Soc. Symp. Proc. Vol. 430 © 1996 Materials Research Society

waveguide is excited and complex permittivities $\varepsilon_n = \varepsilon'_n - j\varepsilon''_n$ of the multi-layered material to be tested have been established with following simplifying assumptions: waveguide, flange and metal backing are assumed to act as a perfect electric wall. The metallic flange radially increases to infinity at $z=0$. The layered media are assumed to be homogeneous, isotropic with parallel faces. The choice of the TE_{01} mode in the circular waveguide is due to its symmetrical axial configuration of the electric field (E_θ) which makes it less sensitive to metallic losses and the air gap effect than the fundamental TE_{11} mode [1]. The discontinuity at the waveguide-material interface generates higher-order TE_{0q} modes [2]. In the operating frequency band, the only propagating mode in the waveguide is TE_{01}, all others are evanescent modes. The fundamental TE_{11} mode is not excited. The modeling of one [3], two[2], three or four media has been achieved and can be done for a higher number of media, only one medium being unknown. In each region, the field can be expressed in terms of electric longitudinal vector potential. In the waveguide, electric vector potential (1) can be expressed as the sum of the incident mode TE_{01}, the amplitude of which is U^+_0 and reflected modes TE_{0q}, the amplitudes of which are U^-_{0q}. For a positive z, electric vector potentials (2,3) are expressed as the sum of the incident potential and the reflected potential except for the last medium (4).

Circular waveguide: $(z<0)$

$$\psi_g(r,z) = U^+_0 J_0(x'_{01} r / a) e^{-\alpha_{01} z} + \sum_{q=1}^{\infty} U^-_{0q} J_0(x'_{0q} r / a) e^{\alpha_{0q} z} \tag{1}$$

First layer: $(0<z<d_1)$

$$\psi_1(r,z) = \psi_{1i}(r) e^{-\lambda_1(s) z} + \psi_{1r}(r) e^{\lambda_1(s)(z-d_1)} \tag{2}$$

Intermediate layer p: $(d_{p-1}<z<d_p)$

$$\psi_p(r,z) = \psi_{pi}(r) e^{-\lambda_p(s)(z-d_{p-1})} + \psi_{pr}(r) e^{\lambda_p(s)(z-d_p)} \tag{3}$$

Last medium: $(d_n<z<\infty)$

$$\psi_n(r,z) = \psi_n(r) e^{-\lambda_n(s)(z-d_{n-1})} \tag{4}$$

were Ψ is the electric vector potential in each region, α_{0q} are the propagation constants of TE_{0q} modes in the waveguide, $\lambda_p(s)$ are the propagation constants in layer p, a is the radius of the waveguide and x'_{0q} are the qth roots of Bessel function $J'_0(x)$.

Applying the continuity conditions on the Hankel [4] transformed field components for $z=0$ and $z=d_p$ and using the orthogonality properties of Bessel functions, amplitudes of incident and reflected modes can be written as a system of matrix equations. The resolution of this system gives the refection coefficient of each mode and in particular the TE_{01} mode (U^-_{01} / U^+_0). Computations have been compared with those obtained for two media with the same dielectric medium for each layer. Results are similar and allow the characterization of a multi-layered medium.

COMPUTATIONS ON REFLECTION COEFFICIENTS

In order to determine the influence of an air gap between the sample and the metallic flange on the reflection coefficient, we have made some computations with three different media: The first layer medium is an air gap were the thickness is increased, the second is the sample under test with dieletric losses and the third is air or metal sheet. Table I presents the evolution of the reflection coefficient

246

in terms of the intermediary air gap thickness for two materials at 10 GHz. These results show the small influence of the magnitude of the reflection coefficient on contact flaws whatever the third medium. However, the phase of the reflection coefficient is more sensitive.

Table I: evolution of the reflection coefficient in term of the air gap thickness

Materials	air gap (mm)	last medium : air					last medium : metal backing				
		0	0,01	0,05	0,1	0,2	0	0,01	0,05	0,1	0,2
9,4 - j1,6 e = 3 mm	|S11|	0,89	0,89				0,591	0,591		0,592	0,593
	arg S11°	175,2	175,0	174,6	174,0	172,8	225,5	225,4	224,9	224,2	223
4,4 - j 0,2 e = 3 mm	|S11|	0,772	0,772		0,771		0,936	0,936			
	arg S11°	193,0	192,9	192,5	192,0	191,1	112	111,8	111,3	110,7	109,4

For the reflection coefficient obtained for an air gap of 0.1 mm, permittivities were determined with an increase of the uncertainty of arround 7%.

The influence of an adapter layer isolating the flanged waveguide has been studied. This layer allows the characterization of the liquid medium. Charts have been plotted giving the permittivities in terms of the reflection coefficient (Figure 2).

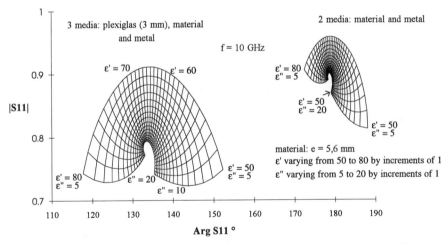

Figure 2: Charts plotting permittivity versus reflection coefficient for 2 and 3 media

The real permittivity varying from 50 to 80 by increments of 1 and the imaginary pemittivity varying from 5 to 20 by increments of 1. Two charts have been computed. The first one with 2 media and the second one for 3 media where the first layer is a sheet of plexiglas of 3 mm thickness. An increase of the reflection coefficient variation has been observed on the second chart which leads to an increase of the permittivity accuracy with the same error on the reflection coefficient. Accuracies computations have been made for an uncertainty arround ±0.005 U for |Γ|

and ±0.5° for ∠Γ. For two materials with high dielectric losses, table II shows that permittivity uncertainties are halved.

table II : Accuracies for 2 or 3 media configurations

Material	Number of media	|Γ|	∠Γ°	±Δε'/ε'	±Δε''/ε''
79 - j 12 e = 5,6 mm	2	0,89	176,75	1,8 %	24%
	3	0,755	129,57	0,95 %	11,6 %
60 - j 8 e = 5,6 mm	2	0,924	181,64	4,2 %	30 %
	3	0,858	138,41	2 %	14,3 %

The knowledge of the electric field in the material is essential to give a finite lateral dimension of the flange and the sample. Figure 3 plots the electrical field (E_θ) on a diametral plan for z=0, 3, 10 and 30 mm. These results are computed for a material ε=4.4-j0.2 (3 mm) and are compared with those obtained by the Finite Element Method (F.E.M.) in 2 Dimensions axisymmetric in forced oscillations. The infinite medium (air) is simulated by the absorbing boundary conditions (ABC).

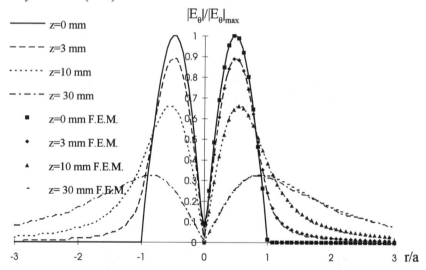

Figure 3: Electric field in term of the waveguide radius

For a radius higher than twice the waveguide radius, the electric field is assumed to be close to zero.

248

EXPERIMENTAL RESULTS

The inverse problem is resolved by two methods. The first one computes vectorial reflection coefficients Γ and permits the construction of ε' and ε'' charts as a function of Γ [5]. These charts are interpolated using bidimensional cubic spline functions; the values of ε' and ε'' can be computed from a Γ measurement with their associated accuracies. The second one is a direct search method. In this case, an arbitrary permittivity is necessary to begin the computations and converge by sucessive steps to the measured reflection coefficient. A solution to get round the problem of multiple solutions is the measurement of the same material first with air as second medium and then with a short-circuit. Only one permittivity value can simultaneously satisfy these two test conditions.

A vector analyzer is connected to a coaxial-rectangular adapter (Figure 4), which is connected to a Marie transition which transforms the TE_{10} mode of the rectangular waveguide to the TE_{01} mode of the circular waveguide. A flanged waveguide is connected to the Marie transition. The network analyzer calibration is made with three shifted short-circuits. The waveguide diameter is chosen to propagate the TE_{01} mode. After this calibration, error measurements are assumed to be ± 0.005 U for $|\Gamma|$ and $\pm 0.5°$ for $\angle\Gamma$.

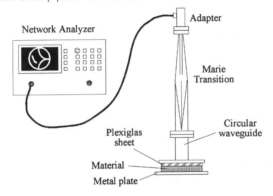

Figure 4: test set-up

Table III presents results obtained with an air gap (0.98 mm) between the flange and the sample.

Table III: Permittivities and accuracies for the two configurations

$\varepsilon' - \varepsilon''$ $\pm\Delta\varepsilon'/\varepsilon'$ $\pm\Delta\varepsilon''/\varepsilon''$	Open-Ended Waveguide 2 media		Open-Ended Waveguide 3 media with air gap	
Materials	**air**	**short-circuit**	**air**	**short-circuit**
Y1	4,49 - j 0,18 $\pm 3,9\%$ $\pm 83\%$	4,54 - j 0,1 $\pm 1,2\%$ $\pm 35\%$	4,51 - j 0,14 $\pm 2,6\%$ $\pm 61\%$	4,48 - j 0,11 $\pm 1,2\%$ $\pm 36\%$
Z1	9,69 - j 1,83 $\pm 7,5\%$ $\pm 34\%$	9,54 - j 0,1,6 $\pm 1\%$ $\pm 5\%$	9,58 - j 1,64 $\pm 7,4\%$ $\pm 36\%$	9,56 - j 1,58 $\pm 1\%$ $\pm 4,7\%$

Two materials with dielectric losses have been measured for the two configurations (2 and 3 media). Permittivities are similar with a slight improvement of the precision for the Y1 material.

CONCLUSION

The circular waveguide radiating into a multi-layered medium permits the change of the experimental conditions without modifying the thickness of the sample or changing the frequency measurement.

Taking into account an air gap between the flanged waveguide and the sample under test, permits an increase in the accuracy of the permittivity determination. The characterization of the dielectric liquid medium becomes possible with the insertion of a plexiglas sheet as a first medium isolating it from the flanged waveguide. A dilatation of the chart plotting the reflection coefficient against permittivity has been noted. This means a higher accuracy for the determination of ε^* for materials with high dielectric parameters. The accurancies of real and imaginary parts of the permittivity are respectively about a few percentage points and more than ten percentage points.

REFERENCES

1   O. Tantot, thesis, Limoges University, France, 1994.

2   B. Chevalier, M. Chatard-Moulin, J.P. Astier, P. Guillon, presented at the 1991 J.C.M.M., Limoges, France, pp. 275-291.

3   C. Fray, K. Chandrasekhar, A. Papernik, Electronic letters, **17**, 718-720, 1981.

4   D. Colombo, in Les Transformées de Mellin et de Hankel, edited by C.N.R.S., Paris, 1959

5   J. F. Villemazet, M. Chatard-Moulin, P. Guillon, H. Jallageas, presented at the 1991 IEEE Instr. and Meas. Conference, Atlanta, 1991 (Poster Session).

A NEW APPROACH FOR DIELECTRIC MEASUREMENTS AT HIGH TEMPERATURES

C. Groffils* and P. J. Luypaert*, Y. Zhou**, O. Van der Biest**

*MEAC, I&I K.U.Leuven, Kapeldreef 60, 3001 Heverlee, Belgium
**Dept. of Metallurgy and Materials Engineering, Katholieke Universiteit Leuven, De-Croylaan 2, B-3001 Heverlee, Belgium.

ABSTRACT

Most of the algorithms for cavity dielectric measurements use a full network frequency scan for the perturbation method. At high temperature the measurement points in one frequency scan will have different temperatures due to the fast cooling of the sample inside the cold cavity. This paper presents a new approach for the perturbation method to measure dielectric constants at high temperatures with a two port, two piston cylindrical cavity and external furnace. Special features of this cavity are the possibility to work with different modes and different relative positions of the coupling aperture.
The dielectric properties are extracted from the full sets of S-parameters for two successive frequency points corresponding to approximately the same temperatures. An extraction algorithm is developed and programmed.

INTRODUCTION

Dielectric constant measurements at high temperatures are necessary to study the influences of microwaves on materials such as gels, ceramics, etc. [1]. Therefore different preheating or heating techniques can be used namely microwave heating, electrical heating, laser heating ... [2,3,4].

The generally excepted measurements technique is based on a perturbation method [5] where we heat only the sample. To extract the resonance frequency and quality factor from the data we have developed a first order approximation of the perturbation method and a first order differential model for the cooling process [2,3].

Together with a cavity network model [6] for each cavity mode and temperature we can calculate the influence of the inductive coupling in order to get an accurate measurement with two successive frequency points.

The temperature measurements of the sample is one of the major practical problems which need to be overcome in order to make a realistic measurement and obtain accurate results. In this paper we use an approximation for the temperature cooling curve obtained for a sample preheated in an external electrical furnace.

In the following sections the set-up is described, together with the perturbation method, cavity model and the dielectric properties extraction algorithm. Results of measurements are presented.

2. SET-UP

Techniques such as metallic probe measurement and fiber optic black body measurement being contact methods, give extra problems due to the field distortion. Therefore non contact methods are suggested in real-time measurements. In our case we use a cooling model to predict the temperature.

The sample is pre-heated outside the cavity and brought into the cavity. Then the two port S-parameters are measured with and without the sample (a cylindrical rod[k1][1]). This cycle is repeated for each frequency scan (from low to high temperatures).

The apparatus consists of:
1. Vector network analyzer (HP8510 C)
2. Resistance electrical heating unit with metallic probe temperature sensor.
3. Microwave mono mode, two port, two piston cavity
4. Computer (PC 80486dx2, 66 Mhz), and software (Lab VIEW)

The advantages of the set-up are a new approach to a coupling study and the possibility to measure different modes with optimum coupling.

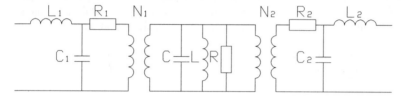

Figure 1: network cavity model for one temperature and one mode.

3. PERTURBATION THEORY
Suppose we have identical cavities with the sample inside one of them. In case of a high quality cavity the microwave parameters can be calculated using the following formulas [5]:

$$\epsilon_r \approx 1 + \left(\frac{\int |E_e|^2 dV_c}{\int E_e \cdot E_f \, dV_s} \right) \cdot \left(\frac{2 . \Delta f}{f^f_{n,m,l}} \right)$$

$$tg\delta \approx \frac{O_c}{k . \epsilon_r O_s} \cdot \left(\frac{1}{Q_f} - \frac{1}{Q_e} \right)$$

The different symbols stands for E_e and E_f the electrical fields, O_s the sample surface, O_c the cavity surface, Δf the resonant frequency shift, k a constant, $f_{n,m,l}$ the resonance frequency of the empty cavity and Q the quality factor. Subscripts "e" and "f" stands for the empty cavity and partially filled cavity, respectively.

---

[1] for the TM010 and TM012 mode

4. CAVITY MODEL

The model used for the cylindrical cavity with loop coupling to TM $_{0,1,1}$-mode is shown in figure 1. The parameters of the model are: L,C and R (unloaded cavity), R1 and R2 (loss in the connections antenna-connector), N1, N2 (loop antenna) and C1,C2, L1, L2 (connector transmission lines).

$$R_1 = \left(Z_{11}^r - \sqrt{\frac{Z_{11}^i}{Z_{22}^i}} \cdot Z_{12}^r \right) \omega = \omega 0$$

$$R_2 = \left(Z_{22}^r - \sqrt{\frac{Z_{22}^i}{Z_{11}^i}} \cdot Z_{12}^r \right) \omega = \omega 1$$

Where $Z_{ij}^{r,i}$ are real and imaginary parts of the impedance matrix of the two port. We have found the internal quality factor and the resonant frequency as a function of the impedance matrix.

$$Q = \left| \frac{-Z_{22}^r}{\left(Z_{22}^r - R_2 \right) \cdot \left(\dfrac{\omega}{\omega 0} - \dfrac{\omega 0}{\omega} \right)} \right|$$

$$\omega 0 = \sqrt{\frac{\dfrac{U_1}{U_2} \cdot \omega 1 - \omega 2}{\dfrac{U_1}{U_2} \cdot \dfrac{1}{\omega 1} - \dfrac{1}{\omega 2}}}$$

Where U_1 and U_2 stand for:

$$U_1 = \left(\frac{Z_{22}^r - R_2}{-Z_{22}^i} \right) \omega = \omega 1$$

$$U_2 = \left(\frac{Z_{22}^r - R_2}{-Z_{22}^i} \right) \omega = \omega 2$$

To calculate the resonance frequency and the quality factor it's sufficient to measure two frequencies for which the model is valid. The internal quality factor is independent of the resistance's R and the transformation numbers N.

5. EXTRACTION ALGORITHM

The extraction algorithm is programmed in the graphic environment LabVIEW[2] and consists of the following steps:

- choose the frequency range (scan), sample thickness
- measure S_{11}, S_{12}, S_{21} and S_{22}

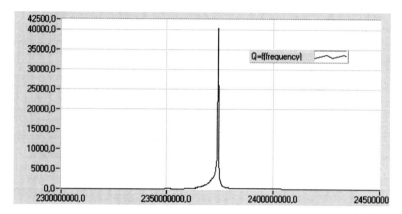

Figure 2: the calculated $Q_{internal}$ as function of the network analyzer frequency scan (zoomed out), at resonance formula (7) becomes infinite

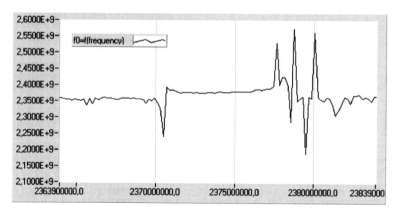

Figure 3: the calculated $f_{resonance}^{internal}$ as function of the network analyzer frequency scan, resonance is at 1720750000 Hz

---

[2] from National Instruments

- change the reference plane relative to the loopantenna position (connector), and find new S_{11}, S_{21}, S_{12}, S_{22}.
- transform S-matrix to Z-matrix
- calculate the internal quality factor (formula (7)) for successive frequency points (see Figure 2)
- calculate of resonant frequency (formula (8)) for successive frequency points (see Figure 3)
- perform a one dimensional curve fitting to find the resonance frequency
- perform a one dimensional curve fitting to find $Q_{internal}$
- calculate ϵ and $\mathrm{tg}\delta$ using formulas (1) and (2)
- link the cooling curve T(t) with the dielectric properties data $\epsilon_r(t)$ and $\mathrm{tg}\delta$ (t)

6. MEASUREMENTS

Two modes are used (TM_{010} and TM_{012}). For both modes we chose the optimum relative antenna position. Measurements are performed to verify the stability and accuracy of the algorithm. The temperature is estimated by the cooling curve. The errors (see Table 1) are determined by an expansion of the formulas (1,2,3,4,5,6,8) into the Taylor series. The value of Δ_{ϵ_r} increases with the temperature due to the rise of error for the resonant frequency because of deterioration of the quality factor (see Table 2). So the size of the sample must be chosen smaller in order to decrease the width of the resonant curve. The value of ϵ_r.

Temp	230	400	450	715	730
ϵ_r	7.18	7.55	8.30	8.44	8.65
Δ_{ϵ_r}	0.07	0.12	0.43	0.64	0.91

Table 1: Tabulation of the real part of the dielectric constant and its accuracy for different temperatures, the material is a DURAN glass rod of 7 mm

is the smallest for lower frequencies as we expected.

Temp.	230	450	715	730
Q	51600	36000	11000	4200

Table 2: Tabulation of the internal quality factor for different temperatures, the material is a DURAN glass rod of 7 mm.

The results of measurements confirm that the transition temperature of glass is at 750 °C.

7. CONCLUSIONS AND REMARKS

In conclusion, the importance of the method can be summarized as follows:
- dielectric measurements at high temperatures can be done and are reproducible
- the influence of the coupling can be neglected
- different modes can be used
- different antenna positions are possible

In the future we intend to measure the temperature by means of an infrared measurement system. A sample correction for more irregular shapes will be investigated.

8. ACKNOWLEDGMENTS

I wish to thank the IWT (Institute for Scientific and Technology Research) for their financial support.

REFERENCES

[1]   C.B. Groffils, P.J. Luypaert, "Microwave Processing Method applied to Glass melting", MHF sept.1993

[2]   R. Hutcheon, M. de Jong, F. Adams, G. Wood ,J. Mc. Gregor, B. Smith, " A System For Rapid Measurements Of RF And Microwave Properties, Up To 1400°C " J. Microwave Power and Electromagnetic Energy Vol. 27 No. 2.1992

[3]   C.B. Groffils, P.J. Luypaert, " An analysis cavity for dielectric measurements", MIOP93, Hagenburg, May 1993

[4]   V. Pohl, D. Fricke, and A. Mühlbauer, "Correction Procedures for the Measurement of Permittivities with the Cavity Perturbation Method," IMPI, vol. 30, No 1, 1995

[5]   M. Sucher, J. Fox, Handbook of Microwave Measurements, Vol. II (John Wiley and Sons, INC. New York, 1963)

[6]   A.C. Metaxas, R.J. Meredith R.J. Industrial Microwave Heating (Peter Pelegrims, London 1983)

COMPLEX PERMEABILITY MEASUREMENTS OF MICROWAVE FERRITES[1]

RICHARD G. GEYER* AND JERZY KRUPKA**
*Electromagnetic Fields Division, National Institute of Standards and Technology, MS 813.08, Boulder, CO 80303-3328, geyer@boulder.nist.gov
**Instytut Mikroelektroniki i Optoelektroniki Politechniki Warszawskiej, Warszawa, Poland

ABSTRACT

A rigorous and accurate method for the experimental determination of the complex permeability of demagnetized ferrites at microwave frequencies is presented. The measurement uses low- loss dielectric ring resonators, is nondestructive, and allows complex permeability characterization of a *single* ferrite sample to be performed at frequencies from 2 GHz to 25 GHz. A wide variety of ceramic microwave ferrites having various compositions and differing saturation magnetizations were measured in the demagnetized state. Generally, at any frequency greater than gyromagnetic resonance, the real part of the complex permeability increases as saturation magnetization increases. For the same frequency magnetic losses increases as saturation magnetization increases. The real permeability results are compared with magnetostatic theoretical predictions. Measurement data show excellent agreement with theoretical predictions, but only when $2\pi\gamma M_s/\omega < 0.75$, where γ is the gyromagnetic ratio, M_s is saturation magnetization, and ω is the radian rf frequency.

INTRODUCTION

Modern microwave systems use many ferrite devices to control the transmission path (circulators, isolators, switches), frequency (YIG-tuned oscillators), frequency bandwidth (filters), amplitude (attenuators, limiters) and phase (phase shifters) of microwave signals. For optimal design of many of these devices, accurate measurements of the relative magnetic permeability and loss factor are needed at the operational frequency.

At frequencies much greater than gyromagnetic resonance, magnetic losses are on the order of 10^{-2} to 10^{-5} and cannot be accurately measured by transmission line methods. The most accurate methods for measuring ferrite magnetic properties at microwave frequencies are techniques that use resonant cavities or techniques incorporating the ferrite sample as part of a dielectric resonator [1-9]. In the past, TM_{mn0} cavities have been employed to measure the complex permeability of demagnetized or partially saturated ferrites at microwave frequencies. In this case, right- and left-circularly polarized mode structure is used to evaluate the component of the complex magnetic permeability tensor transverse to the applied static magnetic field bias. For demagnetized samples, the right- and left-circularly polarized resonant modes are degenerate, and the measurement reduces to a scalar permeability determination. The complex permeability is evaluated from the resonant frequencies and unloaded quality factors that are measured with and without sample insertion.

Here a dielectric resonator measurement system composed of low-loss dielectric sleeves operated with H_{011} mode structure is proposed. This system allows broadband permeability characterization of a single ferrite sample at frequencies both near and far from gyromagnetic resonance. The large measurement bandwidth is realized by the use of commercially available ultra-high Q electroceramic or oriented single-crystal quartz resonators. The aspect ratios and permittivities of the sleeve resonators are chosen so as to enable spectral characterization of a single ferrite sample from 2 GHz to 25 GHz.

---

[1] U.S. Government work not protected by U.S. copyright

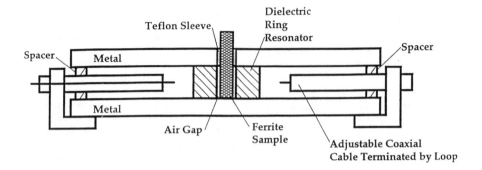

Figure 1. Dielectric resonator system used for microwave permeability measurements.

MEASUREMENTS

The resonant system that was used to measure the microwave complex permeability of ferrite samples is shown in Fig. 1. In the absence of an applied static field, the ferrite in the demagnetized state is isotropic and is described by a scalar frequency-dependent relative permeability, $\mu_d^* = \mu_d' - j\mu_d''$, and complex relative permittivity, $\epsilon_f^* - j\epsilon_f''$. The resonators are coupled to the external microwave source through two loop-terminated coaxial cables that are adjustable so that the measured loaded Q factor is equal to the unloaded Q factor within any prescribed accuracy.

The measurement procedure follows several steps. First, the complex permittivity of the ferrite, as specified by the manufacturer, is verified with a TM_{0n0} cavity. Second, the complex relative permittivity, $\epsilon_r^* = \epsilon_r' - j\epsilon_r''$, of each dielectric ring resonator is found from measurements of the resonant frequencies and unloaded Q factors of the empty ring resonators operating in the H_{011} mode, taking into account conductive microwave losses of the upper and lower metal ground planes. Values of the dielectric loss factors, ϵ_f'' and ϵ_r'', at other measurement frequencies are calculated assuming a linear increase with frequency.

The electromagnetic fields in the ferrite, in the dielectric resonator, and exterior to the dielectric resonator may be expressed as general cylindrical functions which are solutions to the Helmholtz equation [2, 10]. The continuity conditions of the tangential electric and magnetic fields at the interfaces between the ferrite and dielectric resonator are applied to obtain a system of equations that has nontrivial solutions only if the corresponding determinant vanishes. Then the scalar permeability μ_d' is determined from measurements of the resonant frequency of the ring resonator containing a completely demagnetized ferrite sample and solving

$$F(\epsilon_f', \mu_d', f_{r0}) = 0, \tag{1}$$

where f_{r0} is the measured H_{011} mode resonant frequency and F is the operator H_{011} eigenvalue equation. After determining μ_d', the magnetic loss factor μ_d'' is found as a solution to the equation

$$\frac{1}{Q} = \frac{1}{Q_c} + p_{\epsilon_r'} \tan \delta_{e,r} + p_{\epsilon_f'} \tan \delta_{e,f} + p_{\mu_d'} \tan \delta_{m,f}, \tag{2}$$

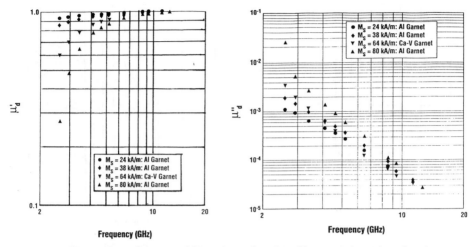

Figure 2. Measured complex permeability, μ_d^*, as a function of frequency for various doped garnets having differing saturation magnetizations.

where Q is the unloaded Q factor for the H_{011} mode, Q_c is the Q factor representing conductor losses in the metal plates, $p_{\epsilon_r'}$ is the electric energy filling factor for the dielectric ring resonator, $p_{\epsilon_f'}, p_{\mu_d'}$ are the ferrite sample electric and magnetic energy filling factors, $\tan \delta_{e,r}$ and $\tan \delta_{e,f}$ are the dielectric loss tangents of the ring resonator and ferrite, and $\tan \delta_{m,f} = \mu_f''/\mu_f'$ is the magnetic loss tangent of the ferrite.

Example complex permeability data for garnets doped with aluminum and calcium-vanadium are given in Fig. 2. Data for lithium and magnesium spinel polycrystallites are given in Fig. 3. Saturation magnetizations of the measured ferrite specimens are also indicated. Gyromagnetic resonance is determined by $2\pi\gamma M_s$, where the gyromagnetic ratio γ is 35.19 MHz·m/kA. The data clearly show the shift in gyromagnetic resonance to higher frequencies as saturation magnetization increases. For small tuning bias fields, ferrites having greater M_s would be more useful in microwave circulator and phase shifter design at frequencies greater than gyromagnetic resonance, albeit at the expense of increased magnetic loss. The aluminum-doped garnet exhibits significantly lower magnetic losses at all measurement frequencies than does the lithium spinel polycrystallite sample having the same saturation magnetization. Magnetic losses rapidly increase nonlinearly as gyromagnetic resonance is approached. The selection of ferrite material is one of the most critical aspects of circulator and phase shifter design. Accurate predictions of bandwidth, tunability and insertion loss that can be expected when using these ceramic materials in component fabrication can be inferred from these measurement data.

Variational uncertainty analyses, which include sources of uncertainty due to measured sample dimensions, sample dielectric loss, surface resistance of the ground planes, resonance frequency, and Q factor, were performed. The estimated total relative uncertainty in μ_d' is ± 0.8 percent. The total uncertainty in μ_d'' is $\pm 1 \times 10^{-5}$. For the same size sample the dielectric resonator technique espoused here is nearly one order of magnitude more sensitive for measurement of ferrite magnetic loss at microwave frequencies than conventionally used cavities.

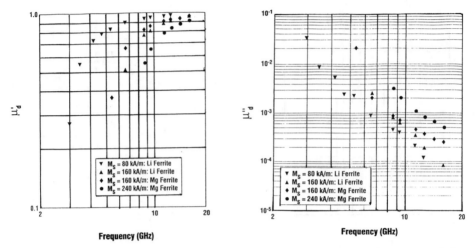

Figure 3. Measured complex permeability, μ_d^*, as a function of frequency for various spinel ceramics having differing saturation magnetizations.

THEORETICAL CONSIDERATIONS

Schlömann [11] treated the high frequency permeability of a cylindrically symmetric domain configuration with a magnetostatic approximation, neglecting exchange effects. In the demagnetized isotropic state this treatment yields the approximate predictive relation,

$$\mu_d' = \frac{1}{3} + \frac{2}{3}\sqrt{1 - (2\pi\gamma M_s/\omega)^2}. \tag{3}$$

With an applied static field the ferrite becomes uniaxially anisotropic with its magnetic permeability described by the well-known tensor,

$$\overline{\overline{\mu}} = \mu_0 \begin{bmatrix} \mu^* & j\kappa^* & 0 \\ -j\kappa^* & \mu^* & 0 \\ 0 & 0 & \mu_z^* \end{bmatrix}, \tag{4}$$

where the applied static field is oriented parallel to the z-axis, $\mu^* = \mu' - j\mu''$ is the principal direction transverse component of the magnetic permeability, $\mu_z^* = \mu_z' - j\mu_z''$ is the parallel component, and $\kappa^* = \kappa' - j\kappa''$ is the off-diagonal transverse component. Green and Sandy [4] empirically fit experimental data of partially magnetized ferrites to the demagnetized permeability and a power law of the ratio of actual average static internal magnetization M to saturation magnetization,

$$\mu' = \mu_d' + (1 - \mu_d')(M/M_s)^{\frac{3}{2}}, \tag{5}$$

$$\mu_z' = \mu_d'^{1-(M/M_s)^{\frac{5}{2}}}, \tag{6}$$

and

$$\kappa' = \frac{2\pi M_s}{\omega}\frac{M}{M_s}, \tag{7}$$

where μ_d' is given by (3).

260

A comparison between the magnetostatic predictive rule with measured microwave μ_d' values determined with the dielectric resonator technique discussed here is given in Fig. 4 for yttrium iron garnet, whose saturation magnetization at 300 K is 142 kA/m. The Schlömann predictive relation shows good agreement with the measurement data, but only when $2\pi\gamma M_s/\omega < 0.75$. At frequencies closer to gyromagnetic resonance, however, significant error in the use of this predictive rule results. Furthermore, measurement data on the relative real permeability at frequencies close to gyromagnetic resonance demonstrate frequency dependence and lower values than the theoretical lower limit of 1/3.

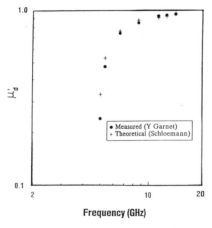

Frequency (GHz)

Figure 4. Comparison of demagnetized permeability measurements using dielectric resonators with theoretical predictions.

CONCLUSIONS

Accurate microwave measurements of the scalar complex permeability of a *single* demagnetized ferrite sample may be performed over a broad range in frequency using low-loss dielectric resonators. Magnetic loss factors as low as 2×10^{-5} can be accurately measured. Magnetic loss factors for lithium spinel polycrystallites are greater than for aluminum-doped garnet having the same saturation magnetization. Because partial magnetic energy filling factors of the sample are generally greater with the H_{011} dielectric resonator system than with TM_{110} mode cavities, relative changes in the Q factor upon introduction of ferrite specimens having low magnetic loss are much greater. In addition, with conventional techniques such broadband measurements would have required the fabrication of many cavities. The scalar permeability data for ceramic ferrites having differing saturation magnetizations are useful in the design of microwave ferrite devices used in microwave systems. Theoretical estimates of the real permeability derived from magnetostatic theory are limited to operational rf frequencies far from gyromagnetic resonance. In general, nonlinearly varying magnetic losses must be measured at microwave frequencies.

ACKNOWLEDGMENTS

The authors gratefully acknowledge the support given to this work by the Polish-American Sklodowska-Curie Foundation.

REFERENCES

[1] R. C. LeGraw and E. G. Spencer, "Tensor permeabilities of ferrites below magnetic saturation," in *IRE Conv. Rec.* (New York), pt. 5, pp. 66-74, 1956.

[2] H. E. Bussey and L. A. Steinert, "Exact solution for a gyromagnetic sample and measurements on a ferrite," *IRE Trans. Microwave Theory Tech.*, vol. MTT-6, pp. 72-76, 1958.

[3] J. J. Green and T. Kohane, "Testing of ferrite materials for microwave applications," *Semicond. Prod. Solid State Technol.*, vol. 7, pp. 46-54, 1964.

[4] J. J. Green and F. Sandy, "Microwave characterization of partially magnetized ferrites," *IEEE Trans. Microwave Theory Tech.*, vol. MTT-22, pp. 641-645, 1974.

[5] N. Ogasawara, T. Fuse, T. Inui, and I. Saito, "Highly sensitive procedures for measuring permeabilities $(\mu_\pm)$ for circularly polarized fields in microwave ferrites," *IEEE Trans. Magn.*, vol. MAG-12, pp. 256-259, 1976.

[6] W. Muller-Gronau and I. Wolff, "A microwave method for the determination of the real parts of the magnetic and dielectric material parameters of premagnetized microwave ferrites," *IEEE Trans. Microwave Theory Tech.*, vol. MTT-32, pp. 377-382, 1983.

[7] P. Le Roux, F. Jecko, and G. Forterre, "Nouvelle methode de determination des parties reeles des parametres electriques et magnetiques des resonateurs a ferrites satures et non satures," *Annales des Telecommunications*, vol. 43, pp. 314-322, 1988.

[8] J. Krupka, "Measurements of all complex permeability tensor components and the effective line widths of microwave ferrites using dielectric ring resonators," *IEEE Trans. Microwave Theory Tech.*, vol. 39, pp. 1148-1157, 1991.

[9] J. Krupka and R. G. Geyer, "Complex permeability of demagnetized microwave ferrites near and above gyromagnetic resonance," *IEEE Trans. Magn.*, vol. 32, no. 3, May, 1996.

[10] P. S. Epstein, "Theory of wave propagation in a gyromagnetic medium," *Rev. Mod. Phys.*, vol. 28, no. 1, pp. 3-17, 1956.

[11] E. Schlömann, "Microwave behavior of partially magnetized ferrites," *J. Appl. Phys.*, vol. 41, pp. 204-214, 1970.

IN—LINE MEASUREMENT OF HIGH TEMPERATURE DIELECTRIC CONSTANT IN THE PROCESS OF SINTERING

Zhou Jian, Cheng Jiping, Tang Yuling and Qiu Jinyu, Advanced Material Research Institute, Wuhan University of Technoloy, Wuhan, 430070, P. R. China.

ABSTRACT

In this paper, a resonant cavity method is developed and some experimental results for measuring dielectric constants of ceramic samples (e. g. Al_2O_3) under different sintering temperatures are reported. The experiments show that this method has higher precision and good prospects of in—line monitoring the high temperature dielectric constant in the process of raising the temperature of the samples. These results provide some scientific experimental basis for physical research of ceramic materials.

BASIC THEORY

There are many methods of microwave measurement of dielectric properties: waveguide reflection method; waveguide transmission method; micro—strip annular resonator method; resontant method and space—wave method, etc[1]. These methods have theirrespective advantages, however an improved resonant cavity method is developed and used in our experiment. Resonant cavity method does not have a special requirement for the geometry and size of the measured sample. Moreover it has higher measuring precision, so that it is used extensively and particularly for ceramic sample.

The perturbation theory is introduced as follows:

Perturbation method generally includes two steps: The first is that the perturbation is neglected, and the solution of electromagnetic field has been known. The second is that the perturbation is considered and its solution is unknown, and required for be defermined.

As shown is Fig. 1, we assume that there is an empty resonant cavity

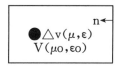

Fig. 1. Resonant cavity with perturbation

filled with air, which volume is V, resonant frequency is ω_0, electric and magnetic intensity is E_0 and H_0 respectively. If we put a dielectric with permittivity ε and permeability μ, volume $\triangle v$ into the empty cavity, so the resonant frequency of the cavity is changed to ω, electric and magnetic intensity in it is changed to E and H. Based on the Maxwell equations and perturbation method, it can be derived that [2].

About this article: The research was supported by 863 plan(NAMC)

$$\frac{\omega-\omega_0}{\omega}=\frac{-\int_{\triangle v}[(\epsilon-\epsilon_0)E_0{}^* \cdot E+(\mu-\mu_0)H_0{}^* \cdot H]dv}{\int v[\epsilon_0 \cdot E_0{}^* \cdot E+\mu_0 \cdot H_0{}^* \cdot H]dv} \qquad (1)$$

Obviously, in order to know the increase of frequency, it is necessary to calculate the electromagnetic field solution of E_0, E and H_0, H. We put a dielectric with very small volume $\triangle v$ into the cavity, so its effect on the outside of $\triangle v$ can be negligible. Moreover we consider approximately that the electric and magnetic energies in the cavity are equal to each other in the region outside of $\triangle v$, So it can be obtained that.

$$\frac{\omega-\omega_0}{\omega}=\frac{-\epsilon_0(\epsilon'-1)\int_{\triangle v} E_0{}^* \cdot E dv}{4W} \qquad (2)$$

$$\frac{1}{Q}-\frac{1}{Q_0}=\frac{\epsilon_0\epsilon''\int_{\triangle v} E_0{}^* \cdot E dv}{2W} \qquad (3)$$

where $\epsilon=\epsilon'-j\epsilon''$, ϵ' and ϵ'' is the real part and imaginery part of dielectric ϵ, Q and Q_0 are quality factor of cavity in the cases of dielectric loaded and unloaded, respectively.

The total stored energies in the cavity is given by

$$W=\frac{1}{4}\int_V (\epsilon_0|E_0|^2+\mu_0|H_0|^2)dv \qquad (4)$$

Some treatment to formulae (2) and (3) (e. g. for TE_{103} mode) are as follows, its field components are:

$$Ey=Ey_0\sin(\pi x/a)\sin(3\pi z/L) \qquad (5-1)$$
$$Hx=Hx_0\sin(\pi x/a)\cos(3\pi z/L) \qquad (5-2)$$
$$Hz=Hz_0\cos(\pi x/a)\sin(3\pi z/L) \qquad (5-3)$$
$$W=2\overline{W}e=1/2\int \epsilon_0|E_0|^2dv=\epsilon_0Ey_0V/8 \quad (6)$$

Where a and b are the width and height of waveguide cross—section, L is the longitudinal length. $\overline{W}e$ is the stored electric energy in the cavity. Substitute the W value into formlae (2) and (3), so can be obtained that

$$\epsilon'=1+v|\triangle f|/(2\triangle vf_0) \qquad (7)$$
$$\epsilon''=v(1/Q-1/Q_0)/(4\triangle v) \qquad (8)$$

It can be seen from above that the part of dielectric constant of loss dielectric results in the change of the resonant frequency and its imaginary part results in the varying of the Q factor. So, by measuring the change of resonant frequency and the varying of quality—factor, it can be determined the dielectric constant ϵ' and ϵ'' of loss dielectric.

EXPERIMENTAL METHODS AND ANALYSIS

1. Preparation of Experimental Samples

A ceramic sample is made of green Al_2O_3 ceramic of square — plain solid with 16mm × 16mm × 5mm. It is heated to the temperature of 600℃, 700℃, 800℃, 900℃, 1000℃, 1100℃, 1200℃ in microwave oven respectively, and the size of the sample at above sintered temperature is determined after cooled, respectively.

The experimental block diagram is shown in Fig. 2.

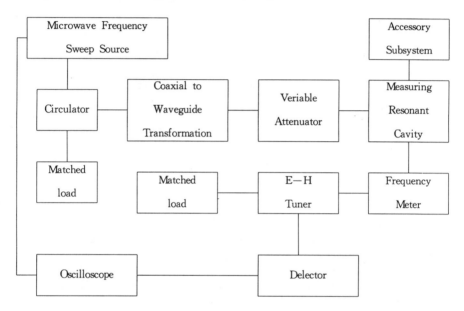

Fig. 2. The experimental block diagram.

2. Experimental Procedure

(1). Assemble the experimental system according to the experimental block diagram and adjust it, measure the parameters f_0 and Q_0 of cavity about five times, and then calculate the average value.

(2). Put a ceramic sample (Al_2O_3) into the resonant cavity, and then sinter it. In the process of sintering, the frequency f_1 and Q factor are measured carefully, when the temperature of sample keep constant at 600℃, 700℃, 800℃, 900℃, 1000℃, 1100℃, 1200℃ respectively.

(3). Repeat step (2), obtain a series of experimental data, then calculate their average values for each temperature by using formulae (2) and (3). The results are shown in Fig. 3

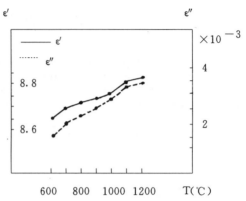

Fig. 3. The curves of dielectric constant
(ϵ' and ϵ'') to sintering temperature.

ANALYSIS OF THE ERROR AND CONCLUSIONS

The experimental results are a little lower than the theory values. The error stem from the following:

1. In the analysis of electromagnetic field, we make use of approximation.
2. Instrument error and reading error.
3. Experimental error is caused by size measure of the sample.

Through the above analysis and discussion, we can conclude as following:

The real part of dielectric constant of Al_2O_3, ceramic material is increasing with the sintering temperature rising, so does the imaginary part. Our experiments show that although Al_2O_3 ceramic material is a kind of lower loss one, monitoring the process of sintering by our experimental system with higher measuring precision is possible. The resonant cavity method is a satisfactory method to measure the dielectric properties of the lower loss material. The experimental results can be provide some scientific experimental basis to physical analysis of ceramic dielectric.

ACKOWLEDGMENTS

We thank Mr. Zhangguo Wan and Lei Chen for the fabrication of the resonant cavity.

REFERENCE

1. J. G. P. Binner, T. E. Cross, et al in High Temperature Dielectric Property Measure ments—An Insight into Microwave Loss Mechanism in Engineering Ceramics. Mat. Res. Soc. Proc. 347(1994)PP. 247—252.
2. Cheng—en Liao, Foundation of Microwave Technology, 1st, (The National Defence Industry Publisher, Bei Jing, 1979), P305.

CHARACTERIZATION OF THIN SEMICONDUCTOR FILMS FOR (OPTO)ELECTRONIC APPLICATIONS

J.R. ELMIGER, H. FEIST, M. KUNST
Hahn-Meitner-Institut, Department Solare Energetik, Glienicker Str. 100, 14109 Berlin, Germany

ABSTRACT

A simple set-up to measure the transient photoconductivity in the microwave frequency range is presented. The effective mobility is derived from the end of pulse transient photoconductivity. This can be used for the characterization of semiconductor films. Examples of measurements on a-Si:H films are given.

INTRODUCTION

The increasing applications of thin films in (opto)electronic devices require reliable and informative characterization techniques for thin films. In particular, contactless techniques are of interest, because they avoid contact problems and they enable in situ characterization. Transient photoconductivity measurements in the microwave frequency range offer the possibility to obtain in a non-invasive way information on charge carrier transport processes. Because these processes control to a large extent the performance of (opto)electronic devices, these measurements are appropriate for the characterization of semiconductor films.

In contributions to preceding symposia several aspects of transient photoconductivity measurements in the microwave frequency range with the Time Resolved Microwave Conductivity (TRMC) method were presented [1-3]. A general framework for the use of this method for the characterization of semiconductors was given [1]. Specific applications to a single crystalline semiconductor, i.e. Si [2], and to in situ optimisation [3], have been reported. In the present work, it will be shown that the end of pulse value of the TRMC signal can be used to characterize thin semiconductor films. To this end an apparatus is described that enables quantitative measurements of the photoconductivity in the microwave frequency range without tailoring of the samples.

EXPERIMENT

Apparatus
The photoconductivity was determined from the change of the reflected power upon pulsed illumination of the sample by a simple reflectometer set up, as described previously [1]. Determination of the conductivity and the photoconductivity is most convenient for a cell, where the sample perfectly fits in a wave guide, closed at one side by a short circuit and at the other side connected to the circulator. Unfortunately, this design requires extensive and cumbersome tailoring of the sample. For this reason we looked for a cell, where no tailoring is necessary and yet in a good approximation the theory for a closed wave guide system can be applied.

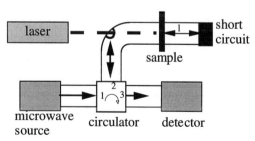

Fig.1: Experimental set-up to measure the photoconductivity from the change of the reflected microwave power

Mat. Res. Soc. Symp. Proc. Vol. 430 © 1996 Materials Research Society

The cell consists of the sample, i.e. a semiconductor film on an isolating (mostly glass) substrate, fixed between the flange of a wave guide connected to the circulator and the flange of a variable short circuit (Fig.1). It was verified that with only the substrate (0.5mm glass) as sample no appreciable loss in the cell was observed. This suggests that the set up can be considered as a closed wave guide system.

The experiments were performed at 10GHz. This frequency was more or less arbitrarily chosen. Additional measurements at other frequencies in the X-band range (8-12.4 GHz) lead to the same results. In general, for low conductivity samples, a high frequency is advantageous, whereas for high conductivity samples low microwave frequencies should be chosen. Illumination of the sample was effectuated through a hole (diameter 3mm) in the wave guide (Fig.1) with 10ns (FWHM) pulses of a NdYAG laser at 532nm. The pulse length limits the time constant of the equipment to 10ns.

Theory

For the present experiments, only the change of the conductivity due to the light induced charge carriers has to be taken into account, as an effect of the illumination. The neglect of other effects of illumination was discussed in a previous paper [1]. If the cell (Fig.1) is modelled as a closed wave guide system, the microwave power reflection coefficient (R) resp. the change of the reflection coefficient upon illumination (ΔR) can be calculated from the conductivity, resp.the photoconductivity, with the theory of the distributed transmission line for a multilayer system [4]. Additionally, it has to be assumed that the cell is uniform in the x/y plane, perpendicular to the propagation direction of the microwaves. Measurements of R and ΔR as a function of the distance l between short circuit and sample can be compared to calculations with this model in order to check, if this model is appropriate to describe the present experiments. It has to be noted that for this calculation the knowledge of the relative dielectric constant (ε) of the sample is not required, because for thin films this property has no appreciable influence on the reflection coefficient within the range of values of ε ($2<\varepsilon<50$) for semiconductors. This leaves only the conductivity, resp. photoconductivity, as unknown parameter.

For a better understanding, it is useful to discuss the theory in a more qualitative way. As a simplification it will be assumed that the influence of the dark conductivity of the sample can be neglected, as it is the case for samples of practical interest. ΔR is equal to the negative microwave absorption due to excess conductivity, $B(\Delta\sigma(t))$:

$$\Delta R = -B(\Delta\sigma(t)) = -\int_0^d \Delta\sigma(z,t)E^2(z,\Delta\sigma)dz \qquad (1)$$

where $E(z,\Delta\sigma)$ is the microwave field in the sample, z refers to the propagation direction of the microwaves and d is the thickness of the sample. For thin films the microwave field in the sample is uniform in the z direction. Consequently:

$$\Delta R = A\int_0^d \Delta\sigma(z,t)dz = A\Delta S(t) \qquad (2)$$

where A, the sensitivity constant, is a function of the microwave field in the sample (Eq. 1) and $\Delta S(t)$ is the excess conductance, i.e. the integral over the excess conductivity in the sample. For not too large values of $\Delta S(t)$, i.e. in the small perturbation approach, A is independent of $\Delta S(t)$ and $\Delta R(t)$ is proportional to the excess conductance. For the measurements presented here, this is always warranted, as it was shown by the model calculation. It must be noted that Eq.2 can be derived more generally [1,4] and so it can be used as the fundamental equation describing TRMC measurements. The model calculation shows also that for thin semiconductor films the TRMC signals depends on the photoconductance and is insensitive to the distribution of the photoconductivity. Consequently, the photoconductance will be determined from the experimental data.

Excess conductance and effective mobility

The excess conductance induced by the laser pulse is due to $\Delta n(t)$ (in cm^{-2}) excess electrons characterized by the mobility μ_n and $\Delta p(t)$ (in cm^{-2}) excess holes with the mobility μ_p:

$$\Delta S(t) = e(\Delta n(t)\mu_n + \Delta p(t)\mu_p) \qquad (3)$$

The zero point for the time is the trigger point of the exciting laser pulse. For our purposes it is convenient to consider only the end of pulse TRMC signal at 10ns, the TRMC amplitude, and the corresponding ΔS. Although this choice minimizes the influence of excess charge carrier decay processes, an influence of these processes during the excitation has to be taken into account. Excess charge carriers can be immobilized during the excitation by three kinds of decay processes:

i)   Higher order electron-hole recombination
ii)   Trapping in defects and recombination via defects
iii)   Surface recombination and surface trapping

The influence of the first process can be eliminated by the use of data at small laser excitation density or extrapolation in the range, where the TRMC amplitude is proportional to the excitation density. The second group of processes is a measure of the (opto)electronic quality of the sample. The third group is problematic and has to be eliminated, because the surface conditions of the samples do not have a clear relation with interface conditions in devices.

ΔS can be expressed in Δn_0, the number of excess electrons induced by the laser pulse, and factors taking into account the three groups of decay processes:

$$\Delta S = e\Delta n_0 \{(1 - {}_n f_{rec} - {}_n f_{tr} - {}_n f_{sr})\mu_n + (1 - {}_p f_{rec} - {}_p f_{tr} - {}_p f_{sr})\mu_p\} \qquad (4)$$

where f_{rec}, f_{tr} and f_{sr} are the relative number of excess carriers that decay within 10ns by respectively the decay processes of the type i),ii) and iii), and the left index indicates the charge carrier involved (n:electrons and p:holes). The effective mobility, m_{eff}, can be introduced:

$$\mu_{eff} = \frac{\Delta S}{e\Delta n_0} = (1 - {}_n f_{rec} - {}_n f_{tr} - {}_n f_{sr})\mu_n + (1 - {}_p f_{rec} - {}_p f_{tr} - {}_p f_{sr})\mu_p \qquad (5)$$

In most cases, one of the terms determining μ_{eff} is much larger than the other one. So only one kind of excess charge carriers is detected. Discarding the influence of the decay processes of the types i) and iii), the effective mobility characterizes a film by the mobility and decay processes via defects. It is clear from Eq.5 that a higher effective mobility refers to a higher quality film, if only one kind of excess charge carriers determines the signal. If no excess charge carrier decay within the pulse takes place, the effective mobility is given by;

$$\mu_{eff} = \mu_n + \mu_p \qquad (6)$$

This value has to be compared to charge carrier mobilities in the same material obtained by other methods, if available.

Evaluation of the data

For the calculation of Δn_0 the absorption coefficient (α) at the excitation wavelength has to be known. This is avoided if the exciting light is completely absorbed in the sample. For 1μm semiconductor films this implies $\alpha > 10^4$cm^{-1}. This is warranted for light with a photon energy larger than the bandgap. For this reason in this work the TRMC signals were excited by 532nm (2.32eV) laser pulses . Most thin semiconductor films show complete absorption of 532 nm light. The calculation of ΔR by the simple model is based on a uniform ΔS in the x/y plane. Although it is possible in this configuration (Fig.1) to illuminate the sample uniformly and so produce a uniform ΔS, this has several serious drawbacks e.g. a very inaccurate knowledge of the excitation density. Therefore the sample is illuminated over a circular spot (0.3 cm diameter) in the middle of the wave guide. The value of ΔS calculated by the model (corresponding to the experimental value of ΔR) will be called ΔS_{eff} and is in the framework of our model related to ΔS, the real (local) change of the conductance, by:

$$\frac{\Delta S_{eff}}{\Delta S} = \frac{\displaystyle\int\int_c \sin^2(\frac{\pi x}{a})dxdy}{\displaystyle\int\int_s \sin^2(\frac{\pi x}{a})dxdy} \qquad (7)$$

where c is the area of the illuminated spot, s the area of the wave guide and a the length of the long side of the wave guide.

ΔS is calculated by Eq.(7) from the value of ΔS_{eff} obtained from the best fit of the experimental data to the model. This value of ΔS is divided by the number of electron-hole pairs induced by the pulse (in cm^{-2}) to yield the effective mobility. The accuracy is mainly controlled by the error in the excitation density (about 20%).

RESULTS

Test of the set up and determination of μ_{eff} in a-Si:H

The data points in Fig.2 show the TRMC amplitude induced by 532nm light pulses in a state-of-the-art a-Si:H film (1μm thick) on a 7059 Corning glass substrate as a function of the distance l between sample and short circuit. The drawn line represents the best fit of the model to the experimental data with ΔS_{eff} as the only fit parameter. The good agreement between the experimental data and the model proves that the model offers an adequate description of distribution of the microwave field in the present set up.

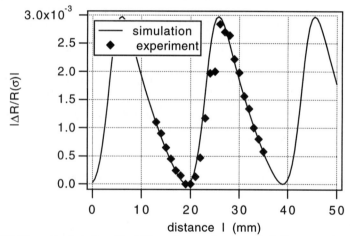

Fig.2: *TRMC amplitude induced by 532 nm laser pulses in a-Si:H film on a glas substrate as a function of the distance between the short circuit and the sample. The drawn line is the calculated curve.*

The experimental data show the movement of the sample through the microwave field in the cell, controlled by a value zero at the short circuit. The TRMC amplitude is maximal if the sample is positioned at the maximum of the microwave field and it is minimal if the sample is positioned at the minimum of the microwave field. The periodicity of the data (2cm) is determined by the half of the wavelength in the empty wave guide. A free standing film would show a square sinus dependence on l [4]. The presence of the glass substrate leads to a distortion of this behaviour.

At the excitation density, where the data points in Fig.2 are determined, the TRMC amplitude (and so ΔS) depends sublinearly on the excitation density. Measurements of the TRMC amplitude as a function of the excitation density at the highest sensitivity enable to determine the amplitude in the range, where it is proportional to the excitation density. From these experiments f_{rec} at the excitation density of interest can be calculated. From additional measurements it was derived that surface recombination can be neglected. It is generally accepted that the hole mobility in intrinsic a-Si:H is about two orders of magnitude smaller than the electron mobility. Consequently, the TRMC signals refer only to electrons and the effective mobility corrected for f_{rec} will be equal to the electron mobility μ_n, if trapping within 10ns can be neglected, as it was observed in this material.

The effective mobility determined from Fig.2 amounts $\mu_{eff}=0.5cm^2V^{-1}cm^{-1}$. This value agrees with the value determined from Time of Flight (TOF) measurements for electrons. This means, that the model not only predicts correctly the distribution of the microwave field but also its strength. So the model can be used to describe experiments performed in our set up.

Effective mobility and quality of intrinsic a-Si:H films
The possibilities to characterize a-Si:H films in the present set up will be shown by the investigation of intrinsic a-Si:H films of different but known quality. To this end films were deposited at different deposition temperature (T_{dep}). In Fig.3 μ_{eff} (corrected for nf_{rec}) is plotted for these films as a function of T_{dep}. The behaviour of μ_{eff} matches the quality closely: the quality is optimal at $T_{dep}=250C$ and decreases with decreasing deposition temperature.

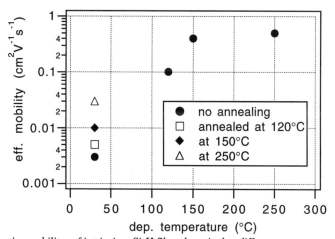

Fig.3: *Effective mobilites of intrinsic a-Si:H films deposited at different temperatures.*

The decrease of μ_{eff} at lower deposition temperature is attributed to a decrease of μ_n (due to increasing disorder of the material at lower deposition temperature) and an increase of f_{tr} (due to an increasing number of deep traps). Separation of these influences is possible, at least partially, by analysis of the dependence of the TRMC amplitude on the excitation density.
It is well known, that the quality of a-Si:H films increases after annealing at higher temperatures. Also this is shown by the increase of μ_{eff} in the film deposited at 30C after annealing at different temperatures (Fig.3)

Other films
Preliminary experiments have shown that also GaAs, WS_2, WSe_2, FeS_2 and AgS films can be characterized by the effective mobility obtained from TRMC measurements. The production process could be optimised by comparison of the effective mobility of films produced under different conditions.

271

CONCLUSIONS

It was shown that in a relatively simple equipment the photoconductivity of semiconductor films can be determined. An example was shown with intrinsic a-Si:H films. The effective mobility derived from the photoconductivity can be used for an optoelectronic characterization of the sample.

REFERENCES

1.	M.Kunst, MRS Symposium Proceedings **189**,75 (1990)
2.	A.Sanders, H.Wetzel and M.Kunst, MRS Symposium Proceedings **189**,403(1990)
3.	C.Swiatkowski, A.Sanders, M.Kunst, G.Seidelmann, C.Haffer and K.Emmelmann, MRS Symposium Proceedings **269**,491 (1992)
4.	C.Swiatkowski, A.Sanders, K.-D.Buhre, and M.Kunst, J.Appl. Phys.**78**,1763 (1995)

THE DIELECTRIC PROPERTIES OF POROUS ZINC OXIDE CERAMICS

J.P. CALAME *, Y. CARMEL *, D. GERSHON *, A. BIRMAN *, L.P. MARTIN †,
D. DADON †, and M. ROSEN †
*Institute for Plasma Research, University of Maryland, College Park, MD 20742
†Dept. of Materials Science, Johns Hopkins University, Baltimore, MD

ABSTRACT

Measurements of the complex dielectric constant of microwave sintered, porous ZnO at
2.45 GHz are presented. The dielectric properties as a function of porosity do not obey the
standard Maxwell-Garnet dielectric mixing law with the ceramic material as the major phase,
but instead behave as if the ceramic grains always remain in relatively poor electrical contact
even at very high densities. Electromagnetic simulations, carried out for a variety of
microstructure geometries, are performed to explore this observation. A model which treats
the ceramic as an array of grains and pores, with the grains separated from each other by non-
or slightly-percolating, fractal-geometry surfaces, provides a good description of the
experimental results.

INTRODUCTION

Microwave sintering of ceramics is a complex process involving the propagation and
absorption of electromagnetic energy, heat transport, and densification that changes both the
macroscopic shape and microstructure morphology. The electrical and thermal properties of
the material that govern this process strongly depend on the local temperature and density,
both of which change during the sintering. A detailed knowledge of these dependencies is
essential for understanding and modeling the sintering process.

In this work the dependence of the permittivity on density is studied, using microwave
sintered zinc oxide (ZnO) as an example. In this paper experimental data is first compared
with common analytic mixing laws. In an effort to gain better insight into the behavior of the
permittivity as a function of density, a series of electromagnetic simulations were performed
for a several microstructure geometries. We find that a model using fractal geometry particle
boundaries to represent aspects of the microstructure geometry provides a physically
meaningful description of the experimental data.

EXPERIMENT AND ANALYTIC THEORY

Cylindrical samples (31 mm dia, 17 mm thick) of ZnO were produced by uniaxially cold-
pressing commercial, 4 μm powder to 37 MPa without the use of a binder. Microwave
sintering was performed in a highly overmoded 2.45 GHz applicator. In order to provide
thermal insulation, the samples were loosely packed in ZnO powder and placed in an alumina
outer enclosure. Type K thermocouples were used to monitor the core and surface
temperatures.

After a slow 1 °C/min heating (to 200 °C) in air for water removal, the microwave power
was increased under computer control to achieve a heating rate of 15 °C/min in air until
various final temperatures (and thus densities) were obtained. Once the desired temperature
was reached, the microwave power was sharply reduced to allow cooling. Density

Mat. Res. Soc. Symp. Proc. Vol. 430 © 1996 Materials Research Society

measurements were performed by dimensional measurement and weighing. They are expressed as a relative density R, which is the fraction of the theoretical X-ray density (5.61 g/cm^3). As a result of this processing procedure, five samples were obtained with relative densities of 53.4, 65.8, 77.0, 86.2, and 94.1%, corresponding to maximum core processing temperatures of 680, 780, 830, 880, and 960 °C, respectively. The complex relative permittivities of the samples at 2.45 GHz were determined using an open-ended coaxial line technique. In Fig. 1 the measured relative complex permittivity ($\varepsilon = \varepsilon'-j\varepsilon''$) of the microwave sintered ZnO at room temperature at 2.45 GHz is plotted versus the relative density R. The most distinguishing feature of the data is the sharp increase in permittivity with relative density, which is particularly evident in the imaginary part.

The two most common theories to describe the dielectric permittivity of a heterogeneous mixture are Maxwell-Garnet (MG) theory and the effective medium approximation (EMA) [1]. Maxwell-Garnet theory is derived based on the polarization induced by an externally applied, uniform electric field on isolated spherical inclusions located within a host material. Depending on whether air or ceramic is considered as the host (major phase), we have two expressions for the permittivity of the mixture ε_m as a function of the relative density of ceramic. When ceramic is considered to be the major phase, one obtains

$$\varepsilon_m = \frac{\varepsilon_{cer}\left(\varepsilon_0(3-2R)+2R\varepsilon_{cer}\right)}{\left(R\varepsilon_0 + \varepsilon_{cer}(3-R)\right)} \quad , \tag{1}$$

and when air is considered to be the major phase the expression is

$$\varepsilon_m = \frac{\varepsilon_0\left(\varepsilon_{cer}(1+2R)+2\varepsilon_0(1-R)\right)}{\left(\varepsilon_{cer}(1-R)+\varepsilon_0(2+R)\right)} \quad , \tag{2}$$

where ε_0 and ε_{cer} are the permittivities of air and ceramic, respectively.

The predictions of MG theory and the EMA [1] are also shown in Fig. 1 so that they may be compared to the data. In this figure MG-Ceramic assumes that the ceramic is the host, MG-Air assumes that air is the host. These calculations assumed that the permittivity of the fully dense, microwave sintered ZnO ceramic at 2.45 GHz was 38.2-j22.1, based on previous work [2]. None of the three approximations adequately describes the experimental data, although the MG-Air theory does at least to some extent produce the concave-upward, nonlinear variation in the real part of the permittivity observed in the data. The agreement between theory and experiment in the case of the imaginary permittivity is poorer still. Even if MG-Air was more compatible with the experimental data, there are strong physical objections to the notion of the ceramic as the minor phase at high relative densities. The theory is derived on the basis of isolated spheres of matter with no quadrupole or higher interactions, which is not the case at high packing densities. Furthermore, only so much volume can be filled by spheres, and once they touch (percolation threshold) all assumptions in developing Maxwell-Garnet theory and spherical-particle EMA break down. These percolation thresholds depend on the arrangement of the spheres, but typical examples are R=0.52 for a simple cubic array, 0.68 for a body centered cubic array, 0.74 for a face centered cubic array, and about 0.6 for a random packing of spheres.

ELECTROMAGNETIC SIMULATIONS AND THE FRACTAL BOUNDARY MODEL

In order to develop a more physically meaningful explanation of the experimentally observed mixing law, a series of electromagnetic simulations were performed on a variety of ceramic microstructure geometries. Finite-difference quasi-electrostatic field solution methods were chosen to perform the computations since the structure sizes are much smaller than a wavelength. In this procedure a 3D model space cube is filled with n^3 small cubical cells that are assigned as either air or ceramic (n is the number of cells along each axis). The top and bottom surfaces of the model space are assigned fixed potentials just like in a parallel plate capacitor, and the ceramic is distributed over the cells in the shape of the microstructure under investigation. Electric fields in each cell are computed using standard relaxation methods. The capacitance of the model space, and thus the effective complex dielectric constant of the mixture with the assumed microstructure, is obtained from the fields. Two important microstructure geometries relevant to the present work have been studied: a random spatial distribution of cubical dielectric cells, and a regular array of dielectric spheres where contact and spherical neck formation are permitted. Example simulation structures are shown in Fig. 2. In the case of the sphere (Fig. 2b) one can allow the diameter to become larger than the size of the model space, at which point the filled volume will look like a truncated sphere with six flat, circular-outline surfaces cut into it. Due to dielectric and potential "mirror" boundary conditions imposed on the model space boundaries, these surfaces represent spherical necks forming connections to neighboring spheres in a simple cubic array. The results of the imaginary permittivity calculations with $n=25$ are shown in Fig. 3.

In the case of the spheres, at low relative density they are isolated from each other so we certainly expect MG-Air to describe the behavior. However, the onset of electrical percolation (direct interconnecting paths linking the top and bottom faces of the model space) creates a transition to MG-Ceramic. At high density the dielectric behavior indicates that the closed off, cusp-like pores that form at the model space corners are well described by MG-Ceramic, in spite of the non-spherical shape. In fact, the random cubical cell simulation exhibits very similar overall behavior when one considers that the percolation threshold for this geometry is very low (about 0.1). Even though at modest densities (0.3 or so) air should clearly still be the host, the percolation of dielectric overshadows this and the system behaves as if ceramic is the major phase. Based on the experimental results and these simulations, one can conclude that zinc oxide behaves as if percolation never occurs to any major extent in spite of high fractional density. This could result, for example, from poor physical contact between particles due to surface geometry mismatches or cracks, from electrically inactive glassy layers formed between the ZnO grains, or due to potential barriers at the grain boundaries created by surface states, impurities, or dislocations.

To address this, a new procedure was developed that attempts to randomly fill cells with dielectric, provided that doing this does not trigger top-to-bottom percolation through the model space. Thus a dielectric cell may be connected to other dielectric cells that eventually connect to either the bottom or top surfaces, but not both. Eventually the model space contains a combination of dielectric cells and critical cells that must remain as air or else percolation would result. The air gap resulting from this procedure turns out to be a fractal surface (dimension between 2 and 3), shown schematically in Fig. 4a. A two-dimensional slice through a representative $n=32$ example is also shown. The particular geometry that is created depends on the random seed given to the computer for an initial shuffling process. For a given number of dielectric cells per axis there is a maximum relative density R_{max} that can be

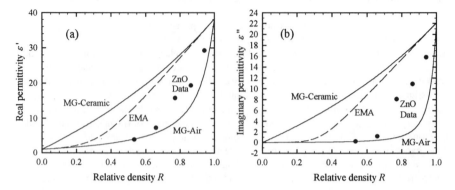

Fig. 1. (a) Real and (b) imaginary permittivity of ZnO vs. density at 2.45 GHz (circles), along with Maxwell-Garnet and EMA analytic theories.

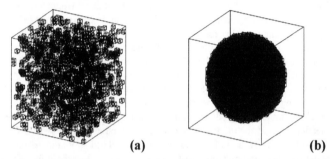

Fig. 2. Geometries for (a) random cubes and (b) spheres in a simple cubic array finite difference calculations. The outer cubical grid defines the model space.

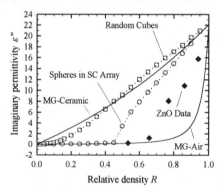

Fig 3. Calculated imaginary permittivity (ε") as a function of relative volume density for the random array of cubes (squares) and the spheres in a simple cubic array (circles) configurations. Also shown are the Maxwell-Garnet theories and the ZnO data (diamonds).

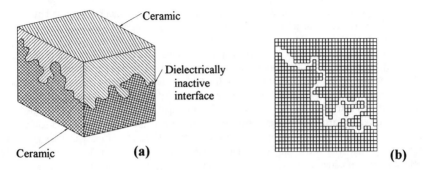

Fig. 4. (a) Schematic diagram of the fractal boundary model representing the interface between two ceramic particles and (b) a representative 2D slice through such a structure.

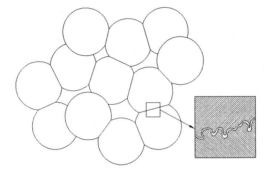

Fig. 5. Diagram of a combination of a mesoscale geometry with the fractal boundary model

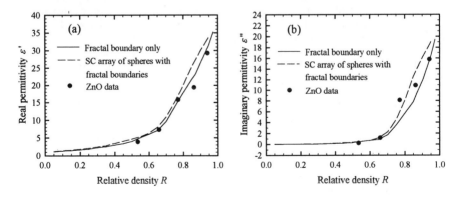

Fig. 6. Computed (a) real and (b) imaginary permittivities using the fractal boundary models, along with the ZnO data.

achieved by this procedure (or else percolation would occur). To achieve less dense disordered structures with the same overall morphology, the air gap can be increased by removing dielectric from the cells bordering the gap, in 3x3x3 chunks to avoid filamentation. In order to extend the results to relative densities higher than R_{max} , the spatial self-similarity property of fractals can be evoked, and the gap can be filled with material having a dielectric permittivity equal to a value computed with the same model at some lower relative density.

An interesting additional application of the fractal boundary mixing law can arise in systems where there is a discernible, larger scale structure created by the overall arrangement of the ceramic particles that make up the green body. For example, there could be a "mesoscale" structure created by large particles either resting against each other or in contact via spherical necks. The contact regions can be modeled using the fractal boundary theory and assigned the appropriate permittivity, while the rest of the mesostructure would be simulated conventionally as cells made up of ceramic or air. A picture of this situation is shown in Fig. 5. An interesting issue associated with this procedure is knowing on first principles what relative density to assign to the microscopic boundary region for a given mesoscale relative density. One possibility is to assume that the surface roughness on both the boundary and mesoscale levels evolves self-similarly. A good way to quantify this roughness-based interface phenomenon in the fractal boundary system for a particular R_{micro} is to define a parameter ζ_{micro} which is the surface area surrounding the gap (both sides) divided by the 2/3 power of the gap volume. For a mesoscopic geometry with relative volume density R, a value of the interface ζ_{meso} can be computed as the total inter-particle contact area divided by the two-thirds power of the total pore volume. The value of R_{micro} for the fractal boundary is selected to force ζ_{micro} to be equal to the value of ζ_{meso}. The results of this two-step procedure with $n=25$, when applied to spheres in a simple cubic array, are shown by the dashed lines in Fig. 6. The behavior of this composite geometry system is very similar to that of the fractal boundary model alone.

CONCLUSIONS

The procedure presented above provides a new means of computing the dielectric properties of a porous body using a fractal boundary model, knowing only the fully dense complex dielectric permittivity and the relative volume density. The model can be extended to two-scale microstructures by self-similarity. Both schemes describe the 2.45 GHz dielectric behavior of microwave sintered zinc oxide with good accuracy. The fractal geometry can arise from poor physical contact due to surface roughness or cracks, the development of glassy intergranular layers, or the existence of surface dislocations or impurities. All of these factors are known to occur in ZnO.

ACKNOWLEDGMENTS

This work was supported by the US Dept. of Energy and by the Army Research Office.

REFERENCES

1. R. Landauer, J. Appl. Phys. **23**, 779 (1952).

2. L.P. Martin, D. Dadon, D. Gershon, A. Birman, B. Levush, Y. Carmel, and M. Rosen, J. Am. Ceram. Soc., (in press).

MICROWAVE DIELECTRIC PROPERTIES, THERMAL ANALYSES AND CHEMICAL COMPOSITION OF HEAVY CLAY BODIES

Garth Tayler, Acme Brick Company, Technical Department, Fort Worth TX;
Michael Hamlyn, Staffordshire University, School of Engineering, Stafford, U K;
Michael Anderson, Staffordshire University, School of Design and Ceramics, Stoke on Trent, U K.

ABSTRACT

To investigate the suitability of heavy clay bodies for microwave processing, a range of material properties needed to be investigated.
This paper presents results of the analyses, chemical, mineralogical and thermal, of a range of heavy clay bodies. The microwave dielectric properties of these bodies were measured, over a range of temperatures, using the cavity perturbation technique. The relationship between the dielectric properties and the results of DTA/TGA analysis is discussed and related to the potential use of microwave energy for the firing of these materials.

INTRODUCTION

The research work described was undertaken at Staffordshire University (UK), as part of a study to examine the feasibility of using microwave energy for the firing in the traditional 'structural clayware' industry. The use of the more efficient volumetric heating effect of microwave energy, has long been recognised as an extremely efficient process in many industrial heating applications. The whiteware ceramic industry has recently begun to consider this method of heating and some successful microwave drying applications have been developed. The use of microwave energy for the firing of traditional ceramic materials, is still to be examined to find out its technical feasibility. This has been the motivation behind the work reported.

RAW MATERIALS

The body component materials used for this research work were selected from the materials being used by a face brick manufacturing operation. The component materials, which include some waste materials, were described as follows :

Clean fireclay	CFC	
Fireclay	BBFC	
Plastic clay	PLC	
Carbonaceous flint clay	CFL	
Pulverised fly ash	LPFA	(inert waste material from a power plant)
Carbon fly ash	CFA	(carbonaceous waste material from a coal processing plant)
Silica sand	SS	

Mat. Res. Soc. Symp. Proc. Vol. 430 ® 1996 Materials Research Society

The two fireclays formed the major components of the brick bodies and provided the required colour characteristic of the face brick product being produced. The plastic clay component was used to improve extrudability.
The carbonaceous flint clay when used in the body, provided a limited source of carbon to the brick body, resulting in an attractive local reduction effect on the surface of the stretcher faces of the bricks produced. The carbon fly ash component (CFA) was added to increase the carbon content of the body.
The inert components were selected for specific purposes. The silica sand was used to provide the necessary stabilising effect on these rather pure kaolinitic bodies, to minimise shrinkage. The PFA was used in conjunction with the silica sand in two of the bodies, in order to try to facilitate the oxidation of the carbon in the body. Each of the raw materials were analysed chemically and by X- Ray Defraction at Staffordshire University.

The results are seen in table 1;

Table 1: Characterisation of component raw materials by Chemical and XRD analyses

Chemical	CFC	BBFC	PLC	CFLC	LPFA	CFA	SS
SiO2	54.49	52.05	57.86	40.35	54.76	34.01	98.97
Al2O3	27.69	32.49	23.19	23.83	31.87	25.09	0.48
Fe2O3	4.42	1.31	7.28	3.18	3.44	2.78	0.37
CaO	0.47	0.12	0.1	0.42	4.49	7.04	0.02
K2O	0.58	0.12	0.61	0.73	0.49	0.5	0.05
Na2O	0.11	0.17	0.09	0.35	0.39	0.3	0.42
TiO2	1.25	1.22	1.33	1.23	1.55	1.36	0.03
Total Carb	0.35	0.13	0.3	7.69	1.77	35.56	0.09
LOI	10.49	11.76	8.89	28.72	1.88	28.1	0.22
Si:Al ratio	1.96	1.6	2.49	1.69	1.72	1.35	n/a
XRD							
Kaolinite	Major	Major	Major	Major	-	-	-
Quartz	Minor	Minor	Minor	Minor	Major	Major	Major
Illite	Minor	Minor	Minor	Minor	-	-	-
Muscovite	Minor	-	Minor	-	-	-	-
Mullite	-	-	-	-	Major	Major	-
Calcite	-	-	-	-	-	-	-
Montm'ite	Minor	-	-	-	-	-	-
Magnetite	-	-	-	-	Minor	-	-
Lime	-	-	-	-	Minor	-	-

Note:" Loss included" results shown in chemical analyses.

METHODOLOGY

The raw material components were sampled from bulk stockpiles and were thus representative of the material used in the production process. Each material was analysed chemically and by XRD. Thermal analyses using simultaneous DTA/TGA were conducted on each material. From these results, four characteristic composite bodies were formulated. Simultaneous DTA/TG analyses were conducted on each of these bodies. The dielectric properties of these composite bodies over a range of temperatures up to 1200°C, were then examined using the Cavity Perturbation Technique at Staffordshire University. The characteristic bodies are shown in table 2.

Table 2 : Characteristic body types

Control body (MC)	a body consisting of a combination of clean fireclays and silica sand.
Carbonaceous body (M4)	body containing 40% carbonaceous flintclay and 10% PFA.
CFA body (M5)	body containing clean fireclays and containing 10% CFA.
CFA Works body (MT)	a body consisting of a combination of fireclays, (M5) silica sand and containing 10% CFA.

Samples of the individual bodies were finely ground and dried. Sufficient moisture was then mixed into the materials to facilitate extrusion. Small 3 mm diameter by 12 mm long, "stick" samples were extruded using a specially manufactured ram extruder (see figure1). The samples were sized to fit into the sample holder of the cavity perturbation apparatus. Nelson (4), has shown that density critically influences dielectric properties. The ram extruder was therefore employed to achieve a similar density to a typical brick extrusion process. The inert materials, which could not be extruded, were tested in the powder form. The apparatus is shown in figure 1.

3mm sample extrusion   extrusion ram

extrusion ram cylinder

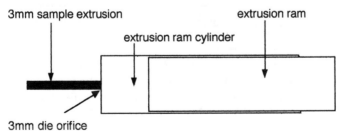

3mm die orifice

Figure 1: Extrusion Ram for Sample Preparation

TOTAL CARBON, SILICA AND ALUMINA RESULTS.

The characterisation of the selected bodies according to their carbon contents and clay contents, expressed by silica and alumina percentages, is seen in table 3.

Table 3: Characterisation by total carbon, silica and alumina contents

CHARACTERISTIC FEATURES	MIX	TOT C	SiO2	Al2O3	LOI
Lower carbon,fireclay and silica sand body	MC	0.42	59.37	26.95	8.71
Higher carbon, fireclay, carb flint, s sand and PFA body	M4	2.71	55.44	28.98	11.31
Higher carbon,fireclay,s sand and CFA body	M5	2.87	59.61	26.51	11.62
High carbon,fireclay,high Fe2O3 CFA, clamp kiln works body	MT	4.01	54.01	28.84	12.65

THERMAL ANALYSES RESULTS

Referring to the chemical and XRD analyses carried out, the DTA/TGA results of the characterised bodies MC, M4, M5 and MT, as set out in Table 3, may be considered. Mixes MC and M4 have total carbon contents of 0.42% and 2.71% respectively. The MC mix contained no carbonaceous addition but still possessed a total carbon content of 0.42%. This carbon was contributed by the fireclay and plastic clay which contain small traces of carbon which was co-deposited with the clay during its formative stage. The carbon in the M4 mix is contributed mainly by the carbonaceous flint clay component. Mixes M5 and MT contained 2.87% and 4.01% total carbon respectively. The contribution of the carbon in these mixes was from the CFA (carbon fly ash) additive in the body. Table 4 shows the comparative DTA/TGA results.

Table 4: Comparison of DTA/TGA results, MC, M4, M5 and MT

MIX	% T C	COMPARATIVE COMMENTS
MC	0.42	Dehydroxilation endothermic peak at 560 ºC. Small carbon sourced exothermic peak at 480 ºC.
M4	2.71	Dehydroxilation endotherm at 560 ºC . Large exotherm due to volatile hydrocarbon release at 440 ºC.
M5	2.87	Dehydroxilation endotherm at 560 ºC. Medium carbon exo - therm at 460 ºC. Large exotherm due to the oxidation of fixed carbon at 680 ºC. Oxidation completed at 800 ºC.
MT	4.01	Dehydroxilation endotherm at 560 ºC.Small exotherm at 420 ºC Large carbon exotherm from 640ºC to 825ºC.

DIELECTRIC PROPERTY MEASUREMENT RESULTS

Dielectric properties from ambient temperature to 1200 oC were measured at Staffordshire University, using the Cavity Perturbation test equipment which has been calibrated to operate at different frequencies. The frequency of 2468 MHz was the test frequency used.

Dielectric properties : Component materials

The dielectric property measurements at the initial lower temperatures, indicated slightly higher Permittivity and the Loss Factor readings, as a result of the hydroscopic moisture present in the material. Most of the curves reflected higher permittivity levels, at the lower temperatures of 20oC to 160 oC. Once this moisture was driven off, the permittivity reduced rapidly.

It is verified by Metaxis and Meridith (2), and Hamlyn (3), that permittivity begins to decrease as the moisture is driven off and as the temperature further increases in the heating cycle, so the permittivity begins to increase.

During the dielectric testing of the body components, it was found that materials with higher total carbon contents, had higher permittivity values.

Comparing the ε' and ε'' values of Clean Fire Clay (CFC), with Carbonaceous Flint Clay (CFLC) and carbonaceous fly ash (CFA), in table 5 for example, it can be seen that permittivity and loss factor values for the higher carbon components are much higher. The real and imaginary parts of the complex permittivity of body components at 100, 600 and 1000 oC are compared as follows:

Table 5: Comparative permittivity and loss factor values of raw material components

MATERIAL	TOT C %	100 oC		600 oC		1000 oC	
		ε'	ε''	ε'	ε''	ε'	ε''
CFC	0.35	5.4	0.69	3.8	0.10	4.4	0.50
BBFC	0.13	4.6	0.34	3.6	0.05	3.75	0.24
PLC	0.3	6.2	1.80	3.8	0.01	4.5	0.60
CFL	7.69	6.8	2.20	5.10	0.80	5.50	0.80
CFA	41.9	8.5	1.60	8.25	1.60	10.80	6.00
PFA	1.77	2.6	0.50	2.68	0.08	2.61	0.026
SS	0.09	2.7	0.005	2.8	0.012	2.81	0.026

Dielectric properties : Composite bodies

The selected composite bodies MC, M4, M5 and MT and their corresponding values of ε' and ε'' at 100 ºC, 500 ºC, 800 ºC and 1000 ºC are compared:

Table 6: Comparative permittivity and loss factor values of composite bodies

MIX	TOTAL %C	100ºC		500ºC		800ºC		1000ºC	
		ε'	ε''	ε'	ε''	ε'	ε''	ε'	ε''
MC*	0.35	5.4	0.69	4.2	0.05	4.00	0.30	4.4	0.50
M4	2.71	5.23	0.50	4.57	0.20	4.9	0.38	4.58	0.57
M5	2.87	7.5	0.65	6.51	0.15	6.10	0.36	5.3	0.50
MT	4.01	8.00	0.30	8.50	0.60	8.75	0.60	8.3	1.10

* Dielectric properties of MC were not measured. Component CFC was taken as the equivalent material for comparison purposes. MC and CFC are similar in composition.

CONCLUSION

From the results obtained to date, a preliminary conclusion can be drawn that the complex permittivity values increase with increased body carbon, as seen in the composite bodies, M5 and MT.
The higher carbon bodies are therefore seen to be susceptible to more efficient coupling with the microwave field due to their increased lossiness. The carbon in the bodies, being a good dielectric material, is responsible for this. Further work is in progress to investigate the influence and effects of this phenomenon, when considering microwave firing of heavy clay products.

ACKNOWLEDGMENTS

The author wishes to acknowledge the assistance of Staffordshire University Microwave Processing Unit and Eskom - TRI, South Africa, who made this research possible.

REFERENCES

1.   Ahmad I, Dalton R, and Clark D.
      "Unique Application of Microwave Energy to the Processing of Ceramic Materials."
      Journal of Microwave Power and Electromagnetic Energy. Vol.26No.3,1991.

2.   Hamlyn M. G. Bowden A. L. Jones G. and Jackson S. M.
      "Microwave Firing Process for Earthenware Ceramics"
      High Frequency and Microwave Processing and Heating Conference
      Arnhem. (1989).

3.   Mataxas, A. C. and Meridith, R. J.
      Industrial Microwave Heating .(Peter Peregrinus, London 1983).

4.   Nelson S. O.
      "Observations on the Density Dependence of Dielectric Properties of
      Particulate Materials."
      Journal of Microwave Power.No.18(2). (1983).

MEASUREMENT OF THE DIELECTRIC PROPERTIES OF POLYSTYRENE PARTICLES IN ELECTROLYTE SOLUTION

C.GROSSE*# , M.C.TIRADO*
* Instituto de Física, Universidad Nacional de Tucumán, Av. Independencia 1800,
(4000) Tucumán, Argentina. julio@untmre.edu.ar
Consejo Nacional de Investigaciones Científicas y Técnicas.

ABSTRACT

The permittivity and conductivity of suspensions of spherical sub micron size polystyrene particles were measured in the 10 kHz to 10 MHz frequency range using a HP 4192A Impedance Analyzer. Due to the high conductivity of the samples (0.1 to 0.6 S/m) every detail of the measurements had to be optimized. For each sample, the instrument was calibrated as an electric quadrupole, by means of short, open, and reference measurements, using for this last a standard electrolyte with a conductivity similar to that of the sample. The results obtained were combined with measurements in the 1 MHz to 28 GHz range made in Goettingen, Germany, using precisely the same samples. The final spectra cover the broadest frequency range to date for these systems.

INTRODUCTION

Suspensions of monodispersed polystyrene spherical particles are ideal model systems for the study of low frequency dielectric relaxation mechanisms. In order to compare the validity of existing theoretical models [1-9] it is necessary to perform high precision measurements in a broad frequency range. This kind of data is currently unavailable.

This situation prompted the Research Group of Prof. R. Pottel at Goettingen, Germany, and our own at Tucumán, Argentina, to an agreement which is supported by the Volkswagen Foundation. The goal is to use the specialized equipment of both Laboratories to measure the dielectric response of precisely the same samples in the range of 10 kHz to 10 MHz (Tucumán) and 1 MHz to 28 GHz (Goettingen).

Here we present the first results obtained in the whole frequency range, and discuss the experimental details corresponding to the low frequency part of the spectrum.

LOW FREQUENCY DIELECTRIC MEASUREMENTS OF CONDUCTING LIQUIDS

These measurements were made using a HP 4192 A Impedance Analyzer, which has a broad frequency range (5 Hz to 13 MHz) and can be fully computer controlled. The determination of the dielectric properties of conducting liquids presents serious difficulties.

The impedance Z_C of the measuring cell is made of the impedance of the sample Z_S, in series with the polarization impedance of the electrodes Z_P.

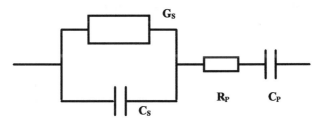

For a cylindrical cell with flat electrodes:

$$Z_C = Z_S + Z_P = \frac{4}{\pi\,D}\left[\frac{d}{\sigma\,D}\frac{1 - i\,\omega\,\varepsilon/\sigma}{1 + \left(\omega\,\varepsilon/\sigma\right)^2} + \frac{2}{D}\left(r_P - \frac{i}{\omega\,c_P}\right)\right] \qquad (1)$$

where ω is the angular frequency, D the diameter of each electrode, d the spacing between them, σ the conductivity of the liquid, ε its absolute permittivity, and r_P, c_P are the electrode polarization resistance and capacitance, respectively, corresponding to a single electrode of unit area.

If the electrodes had zero impedance, the phase angle would be

$$\operatorname{tg}\theta_S = -\omega\,\varepsilon\!/\!\sigma \approx 5\;10^{-8}\,f \qquad (2)$$

where f is the frequency, and the last term corresponds to the typical case:
$\varepsilon = 80\;\;8.85\;10^{-12}$ F/m, $\sigma = 0.1$ S/m. The resolution of the HP 4192 A for the phase angle is 0.0001 radians under optimum conditions so that, for the considered example, no permittivity measurement can be performed at frequencies lower than 2 kHz.

In the real case, taking into account the electrode impedance, eq (1) shows that:
a. Precise measurements of the sample properties can only be achieved when Z_S is greater than, or at least comparable to Z_P. Therefore, the spacing d should be as large as possible.
b. The ratio d/D cannot be made arbitrarily large because of fringing field effects. Keeping it at a constant value, the first term in eq (1) becomes independent of D. Therefore, the cell size should be as large as possible.
c. The electrode material must have the lowest possible impedance (low resistance and high capacitance). In practice this means platinum black electrodes whenever possible.
d. The electrode impedance Z_P is an unknown which depends on the electrode material and on the sample. Therefore, two or more measurements performed at different spacings are needed to determine the sample properties.
e. The electrode impedance sets a further limit on low frequency dielectric measurements as can be seen from eq (1) written for low frequencies ($\omega \ll \sigma/\varepsilon$, or $f \ll 140$ MHz for the considered example):

$$Z_C = \frac{4}{\pi\,D}\left[\frac{2\,r_P}{D} + \frac{d}{\sigma\,D} - i\left(\frac{2}{\omega\,c_P\,D} + \frac{d\,\omega\,\varepsilon}{\sigma^2\,D}\right)\right] \qquad (3)$$

The electrode properties are frequency dependent (r_P and c_P decrease with f) so that, for sufficiently low frequencies, Z_P becomes always bigger than Z_S. For f = 10 kHz and platinum black electrodes in aqueous NaCl 0.1 S/m electrolyte: $r_P \approx 10^{-4}$ Ω and $c_P \approx 1$ F [10]. Considering a cylindrical cell with d = 1 cm and D = 1.5 cm, eq (3) becomes:

$$Z_C\,[\Omega] = 1.1 + 570 - i\left[\,0.18 + 0.25\,\right] \qquad (4)$$

showing that electrode polarization starts to be a limiting factor for reactance measurements at frequencies below 10 kHz.

In view of these difficulties, every possible measure to improve the precision of the measurements must be taken into account [11-12], the most important of which is the calibration of the instrument.

INSTRUMENT CALIBRATION

The method recommended by the manufacturer consists in representing the parasitic elements as a series impedance Z_0 and a parallel admittance Y_0 connected as shown on the left:

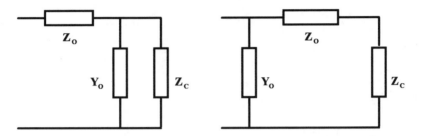

Their values are determined performing a Short and Open circuit calibration leading to the following expresion for the cell impedance:

$$Z_C = \frac{(Z_{Me} - Z_{Sh})(1 - Z_{Sh} Y_{Op})}{1 - Z_{Me} Y_0} \tag{5}$$

where Z_{Me} is the measured value whereas Z_{Sh} and Y_{Op} are the calibration results.

This calibration procedure has two major shortcomings:

a. It is based on a rather arbitrary assumption of a particular configuration of the parasitic elements. If, for instance, the right hand side configuration were used, and the same calibration performed, the expression for the cell impedance would be:

$$Z_C = \frac{Z_{Me} - Z_{Sh}}{(1 - Z_{Me} Y_{Op})(1 - Z_{Sh} Y_{Op})} \tag{6}$$

b. It assumes that the instrument itself is perfectly calibrated, so that the displayed values actually represent the impedance parameters of the equivalent circuit connected to the terminals.

The proposed way to avoid these shortcomings consists in representing the instrument as an electric quadrupole:

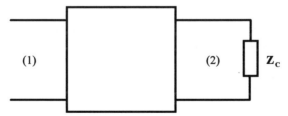

This element is characterized by four coefficients Z_{ij} which relate the voltage to current ratio $Z_i = V_i / I_i$ on its input (2) and output (1) sides:

$$(Z_1 - Z_{11})(Z_2 - Z_{22}) = Z_{12} Z_{21} \tag{7}$$

It is assumed that the system is linear, so that Z_{ij} do not depend neither on the voltages nor on the currents. Their values are obtained performing three calibration measurements:

Short circuit: $\mathbf{Z}_2 = 0$ $\mathbf{Z}_1 = \mathbf{Z}_{1Sh}$

Open circuit: $\mathbf{Y}_2 = 0$ $\mathbf{Y}_1 = \mathbf{Y}_{1Op}$

Reference: $\mathbf{Z}_2 = \mathbf{Z}_{2Re}$ $\mathbf{Z}_1 = \mathbf{Z}_{1Re}$

leading to:

$$\mathbf{Z}_{11} = 1 / \mathbf{Y}_{Op} \tag{8}$$

$$\mathbf{Z}_{22} = \frac{\mathbf{Z}_{2Re}\left(\mathbf{Z}_{1Re}\,\mathbf{Y}_{1Op} - 1\right)}{\mathbf{Y}_{1Op}\left(\mathbf{Z}_{1Re} - \mathbf{Z}_{1Sh}\right)} \tag{9}$$

$$\mathbf{Z}_{12}\,\mathbf{Z}_{211} = \frac{\mathbf{Z}_{2Re}\left(\mathbf{Z}_{1Re}\,\mathbf{Y}_{1Op} - 1\right)\left(1 - \mathbf{Z}_{1Sh}\,\mathbf{Y}_{1Op}\right)}{\mathbf{Y}_{1Op}^2\left(\mathbf{Z}_{1Re} - \mathbf{Z}_{1Sh}\right)} \tag{10}$$

The value of the cell impedance Z_C can now be related to its measured value Z_{1Me} by the following expression:

$$\mathbf{Z}_C = \mathbf{Z}_{2Re}\frac{\left(1 - \mathbf{Z}_{1Re}\,\mathbf{Y}_{Op}\right)\left(\mathbf{Z}_{1Me} - \mathbf{Z}_{1Sh}\right)}{\left(\mathbf{Z}_{1Re} - \mathbf{Z}_{1Sh}\right)\left(1 - \mathbf{Z}_{1Me}\,\mathbf{Y}_{1Op}\right)} = Q\frac{\mathbf{Z}_{1Me} - \mathbf{Z}_{1Sh}}{1 - \mathbf{Z}_{1Me}\,\mathbf{Y}_{1Op}} \tag{11}$$

A convenient reference impedance can be made using the measuring cell with a known electrolyte. This is possible because the coefficient Q in eq (11) does not depend on the spacing d of the reference cell since, from eq (9):

$$Q = -\mathbf{Z}_{22}\,\mathbf{Y}_{1Op} \tag{12}$$

The cell impedance for the reference can be written as

$$\mathbf{Z}_{2Re} = \mathbf{Z}_{PRe} + \mathbf{k}_{Re}\,\mathbf{d} \tag{13}$$

where Z_{PRe} is the electrode impedance for the cell filled with the reference electrolyte. Fitting a straight line to experimental values of the function:

$$\frac{\mathbf{Z}_{1Re} - \mathbf{Z}_{1Sh}}{\left(1 - \mathbf{Z}_{1Re}\,\mathbf{Y}_{1Op}\right)} = \frac{\mathbf{Z}_{PRe} + \mathbf{k}_{Re}\,\mathbf{d}}{Q} \tag{14}$$

obtained measuring the reference at different spacings leads to a slope equal to k_{Re}/Q, from which the coefficient Q is determined.

Analogously, the impedance of the cell can be expressed as:

$$\mathbf{Z}_C = \mathbf{Z}_{PMe} + \mathbf{k}_{Me}\,\mathbf{d} \tag{15}$$

Making a series of measurements at different spacings and fitting a straight line to the function:

$$\frac{Z_{1Me} - Z_{1Sh}}{\left(1 - Z_{1Me}\, Y_{1Op}\right)} = \frac{Z_{PMe} + k_{Me}\, d}{Q} \tag{16}$$

makes it possible to determine k_{Me} as Q multiplied by the slope.

EXPERIMENTAL RESULTS

The measured systems were based on the Basf #25237/27-1 aqueous suspension of polystyrene spherical particles with the following characteristics. Composition: 97.5% styrol and 2.5% acrylic acid. Concentration by weight: 38.3%. Average radius and standard deviation: 0.0265 and 0.007 micro meters. Conductivity: 0.690 S/m, pH: 6.5. This system was measured as is, and diluted with destilled water to 30, 20, and 10% by weight concentrations. The temperature was $25 \pm 0.1\,^{\circ}C$.

An example of our results for the real and imaginary parts of the complex conductivity:

$$\sigma^{*}(\omega) = \sigma(\omega) + i\,\omega\,\varepsilon(\omega) \tag{17}$$

are presented in the following figures, together with high frequency data measured at Goettingen by R. Pottel and co workers W. Pieper and M. Lange, using a Network Analyzer and an Interferometer.

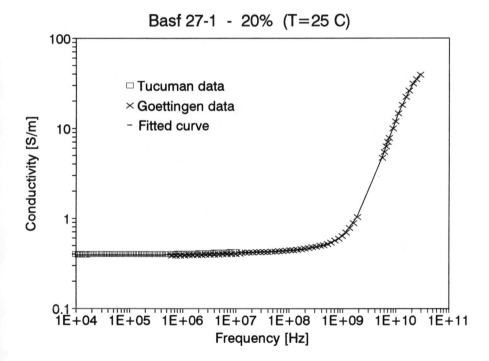

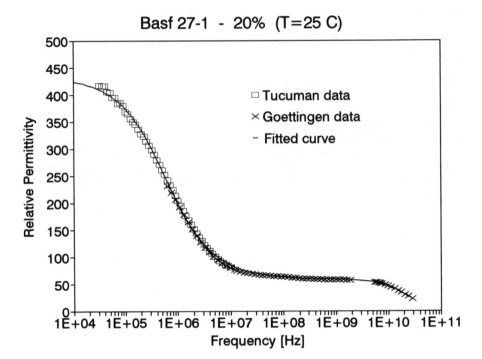

Basf 27-1 - 20% (T=25 C)

The combined data was used to perform non linnear fits to a sum of two Cole-Cole relaxation terms, by means of the Marquardt [13] algorithm. The obtained results appear on the table.

TABLE

Weight Concentration	38.3%	30%	20%	10%
$\delta\varepsilon_1/\varepsilon_0$	170	310	380	350
τ_1 [s]	$1.5\ 10^{-7}$	$2.4\ 10^{-7}$	$2.7\ 10^{-7}$	$4.1\ 10^{-7}$
α_1	0.30	0.31	0.30	0.31
$\delta\varepsilon_2/\varepsilon_0$	40	48	56	64
τ_2 [s]	$7.8\ 10^{-12}$	$8.1\ 10^{-12}$	$7.9\ 10^{-12}$	$8.2\ 10^{-12}$
α_2	0.05	0.06	0.04	0.02
$\varepsilon_\infty/\varepsilon_0$	2.5	2.7	2.9	4.2
σ [S/m]	0.69	0.58	0.39	0.19

CONCLUSION

These first results show that careful cell design combined with quadrupole calibration make it possible to perform satisfactory low frequency dielectric measurements of highly conductive systems using an Impedance Analyzer. The values obtained present an overlap in the 1 MHz to 10 MHz range with measurements made with high frequency equipment, which is sufficiently good to build a full spectrum.

The combined measurements will be pursued using systems with different particle radii and electrolyte compositions which will hopefully either permit an interpretation using existing theoretical models or provide the necessary data for their improvement.

REFERENCES

1. G. Schwarz, J. Phys. Chem. **66**, 2636 (1962).
2. J.M. Schurr, J. Phys. Chem. **68**, 2407 (1964).
3. S.S. Dukhin, V.N. Shilov, Dielectric Phenomena and the Double Layer in Disperse Systems and Polyelectrolytes, Wiley, New York (1974).
4. M. Fixman, J. Chem. Phys. **72**, 5177 (1980).
5. E.H.B. DeLacey, L.R. White, J. Chem. Soc. Faraday. Trans. 2, **77**, 2007 (1981).
6. W.C. Chew, P.N. Sen, J. Chem. Phys. **77**, 4683 (1982).
7. J. Lyklema, S.S. Dukhin, V.N. Shilov, J. Electroanal. Chem. Interfacial Electrochem. **1**, 143 (1983).
8. C. Grosse, K.R. Foster, J. Phys. Chem. **91**, 3073 (1987).
9. C. Grosse, J. Phys. Chem., **92**, 3905 (1988).
10. C. Grosse, M. Tirado, submitted to Anales AFA (1994).
11. D.E. Dunstan, L.R. White, J. Colloid. Interfac. Sci. **152**, (2), 308 (1992).
12. G. Blum, H. Maier, F. Sauer, H.P. Schwan, J. Phys. Chem. **99**, 780 (1995).
13. D.W. Marquardt, J. Soc. Industrial Applied Mathem. **11**, 431 (1963).

OPEN APPLICATOR ANALYSIS FOR MATERIAL JOINING AND DIELECTRIC MEASUREMENTS

S.P.CHEN, W.R.TINGA and F.E.VERMEULEN, E.E. Dept., University of Alberta, Edmonton, Alberta, Canada, T6G 2G7

ABSTRACT

Resonant internal and radiated external electromagnetic fields of an open coaxial structure are analyzed using FDTD. These fields are characterized for two structures, with and without a quarter-wave choke. Techniques utilized to obtain a steady state solution are discussed. Good agreement with field measurements is obtained. This moveable open applicator is convenient for creating intense hot zones as required in ceramic joining. It is also suitable for the measurement of surface dielectric properties of planar materials at room or elevated temperatures.

INTRODUCTION

To allow microwave heat treatment of large surface area materials, an open coaxial microwave structure was analyzed and designed in an earlier paper [1] but radiation losses and choke effects were not considered. To improve the model we must not only consider the internal resonator field but also its radiated field through the gap between the applicator and the treated lossy material .
For this model analysis we choose the Finite Difference Time Domain (FDTD) method which has been widely used for solving electromagnetic radiation problems and eigenvalue problems associated with resonator structures [2,3]. We apply the FDTD method using pulse excitation and Mur's Absorbing Boundary Condition (ABC). By applying the Discrete Fourier Transform (DFT) to the solution we obtain the resonant frequency and the corresponding time harmonic solution. We also obtain the time harmonic solution by applying sinusoidal excitation. The field distribution calculated for the cavity-material interaction region is compared to the results of the previously used finite element method [1] and our new experimental results. Effects of a 2450 MHz, quarter-wave choke on the energy distribution inside and outside the cavity are calculated and discussed. Further, the resonant frequency shift caused by different materials is determined so that this device can be used to measure dielectric properties of planar materials over a wide temperature range.

CAVITY ANALYSIS VIA THE FDTD METHOD

Since the open cavity, shown in Fig. 1, is rotationally symmetric, only the TM modes are

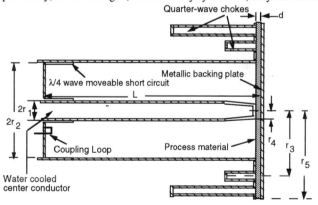

Figure 1. Cross-sectional view of the open coaxial cavity. $r_1 = 10$mm, $r_2 = 38$mm, $r_3 = 47$mm, $r_4 = 4$mm, $r_5 = 74$mm, $L = 198$mm and $d = 6$mm.

considered. The relevant form of Maxwell's equations is

$$\frac{\partial(-\mu H_\varphi)}{\partial t} = \frac{\partial(E_r)}{\partial z} - \frac{\partial(E_z)}{\partial r}$$

$$\frac{\partial(\varepsilon E_r)}{\partial t} = \frac{\partial(-H_\varphi)}{\partial z} + \sigma E_r \tag{1}$$

$$\frac{\partial(\varepsilon E_z)}{\partial t} = \frac{1}{r}\frac{\partial(r H_\varphi)}{\partial r} + \sigma E_z$$

Using a central difference scheme similar to that used by Yee [4], these can be discretized as:

$$H_\varphi^{n+\frac{1}{2}}(i,k) = H_\varphi^{n-\frac{1}{2}}(i,k) - \frac{\Delta t}{\mu \Delta z}\left[E_r^n(i,k+\tfrac{1}{2}) - E_r^n(i,k-\tfrac{1}{2})\right] + \frac{\Delta t}{\mu \Delta r}\left[E_z^n(i+\tfrac{1}{2},k) - E_z^n(i-\tfrac{1}{2},k)\right]$$

$$E_z^{n+1}(i+\tfrac{1}{2},k) = \frac{\varepsilon}{\varepsilon + \sigma \Delta t}E_z^n(i+\tfrac{1}{2},k) + \frac{\Delta t}{(\varepsilon + \sigma \Delta t)\Delta r}\cdot\frac{1}{i+\tfrac{1}{2}}\left[(i+1)H_\varphi^{n+\frac{1}{2}}(i+1,k) - iH_\varphi^{n+\frac{1}{2}}(i,k)\right] \tag{2}$$

$$E_r^{n+1}(i,k-\tfrac{1}{2}) = \frac{\varepsilon}{\varepsilon + \sigma \Delta t}E_r^n(i,k-\tfrac{1}{2}) - \frac{\Delta t}{(\varepsilon + \sigma \Delta t)\Delta z}\left[H_\varphi^{n+\frac{1}{2}}(i,k) - H_\varphi^{n+\frac{1}{2}}(i,k-1)\right]$$

where i and k are the spatial indices, n is the timestep index and σ is the conductivity of the lossy dielectric.

Using the first order Mur's approximation [5] we obtain

$$\left(\frac{\partial}{\partial n} + \frac{1}{v_p}\frac{\partial}{\partial t}\right)E_T = 0 \tag{3}$$

where $\hat{n}$ is the normal direction, E_T is the tangential electric field component on a boundary wall and v_p represents the phase velocity of the field. This equation is easily discretized using only field components on and just inside the mesh boundary, yielding

$$E_T^{n+1} = E_{T-1}^n + \frac{v_p \Delta t - \Delta n}{v_p \Delta t + \Delta n}\left[E_{T-1}^{n+1} - E_T^n\right] \tag{4}$$

where E_{T-1} represents the tangential electric field component one node from the boundary. In treating the corners, an average can be used,

$$\frac{1}{2}\left[\left(\frac{\partial}{\partial n_1} + \frac{\partial}{\partial t}\right) + \left(\frac{\partial}{\partial n_2} + \frac{\partial}{\partial t}\right)\right]E_T = 0 \tag{5}$$

where $\hat{n}_1$ and $\hat{n}_2$ are the two limiting cases for the normal directions at the corner. In this analysis, two types of excitation are used. One is a point source of a continuous sinusoidal wave of frequency f_0 [6], located near the short circuited end of the cavity, to simulate the coupling loop. By sampling the fields after the steady state has been reached, the amplitudes of the fields are obtained. This steady-state field distribution, calculated by taking the time average of the time-domain solution, A, at each mesh point, is given by [7]

$$A(i,k) = \sum_n \left|A^n(i,k)\right| \Big/ N \qquad (6)$$

where N is the total number of time steps.

A pulse is used as another source thereby exciting a large number of modes. By taking the Fourier transform of this computed time domain response, the resonant frequency can be obtained. The Fourier transform of the field shows a local absolute maximum at the frequency f_m nearest the resonant frequency f_0. Thus, the field distribution at any particular frequency is obtained by performing the Fourier transform of the time-domain solution at each grid point [8] resulting in

$$F_{f_m}(i,k) = \sum_{n=0}^{N_{DFT}-1} A^n(i,k,n\Delta t) e\left[-\frac{j2\pi mn}{N_{DFT}}\right]\Delta t \qquad (7)$$

$$f_m = \frac{m}{N_{DFT}\Delta t} \qquad m = 0,1\ldots\ldots\frac{N_{DFT}}{2} \qquad (8)$$

where A^n is the time domain solution and m is the frequency index. Once the field distribution is known, the energy distribution can be obtained from

$$W_e = \frac{1}{2}\int_v \varepsilon|E|^2 dv \qquad \text{and} \qquad W_m = \frac{1}{2}\oint_s \mu|H_t|^2 ds \qquad (9)$$

CALCULATION AND EXPERIMENTAL RESULTS

A 2-dimensional FDTD model of the cavity is used and discretized into a $91\Delta r \times 143\Delta z$ grid, including the external radiation region, where

$$\Delta r = \Delta z = 2mm \qquad \text{and} \qquad \Delta t = \frac{1}{v_p\sqrt{\left(\frac{1}{\Delta r}\right)^2 + \left(\frac{1}{\Delta z}\right)^2}}.$$

The tangential electric field components and the electromagnetic field inside the conductor are set to zero during the whole computation cycle. Other initial electric and magnetic fields are set to zero throughout the grid except at the source point. Electric fields at the interface between different materials are calculated by using an average permittivity of $(\varepsilon_0 + \varepsilon_r)/2$. For the dielectric in the gap region, the material parameters are $\varepsilon_r = 2.0$, $\sigma = 0.02$, $\mu_r = 1$. The resonant frequencies are calculated for two structures one with and one without a quarter-wave choke. For these two cases, Fig. 2 shows the calculated Fourier amplitude spectrum with resonant frequencies in the ISM band at 2.445 or 2.484 GHz and Table I lists both the measured and calculated resonant frequencies.

Finally, the Fourier coefficients associated with the resonant frequency at different points in the cavity give the spatial distribution of the electromagnetic field in the gap region with or without a choke. These results, shown in Fig. 3a and 3b, are compared with previous results obtained with the finite element method (FEM)[1]. These earlier results were obtained by setting H_ϕ on the center line of the inner choke to zero (magnetic wall).

Table I. Choke Effect On Resonant Frequency

Structure	Calculated Frequency, GHz	Measured Frequency, GHz
With choke	2.445	2.451
Without choke	2.484	2.500

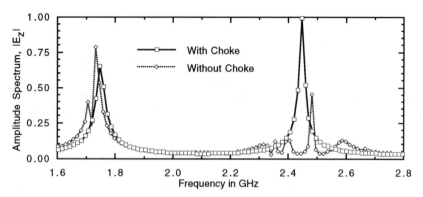

Figure 2. Amplitude spectrum of the electric field showing a resonant frequency at 2.445 or 2.484 GHz with and without a choke, respectively.

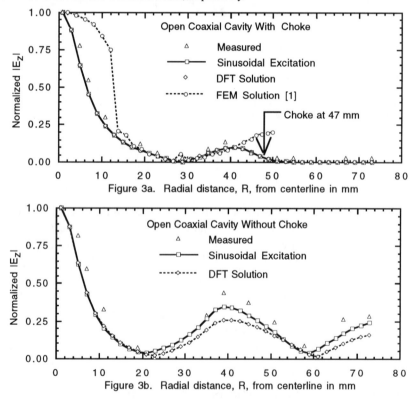

Figure 3. Internal radial field distribution of $|E_z|$ measured at the sample surface adjacent to the metallic backing plate. FEM results are for a uniform 24 mm diameter center conductor. All other results are for a center conductor tapered over a 5 cm length from 20 mm to 8 mm diameter. a) with choke at 2.451 GHz, b) without choke at 2.500 GHz.

It is clear from Fig. 3a that the FDTD results are closer to the experimental results than the FEM results. With FDTD we accounted for energy radiated via the cavity's interaction gap, but not with FEM. Measurements were made with a 0.8 mm diameter coaxial probe 1 mm long. Comparing Figures 3a and 3b clearly shows the effectiveness of the choke in blocking energy leakage to the external region. In fact, with the choke only about 3% of the energy appears outside the cavity but for the cavity without a choke, preliminary results show that most of the energy appears outside the open cavity. External field distributions are shown in Figures 4a and 4b, normalized with respect to two different values of $|E_Z|$ at the cavity's outside edge at R= 74 mm.

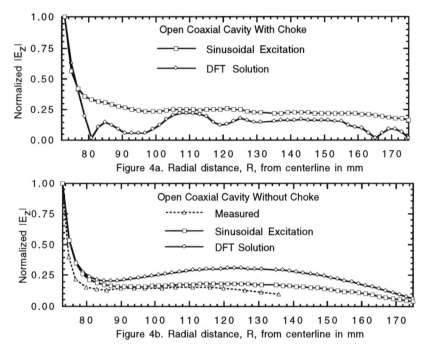

Figure 4. External radial field distribution of $|E_Z|$, measured at the same level as that used in Fig. 3. a) with choke at 2.451 GHz, b) without choke at 2.500 GHz.

The field intensity at the external cavity boundary, at R=74mm, is about 30 dB lower than the value of $|E_Z|$ in the center of the internal gap region. The high energy density in the cavity's gap region, see figure 3, will lead to high temperatures in a lossy material placed in this region.

If the material properties in this region are changed, the field and energy are changed greatly causing the resonant frequency to shift. From the calibration curve, shown in Figure 5, it is clear that measurement of the microwave dielectric constant of a material is possible by measuring the cavity's resonant frequency shift.

CONCLUSION

Internal and external electromagnetic fields of an open coaxial structure, with and without a quarter-wave choke, are calculated using FDTD and agree with field measurements. This moveable open applicator is convenient for creating intense hot zones as required in ceramic joining and for measurement of the dielectric constant of a planar material in the gap region. Use of a choke is shown to be effective in containing most of the energy in the open resonator.

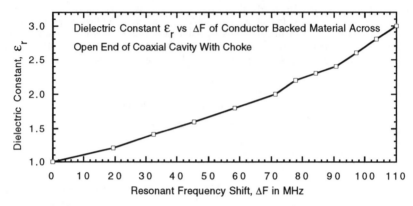

Figure 5   Dielectric constant of the material in the gap region versus the resonator's frequency shift, Δf, in MHz. Frequency resolution in the Fourier transformation was limited to 6.5 MHz.

ACKNOWLEDGEMENTS

This work was supported in part by grants from the Natural Science and Engineering Research Council of Canada.

REFERENCES

1. W.R. Tinga, J.D. Xu, and F.E. Vermeulen, "Open Coaxial Microwave Spot Joining Applicator", Ceramic Trans. vol. 59, pp 347-356, 1995.
2. A. Navarro, M.J. Nunez, and E. Martin, " Study of TE_0 and TM_0 Modes In Dielectric Resonators By A Finite Difference Time-Domain Method Coupled With the Discrete Fourier Transform", IEEE Trans. Microwave Theory Tech., vol MTT-39, pp 14-17, Jan. 1991.
3. J. Litva, Z. Bi, K. Wu, R. Fralich and C. Wu, " Full-wave Analysis of an Assortment of Printed Antenna Structures Using the FD-TD method", in IEEE AP-S Int. Symp. Dig., pp 410-413, June 1991.
4. K.S. Yee, "Numerical Solution of Initial Boundary Value Problems Involving Maxwell's Equations in Isotropic Media", IEEE Trans. Antenna Propagat., vol AP-14, pp 302-307, May, 1966.
5. G. Mur, "Absorbing Boundary Conditions for the Finite Difference Approximation of the Time-Domain Electromagnetic Field Equations", IEEE Trans. Electromagn. Compat., vol. EMC-23, no. 4 pp 377-382, Nov., 1981.
6. A. Taflove, in Computational Electrodynamics: the Finite Difference Time-Domain Method, Artech House, Boston, London, 1995.
7. D.H. Choi and W.J.R. Hoefer, "The Finite Difference - Time-Domain Method and Its Application To Eigenvalue Problems", IEEE Trans. Microwave Theory Tech., vol MTT-34, pp 1464-1469, Dec. 1986.
8. M.F. Catedra, R.P. Torres, J. Basterrechea and E. Cago, in The CG-FFT Method: Application of Signal Processing Techniques To Electromagnetics, Artech House, Boston, London, 1995.

ESTIMATION OF COMPLEX PERMITTIVITY OF PRINTED CIRCUIT BOARD
MATERIAL USING WAVEGUIDE MEASUREMENTS

C. J. REDDY*, M. D. DESHPANDE** AND G. A. HANIDU*
* Department of Electrical Engineering, Hampton University, Hampton VA 23668
** ViGYAN Inc., Hampton VA 23681

ABSTRACT

A simple waveguide measurement technique is presented to determine the complex permittivity of printed circuit board material. The printed circuit board with metal coating removed from both sides and cut into size which is the same as the cross section of the waveguide is loaded in a short X-band rectangular waveguide. Using a network analyzer, the reflection coefficient of the shorted waveguide(loaded with the sample) is measured. Using the Finite Element Method(FEM) the exact reflection coefficient of the shorted wavguide(loaded with the sample) is determined as a function of dielectric constant. Matching the measured value of the reflection coefficient with the reflection value calculated using FEM and utilizing Newton-Raphson Method, an estimate of the dielectric constant of a printed circuit board material is obtained. A comparison of estimated values of permittivity constant obtained using the present approach with the available data.

INTRODUCTION

Due to the revolution in communication technology in the recent past, there is an increasing use of high frequency circuits for various applications. The design of printed circuits at high frequencies require the knowledge of the dielectric properties of the material used in manufacturing the printed circuit board. Over the years many methods have been developed and used for measuring permittivity $(\varepsilon_r', \varepsilon_r'')$ and permeability (μ_r', μ_r'') of materials [1]. The most accurate measurement at high frequency can be done using the high Q resonant cavity technique [2]. However, the main disadvantage of the cavity method is that the measured results are applicable only over a narrow frequency band[2]. Electric properties of material over wide range of frequencies can be done with less accuracy using the transmission line methods. The material sample used in these measurements is usually of a cross section which is the same as that of the transmission line. However, when the sample selected is not of uniform cross section or the sample occupies only a part of transmission line cross section, then the complete modal analysis is required to measure accurately the material properties. The complete modal analysis is quite complicated, if not impossible. In such cases, when the sample cross section is different from that of the transmission line, a numerical method such as the FEM instead of the modal analysis is much easier to implement to obtain material properties.

In this paper, FEM is used to estimate complex permittivity of material using a terminated rectangular waveguide. The material sample of specific length but of arbitrary cross section is assumed to be present in a shorted rectangular waveguide. The reflection coefficient at some arbitrary selected reference plane in the rectangular waveguide is measured at a given frequency. Since only determination of permittivity is required, a single reflection coefficient measurement suffices. The reflection coefficient at the given frequency is also calculated as a function of $(\varepsilon_r', \varepsilon_r'')$ using the FEM. From the calculated and measured values of reflection coefficients and the use of Newton-Raphson Method [3], the complex permittivity of the given material is then

301

determined. The complex permittivity of Teflon, obtained using the present techniques are compared with the values obtained using the standard software available with the HP-8510c Network Analyzer [4,5].

THEORY

(a) Direct Problem

In this section the FEM is used to determine the reflection coefficient of a short circuited rectangular transmission line loaded with an arbitrary shaped dielectric material. Figure 1 shows a terminated rectangular waveguide with a dielectric sample of arbitrary cross section. It is assumed that the waveguide is excited by a dominant TE_{10} mode from the right and the reflection coefficient is measured at the reference plane P_1 as shown in figure 1(a). For the purpose of analysis the problem is divided into two regions: Region I ($z < 0$) and Region II ($z > 0$). Using the waveguide vector modal functions, the transverse electromagnetic fields in the region I are expressed as a sum of incident field and reflected fields. The reflection coefficients can be found using orthogonality properties of the modes[6].

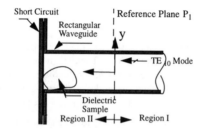

(a) Longitudinal view of rectangular waveguide
loaded with a dielectric sample.

(b) Cross sectional view of rectangular waveguide

Figure 1 Geometry of rectangular waveguide excited by TE_{10} mode

The electromagnetic field inside Region II is obtained using the FEM formulation [7]. The electric field inside the Region II satisfies the vector wave equation. Following the procedure described in [7] and [8] and using vector edge basis functions a matrix equation can be written as[5]

$$[S] \cdot [b] = [v] \qquad (1)$$

The solution vector $[b]$ of the matrix equation (1) is then used to determine reflection coefficient at the reference plane P_1.

(b) Rectangular Waveguide Measurement System

The measurement system for single port waveguide measurements is shown in figure 2. The reflection coefficient a_0' due to a terminated rectangular waveguide loaded with a given material sample piece can be measured using the procedure described in[4].

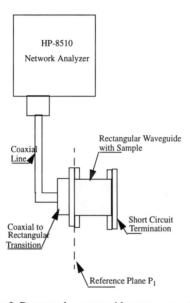

Figure 2 Rectangular waveguide measurement system.

(c) Inverse Problem

This section presents computations of complex dielectric constant of a given sample piece from a one port measurement of the reflection coefficient. From the given geometry of the sample and its position in the terminated rectangular waveguide, the reflection coefficient $a_0 (\varepsilon_r', \varepsilon_r'')$ is calculated using the FEM for assumed values of $(\varepsilon_r', \varepsilon_r'')$. If a_0' is the measured reflection coefficient then the error in the calculated value of reflection coefficient is $a_0 (\varepsilon_r', \varepsilon_r'') - a_0'$. Writing the error in real and imaginary part we get

303

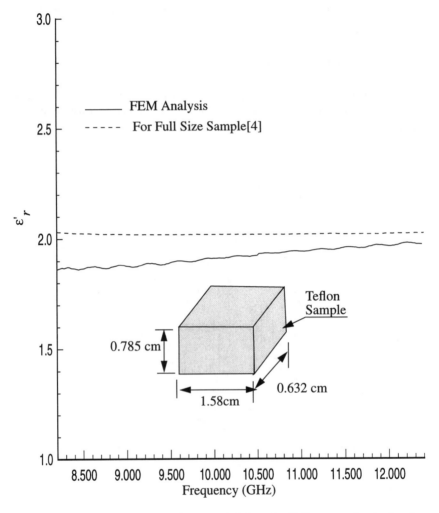

Figure 3   Real part of complex permittivity of a Teflon sample obtained
using undersize sample.

$$f_1(\varepsilon_r', \varepsilon_r'') = \text{real}(a_0(\varepsilon_r', \varepsilon_r'') - a_0') \tag{2}$$

$$f_2(\varepsilon_r', \varepsilon_r'') = \text{imag}(a_0(\varepsilon_r', \varepsilon_r'') - a_0') \tag{3}$$

If $(\varepsilon_r', \varepsilon_r'')$ are incremented by small values to $(\varepsilon_r' + d\varepsilon_r', \varepsilon_r'' + d\varepsilon_r'')$ such that $f_1(\varepsilon_r' + d\varepsilon_r', \varepsilon_r'' + d\varepsilon_r'')$ and $f_2(\varepsilon_r' + d\varepsilon_r', \varepsilon_r'' + d\varepsilon_r'')$ are simultaneously zero. The correct value of $(\varepsilon_r', \varepsilon_r'')$ is finally arrived by an iterative procedure described in [8].

NUMERICAL RESULTS

To validate the analysis carried out in this paper, first an arbitrary shaped sample of Teflon with dimensions 1.58cmX0.785cmX0.632cm is placed in a rectangular waveguide for measurement of reflection coefficient. From the measurement of reflection coefficient and following the procedure described in the above section, the complex permittivity is calculated and presented in figure 3. For comparison, the complex permittivity obtained using full size (2.29cmX1.0cmX0.947cm) is also presented in figure 3. From figure 3, it can be concluded that FEM procedure can be used to determine the complex permittivity of dielectric material using an arbitrarily shaped sample.

CONCLUSIONS

FEM has been successfully used to characterize dielectric materials using waveguide reflection measurements. This method is very useful when the dielectric sample to be characterized is available only in an arbitrary shape. The numerical data presented in this papers validate the analysis presented.

REFERENCES

[1] James Baker-Jarvis, Dielectric and magnetic measurement methods in transmission lines: an overview, (Proceedings of the 1992 AMTA Workshop, 1992).

[2] Leo P. Ligthart, IEEE Trans. on MTT, **31**, 249-254 (1983).

[3] W. H. Press, Numerical recipes, The art of scientific computing (Fortran version), (Cambridge University Press, Cambridge, 1989) Chap. 9.

[4] R. L Cravey et al, NASA TM 110147, April 1995.

[5] Dielectric material measurement forum, (Hewlett Packard, 1993).

[6] R. F. Harrington, Time-harmonic electromagnetic fields, (McGraw-Hill Book Company, New York, 1961).

[7] C. J. Reddy, M.D. Deshpande, C.R. Cockrell and F.B.Beck, NASA TP 3485, December 1994.

[8] M. D. Deshpande, C. J. Reddy, NASA CR 198203, August 1995.

Part VI

Modeling of Microwave Heating

AN OVERVIEW OF ELECTROMAGNETIC MODELING OF SINGLE- AND MULTI-MODE APPLICATORS AT EMR MICROWAVE TECHNOLOGY CORPORATION

J.M. TRANQUILLA, H.M. AL-RIZZO, AND K.G. CLARK
EMR Microwave Technology Corporation
64 Alison Blvd., Fredericton, New Brunswick, Canada E3C 1N2

ABSTRACT

This paper examines the design aspects and numerical modeling of single- and multi-mode, high power, microwave resonant cavity applicators. A novel numerical electromagnetic modeling approach, based on the finite difference solution of Maxwell's equations in the time domain, is utilized taking into account the finite conductivity and thickness of the cavity walls. Results obtained from the model enable an unambiguous choice of the optimum iris size for critical coupling; shape, dimension, orientation and number of radiating elements for a prescribed irradiation pattern of slotted waveguide feeds as well as dimensions and excitation mechanism of multi-mode cavities. The model also allows the evaluation of the volumetric distribution of the microwave energy at any point inside the load and may eventually be used to determine the dynamic transient temperature profiles induced during the curing process.

INTRODUCTION

Rapid heating rates and high processing speed requirements in industrial microwave heating applications dictate efficient applicator designs exhibiting uniform and high energy densities. Various factors need to be considered in a normally costly and time consuming, trial-and-error, experimental process due to the several complex and interacting physical mechanisms involved. It is therefore important to devise a sophisticated predictive microwave modeling capability in order to guide and optimize full-scale industrial designs and to ensure that most of the available microwave energy is efficiently and appropriately coupled to the load.

Our purpose here is to report on current research and development at EMR Microwave Technology Corp. which utilizes, among other software tools, the Finite Integration Technique (FIT) for the 3-D modeling and innovative design of novel and versatile types of multislotted, multimode circular and rectangular as well as single-mode resonant microwave applicators. It is believed that such an advanced expertise in the numerical electromagnetic aspects of microwave heating is currently lacking and urgently needed in industrial applications.

Mat. Res. Soc. Symp. Proc. Vol. 430 © 1996 Materials Research Society

NUMERICAL RESULTS

Slotted Waveguide Applicators

As far as design and analysis of general purpose, multislotted waveguide applicators of arbitrary length is concerned, our work has been directed toward optimizing the shape, number, location, orientation and distribution of the individual radiating apertures to produce a high coupling efficiency and a prescribed energy distribution under a variety of loading conditions.

Figure 1 depicts the slot distribution for a 5-foot section of a WR-975 waveguide with seven rectangular apertures, positioned in the broad face, the length of each being in the neighborhood of one-half free-space wavelength. The width of the individual slot, which is taken as 2 cm, is chosen such that breakdown problems are avoided. The structure is excited by the dominant TE_{10} mode at 915 MHz with an incident power of 10 kW at one end and terminated by a short circuit on the other. The inclusion of the two transverse slots ensures that 95.25% of the input power has been effectively coupled into the load.

The eigensolver employs 17x11 orthogonal mesh lines in Cartesian coordinates along the wide and narrow dimensions of the waveguide cross section, respectively. A total of five eigenmodes was evaluated and used in the 3-D time-domain solver as incoming time-harmonic, boundary-condition fields. A total of 51x44x92 mesh cells is used along the x, y and z directions. The induced surface current density, $\hat{n} \times H$, near the sixth and seventh slots is shown in Figure 2, where $\hat{n}$ is a unit vector normal to the surface of the guide and H is the magnetic field intensity vector. In Figure 3 the magnitude of the electric field intensity vector inside the applicator is displayed along the center line of the guide broad face. Finally, the near-field behavior of the exterior electric field intensity vector is shown for two plane cuts, across and along the waveguide axis, in Figures 4 and 5, respectively.

Cylindrical Multimode Cavity Applicator

A cylindrical microwave applicator, 1.5 m in diameter and 0.56 m in height, filled with a representative load of permittivity, $\varepsilon_r = 5$ up to a height of 0.5 m is used to carry out the simulation. The problem space is modeled by a total of 65 x 33 x 85 mesh cells along the x, y and z directions, respectively. Energy is coupled to the cavity via the slotted waveguide section, described in the previous section, placed horizontally along the top of the cylinder and energized by means of a 50kW, 915MHz dominant TE_{10} mode. The geometry of the applicator is shown in Figure 6. Figure 7 displays the absorbed power density (W/m^3) in a plane located at a depth of 5 cm from the top layer of the sample. Finally, in Figure 8 the power density distribution is shown in a plane cut along the center line of the slotted-waveguide section.

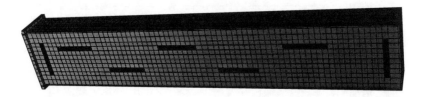

Figure 1: Geometry for the multislotted rectangular waveguide applicator

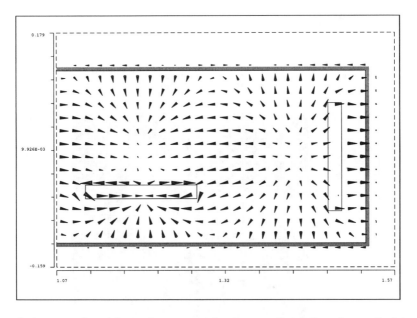

Figure 2: A vector plot of the surface current desnity near the sixth and seventh slots of the applicator shown in figure 1.

Guide Axis E-Field

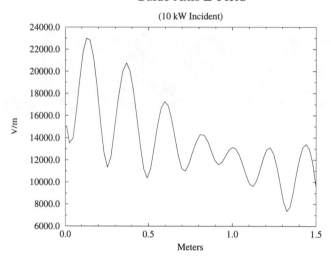

Figure 3: Magnitude of the electric field intensity inside the applicator of Figure 1 evaluated at the center line

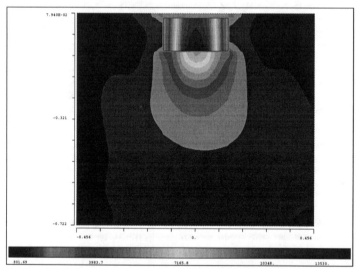

Figure 4: Near-field characteristics of the applicator shown in Figure 1 displayed in terms of the magnitude of the total electric field in a plane normal to the axis of the waveguide

312

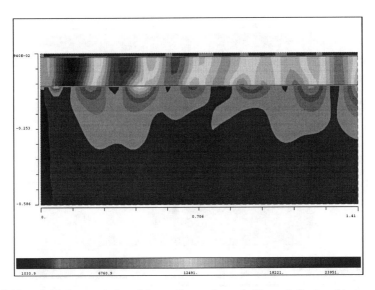

Figure 5: Near-field characteristics of the applicator shown in Figure 1 displayed in terms of the total electric field in a plane along the axis of the waveguide

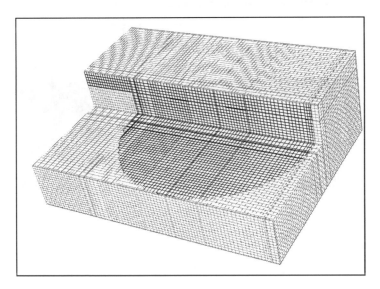

Figure 6: Geometry for the cylindrical multimode cavity applicator with the slotted waveguide feed structure symmetrically oriented on the top

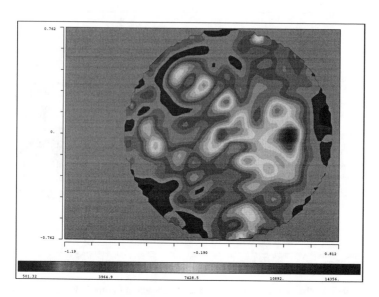

Figure 7: A cross-sectional view illustrating the power density distribution in W/m³ in a plane located at a depth of 5 cm from the top of the sample

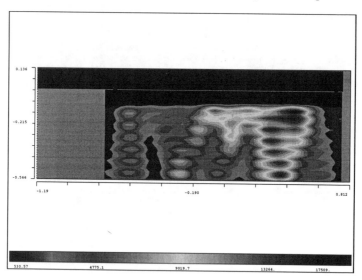

Figure 8: A longitudinal cross-section view illustrating the power density distribution in W/m³ in a plane located at the center line of the feed waveguide

314

Single-Mode Cavity Resonators

The resonant cavity consists of a straight section of WR-975 waveguide terminated by a computer-controlled, adjustable short circuit. Microwave energy is introduced via a circular coupling iris located at the center of the transverse wall. The sample is introduced via two choke tubes passing through the broad wall and located at a point of maximum electric field strength. A 10-inch long circular tube of 2.75-inch inner diameter ensures a worst case microwave leakage of 0.75 μW when the generator operates at its rated 50 kW output power. Critical coupling of the empty cavity occurred at an aperture of 2.45 inches in diameter with S_{11}=-15.75 dB. Due to symmetry, perfect magnetic and electric sheets were placed along and across the center of the broad wall, respectively, resulting in only a quarter section, the geometry of which is shown in Figure 9. The total computational domain is modeled using 500,000 mesh cells with the simulation conducted for iris diameters ranging from 4.5 to 8 cm. Calculations were performed for the unloaded cavity and the actual field strength inside the cavity is shown in Figure 10. For the TE_{103} mode both measured and computed results indicate an optimum iris diameter of about 6.25 cm. As the cavity could be exposed to substantial power levels, the potential for dielectric breakdown, as well as the amount of attenuation provided by the choke tube, should be considered. Figure 11 depicts a contour plot of the electric field intensity near the intersection of the sample-introduction tube and the guide wall, normalized to its peak value in the cavity. A plot of the field along the choke tube is shown in Figure 12 drawn from the center of the guide to the region outside the cavity.

Figure 9: Geometry for the single-mode rectangular waveguide applicator

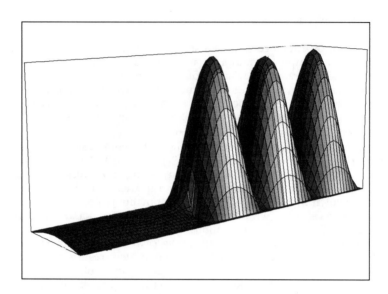

Figure 10: Three-dimensional electric field distribution for the empty cavity

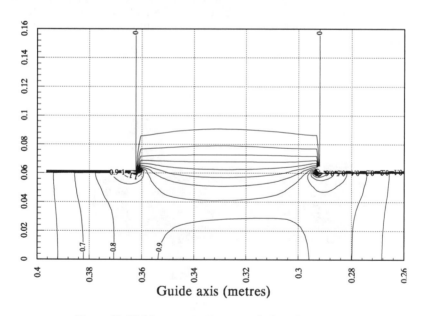

Guide axis (metres)

Figure 11: Field concentration near choke tube corners

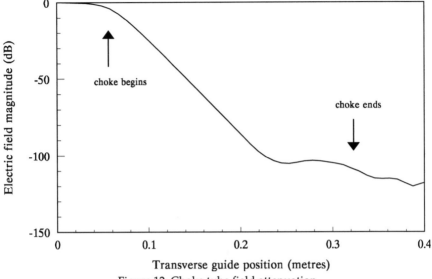

Figure 12: Choke tube field attenuation

SUMMARY

This paper presents a few cases of focused research effort at EMR Microwave Technology Corp. to model, design and optimize various high-power microwave applicators for potential new applications that enhance the extraction efficiency and processing costs of minerals, petroleum and forest products. A wide range of traveling wave, single and multimode heating applicators as well as fluidized-bed reactors has been analyzed. The existence of such modeling capabilities is urgently needed in industry to explore the performance of new applicator geometries, predict and control processing conditions and reduce the empiricism of experimental work. It is believed that the developed simulations, which can provide precise quantitative data on absorbed power, scattering parameters, eigenmodes, quality factors and energy density patterns, taking into account the finite conductivity and thickness of waveguide walls is the most accurate available to date.

A HYBRID NUMERICAL METHOD FOR MODELING MICROWAVE SINTERING EXPERIMENTS

C. V. Hile and G. A. Kriegsmann
Department of Mathematics
Center for Applied Mathematics and Statistics
New Jersey Institute of Technology
Newark, NJ 07102

ABSTRACT

We describe a hybrid numerical method for modeling the electromagnetic interaction of a low-loss material in a single-mode waveguide applicator. The method we propose utilizes a combination of asymptotic and numerical techniques. The interaction between the applicator and the electromagnetic fields is described using scattering matrix theory and the interaction between the electromagnetic fields and the ceramic is determined numerically. Several simulations are presented to show the accuracy and simplicity of this method along with the relatively small amount of computer resources it requires.

INTRODUCTION

Over last several the years, the use of microwave energy to sinter or join a wide variety of ceramic materials has become an important technology. However, this technology can not be utilized to its full potential without a more fundamental understanding of the complicated interaction between the waveguide applicator, the electromagnetic fields and the ceramic. Currently, the modeling of the interaction between the waveguide applicator, the electromagnetic fields, and the ceramic focuses on full numerical simulations of the heating process [1]-[3]. The numerical technique used most frequently is the finite difference time domain (FDTD) method [4]. The convergence rate of this technique depends upon the amount of electromagnetic energy absorbed by the ceramic material and upon the time required for the electromagnetic fields to radiate out of the waveguide applicator. Consequently, this technique works effectively for lossy ceramics, which readily absorb electromagnetic energy, and for low Q cavities, where the transient electromagnetic fields radiate out of the cavity rather quickly.

When the ceramic is a low-loss material, only a small amount of electromagnetic energy is absorbed. Consequently the cavity must have a high Q to allow the electromagnetic fields to build up sufficient strength to heat the ceramic. This adversely affects the convergence rate of the FDTD method and dramatically increases the convergence time since the electromagnetic fields become trapped within the cavity and the transients are forced to linger for many periods before radiating out of the structure. Thus, although the FDTD method provides a detailed description of the heating process, it can require extensive computer resources. This is especially true when parameter studies are needed to deduce trends and functional relationships.

In this paper we develop a hybrid numerical method for describing the interaction of an electromagnetic wave with a low-loss ceramic material in a high Q cavity. The method we propose utilizes a combination of asymptotic and numerical techniques: the asymptotic aspects are based upon scattering matrix theory, which allows the evanescent modes excited by the iris and ceramic material to be neglected, and the numerical aspects are based upon the FDTD method. The resultant hybrid method is much faster than a straightforward application of the FDTD method.

This hybrid method assumes a prior knowledge of the resonant frequency of the cavity, taking into account the effect of the sample and the dimensions of the cavity and iris. This resonant frequency is obtained at the outset by applying the FDTD mehtod throughout the entire waveguide applicator as described in reference [1]. Specifically, the cavity is excited at a frequency equal to the resonant frequency of the empty cavity and the period of the time harmonic steady state solution is used to calculate the resonant frequency of the loaded cavity.

FORMULATION

A ceramic sample of arbitrary cylindrical shape occupies a portion of a TE_{103} waveguide. The applicator is comprised of a waveguide, a symmetric iris and a movable back wall, called a short. The cavity and its load are excited by the incident TE_{10} mode

$$\mathbf{E}_{inc} = E_0 e^{iK_1 Z} \sin(\pi X/W)\hat{\mathbf{Y}} \tag{1}$$

where E_0 is the strength of the incident mode, $K_1 = \sqrt{\omega^2/c^2 - \pi^2/W^2}$ is the wave number of the propagating mode, and W is the width of the guide.

The time harmonic electromagnetic fields in the applicator are described by Maxwell's equations, where the effective permittivity, ϵ, is a function of position. Outside the sample $\epsilon = \epsilon_0$, and inside the sample $\epsilon = \epsilon_R + i\epsilon_I$ where the imaginary part takes the electrical losses of the material into account. The electromagnetic fields satisfy boundary conditions which follow from Maxwell's equations and from the assumption that the waveguide walls and iris are perfectly conducting. In addition, the tangential components of the electric and magnetic fields must be continuous across the surface of the sample.

The waveguide feeding the cavity is assumed to support only a TE_{10} mode. Thus, far away from the iris ($Z << 0$) the electric field takes the asymptotic form

$$\mathbf{E} = E_0[e^{iK_1 Z} + R_0 e^{-iK_1 Z}]\sin(\pi X/W)\hat{\mathbf{Y}} \tag{2}$$

where R_0 is the amplitude of the reflected mode.

For convenience in the analysis which follows, we introduce the dimensionless variables and parameters,

$$\tilde{\mathbf{E}} = \mathbf{E}/E_0, \quad \tilde{\mathbf{H}} = \mathbf{H}/(\sqrt{\epsilon_0/\mu_0}\, E_0), \quad (x, y, z) = (X, Y, Z)/W,$$

$$(l, h, a) = (L, H, A)/W, \quad t = \omega T, \quad k_1 = K_1 W.$$

The problem now at hand is to find $\tilde{\mathbf{E}}$ and $\tilde{\mathbf{H}}$ that satisfy Maxwell's equations in the region $(-\infty < z < l)$ and satisfy the asymptotic expression in equation 2 in the region far away from the iris $(z \ll 0)$.

ASYMPTOTIC METHOD

The basic idea behind the analysis that follows is to mathematically remove the iris from the problem. We do this by applying scattering matrix theory to describe the interactions of the iris with the incident electromagnetic field and with the electromagnetic field reflected by the ceramic material and the short [5]. This theory is asymptotic in character, as it neglects the evanescent modes excited by the iris and ceramic material.

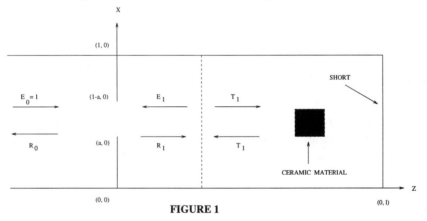

FIGURE 1

In Figure 1, we let E_j and R_j denote the amplitudes of the incident and reflected modes, respectively, for $j = 0, 1$. Note that $E_0 = 1$ with the present scaling. These amplitudes are related by the linear equations

$$R_0 = r_1(a) + (1 + r_1(a))E_1, \tag{3a}$$

$$R_1 = (1 + r_1(a)) + r_1(a)E_1, \tag{3b}$$

where $r_1(a)$ is the reflection coefficient for an iris in an infinite and homogeneous waveguide. This reflection coefficient depends upon the iris height a and the source frequency ω. Approximations for r_1 are given in many standard texts [6].

The amplitude T_1 of the propagating mode that impinges upon the ceramic material, is calculated using the connection equations, $T_1 = R_1$ and $E_1 = \gamma_1 T_1$, along with equation (3). This yields

$$T_1 = \frac{1 + r_1(a)}{1 - r_1(a)\gamma_1}.$$ (4)

Similarly the amplitude of the reflected mode is given by

$$R_0 = r_1(a) + (1 + r_1(a))\gamma_1 \left[\frac{1 + r_1(a)}{1 - r_1(a)\gamma_1} \right],$$ (5)

which shows the explicit dependence of T_1 and R_0 upon the iris height, the amplitude of the incident mode, and γ_1. The reflection coefficient γ_1 is the amplitude of the reflected wave caused by a unit incident field upon the ceramic material and the short **without the iris**.

According to this scattering matrix approximation, the ceramic material is essentially irradiated by a TE_{10} mode with strength T_1 producing a reflected wave of strength $\gamma_1 T_1$. The effects of the iris are contained in T_1 and R_0. We now define

$$\mathbf{e} = \tilde{\mathbf{E}}/T_1, \quad \text{and} \quad \mathbf{h} = \tilde{\mathbf{H}}/T_1,$$ (6)

which satisfy Maxwell's equations. The boundary conditions on the lateral walls of the waveguide, on the short, and across the ceramic material are the same as before. However, we now have for $z \ll 0$

$$\mathbf{e} = [e^{ik_1 z} + \gamma_1 e^{-ik_1 z}] \sin(\pi x)\hat{\mathbf{y}}.$$ (7)

The problem now at hand is to solve Maxwell's equations for $\mathbf{e}$ and $\mathbf{h}$ throughout the entire waveguide without the iris. Once this is done, γ_1 is deduced from equation (7) and T_1 is obtained from equation (4). The electric field $\tilde{\mathbf{E}}$ is then given by equation (6). Finally, the reflection coefficient R_0 is given by equation (5).

NUMERICAL METHOD

We use the finite difference time domain (FDTD) method to solve Maxwell's equations throughout the entire waveguide without the iris. The electromagnetic fields are written as

$$\begin{aligned} \mathbf{e} &= E_y(x, z, t)\mathbf{y}, \\ \mathbf{h} &= H_x(x, z, t)\mathbf{x} + H_z(x, z, t)\mathbf{z}. \end{aligned}$$

At the left boundary $z = -z_\infty$, the incident mode is prescribed as [7]

$$\frac{\partial E_y}{\partial z} - k_1 \frac{\partial E_y}{\partial t} = 2ik_1 e^{-i(k_1 z_\infty + t)} \sin(\pi x).$$ (8)

This formulation allows incident waves to propagate to the right (into the computational domain) and annihilates waves propagating to the left (out of the computational domain). The size of z_∞ is chosen to ensure that the ceramic material is far enough away from the left boundary that the evanescent modes excited by the ceramic material can be neglected.

On the surface of the ceramic, the boundary conditions require that the tangential components of the electric and magnetic fields be continuous. We satisfy these conditions by using average values for the permittivity on the boundary points of the ceramic.

There are two possible ways of handling the short at $z = l$. The simplest is to prescribe $E_y = 0$ there. However, we can take this condition into account and make the computational domain smaller by setting

$$\frac{\partial E_y}{\partial z} + ik_1 \cot[k_1(l - z)]\frac{\partial E_y}{\partial t} = 0. \tag{9}$$

at $z = z_r$ [5]. The size of z_r is chosen to insure that the evanescent modes excited by the ceramic material can be neglected. Using these equations and boundary conditions, the electromagnetic fields are updated recursively in time until a time harmonic steady state solution is reached.

SIMULATIONS

To demonstrate the accuracy and computational efficiency of this hybrid numerical method, we consider a TE_{103} waveguide whose cavity contains a low-loss slab which covers the entire waveguide cross section. The incident TE_{10} mode is prescribed at the left boundary according to equation (9) and is located at $z = 0$. The right boundary, given by equation (9), is located at $z = l$. The dimensions of the cavity are $L = 220.71$ mm ($l = 2.592$) and $W = 109.22$ mm. This value for the length corresponds to a resonate cavity with an aperture opening of 17.5 mm ($a = 0.42$). The source frequency corresponds to the industrial microwave frequency, $f = 2.45 \times 10^9$ GHz. The numerical space step is $\Delta = l/600$ and the time step is $\Delta t = k\Delta/\sqrt{2}$. Figure 2 shows the maximum amplitude of the electric field for three different simulations of a cavity which contains varying degrees of loss ($\epsilon_I/\epsilon_0 = 0.001$, $\epsilon_I/\epsilon_0 = 0.01$, and $\epsilon_I/\epsilon_0 = 0.1$). The convergence times (in CPU minutes on a HP 735 workstation) for these simulations are 305 min, 160 min, 60 min. The convergence times for the simulation with $\epsilon_I/\epsilon_0 = 0.01$ obtained by a straightforward application of the FDTD method throughout the entire waveguide applicator (including the iris) is 900 min.

FIGURE 2

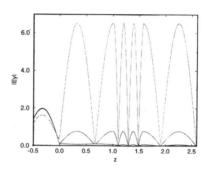

CONCLUSIONS

A hybrid numerical method was developed for describing the interaction of an electro-magnetic wave with a low-loss ceramic material in a high Q cavity. The method utilizes a combination of asymptotic and numerical techniques. The resultant hybrid method is much faster than a straightforward application of the FDTD method. It is worthwhile to point out that this hybrid method is not bound to the FDTD method. Other numerical methods, such as finite element methods or any elliptic solver, can be incorporated into this hybrid method.

ACKNOWLEDGEMENTS

This work was supported by grants from the Air Force Office of Scientific Research (Grant No. 94-1-0338), the National Science Foundation (Grant No. DMS-9407196 and Grant No. DMS-9510492), and the New Jersey Institute of Technology (Grant No. 421070).

REFERENCES

1. B. Chapman, M.F. Iskander, R.L. Smith, and O.M. Andrade, Microwave Processing of Materials III, MRS Symposium Proceedings, **269**, 53 (1992).
2. J. Wei, M. Hawley and J. Asmussen, J. Micro. Power, **28** (4), 234 (1993).
3. M.F. Iskander, IEEE Trans. Micro. Theory, **42** (5), (1994).
4. K.S. Yee, IEEE Trans. Ant. Prop., **AP-14**, 302 (1966).
5. G.A. Kriegsmann, SIAM J. Appl Math., to appear.
6. L. Lewin, Theory of Waveguides, (John Wiley, 1975).
7. G.A. Kriegsmann, SIAM J. Sci. Stat. Comput., **3** (3), 318 (1982).

DEVELOPMENT OF A MULTI-GRID FDTD CODE FOR THREE-DIMENSIONAL SIMULATION OF LARGE MICROWAVE SINTERING EXPERIMENTS

Mikel J White[1], Magdy F. Iskander[1], and Hal D. Kimrey[1]
[1]University of Utah, Electrical Engineering Department, Salt Lake City, UT 84112
[2]Oak Ridge National Lab, Oak Ridge, TN 37831

ABSTRACT

The Finite-Difference Time-Domain (FDTD) code available at the University of Utah has been used to simulate sintering of ceramics in single and multimode cavities, and many useful results have been reported in literature [1-4]. More detailed and accurate results, specifically around and including the ceramic sample, are often desired to help evaluate the adequacy of the heating procedure. In electrically large multimode cavities, however, computer memory requirements limit the number of the mathematical cells, and the desired resolution is impractical to achieve due to limited computer resources. Therefore, an FDTD algorithm which incorporates multiple-grid regions with variable-grid sizes is required to adequately perform the desired simulations. In this paper we describe the development of a three-dimensional multi-grid FDTD code to help focus a large number of cells around the desired region. Test geometries were solved using a uniform-grid and the developed multi-grid code to help validate the results from the developed code. Results from these comparisons, as well as the results of comparisons between the developed FDTD code and other available variable-grid codes are presented. In addition, results from the simulation of realistic microwave sintering experiments showed improved resolution in critical sites inside the three-dimensional sintering cavity. With the validation of the FDTD code, simulations were performed for electrically large, multimode, microwave sintering cavities to fully demonstrate the advantages of the developed multi-grid FDTD code.

INTRODUCTION

In the past few years there have been reports of benefits associated with sintering ceramics using microwaves [5-7]. With the significant interest and continued publication of results inferring advantages of using microwave sintering rather than conventional sintering of ceramics, there has been a significant need to improve modeling of microwave-sintering processes. Various geometrical arrangements and material-compatibility aspects are often involved in sintering ceramics using the new technology for microwave processing of materials. To help address these issues, microwave sintering in single- and multi-mode microwave cavities was recently simulated using the FDTD technique [1-4].

In previous simulations of microwave sintering experiments, however, regions of interest (e.g., samples, SiC rods or plates, etc.) often occupy a limited volume in the FDTD computation domain. The available FDTD code utilizes uniform cell sizes which restricts the ability to refine the mesh to the desired dimensions due to computational memory requirements and excessive execution times. Due to this limitation, the ability to accurately calculate the field distribution in the sample, surrounding insulation, and in the SiC rods when hybrid heating is used, becomes computationally limited. To overcome this difficulty, and in particular to achieve high accuracy in the region of interest with a reasonable memory and computational run time requirements, we developed an FDTD code which incorporates multi-grid regions with a variable-grid size. This code makes it possible to simulate large models with localized structures, thus allowing the concentration of FDTD

Mat. Res. Soc. Symp. Proc. Vol. 430 © 1996 Materials Research Society

cells in regions of interest such as those around the sample, the SiC rods, and the plates of high loss materials which are often used to stimulate the heating process.

Other papers have been published on sub-gridding the FDTD algorithm [8-10]. In reviewing the literature we decided to extend the method developed by Prescott and Shuley [10] to three dimensions. In extending the calculation to 3D, we found it necessary to separate the fine/coarse grid boundary from dielectric boundaries by a "buffer" zone of approximately four coarse cells. This helps achieve a stable solution and obtain accurate results. In addition to this extension we may also mention that Prescott and Shuley did not describe in their paper how the results from the fine grid were used to affect changes in the coarse grid data. Furthermore, we were unable to obtain a stable solution by updating the coarse magnetic fields at the proper point in time as is done by Zivanovic in the variable step size method (VSSM) [9]. However, we were able to obtain stability in the 3D FDTD code by updating the coarse electric field values which are completely contained inside the fine region, but not on the boundary. Tangential fields on the boundary are replaced by the average of the coarse region field value and the fine region field value. Our update method resulted in increased stability and accuracy in the coarse region. A detailed solution procedure is described in a paper submitted for publication in *IEEE Antennas and Propagation*. We also programmed the VSSM code [9] for comparison purposes. The obtained results from the newly developed Multi-Grid Displacement Method (MGDM) code and a comparison with the VSSM code, will be presented.

SIMULATION RESULTS IN AN AIR-FILLED WAVEGUIDE

To evaluate the developed 3D code we performed amplitude and phase error calculations in a waveguide operating at 10 GHz excited by a TE_{10} mode. A cell size of 1.27 mm (0.05") was chosen. This corresponds to a cell size of approximately 24 cells per wavelength. Model dimensions were 48x28x60 cells, with one cell of metal surrounding the waveguide. A (+z) directed transparent source was placed at $k=10\Delta x_c$. A schematic of the waveguide with the excitation and absorbing boundary planes is shown in Figure 1. Fine-grid regions of different sizes were modeled. All models were simulated with four different values of nfact (ratio between the cell sizes in the fine and coarse regions): 2, 3, 4, and 8. We performed simulations on our code and another FDTD

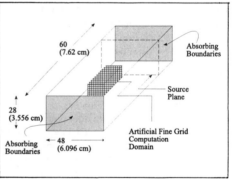

Figure 1, Schematic of waveguide used for evaluating magnitude and phase errors introduced by the fine-grid computation domain

code based on the Variable Step Size Method (VSSM) [9]. Since there is typically some finite error involved with FDTD simulations, errors from the developed multi-grid code with the VSSM code are reported with respect to results from an available uniform-grid FDTD. The uniform-grid model was developed with a cell size of 0.635 mm (0.025"), half the cell size of the coarse grid of the multi-grid code, for comparison. Magnitude and phase at various points in the fine grid region were compared to the corresponding values determined by the uniform grid code. A fair representation of the accuracy of the multi-grid code can be made by comparing E_y at the center of the waveguide in the three dimensions ($24\Delta x_c$, $14\Delta y_c$, $30\Delta z_c$). Amplitude errors for a fine grid of dimension $9\Delta x_c$ x $9\Delta x_c$ x $10\Delta x_c$ are shown in Figure 2. Amplitude error was computed as

$$\varrho_{MAG} = \left| \frac{E_{UNIFORM} - E_{MG}}{E_{UNIFORM}} \right| \times 100 \ \% \tag{10}$$

Phase errors were computed by using the difference in time of the zero crossings between the uniform and multi-grid codes (Δt) as

$$\varrho_{PHASE} = \left| \frac{\Delta t}{T} \right| \times 2\pi \tag{11}$$

The errors are comparable to those reported in [9], even though the number of nodes in our simulations are much larger. We were also unable to obtain long term stability for the VSSM for these models when larger dimensions were used. When the refinement factor, nfact, was increased to 8, it was not possible to achieve a stable solution using VSSM [9]. The developed (MGDM) code, however, was stable at least three times longer than the VSSM code, and we were able to reach a steady-state response for nfact=8 which had acceptable magnitude and phase errors compared to the uniform grid results.

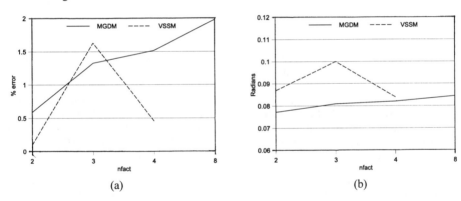

(a)   (b)

Figure 2, Error calculations for an air filled waveguide in which an artificial fine grid with dimensions of $9\Delta\ell_c$ x $9\Delta\ell_c$ x $10\Delta\ell_c$ was introduced. (a) Magnitude percentage error and (b) phase error in radians. Results are shown for both the developed MGDM code and the VSSM code available in literature [9].

SIMULATION OF SINTERING OF A CERAMIC SAMPLE IN A WAVEGUIDE USING THE MGDM CODE

The next objective was to obtain guidelines for creating models using the MGDM method for microwave sintering of ceramics. Concerns existed regarding possible instabilities in the code when the boundary between the fine and coarse grid regions was placed at the air-dielectric interface of the ceramic sample. To examine the presence of such a possibility and to further identify the number of coarse grid air nodes that are needed to separate the boundary of the two grid sizes from the dielectric boundary, and hence to achieve stable results, we created a model of a WR 340 waveguide, operating at 2.45 GHz. The waveguide was excited with the dominant TE_{10} mode, propagating in the (+z) direction. A ceramic sample of dimensions 2.8 cm x 1.3 cm x 5.1 cm was assumed suspended at the center of the waveguide. Dielectric properties were chosen which are typical of those used in sintering experiments at Oak Ridge National Lab ($\epsilon_r' = 4.3$, and $\sigma = 6.4$ x 10^{-5} S/m). The sample was placed at the center of the waveguide and five simulations were performed. The first simulation placed the multi-grid boundary on the surface of the ceramic sample. In subsequent simulations the boundary between the coarse- and fine-grid regions was placed one, two, four, and six coarse nodes away from the air/dielectric interface. When the boundary is placed at the dielectric interface it was observed that the code was not stable, and the obtained results diverged. This is expected since the displacement calculation uses the speed of light in the medium, and there are two media at the boundary between these two grids. When the boundary was placed one cell away from the dielectric, the solution was stable but not highly accurate. Results were acceptable when the interface was placed two or more cells away from the dielectric. However, the best results were obtained when the boundary between the multiple grids was placed four cells from the ceramic sample. Increasing the boundary to six cells showed no additional improvement in the accuracy of the results. Therefore, a region of four coarse cells between the grid boundary and any change in dielectric properties is suggested to ensure a resulting stable and accurate solution.

MICROWAVE SINTERING IN A MICROWAVE CAVITY

A multimode cavity used by Oak Ridge National Lab (ORNL) was modeled using a uniform-grid FDTD code and the MGDM code. The cavity had dimensions of 40 x 30 x 40 cm³. The source waveguide, which was placed near the bottom of the cavity, has a cross-section of 8.5 x 4.5 cm². The waveguide was excited with a TE_{10} mode operating at 2.45 GHz. The simulation contained a 2 x 2 x 2 cm³ ceramic sample ($\epsilon_r' = 6.5$, $\sigma = 8.25$ x 10^{-3}), surrounded by 2 cm thick insulation ($\epsilon_r' = 1.655$, $\sigma = 1.76$ x 10^{-4}), and a 1 cm thick SiC box ($\epsilon_r' = 29.4$, $\sigma = 0.9$) which was open at the top and bottom. The overall dimensions of the sample and insulation model are 8 x 6 x 8 cm³. This arrangement containing the ceramic was modeled with the MGDM code, with 4 coarse cells of air (8 fine cells for a refinement factor nfact=2) surrounding it. As shown in Figure 3, the simulation results using the uniform grid FDTD code and the newly developed MGDM code are in good agreement. However, the MGDM simulation required only half the CPU time as compared to the uniform-grid code. This, along with the associated memory savings, represent the advantages of using the MGDM code. It should also be noted that we attempted to model this problem using the VSSM code, but the code was not stable and failed to reach steady-state. It is observed that the VSSM method works well for small, refined regions, and for simple geometries, however, as the dimensions of the refined region increase, and the model becomes more complex, the VSSM method generally fails.

An important objective of modeling microwave sintering of ceramics is to investigate the feasibility of scale-up and commercial use of microwave sintering technology. To this end we developed the multi-grid displacement-method code. This code is capable of modeling microwave sintering in large multimode cavities, and hence is being used to help guide the on-going experimental effort at ORNL, which is focused on addressing commercialization and scale-up issues. The experimental activities at ORNL involve the use of 500 and 4000 liter multimode microwave cavities. Scale-up experimental procedure involves the sintering of bucket tappet samples with wall

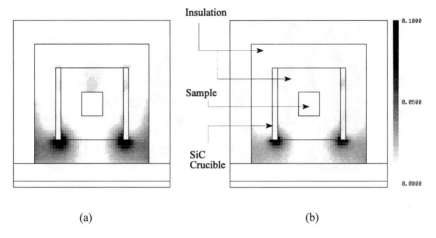

(a)   (b)

Figure 3, Electric field simulation results in a small microwave cavity. (a) Uniform grid FDTD simulation results with $\Delta\ell = 0.25$ cm and (b) MGDM results in the coarse region with cell sizes of $\Delta\ell_c = 0.5$ cm and $\Delta\ell_f = 0.25$ cm

thicknesses of 0.381 cm. Modeling such experimental arrangements using a uniform grid FDTD code will result in prohibitively large number of cells. We were able, however, to model this cavity using the developed MGDM code. The coarse grid was modeled with a 1.27 cm cell size, and the fine grid used a 0.3175 cm cell size. Dimensions of the coarse and fine regions are 63 x 63 x 113 cells, and 98 x 86 x 98 cells, respectively. Four layers, each containing a 4 x 4 array of samples, were modeled. The bucket-tappet samples had a nominal composition of 30 wt. % Si_3N_4 and 5 wt. % each of Al_2O_3 and Y_2O_3 ($\epsilon_r' = 16.83$ $\sigma = 0.3132$ S/m at 1000 °C) and had dimensions of 4.76 cm diameter, 3.81 cm height, and a 0.32 cm thickness. Each layer sat on a Boron Nitride shelf ($\epsilon_r' = 2.11$ $\sigma = 0.0076$ S/m at 1000 °C) with an area of 20.0 x 20.0 cm^2 and a thickness of 0.64 cm. The sintering arrangement was then surrounded by an Alumina fiberboard insulation box ($\epsilon_r' = 9.67$ $\sigma = 0.0388$ S/m at 1000 °C) with a thickness of 0.32 cm, and a height of 21.0 cm. Four rings of Silicon Carbide ($\epsilon_r' = 29.0$, $\sigma = 0.90$ S/m at 1000 °C) were placed around the fiberboard box on each layer to aid the hybrid heating process. The rings had a thickness of 0.32 cm, and a height of 2.22 cm. Results of the electric field patterns in the cavity modeled by the fine region are shown in Figure 4. The MGDM code allowed us to model this most realistic experimental arrangement in an electrically large, multimode, microwave cavity and obtain reasonable resolution, while maintaining reasonable computer requirements. Additional results illustrating the use of the developed MGDM code in modeling and simulating electrically large, multimode, microwave cavities are reported in a companion paper published in this proceedings [11].

CONCLUSIONS

In this paper we described a procedure to extend the mesh refinement algorithm (MRA) proposed by Prescott and Shuley [10] to three dimensions. A procedure to update the coarse-grid results based on fields updated in the fine region was also implemented. It was shown that when

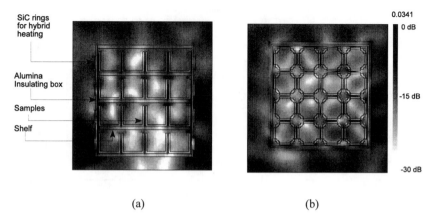

SiC rings
for hybrid
heating

Alumina
Insulating box

Samples

Shelf

0.0341

0 dB

-15 dB

-30 dB

(a)   (b)

Figure 4, $|E|^2$ field patterns in the fine region of a 500 liter multimode cavity modeled using the developed MGDM code. (a) Vertical cross section through the samples, and (b) horizontal cross section

replacing the field values of the coarse grid region inside the fine region with the average of the field values calculated from the fine region and that calculated from the coarse region, stable steady state results are obtained from the developed code. Electric field calculations were performed in an empty waveguide. Fine-grid sizes of $3\Delta\ell_c$ x $3\Delta\ell_c$ x $10\Delta\ell_c$, $6\Delta\ell_c$ x $6\Delta\ell_c$ x $10\Delta\ell_c$, and $9\Delta\ell_c$ x $9\Delta\ell_c$ x $10\Delta\ell_c$, with refinement factors of 2, 3, 4, and 8 were modeled. Comparisons were made between the developed multi-grid displacement method (MGDM) and the variable step size method (VSSM) [9]. In an air filled waveguide with a refinement factor up to 4, the two codes gave very comparable results. However, for larger refinement factors, or more complex geometries of the fine grid region, the VSSM code became unstable, while the developed MGDM remained stable and produced accurate results.

Additional simulations which modeled the placement of a ceramic dielectric inside a rectangular waveguide were also performed. It is shown that a "buffer" region of four cells between the fine/coarse grid boundary and the dielectric interface is necessary to obtain acceptable accuracy and a stable solution. As the "buffer" zone (number of coarse cells in the zone) is decreased, numerical errors associated with the calculation of the displacement term cause inconsistent results. Once results from the code were validated and the minimum required size of the "buffer" zone (four coarse cells) was identified, the code was used to model realistic sintering experiments in a multimode cavity. The results were compared to those from a uniform-grid code and acceptable agreement was observed. In addition, the MGDM resulted in a significant reduction in the CPU time. When the VSSM method was applied to this model, the solution was unstable. Based on these validation results, we intend to continue to use the developed MGDM code to study microwave sintering of ceramics, and in particular, to simulate realistic sintering experiments in an effort to provide general guidelines for newly developed, hybrid-heating techniques, as well as to help guide the scale-up and commercialization of microwave sintering technology.

ACKNOWLEDGMENT

Research sponsored (in part, if applicable) by the US. Department of Energy, Assistant Secretary for Energy Efficiency and Renewable Energy, Office of Industrial Technologies, Advanced Industrial Concepts (AIC) Materials Program, under contract DE-AC05-84OR21400 with Martin Marietta Energy Systems, Inc.

REFERENCES

[1]   M.A. Janney and H.D. Kimrey, in Ceramic Power Science II, edited by G. L. Messing, E.R. Fuller, and H. Hausner, American Ceramic Society, Westerville, Ohio, 1988, pp. 919-924.

[2]   Y.L. Tian, D.L. Johnson, and M.E. Brodwin, in Ceramic Power Science II, edited by G. L. Messing, E.R. Fuller, and H. Hausner, American Ceramic Society, Westerville, Ohio, 1988, pp. 925-932.

[3]   H.D. Kimrey and M.A. Janney, Microwave Processing of Materials, edited by W.H. Sutton, M.K. Brooks, and I.J. Chabinsky, (Mater. Res. Soc. Proc. 124, San Francisco, CA, 1988) p. 367

[4]   M.F. Iskander, R.L. Smith, O. Andrade, H. Kimrey, and L. Walsh, *IEEE Trans. on Microwave Theory and Techniques*, Vol. 42, pp. 793-800, May 1994.

[5]   J.D. Newman, L. Walsh, R. Evans, T. Tholen, O.M. Andrade, M.F. Iskander, K.J. Bunch and H. Kimrey, in Microwave: Theory and Application in Materials Processing, *Ceramic Transactions*, Vol. 36, pp. 229-237, 1993.

[6]   M.F. Iskander, O. Andrade, A. Virkar, and H. Kimrey, in Microwave: Theory and Application in Materials Processing, *Ceramic Transactions*, Vol. 21, pp.35-48, 1991.

[7]   M.F. Iskander, Microwave Processing of Materials II, edited by W.B. Snyder, Jr., W.H. Sutton, M.F. Iskander, and D.L. Johnson (Mater. Res. Soc. Proc. 189, San Francisco, CA, 1990) pp. 149-171.

[8]   I.S. Kim and W.J.R. Hoefer, *IEEE Transactions on Microwave Theory and Techniques*, Vol. 38, pp. 812-815, June 1990.

[9]   S.S. Zivanovic, K.S. Yee, and K.K. Mei, *IEEE Transactions on Microwave Theory and Techniques*, Vol. 39, pp. 471-479, Mar. 1991.

[10]   Deane T. Prescott, and N.V. Shuley, *IEEE Microwave and Guided Wave Letters*, Vol. 2, No. 11, pp. 434-436 Nov. 1992.

[11]   M. Subirats, M. F. Iskander, M.J White, and J. Kiggans "FDTD Simulation of Microwave Sintering in Large (500/4000 liter) Multimode Cavities", presented at the 1996 MRS Spring Meeting, San Francisco CA, April 1996 (unpublished).

COMPARISON OF FINITE-DIFFERENCE AND ANALYTIC MICROWAVE CALCULATION METHODS

F.I. Friedlander*, H.W. Jackson, M. Barmatz, and P. Wagner**
*Communications and Power Industries, PO Box 50750, Palo Alto, CA 94303
** Jet Propulsion Laboratory, California Institute of Technology, Pasadena, CA 91109

ABSTRACT

Normal modes and power absorption distributions in microwave cavities containing lossy dielectric samples were calculated for problems of interest in materials processing. The calculations were performed both using a commercially available finite-difference electromagnetic solver and by numerical evaluation of exact analytic expressions. Results obtained by the two methods applied to identical physical situations were compared. Our studies validate the accuracy of the finite-difference electromagnetic solver. Relative advantages of the analytic and finite-difference methods are discussed.

INTRODUCTION

Use of exact analytic expressions provides accurate and relatively fast results for electromagnetic problems with sufficiently simple geometries. Numerical electromagnetic solvers can treat complex geometries, but with less certain accuracy. The cooperative effort discussed in this paper applied and compared both methods to the representative problem of a lossy ceramic rod situated along the axis of a cylindrical cavity reactor to validate the numerical approach, and to begin a longer-term effort to take advantage of both approaches (and possibly also of approximate analytic methods) in solving problems related to microwave processing of materials.

DESCRIPTION OF THE STEADY-STATE MODEL

The relatively simple model consists of a closed metal cylindrical cavity reactor with an alumina-like ceramic rod positioned along the entire length of the axis of symmetry. The cavity radius, ρ_c=4.69 cm, and length, 6.63 cm, correspond to a 2.45 GHz resonance frequency for the empty cavity TM010 mode. Rods with several different values of radius were modeled.

DESCRIPTION OF THE ANALYTIC METHOD

Electromagnetic properties of the microwave cavity reactor represented in the model described above were treated by analytic methods at JPL [1,2]. To account for radial variations of the complex dielectric constant the lossy dielectric sample (the rod) is partitioned into thin cylindrical zones, or shells, each of which is assumed to have uniform properties. The vacuum (or air) space around the rod is one additional zone, with relative dielectric constant of unity. The curved cavity wall may also be taken as one zone in order to include wall losses. Treated as a perfect electrical conductor, that zone is omitted and an appropriate boundary condition is imposed.

Maxwell's equations can be solved exactly for this shell model using a 4x4 matrix formalism originally due to Sphicopoules, Bernier, and Gardiol [3]. The normal mode frequencies are calculated as complex-valued roots of a complex-valued determinant, using a root finder developed at JPL. Ordinary matrix methods were then used to determine certain expansion coefficients that occur in expressions for normal mode fields. The formulas for this procedure provide rapid and efficient means for evaluating normal mode electric and magnetic fields and power absorption density.

Mat. Res. Soc. Symp. Proc. Vol. 430 © 1996 Materials Research Society

In the exact solution method just described, the flat end plates are treated as perfect electrical conductors. If desired, microwave power absorption in the end plates can be calculated from the adjacent magnetic fields and a surface resistance approximation. The end-plate losses can then be combined with the results of the exact solution (assuming lossless end plates) to calculate the Q of the cavity that includes all the losses. In addition, the field strengths inside the cavity are renormalized so that they correspond to a specified value of total power dissipated in the sample and cavity walls, including the end plates.

DESCRIPTION OF THE FINITE-DIFFERENCE CODE

The finite-difference code used in this study was the frequency domain module for modes of resonance contained within the electomagnetic/plasma-physics code known as MAFIA (for "MAxwell's equations by Finite-Integral Algorithm"). Maxwell's equations are transformed into a set of fully self-consistent discrete matrix equations and solved (without sources). Two- or three-dimensional geometries may be treated. With the latest release (MAFIA 3.20), no-loss, low-loss, and lossy solvers are all available for solving the resonance frequency and field patterns of the modes. The solvers which treat lossy dielectrics account for finite conductivity. The post-processor is also able to compute the approximate losses from the no-loss field solution, given the relevant material properties. Skin-effect losses in conducting walls are evaluated using the magnetic field adjacent and parallel to the wall surface, as in the treatment of end walls by the analytic method.

Time domain modules that account for sources of microwave power are also available within MAFIA. An extension of these modules that will treat nonlinear (temperature-dependent) materials is now being tested by the code developer, Computer Simulation Technology (CST) in Darmstadt, Germany.

COMPARISON OF RESULTS

The frequency vs. normalized rod radius for the TM010 and TM110 modes using the two methods are compared in Fig. 1. For the numerical approach, a 100x100 uniform rectangular mesh was used. The agreement is excellent, the maximum deviation being about 0.1%. Fig. 2 shows the comparison of results from the two methods for the quality factor due to losses in the dielectric using the TM010 mode. Similarly excellent agreement was obtained with the TM110 mode.

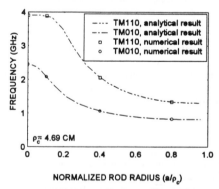

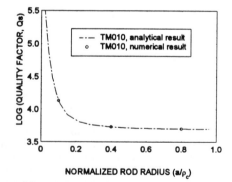

Figure 1. Resonance frequency vs. normalized rod radius for TM010 and TM110 modes.

Figure 2. LOG10(Quality Factor) vs. normalized rod radius for TM010 mode.

In Fig. 3, results are compared for power density vs. radius with three different rod sizes, using the TM010 mode. Fig. 4 shows equivalent results for two rod sizes for the TM110 mode.

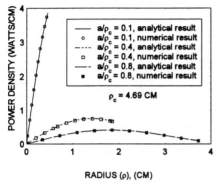

Figure 3. TM010 mode power absorption profile for various rod radii, with 1 watt absorbed.

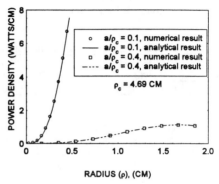

Figure 4. TM110 mode power absorption profile for various rod radii, with 1 watt absorbed.

Table I compares results for the lower frequency Hybrid-111 mode. For this case the relative dielectric constant is 8.5459 and the loss tangent is 0.000559. With such a low loss case, the results from the lossless and lossy solvers are found to be in good agreement, as expected.

Table I. Comparison of Results for a Hybrid-111 Mode

APPROACH	REAL FREQUENCY	QUALITY FACTOR
analytic	1.8757 GHz	2576.4
numeric, lossless	1.8781 (.13% high)	2608.0 (1.2% high)
numeric, lossy	1.8781 (.13% high)	2620.7 (1.7% high)

Table II compares resonance frequency and quality factor results for the TM010 mode with a relatively high loss tangent, 0.4. With such high loss the results from the lossy solver should be more accurate than those from the lossless solution. The comparison of results from the numeric approach to those from the analytic approach confirms this assumption.

Table II. Comparison of Results for TM010 Mode, Loss Tangent=0.4

APPROACH	REAL FREQUENCY	QUALITY FACTOR
analytic	1.0146 GHz	2.7740
numeric, lossless	1.0681 (5.2% high)	2.7036 (2.5% low)
numeric, lossy	1.0177 (.31% high)	2.7749 (.03% high)

Table III compares results for the TM_{010} mode with linearly varying complex dielectric constant as a function of radius. Forty shell regions were used to represent the rod with both approaches. The relative real dielectric constant varies from 15 on axis to 10 at the surface; while the imaginary part varies from 0.75 to 0.5. This choice results in a constant loss tangent of 0.05. The excellent agreement between the calculation methods validates the approaches for treating a spatially varying dielectric constant.

Table III. Comparison of Results for TM_{010} Mode, Linearly Varying Dielectric Constant

APPROACH	REAL FREQUENCY	QUALITY FACTOR
analytic	.93252 GHz	21.079
numeric, lossless	.93179 (.078% low)	21.053 (.12% low)

The good-to-excellent agreement typical of the results provides mutual validation of the two basic approaches. Either method can provide detailed results for parametric variations, even for quite lossy materials. Since the analytic model gives exact results, the differences between the two approaches due to different sample sizes or properties, modal field patterns and/or frequencies can be evaluated. This is very beneficial to adjusting numerical modeling parameters to best obtain desired accuracy with more complex geometries. For example, when the mesh was changed from 100x100 to 50x50, results for the TM_{010} mode quality factor due to dielectric loss in a rod with $a/\rho_c=0.4$ changed by 0.16%, and for the TM_{020} mode by only 0.03%. Results for the 100x100 mesh had been validated by the analytic approach. However, using a 30x30 mesh, the deviation from the validated results for the TM_{010} mode quality factor jumped to 0.5%. Similarly, the resonance frequency of the two modes changed less than 0.1% going to a 50x50 mesh, but the deviation for the TM_{020} mode frequency jumped to 0.48% for a 25x25 mesh.

ADVANTAGES OF THE ANALYTIC APPROACH

The analytic approach can typically provide highly accurate solutions to simple electromagnetic problems much faster than numerical methods such as the finite-difference approach. This makes extensive parametric studies more economically feasible. The exact formulas provide a ready means for detailed evaluation of the solution, as desired. This approach is also better adapted to solving problems where both electromagnetic and thermal properties should be calculated self-consistently, e.g. when realistic thermal emissivities of both the sample and the cavity wall are required in the model.

ADVANTAGES OF THE NUMERICAL APPROACH

Whereas the analytic approach is limited to problems with relatively simple geometry and symmetry, the numerical approach considered here is limited primarily by the ability of a computer to solve the problem discretized on a mesh, and to accomplish this in a "reasonable" time. This means that the overall problem size should not be extremely large compared to the smallest detail (including electromagnetic field variations as well as geometrical model variations) that must be resolved on the mesh. Geometries can be quite complex in spite of that limitation. The multi-solver numerical code is also capable of solving many aspects of complex problems and a great variety of simple problems. For example, in the analytic approach for a cavity reactor a source of excitation cannot be readily included in the model, whereas the time-domain numerical code solver can provide the multi-mode solution for a reactor large compared to the wavelength of the drive signal, and even permit the choice of pulsed, transient, or sinusoidal microwave drive.

BENEFITS OF TWO-METHOD APPROACH TO SOLVING REAL PROBLEMS

Through appropriate application of each of the methods described, the advantages of both methods can be exploited. For example, parametric studies of a simple model can be rapidly performed using the analytic approach, and an approximate optimum set of parameters extracted from the results. This can then be used as a good point-of-departure design for rigorous analysis using the numerical approach. In general, this will result in better designs achieved in a shorter time. For solution of problems of unusual nature or material properties, it is valuable to start with mutual validation of the two approaches by applying them to a representative, sufficiently simplified version of the problem. In this way any significant errors in either approach may be detected.

APPLICATION TO PROBLEMS WITH CYLINDRICAL GEOMETRY

In general, the two-method approach is useful for validation of new variations in method, mutual validation of analytic and numerical solutions to specific problems, and evaluation of numerical error. The following is a representative list of possible problems for using the two-method approach discussed above.
1. Dielectric constant measurements for a large range of sample sizes and a large range of lossy dielectric materials (not limited to small sample size and low-loss materials needed for cavity perturbation theory).
2. Inverted temperature profile in materials processing.
   a. chemical vapor infiltration
   b. improved grain characteristics for high critical current in high T_c superconductors.
3. Fiber coating: determining optimum experimental processing conditions for heating the fiber as its diameter increases.
4. Uniform heating of a fluid flowing in a cylindrical tube.

FUTURE JOINT EFFORTS

The following is a list of joint efforts that are planned or under consideration: (1) comparison of analytic and finite-difference results for transient conditions, (2) validation of a normal mode expansion method, particularly for practical application to batch microwave processing, and (3) application to the design of a batch-processing reactor.

ACKNOWLEDGMENT

The research described in this paper was carried out at the Microwave Power Tube Products Division of Communications and Power Industries, Inc., and at the Jet Propulsion Laboratory (JPL), California Institute of Technology, under a Joint Technology Cooperation Agreement. The research at JPL was performed under contract with the National Aeronautics and Space Administration.

REFERENCES

1. H.W. Jackson, M. Barmatz, and P. Wagner, Mater. Res. Soc. Proc. **347**, 317 (1994).

2. H.W. Jackson, M. Barmatz, and P. Wagner, Ceramic Transactions, **59**, 279 (1995).

3. T. Sphicopoules, L.-G. Bernier, and F. Gardiol, IEEE Proc. **131**, Pt. H, No. 2, 94 (1984).

NUMERICAL SIMULATION AND EXPERIMENTAL VALIDATION OF RF DRYING

S. BRINGHURST, M. J WHITE, and M. F. ISKANDER
Electrical Engineering Department, University of Utah, Salt Lake City, UT 84112

ABSTRACT

The Finite-Difference Time-Domain (FDTD) method has been used by our group to simulate a wide variety of Radio Frequency (RF) and induction-drying processes and realistic, microwave-sintering experiments. Many results were presented and some guidelines towards the effective use of the microwave and RF heating technologies of ceramic ware were developed.

In this paper we describe an experimental effort which was used to validate the FDTD simulation results. Specifically an experimental RF dryer, Thermax Model No. T3GB operating at 25 MHz, was used to dry ceramic ware of various materials, sizes, and shapes and the temperature distribution pattern was monitored using six fiber-optic temperature probes. The measured heating patterns were then compared with the FDTD simulation results. Many of the guidelines developed using the numerical simulations were confirmed experimentally.

Results from various comparisons between simulation and experimental data will be presented. Additional results from the simulation efforts illustrating possible procedures for improving the efficiency and the uniformity of RF drying will also be described

INTRODUCTION

Radio Frequency (RF) drying has been in commercial use for many years, but still many aspects of the drying processes are not fully understood. Changes in the material, shape, or dimensions of the ware being dried often require significant adjustment in the drying equipment. This sometimes requires extensive experimentation so that satisfactory drying performance can be achieved. Many drying defects such as cracks, blisters, end flare, and warp are still unpredictable.

Ways for optimizing equipment designs are needed so as to make available dryers more efficient, more stable under varying loading conditions, and provide an improved product quality. Experimentation can be both costly and time consuming, therefore it is advantageous to use numerical simulation to help develop a fundamental understanding of the RF drying process. Using numerical simulations it is possible to examine design tradeoffs and suggest modifications to improve efficiency of drying and quality of the resulting products. It is also possible to provide a cost-effective means for developing an understanding of the electromagnetic interactions with materials. The electromagnetic interactions will change based on: materials properties, shapes and dimensions of ware, and drying rates. The final goal of the numerical simulations is to guide the design, characterization, and optimization of various electromagnetic processing systems so that efficiency of the drying process can be improved, the heating patterns can be more uniform, and so that the available and possible controls can be identified and effectively used. An important part of this numerical simulation effort is the experimental validation of the obtained results and the verification of suggested changes and modifications in the design and the use of the drying equipment. In this paper we describe the results of the experimental verification of some of the FDTD simulation results.

Mat. Res. Soc. Symp. Proc. Vol. 430 © 1996 Materials Research Society

NUMERICAL SIMULATIONS

For the numerical simulations, a 3D Finite-Difference Time-Domain (FDTD) Code was used. The typical model size is 200 x 100 x 50 computational cells, with the cell size of 2.54 cm³. At 25 MHz the typical run time for this size model was 3 hours. The FDTD code was integrated with a 3D Finite Difference heat transfer code to calculate the heating patterns. The heat transfer code was developed at the University of Utah, and it accounts for heat transfer from conduction, convection, and radiation. The heat transfer code was validated by comparing results with ANSYS® which is a commercial finite element heat transfer code. As the ware continues to heat and dry the material properties are expected to significantly change. Therefore, the FDTD and Heat Transfer codes have been integrated so that they can provide updated heating patterns with the changes in the materials properties versus temperature and moisture content.

Two RF dryers were modeled, a dryer with vertical parallel plate electrodes, and a dryer with horizontal parallel plate electrodes. The dryers are shown in Figure 1.

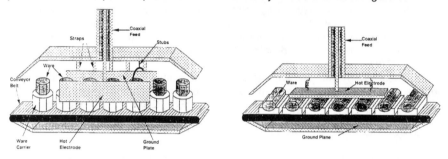

(a)   (b)

Figure 1: Schematic of two RF dryers that were modeled using a 3D FDTD code. (a) a vertical parallel plate dryer, and (b) a horizontal parallel plate dryer.

The numerical modeling was used to study the effect of the orientation of ware on the uniformity of drying, examine the complex role of stubs on the performance of the dryer and examine the feasibility of using auxiliary electrodes to improve uniformity of drying. It was found that stubs can be used to improve the field distribution along the dryer, and also to adjust the input impedance of the dryer, and hence the overall efficiency of the system. A drawback to using stubs, is that the fields tend to concentrate around the stubs and cause arcing. FDTD results also showed that auxiliary electrodes can be used to improve the uniformity of the fields in the ware, but they also create a large capacitance between the hot and the auxiliary electrodes which may significantly impact the impedance matching and the efficiency of the dryer. Some of these results from the numerical simulations were validated experimentally as will be described in the next section.

EXPERIMENTAL VALIDATION

Results from the numerical simulations were validated experimentally using an experimental dryer at the University of Utah. The experimental dryer which operates at 25

MHz is shown in Figure 2. The drying chamber is 26" x 10" x 15", and the drying is done with a height adjustable horizontal electrode with dimensions of 10" x 6". There is a scale at the bottom of the dryer so that water loss can be measured.

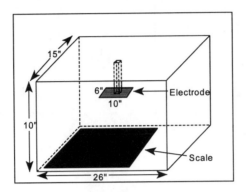

Figure 2: A schematic of the experimental drying chamber used to validate the numerical simulations

The experiments which were done in the experimental dryer were used to confirm many of the critical suggestions made to improve the drying process. For the experimental arrangement the sample was placed in the dryer on top of the scale, and the temperature was then monitored in six locations using LUXTRON fiber-optic temperature probes. The arrangement of the temperature probes is shown in Figure 3.

As can be seen from Figure 3 probes 1-3 are in a horizontal plane directly through the center of the sample. Probes 4-6 are in a horizontal plane about 1/4" below the top surface of the sample. Probes 1 and 4 were placed along a vertical line through the center of the sample. Probes 3 and 6 were placed about 1/4" from the edge of the major axis, while probes 2 and 5 were placed about 1/4" from the edge of the minor axis. The temperature was updated and recorded every 30 seconds during the drying process which usually lasted about 5 minutes.

Figure 4 shows a comparison between the numerical simulation of a sample and the temperature profile given by the temperature probes for a sample dried in the experimental dryer. It is shown in both the numerical simulation and the experimental results that there are more electromagnetic fields and higher temperatures along the major axis at the cylindrical surface of the ware, and then less in the middle section. In the temperature profile the highest temperatures corresponds to the highest concentration of electromagnetic fields. This is observed by comparing results from probes 3 and 4.

Figure 5 shows the same sample but this time with an auxiliary electrode on top of the sample. This is done in an attempt to improve the uniformity of the field distribution inside the ware. Figure 5 shows that the fields are much more uniform and the temperature distribution confirms the numerical results and also shows that the sample heats up much quicker, and reaches a temperature of about 95 ` C, which is about 20 ` C higher than the sample in Figure 4, in 250 seconds instead of 300 seconds.

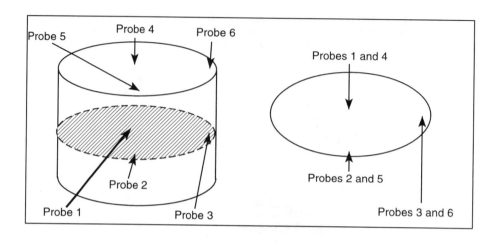

(a)   (b)

Figure 3: Arrangement of the fiber optic temperature probes in the ceramic sample (a) front view of the sample and (b) top view of the sample.

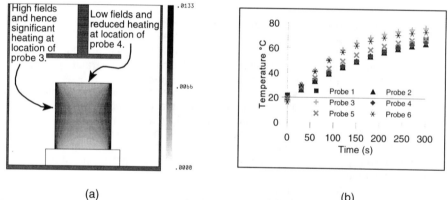

(a)   (b)

Figure 4: Comparison of (a) numerical simulation, and (b) actual experimental results of drying of a single sample.

Figure 6 shows a smaller sample that has the same material properties as the sample in Figure 4. It is shown that the electromagnetic field distribution once again matches the temperature profile from the actual experiment. By comparing the results in Figures 4 and 6, however, it may be observed that the different size sample significantly affected the loading and hence the input impedance of the dryer, which resulted in a significant reduction in the efficiency of the dryer. This is clearly demonstrated by the reduced final temperature values in Figure 6. The sample in Figure 6 is smaller and thus produces less loading on the dryer. Therefore the magnitude of the electromagnetic fields is smaller in the sample and therefore the maximum temperature in the sample is about 55° C which is 20° C less than the maximum temperature of 75° C which is shown in the larger sample case of Figure 4.

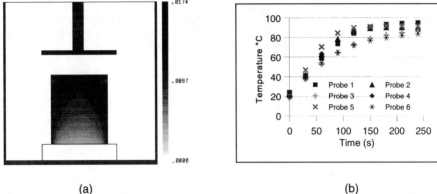

(a)   (b)

Figure 5: Comparison of (a) numerical simulation, and (b) actual experimental drying of a single sample with an auxiliary electrode on top of it.

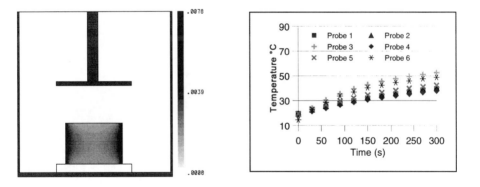

(a)   (b)

Figure 6: Comparison of (a) numerical simulation and (b) actual experimental drying of a short single sample. This is to determine loading effects as compared to a larger sample with the same material properties. The highest value of electromagnetic fields in Figure 6(a) is 0.0078, which is almost half the corresponding value of Figure 4(a).

The results of Figure 6 show, once again, that the FDTD results and the obtained trends in the field variation follow the experimental observations made using temperature measurements. These as well as other correlations between the numerical and experimental data reaffirmed the confidence in the results from the FDTD code.

CONCLUSIONS

A 3D FDTD code was used to simulate realistic RF dryers in commercial use. Both a horizontal plate dryer and a vertical plate dryer were modeled. The numerical simulations were used to calculate the electromagnetic power distribution patterns which were then used in the heat-transfer code to calculate the heating patterns. The heating patterns were used to make many important suggestions to help improve uniformity and efficiency of the drying process. Many of the numerical observations were then validated experimentally using the experimental dryer at the University of Utah.

3-D TEMPERATURE DISTRIBUTIONS PRODUCED BY
A MICROWAVE SPOT HEATING APPLICATOR

R. M. ROOS, J. R. THOMAS, JR.

Mechanical Engineering Department, Virginia Polytechnic Institute and State University, Blacksburg, VA 24061

ABSTRACT

Numerical simulations were used to investigate the effectiveness for materials processing of recently announced microwave spot-heating applicators . These devices can produce heating in depth over a region a centimeter or two in diameter. Computed 3-D time-dependent temperature distributions for both stationary and moving samples suggest that such applicators could be very useful for joining of ceramics, such as alumina. Moving the beam relative to the workpiece at an appropriate rate avoids thermal runaway.

INTRODUCTION

Recent publications [1,2] have described microwave devices with the capability of producing a concentrated spot of microwave energy with an approximate Gaussian spatial distribution on the surface of a workpiece. These devices include quasi-optical focusing gyrotrons [1], and a spot heating applicator specifically designed with ceramic joining in mind [2]. If the microwave energy penetrates sufficiently, such a device would be useful for specialized processing applications, such as joining of ceramics. The effectiveness of such applicators will depend on the temperature distributions they can produce within a workpiece. In an attempt to assess this capability, we have developed a numerical model of the temperature profile produced by a spot of applied electromagnetic energy on a bar of ceramic material in both stationary and moving-sample processing scenarios. This model reveals the extent of the heated volume, and the microwave field strength at the surface of the bar required to produce a maximum temperature of 1500K in depth. A possible control strategy is also described.

MATHEMATICAL MODEL

We consider a bar of ceramic material of thickness h in the z-direction, width w in the y-direction, and length ℓ in the x-direction. The microwave beam is focused on the x-y plane at $z = h$. The microwave energy will penetrate the workpiece, producing heating in depth and resulting in a temperature distribution $T(x, y, z, t)$ described by the equation of energy conservation

$$\nabla \cdot (k \nabla T) - \rho C_p u \frac{\partial T}{\partial x} + \dot{q}(x, y, z, T) = \rho C_p \frac{\partial T}{\partial t}, \tag{1}$$

where $k(T)$ represents the thermal conductivity, ρC_p the density and specific heat, and u the velocity of the beam in the x-direction relative to the workpiece. The volumetric heat source $\dot{q}(x, y, z, t)$ produced by the absorption of microwave energy has the maximum value

$$\dot{q}_0(x, y, z, T) = 2\pi f \epsilon''(T) |E(z)|^2, \tag{2}$$

where f is the microwave frequency, $\epsilon''(T)$ is the dielectric loss, and $E_f(z)$ is the field strength at depth z. The field strength is attenuated exponentially through the depth h of the material using the skin-depth approach, while in the x-y plane, the field is assumed to have a Gaussian distribution

$$\dot{q}(x, y, z, T) = \dot{q}_0(x, y, z, T) \exp\left(-\frac{x^2 + y^2}{r_0^2}\right), \qquad (3)$$

where r_0 is the beam radius.

The boundary conditions were chosen to represent convection and radiation heat loss from the top and all lateral sides, while the bottom is assumed to contact a support surface through a contact conductance h_c.

SOLUTION

Eqs(1-3) were discretized using the fully implicit control-volume finite-difference technique, allowing for temperature dependence of all material properties. The temperature dependence was represented by curve fits to literature data. After performing some numerical experiments on the effect of grid size, a total of 41 grid points were used in the x-direction, 21 in the y-direction, and 11 in the z-direction. A time step of 0.1s produced sufficiently accurate results in the temporal domain.

The computational requirements of the model are significant. A typical run for a moving sample case (described below) requires about 10 hours of CPU time on a 100 Mflop processor. It must be admitted, however, that no attempts were made to optimize the code.

RESULTS

All results presented are based on calculations using the physical properties of alumina. Measured data for 99.8% pure alumina [3] were fit to a polynomial in temperature to provide a continuous representation. This relation takes the simple form

$$\epsilon''(T) = \begin{cases} 3.114 \times 10^{-4}T - 0.1039, & T \geq 430K, \\ 0.030, & \text{otherwise.} \end{cases}$$

In the first set of numerical experiments we explored the time required to heat the sample surface to 1500K for various values of field strength and contact conductance h_c. The results of this excercise are displayed in Fig. [1], where the time required is plotted $vs.$ field strength for two values of h_c. To reach 1500K in one minute requires a maximum field of about 1.65×10^5 V/m for $h_c = 100W/m^2K$, and slightly less for $h_c = 10W/m^2K$.

Temperature distributions in the x - z plane for a field strength of $10^5 V/m$ are shown in Fig. [2] for $h_c = 100W/m^2K$, and for $h_c = 1000W/m^2K$ in Fig. [3]. The uniformity of the temperatures in the z-direction is notable for $h_c = 100W/m^2K$, but is less than desirable for $h_c = 1000W/m^2K$. This suggests the advisability of providing insulation between the sample and any highly conductive support surface to minimize thermal contact. We turned next to numerical experiments to explore the effect of moving the beam relative to the workpiece in the x-direction at various velocities. This effect could be achieved, of course, with a stationary beam and a moving workpiece. For these calculations, the contact conductance h_c between the workpiece and the support was held constant at $h_c = 100W/m^2K$ to insure a

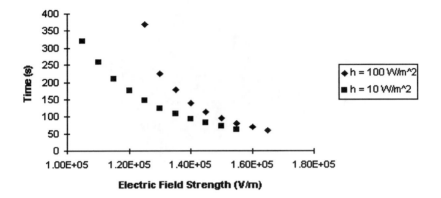

Figure 1: Time required to raise the sample maximum temperature to $1500K$ for two values of contact conductance h_c. Larger h_c (better thermal contact) drains energy from the workpiece and lengthens heating time.

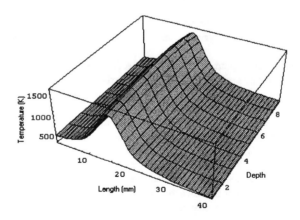

Figure 2: Temperature distribution in the x - z plane at the time when the maximum temperature reaches $1500K$ for $h_c = 100W/m^2K$.

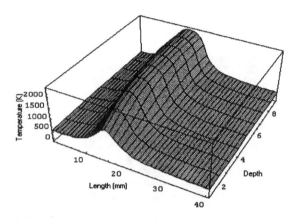

Figure 3: Temperature distribution in the x - z plane at the time when the maximum temperature reaches $1500K$ for $h_c = 1000W/m^2K$.

uniform temperature through the thickness of the material. The microwave beam was held in a fixed position long enough to develop a maximum temperature of 1500K with a maximum field of $10^5V/m$, and the workpiece was then moved steadily in the $+x$-direction at a speed of $0.1cm/s$. The starting temperature distribution in the x-y plane at the top surface is that shown in Fig. [3]. After 10 seconds, the sample has moved $1cm$ and has the temperature distribution in the x-y plane shown in Fig. [4], and in the x-z plane at $y = w/2$ in Fig. [5]. The temperature is still uniform in the z-direction, but the maximum temperature has dropped to 1200K and is not centered under the beam. Evidently this speed is too great for the applied field of $10^5V/m$.

Further numerical experiments with various speeds reveal that at $0.02cm/s$ thermal runaway sets in, and $0.05cm/s$ produces only a very gradual drop in temperature. Thus a speed of approximately $0.04cm/s$ is apparently approximately optimum for a field strength of $10^5V/m$.

The absorbed power for all three speeds considered, as a function of time, is shown in Fig. [6]. When the speed is too great $(0.1cm/s)$ the absorbed power gradually decreases, while if it is too small, it gradually increases, eventually leading to thermal runaway. Thus controlling the absorbed power is a possible control strategy if power can be measured more conveniently or accurately than local temperatures.

Similar numerical experiments were attempted using the thermal and dielectric properties of ordinary silica glass, but were unsuccessful. Thermal runaway always developed for the dielectric loss data used [4]. Since the available data for $\epsilon''(T)$ for glass includes only 2 data points for temperatures above $1000°C$, it is not clear that these results are realistic.

CONCLUSIONS

A microwave applicator capable of producing a focused beam of microwave radiation such

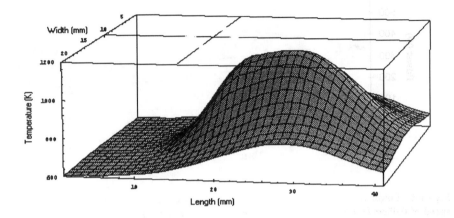

Figure 4: Temperature distribution in the x - y plane after 10 s of sample motion in the $+x$ - direction at a speed of $0.1 cm/s$ for $h_c = 100W/m^2K$. The beam is centered at the cross-hairs.

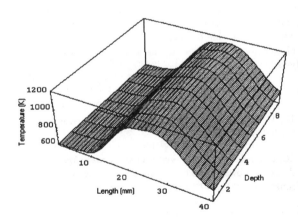

Figure 5: Temperature distribution in the x - z plane after 10 s of motion in the $+x$ - direction at a speed of $0.1 cm/s$ for $h_c = 100W/m^2K$.

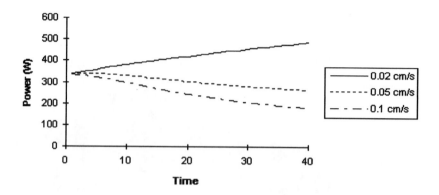

Figure 6: Power absorbed by the workpiece during heating for three speeds. Apparently a speed of $0.05cm/s$ will just about produce constant temperatures.

as those described in Refs. [1,2] could be an effective tool for many materials processing requirements, such as joining of ceramics. Our mathematical model shows that it should be possible to heat alumina to the temperature necessary for joining while advoiding thermal runaway, and to maintain this temperature over a local area while the workpiece is being moved. Care must be exercised, however, to insulate the workpiece on the surface opposite the applied beam in order to produce uniform temperatures through the thickness. The processing speed and field strength must be chosen carefully and coordinated to avoid thermal runaway and to maintain the desired temperatures. Controlling the absorbed power is a viable approach for achieving this balance.

It would be of considerable interest to the authors to see some experimental results to verify their analysis.

REFERENCES

1. Paton, B. E., V. E. Sklyarevich, and M. M. G. Slusarczuk, "Gyrotron Processing of Materials," MRS Bull. **18**, 58 (1993).

2. Tinga, W. R., J. D. Xu, and F. E. Vermeulen, "Open Coaxial Microwave Spot Joining Applicator," in Microwaves: Theory and Applications in Materials Processing III, Ceramic Transactions **59**, 347 (1996).

3. Cross, T. E. (Personal Communication), 1995.

4. Westphal, W. B., and A. Sils "Dielectric Constant and Loss Data," Report AFML-TR-72-39, Air Force Materials Laboratory, 1972.

ENHANCED COMPUTER MODELLING FOR HIGH TEMPERATURE MICROWAVE PROCESSING OF CERAMIC MATERIALS

M.P.CRAVEN*, T.E. CROSS*, J.G.P. BINNER**
University of Nottingham, University Park, Nottingham, NG7 2RD, U.K.
*Department of Electrical and Electronic Engineering, mpc@eee.nott.ac.uk
**Department of Materials Engineering and Materials Design

ABSTRACT

A new software package is described for 2-Dimensional modelling of microwave heating. It is particularly suited to the high temperature heating of ceramics, where electrical and thermal properties are strongly dependent on temperature. The described program, x_tlm, runs under X-windows, which allows a microwave cavity and its contents to be entered graphically and stored as a schematic diagram. The diagram is automatically converted into a mesh format for simulation using a topology independent Transmission Line Modelling formulation. Temperature dependencies of material properties are specified as polynomial functions of temperature. Surface heat loss by the radiation mechanism, which dominates at higher temperatures, is also introduced into the model.

INTRODUCTION

The increasing use of microwave heating in the processing of ceramics, such as drying, sintering and joining, has resulted in the need to characterise the relevant material properties, both electrical and thermal, over a wide range of temperatures. Likewise, in attempts at computer modelling the interaction of such materials with microwave energy, the strong temperature dependence of the electrical properties such as permittivity and loss factor, and thermal properties such as thermal conductivity and heat capacity should not be ignored[1]. There is also a requirement for a user-friendly method of entering the microwave cavity and its contents, which may have a complex shape, into the model. Thus it has become necessary to enhance existing software.

THEORY

Transmission Line Modelling

Transmission line modelling (TLM) is an unconditionally stable time-domain modelling technique, based on the analogy between electrical networks and field networks using equivalent circuits[2], and first developed into a modelling technique by Johns and Beurle in 1971[3]. At the University of Nottingham, work has continued into electro-magnetic (e-m) and thermal problems, and more recently into the coupled electro-heat problem (Figure 1), where both the evolution of the electric field inside a microwave cavity, and the subsequent temperature profile of a material placed inside, can be modelled using similar TLM algorithms running at two different time-scales[4]. The coupling is loose because the real time taken to achieve a stable field is of the order of nanoseconds, whereas the heating process takes place over a time period of many seconds, thus allowing the two processes to be executed sequentially. Unfortunately a feedback process also exists due to the temperature dependence of the electrical and thermal parameters. In particular, the loss factor for ceramics has been found to vary greatly as temperature is

351

Mat. Res. Soc. Symp. Proc. Vol. 430 © 1996 Materials Research Society

increased, which is also true to a lesser extent for the real permittivity and the thermal properties. Thus it is necessary to recalculate the field distribution after a certain change in temperature. Furthermore, heat loss from the material surface due to radiation is very significant at high temperatures (due to the T^4 dependence). Earlier work implemented a 2D mesh which modelled a rectangular microwave cavity, a single rectangular load of one type of material, and sources along one of the boundary edges. The e-m model was coupled to the thermal model only to the extent that power dissipation due to dielectric loss was used as the heat source for the thermal model. Heat loss was modelled as a lumped value subtracted from the heating source.

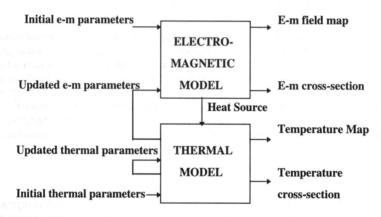

Figure 1 Block diagram of the coupled electro-heat model

Software Design

The software has been redesigned and improved in several ways:
- Hardware independence using X-windows on a UNIX platfrom
- Schematic editor based on overlapping graphical objects
- Multiple types of materials
- Multiple independent sources and sinks along any boundary
- Automatic compilation of schematic to a topology independent TLM model
- Modelling of temperature dependent electrical and thermal parameters
- Modelling of surface heat loss by a radiation mechanism
- Interactive on-line graphing facility

The programming language chosen was standard (ANSI) C linked with an X-windows function library, which makes the software virtually hardware independent. In this work, the program was compiled and executed on a networked Sun workstation, using an IBM compatible personal computer as the X-terminal. X-windows functions can be accessed at various levels. At the lowest level is the Xlib library. Although it is possible to write programs using only Xlib function calls, it is usually better practice (and much easier) to use a toolkit, which makes calls to accomplish standard functions using Xlib. The basic toolkit is the Xt (X Intrinsics) toolkit. However, it is more usual to use a specialised toolkit (built over Xt and Xlib) which incorporates standard functions for buttons and menus etc., such as the proprietary toolkits Motif, OpenWindows and the GNU toolkit Libsx. Libsx was used in this work.

The software design, called x_tlm, is divided into 3 stages each with its associated window(s):
- Schematic capture - schematic drawing window
- Electro-magnetic simulation - e-m 2D map window, e-m cross-section graphing window
- Thermal simulation - thermal 2D map window, thermal cross-section graphing window

In addition a main window is required when the program is first executed, to set up and coordinate the simulation and the other windows.

The schematic window was designed to allow the creation of schematic drawings of a microwave furnace and its contents, based on a set of objects; lines, rectangles, triangles and circles, to be drawn with a computer mouse. Colours are used to represent the different materials; cavity wall, cavity, magnetron sources and materials to be heated (loads). Electrical and thermal parameters, and e-m source/sink parameters are attached to each colour from external databases, as shown in Figure 2. Temperature dependence is described for the properties PERMITTIVITY, LOSS_FACTOR, THERMAL_CONDUCTIVITY and HEAT_CAPACITY as polynomial expressions, the first value denoting the order itself, followed by up to 4 coefficents.

/material.dat	/source.dat
@air	@SINK
PERMITTIVITY 0 1	FREQUENCY 2.45e+09
LOSS_FACTOR 0 0	AMPLITUDE 0
@composite	PHASE 0
PERMITTIVITY 3 4 0 0 0.000000002	REFLECTION_COEFFICIENT 0
LOSS_FACTOR 1 0 0.0003	@A2.45GHz10000
THERMAL_CONDUCTIVITY 1 130 -0.05	FREQUENCY 2.45e+09
HEAT_CAPACITY 0 675	AMPLITUDE 10000
EMISSIVITY 1.0	PHASE 0
DENSITY 3000	REFLECTION_COEFFICIENT -0.9

Figure 2 Example database files containing material and e-m source information

The final schematic consisting of the overlapping objects is super-imposed on to a computational mesh using a multiple pass compilation procedure, in preparation for simulation. The most important feature of the TLM node array after compilation is that it is linear, with each node having knowledge of its own neighbours in the mesh, making the simulation topology independent. The compilation procedure automatically converts the 2D schematic picture (after grabbing it into a 2D pixel array) into its 1D node description in a series of passes as follows:

1. Choose mesh interval and assign a *node index* number to each grid square. Only cavity and load squares are assigned a unique node number (starting with zero). All other grid square are assigned an index number of -1.
2. Around each source line in the schematic, search for nearby boundary grid squares. Any square near enough to the line is assigned a negative index, unique to the source. Thus, all boundary grid squares near Source 0 are given the index -2, squares near Source 1 are given the index -3, and so on.
3. For each grid square with non-negative index, assign four *neighbourhood index* numbers, (denoted n1, n2, n3 and n4), one for each port. A neighbourhood index will be positive if the port is connected to the port of another node in the node array, otherwise it will be negative.
4. Find load thermal surfaces. If a port is the boundary between any load node and a cavity node, or between any load node and a port with a negative index (boundary or source), that port is designated as a thermal surface, and will lose heat by radiation. Each node is given a *surface* value which depends on how many ports are thermal surfaces.

Simulation

Once a design is compiled, the coupled TLM algorithm is executed incorporating a set of simulation options and parameters which are combined with the material and source information described earlier. Some options define how information is displayed to the user during simulation. The other parameters define the time steps for the different parts of a simulation run. A typical coupled microwave heating simulation proceeds as follows:

1. Run the e-m simulation for a number of time-steps until the field is stable.
2. Run the thermal simulation for one time-step
3. Update temperature dependent parameters using the polynomial descriptions
4. Run the e-m simulation until stable again. This may take less time than step 1.
5. Repeat steps 2-4 for as long as the thermal simulation is to run.

The e-m time-step is given by,

$$\Delta t_{em} = \frac{\Delta l}{c\sqrt{2}}$$

where Δl is the mesh spacing and c is the velocity of light. Since the stability of the e-m simulation is dependent on many factors, the program allows the user to decide how many time-steps to allocate to each part of the e-m simulation, before running, thus ensuring maximum flexibility. All time parameters are specified as the integer exponent of a power of 2 in the interests of computational efficiency. The thermal model is not started until $2^{thermal\_start}$ e-m time-steps have elapsed, corresponding to step 1 above. After this, the thermal model is run for one thermal time-step after every $2^{thermal\_update}$ e-m time-steps (step 4 above). The temperature increment ΔT due to absorbed e-m power is calculated for each node every $n = 2^{em\_power}$ e-m time-steps using,

$$\Delta T = K\frac{l}{n}\sum^{n}(_{em}V_{t}^{2})$$

which is an approximation to the integral, summed over n samples. This increment is used as the source term in the TLM thermal model. $_{em}V_t$ is the total node voltage in the TLM e-m model, which is analogous to the instantaneous electric field E. The constant K is found to be,

$$K = G\frac{\Delta l^{2}}{4}\left[\frac{1}{2K_{t}\Delta l} + \frac{2\Delta t_{th}}{C_{p}\rho\Delta l^{3}}\right]$$

where Δt_{th} is the specified thermal time-step and Δl is the inter-node distance as before. Material properties are the thermal conductivity K_t, specific heat capacity C_p and density ρ, all of which depend on the type of material and $G = \varepsilon''\varepsilon_0\omega\Delta l$ is the TLM *shunt conductance* of the node, where ε'' is the dielectric loss factor, ε_0 is the electrical permittivity of free space, and ω is the angular frequency of the source. Thus K will in general be different for each node and will always be zero for materials with a zero loss factor.

Software Verification

A verification procedure was devised to check the correct operation of the schematic entry of a design and its simulation in a microwave heating experiment as follows:

- Frequency Ratio - The ratio of the free space wavelength λ_0 to the wavelength λ_m in a material of relative real permittivity ε' should be $\sqrt{\varepsilon'}$.
- Voltage Standing Wave Ratio (VSWR) - The ratio of the maximum and minimum values of the wave envelope in the cavity should also be equal to $\sqrt{\varepsilon'}$.
- Skin Depth δ - In a material with $\varepsilon'' \neq 0$, the amplitude of the field is reduced to a fraction $1/e$ in distance δ, determined from $\delta=(x_1-x_2)/\log_e(E_1/E_2)$, where E_1 is the field at distance x_1 into the load, and E_2 is the attenuated field at a distance x_2. The measured value may then be compared to the theoretical value, $2/\omega\sqrt{\{2\mu\varepsilon_0\varepsilon'(\sqrt{\{[\varepsilon''/\varepsilon']^2+1\}}-1)\}}$.
- Energy conservation in the thermal model - The equilibrium temperature T of a perfectly insulated material of area A_1+A_2 initially with the area A_1 at temperature T_1 and area A_2 at temperature T_2 is equal to $(A_1T_1+A_2T_2)/(A_1+A_2)$, which can be compared to the simulated value.
- Radiative heat loss - Rate of change of temperature due to surface radiation can be compared to theory using the diffusion equation and black-body radiation law. This can be carried out by setting each node in the thermal model to the same initial value above ambient temperature, and examining the time evolution of the temperature profile as the thermal simulation progresses.
- Energy conservation in coupling - The electrical power of a TLM source incident on to a free space cavity is given by $P_1 = E^2_{rms}/Z_0$ per unit length of source, where Z_0 is the impedance. Power absorbed by the load is $P_2=\rho C_p\Delta T/\Delta t$ per unit area where ΔT is the change in temperature in time Δt. If the load is insulated so there is no radiation loss, P_1 should be equal to P_2.

RESULTS

The EM model was tested using rectangular cavity and a load of 50x50 nodes with a mesh spacing of 0.0025m, divided in half, with air on the left and dielectric with $\varepsilon'=9$ on the right. A source of 2.45GHz was selected, with amplitude 10000V/m and reflection coefficient -1, running along the left hand wall. The right hand wall was defined as a sink with zero reflection coefficient. Initially ε'' was set to zero in both the cavity and the load, so that no heating occurred. Figure 3 shows examples of x_tlm windows used during verification. Reading clockwise, these are the schematic window showing the design, the 2D e-m map, a horizontal cross-section at the centre of the cavity, and finally a thermal map of the dielectric which had had its top left node initialised at four times the ambient temperature of 300K, shown as it approached equilibrium. The difference in wavelength in the cavity and load can clearly be seen. The cavity wavelength was measured as $\lambda_0 = 0.1225$m, to the closest mesh point, confirming the frequency c/λ_0 as 2.449GHz. Wavelength in the dielectric λ_m, measured by averaging peak-to-peak distances, was found to be 0.0408. This gave $\lambda_0/\lambda_m=3.00=\sqrt{\varepsilon'}$ as required. The amplitude of the e-m field in the cavity varied between two extremes, which yielded a VSWR of 3.00, as predicted by theory. Skin depth δ was verified by changing the loss factor in the load to a non-zero value. Using $\varepsilon''=2.5$ the theoretical skin depth is 0.04721, whereas the value measured from data dumped from the simulation was 0.046, an error of less than 3%. A graph of the RMS e-m field inside the cavity showed the familiar exponential decay. The measured average equilibrium temperature was found to be 300.41, close to the theoretical value of 300.36 expected for a single node initialised at 1200K. The second experiment was carried out with a circular load (not shown) which was initialised with all nodes at four times the ambient temperature, this time using the radiative loss mechanism with an emissivity equal to 1. The temperature profile was close to that expected from theory. The final equilibrium temperature was then measured for the same load but without radiative loss, heated from 300K for 15 minutes of thermal model time in order to verify the coupling mechanism, which also behaved as expected.

355

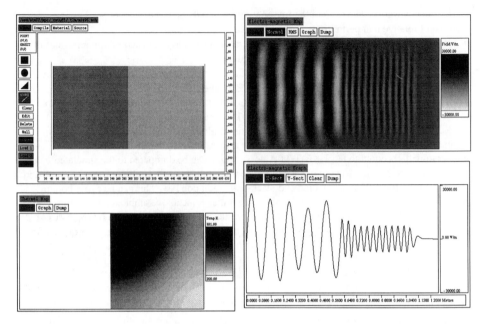

Figure 3 Examples of x_tlm windows (clockwise); schematic, e-m map, e-m graph, thermal map

CONCLUSIONS

New software for implementing the coupled electromagnetic and thermal 2-Dimensional TLM model for microwave heating has been designed, tested and verified. The graphical user interface was simple to use, allowing designs to be entered quickly. Interactive display of 2D map and x or y cross-section graphs enabled the progress of the simulation to be monitored easily. The inclusion of polynomial expressions for electrical and thermal material parameters provides the means to model load materials in a more sophisticated manner. Likewise the radiation model for surface cooling provides a more realistic heat loss mechanism for the heating process.

REFERENCES

1. TIAN Y.L., FENG J.H., SUN L.C. and TU C.J., "Computer modeling of two dimensional temperature distributions in microwave heated ceramics", in BEATTY R.L. et al (Eds.) "Microwave Processing of Materials III", Mat. Res. Soc. Symp. Proc. Vol 269, pp231-243, 1992
2. KRON G., "Equivalent circuit of the field equations of Maxwell", Proc. IRE, Vol. 32, pp289-299, 1944
3. JOHNS P.B. and BEURLE, R.L., "Numerical solution of two-dimensional scattering problems using transmission-line matrix", Proc. IEE, Vol. 188, pp1203-1208, 1971
4. TRENKIC V., CHRISTOPOULOS C. and BINNER J.G.P, "The application of the transmission-line modelling technique (TLM) in combined thermal and magnetic problems", in LEWIS R.W. (Ed.) "Numerical Methods in Thermal Problems", Vol. VIII, Part 2, (Pineridge Press, Swansea, UK), pp1263-1274, 1993

NUMERICAL SIMULATION OF RESONANT CAVITY MICROWAVE SYSTEMS FOR MATERIALS PROCESSING

TIMOTHY A. GROTJOHN, JES ASMUSSEN
Dept. of Electrical Engineering, Michigan State University, East Lansing, MI 48824

ABSTRACT

A single mode microwave resonant cavity used for materials heating is simulated using the FDTD method. The simulation includes both the resonant cavity structure and the input microwave power coupling structure. An overlapping grid method with interpolation between the conformal grid systems developed for each section of the microwave system is used to simulate geometrically complex microwave structures. Comparisons of the FDTD solution for a simplified, lossy-loaded cavity is made with a known analytical solution.

INTRODUCTION

The finite-difference time-domain (FDTD) method is one of the solution techniques for the simulation of microwave heating of materials in single mode and multimode cavities[1]. The FDTD method permits the simulation of microwave power absorption into spatially varying lossy material loads in cavities. The accurate simulation of cavities requires that the cavity geometry, input power coupling structure, and lossy materials shape, size and properties be accurately represented by the FDTD grid structure. To simulate complex shaped electromagnetic structures using the FDTD technique a number of methods[2-4] have been developed including cartesian, cylindrical and spherical coordinate system implementations of the FDTD equations. Additionally, non-orthogonal and non-orthogonal/unstructured grid implementations have also been developed. The non-orthogonal grid systems generally require a solution time approximately 2-4 times that of an orthogonal grid system (cartesian, cylindrical, or spherical). In this paper we apply and extend the overlapping grid method developed by Yee, Chen, and Chang [5] to the solution of complex-shaped, material-loaded, resonant microwave cavities. The FDTD solution developed in this paper represents the various parts of the microwave system with a locally conformal grid structure using either cartesian or cylindrical grid system. The local coordinate systems are constructed so that an overlapping of the grids occurs. In the overlapping regions an interpolation method is used to connected the solutions in the two regions. This paper also illustrates the application of the FDTD method to the simulation of controlled mode microwave cavity tuning characteristics.

MODEL DESCRIPTION

The Maxwell's curl equations, which govern the distribution of microwave fields in the resonant cavity, are given by

$$\nabla \times \vec{E} = -\mu \frac{\partial}{\partial t} \vec{H} \tag{1}$$

$$\nabla \times \vec{H} = \varepsilon \frac{\partial}{\partial t} \vec{E} + \vec{J} \tag{2}$$

where $\vec{E}$ and $\vec{H}$ are the electric and magnetic fields, μ is the permeability, ε is the permittivity, and $\vec{J}$ is the current density. The finite difference method is formulated by discretising the partial differential equations (1) and (2) with a centered difference approximation in both the time and

Mat. Res. Soc. Symp. Proc. Vol. 430 © 1996 Materials Research Society

space domain. The three electric field components (one for each orthogonal direction) and the three magnetic field components are interleaved in space. The two equations are solved in the time domain using a leap-frog method with the E and H values being calculated at alternating half-time steps. Here, both cylindrical and cartesian coordinate system solutions are utilized with the time step and grid spacing along each coordinate direction being chosen appropriately in order to satisfy both the accuracy and stability conditions of the finite-difference equations[6].

The lossy material in a microwave resonant cavity has a conductivity σ that leads to a current density given by

$$\vec{J} = \sigma \vec{E} \tag{3}$$

The current density contributes to the electromagnetic filed solutions according to (2). The current density also leads to power absorption in the lossy material given by

$$P(\vec{r}, t) = \vec{J}(\vec{r}, t) \cdot \vec{E}(\vec{r}, t) \tag{4}$$

Within the simulation region the local values of ε, μ, and σ are assigned at each grid point. Another quantity of interest for resonant cavity simulations is the electromagnetic energy, $U(t)$, stored in the cavity which is given by

$$U(t) = \frac{1}{2} \int_V \{\varepsilon(\vec{r}) [E(\vec{r}, t)]^2 + \mu(\vec{r}) [H(\vec{r}, t)]^2\} dV \tag{5}$$

where the integration is performed for the volume of the cavity V.

IMPLEMENTATION AND RESULTS FOR AN IDEALIZED MICROWAVE CAVITY

The FDTD method is implemented in this section for a material loaded microwave resonant cavity[7]. This first implementation is idealized by using a cylindrical cavity, which is 15.25 cm in diameter, loaded with a lossy homogeneous load of diameter 2.5 cm that is centered in the cavity and that is the same height as the cavity. The electromagnetic fields are excited at selected grid points inside the simulation region based on the expected electromagnetic mode within the resonant cavity, i.e, the input microwave power structure in not considered in this section.

The natural frequency describes the transient response of the cavity to a pulse, or to having a continuous-wave source shut off. For a resonator with a lossy load, the natural frequency is a complex number. The real part of the natural frequency describes the frequency of the field oscillations while the imaginary part of the natural frequency is the exponential decay coefficient due to power absorption. The natural frequency response of a material loaded microwave cavity can be done by first exciting the microwave cavity with a continuous electromagnetic mode, then the source is turned off and the natural oscillating frequency and electromagnetic field decay are monitored. The relationships used include

$$\hat{\omega} = \omega' + j\omega'' \tag{6}$$

$$E(t) = E_0 e^{-\omega'' t} e^{j\omega' t} \tag{7}$$

$$U(t) = U_0 e^{-2\omega'' t} \tag{8}$$

where $\hat{\omega}$ is the complex natural frequency with a real part ω' and an imaginary part ω'', E_0 is the initial electric field amplitude and U_0 is the initial stored electromagnetic energy at the time the source is turned off. By simulating the cavity electromagnetic fields and noting the electric field oscillating frequency and the rate of stored energy decay after the source is turned off, the real and the imaginary part of the natural frequency can be obtained.

An idealized resonant cavity as described above is excited in the TE_{011} mode. To study the effect of the load conductivity on the electromagnetic fields inside the resonating structure, the conductivity of the load was varied from zero to several hundred mho/m. The plot of the simulated total stored energy decay is shown in Figure 1. The decay rate of the stored energy varies with the different conductivities. For example, at $\sigma = 0.1$ mho/m the result was $\omega' = 20.68x10^9$ rad/sec and $\omega'' = 0.09x10^9$ rad/sec, at $\sigma = 1.0$ mho/m the result was $\omega' = 21.13x10^9$ rad/sec and $\omega'' = 0.45x10^9$ rad/sec, and at $\sigma = 10$ mho/m the result was $\omega' = 21.59x10^9$ rad/sec and $\omega'' = 0.16x10^9$ rad/sec. The radial variation of E_ϕ for the three different conductivity loads is shown in Figure 2. For the case of the larger conductivity, the electric field decays rapidly into the 2.5 cm diameter load.

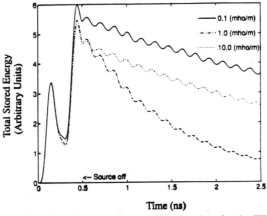

Figure 1: Total stored electromagnetic energy U versus time for the TE_{011} mode for three different load conductivities. The source is turned off at the indicated time.

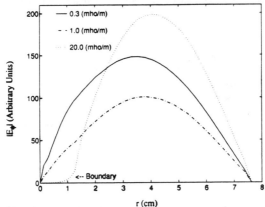

Figure 2. Radial dependence of $|E_\phi|$ (arbitrary units), for the TE_{011} mode.

An s-plane frequency chart can be used to further detail the effect of conductivity variations on the electromagnetic fields. As shown in Figure 3, the horizontal axis represents the imaginary part of the natural frequency, ω'', and the vertical axis is the real part of the complex natural frequency, ω'. For $\sigma = 0$, the cavity load is lossless and hence $\omega'' = 0$. When the conductivity increases from zero, ω'' starts to increase, which is due to power dissipated into the lossy load. After σ is larger than 1.0 mho/m in this example, ω'' starts to decrease, because the cavity load becomes a conducting like material and the electromagnetic fields only penetrate the load a small amount.

For the case of a resonator loaded by a lossy, nondispersive coaxial load as just simulated, analytical solution of ω' and ω'' exist. Figure 3 compares the FDTD method solution to an analytical solution which is possible for this simplified geometry case[7,8]. The analytical solution and the FDTD solution show good agreement. The analytical solution method is good for verifying the FDTD method for simple geometries, but for more complex geometries the analytical solution becomes difficult to formulate and solve.

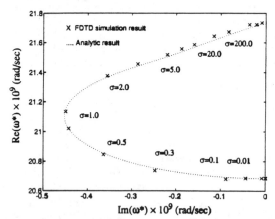

Figure 3. s-plane plot for the complex frequency as a function of load conductivity for the TE011 mode.

IMPLEMENTATION AND RESULTS FOR MICROWAVE CAVITY TUNING

The simulation of the input power coupling structure and the tuning characteristics of a microwave resonant cavity was studied using the structure shown in Figure 4. The cylindrical structure has a coaxial power feed at the center of the top plate of the cavity. The resonant cavity is adjustable in size by moving the upper end of the resonant cavity up or down to change the height of the resonant structure. The specific mode within the resonant cavity is not assumed a priori in this simulation. The only mode assumption is that the coaxial feed is operating in a standard TEM coaxial mode. The plasma load for this simulation was placed at the bottom of the cavity. It has a diameter of 8.85 cm and a height of 2.0 cm. The coaxial mode was excited at the indicated region in the coaxial input power feed. The open boundary of the coaxial power feed was terminated with a non-reflecting boundary. Simulations of the cavity were performed for different cavity heights. Figure 5 shows the amount of power that was reflected back out of the cavity to the non-reflecting boundary. The reflected power reached a minimum for a cavity height of approximately

14 cm which corresponds to the expected empty cavity mode TM_{012}. Plots of the internal fields within the cavity confirm the presence of the TM_{012} mode. Figure 5 shows the reflected power for two different load conductivites. The reflecting power was determined in the simulations using the Poynting vector given by $\vec{E} \times \vec{H}$.

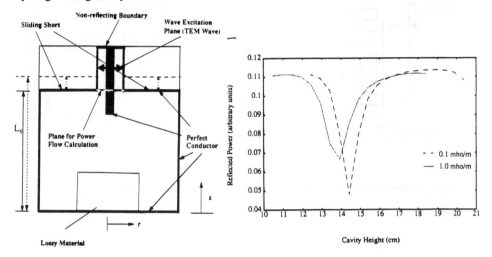

Figure 4: Resonant cavity structure.   Figure 5: Reflected power versus cavity height.

IMPLEMENTATION AND RESULTS FOR A COMPLEX GEOMETERY CAVITY

The FDTD method is applied next to the structure shown in Figure 6. This structure is a microwave resonant cavity with a side-feed power input and a material load shape that does no match the cavity shape. The simulation of this structure is done using three FDTD coordinate systems. The separate grid systems are used to model the main resonant cavity (cylindrical coordinates with the axis up), the input power feed (cylindrical coordinates with the axis to the right), and the materials load (cartesian coordinates). The three grid systems are overlapped as shown for the main resonant cavity coordinate system and the materials load coordinate system in Figure 7. The two grid systems are interpolated using the technique initially described in [5]. The interpolation is performed so that a wave propagating from one grid system to the next will do so with the correct speed, no reflection, and no distortion.

An example simulation for the structure shown in Figure 6 was done with a load conductivity of 0.03 mho/m. The input power was applied for the first 10,000 time steps and then turned off. The natural response of the total stored electromagnetic field energy is shown in Figure 8.

SUMMARY

The simulation of a geometrically-complex microwave resonant cavity used for materials processing has been done using the FDTD method. To model the cavity accurately the grid structure is conformal to the boundaries of the cavity and power coupling components. The conformal grid established for each component is overlapped with grids in neighboring regions and an interpolation is done between grids. This method of implementing the FDTD technique permits time-efficient and accurate solutions of geometrically-complex resonant cavities.

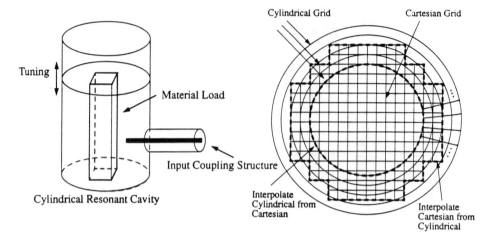

Figure 6: Microwave cavity structure.   Figure 7: Overlapping grid example.

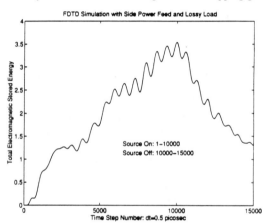

Figure 8: Total stored electromagnetic energy versus time.

REFERENCES

1. M. F. Iskander, R. L. Smith, A. O. M. Adrande, H. Kimrey, Jr., and L. M. Walsh, IEEE Trans. Microwave Theory Tech. 42, 793 (1994). (And references contained within).
2. R. Holland, IEEE Trans. on Nuclear Science 30, 4589 (1983) and IEEE Trans. on Nuclear Science 30, 4592 (1983).
3. J. F. Lee, R. Palandech, and R. Mittra, IEEE Trans. Microwave Theory Tech. 40, 346 (1992).
4. P. H. Harms, J.F. Lee and R. Mittra, IEEE Trans. Microwave Theory Tech. 40, 741 (1992).
5. K. S. Yee, J. S. Chen, and A. H. Chang, IEEE Trans. on Attenas and Propag. 40, 1068 (1992).
6. A. Taflove and M. E. Brodwin, IEEE Trans. Microwave Theory Tech. 23, 623 (1975).
7. W. Y. Tan, PhD Dissertation, Michigan State Univ., (1994).
8. E. B. Manring, PhD Dissertation, Michigan State Univ. (1992).

The MiRa/THESIS3D-Code Package for Resonator Design and Modeling of Millimeter-Wave Material Processing

L. Feher, G.Link, M. Thumm*

Forschungszentrum Karlsruhe, Technik und Umwelt
Institut für Technische Physik-ITP
P.O.Box 3640, D-76021 Karlsruhe, Germany
* and University Karlsruhe,
Institut für Höchstfrequenztechnik und Elektronik,
Kaiserstr. 12, D-76128 Karlsruhe, Germany

Abstract. Precise knowledge of millimeter-wave oven properties and design studies have to be obtained by 3D numerical field calculations. A simulation code solving the electromagnetic field problem based on a covariant raytracing scheme (MiRa-Code) has been developed. Time dependent electromagnetic field-material interactions during sintering as well as the heat transfer processes within the samples has been investigated. A numerical code solving the nonlinear heat transfer problem due to millimeter-wave heating has been developed (THESIS3D-Code). For a self consistent sintering simulation, a zip interface between both codes exchanging the time advancing fields and material parameters is implemented. Recent results and progress on calculations of field distributions in large overmoded resonators as well as results on modeling heating of materials with millimeter waves are presented in this paper. The calculations are compared to experiments.

Millimeter-wave material processing at Karlsruhe Research Center

Microwave technology for industrial material processing is an advantageous, but sophisticated alternative to conventional processing and heating techniques [1]. Precise knowledge and experience is neccessary for a controlled and optimized processing of charges in an industrial scale. Since 1994 R&D investigations have been undertaken at the Karlsruhe Research Center, ITP for processing ceramic samples with millimeter waves [2]. Possible benefits from applying HF radiation to dielectric materials include rapid volumetric heating of ceramics, resulting in enhanced material properties (high densification) or new materials (e.g. plastic nanoceramics). A small and compact 30 GHz gyrotron installation GOTS (Gyrotron Oscillator Technological System) working at the second harmonic TE_{02} mode has been chosen as the radiation source. The installation has been originally developed and manufactured by the Institute of Applied Physics (IAP) in Nizhny Novgorod (Russia). A cylindrical, water cooled applicator with a volume of 100 l (height=50 cm) is used for material processing, especially for sintering of ceramic powders. Advantages of using millimeter waves are better absorption properties of the ceramics and an enhanced field uniformity within an applicator [3]. In fact, a highly uniform electric field is the crucial state that has to be achieved for producing equivalent and stressfree conditions for an industrial high quality ceramic charge processing. Simulations and experiments have shown the need for optimizing the present configuration. Focusing effects due to the resonator geometry and beam reflections within the resonator result only in a limited yield for a processing area. Temperature development and material processing dynamics as well as the resonator design problem and the electric loss interaction of the sample and oven material have to be investigated. Thermal insulation of ceramic samples in addition is a disadvantageous requirement under industrial guidelines increasing costs and processing time. Measurements and simulations showed high temperature gradients within the ceramics leading to high thermal stresses and unsuccessful densification caused by the direct thermal radiation at the ceramics surface.

Optical field calculations with the MiRa-Code

Many computational codes and techniques for numerical field calculations have been developed and are available [4],[5],[6]. Calculation of stationary field distributions for large overmoded resonators imply an enourmous amount of memory allocation [7] and cost intensive computation time. The verification of the simulation results by experiments is essential. The MiRa-Code (Microwave Raytracer) has been developed for this purpose as a gridless (in the sense of not discretizing differential operators on a grid), analytical approach intended to be a tool for engineering support simulating complex resonator geometries and to perform design studies on small workstations.

A ray formalism representing the full properties for the electromagnetic fields in the stationary state provides the theoretical basis for this code [8]. This allows a description as a monochromatic harmonically varying wave field for the vector potential $\vec{A}(\vec{x}, t)$

$$\vec{A}(\vec{x}, t) = \vec{A}(\vec{x})e^{-i\omega t} \quad . \tag{1}$$

Taking advance of gauge transformations one can always impose the condition $\Phi(\vec{x}, t) = 0$. If now $\vec{A}(\vec{x}, t)$ is known, the fields

$$\vec{B}(\vec{x}, t) = \nabla \times \vec{A}(\vec{x}, t) \qquad \vec{E}(\vec{x}, t) = i\omega \vec{A}(\vec{x}, t) \tag{2}$$

can be determined immediately. The stationary vector potential is determined as an eigenvalue problem by the reduced wave equation (Helmholtz equation)

$$\Delta \vec{A}(\vec{x}) + k^2(\vec{x})\vec{A}(\vec{x}) = 0 \quad . \tag{3}$$

The complex wavenumber $k(\vec{x})$ is related to the material and wave properties according to $k(\vec{x})^2 = \frac{\omega^2}{c^2}\epsilon_r^*(\vec{x})$ where ϵ_r^* is the complex dielectric constant. For obtaining determining equations in optical terms, we express the vector potential as a wavefield by a combination of a spatial amplitude $a^\mu(x^\alpha)$ and spatial phase using a covariant vector notation $(\alpha, \mu = 1, 2, 3)$

$$A^\mu(x^\alpha) = a^\mu(x^\alpha)e^{ik_0 S(x^\alpha)} \quad . \tag{4}$$

The covariant derivative in 3 dimensions $T^a_{;n} = T^a_{,n} + \Gamma^a_{nm}T^m$ as well as $T_{a;n} = T_{a,n} - \Gamma^m_{an}T_m$ where T^a is an arbitrary 3 dimensional vector $(a, n, m = 1, 2, 3)$ has to be introduced. The Γ^a_{nm} are defined in [8]. Inserting eq.(4) in eq.(3) yields after evaluation for real and imaginary parts a polynomial of second order in k_0 giving an exact representation of (3) by the coupled differential equation system, second order in space. Matching coefficients gives determining equations for $a^\mu(x^\alpha)$ and $S(x^\alpha)$

$$S^{,\nu}(x^\alpha)S_{;\nu}(x^\alpha) = \epsilon_r'(x^\alpha) \tag{5}$$

$$[a^\mu(x^\alpha)S(x^\alpha)]^{,\nu}_{;\nu} = 0 \tag{6}$$

with the condition

$$[a^\mu(x^\alpha)]^{,\nu}_{;\nu} = 0 \quad . \tag{7}$$

Eq.(5) is equivalent to the result of the scalar Sommerfeld eikonal approach of geometrical optics; eq.(6) determines the amplitude and phase properties of the wave field with $a(x^\alpha)^\mu$ being a function class given by eq.(7). A generalized Fermat's principle leads to ray equations for tracing phase and amplitude in space. As an example, Fig.1 shows a spherically symmetric point source situated in an homogeneous or inhomogeneous material. The propagating rays and wavefronts are distorted by the inhomogeneous dielectric distribution.

The dissipated power within a lossy dielectric can be evaluated from the electric field according to $\vec{E}(\vec{x})$

$$P_{elec} = \frac{1}{2} \iiint \sigma(\vec{x})|\vec{E}(\vec{x})|^2 dV = \frac{1}{2}\omega\epsilon_0 \iiint \epsilon_r''(\vec{x})|\vec{E}(\vec{x})|^2 dV \quad . \tag{8}$$

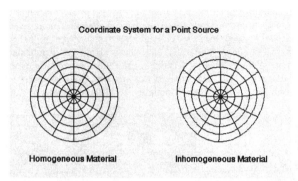

Figure 1: Wavefield distortion in an inhomogeneous medium

Results for the GOTS applicator

A mode stirrer at the top of the applicator, rotating on a conical shape around the cylindrical axis is intended to smooth inhomogeneous field distributions in the time average. For measuring

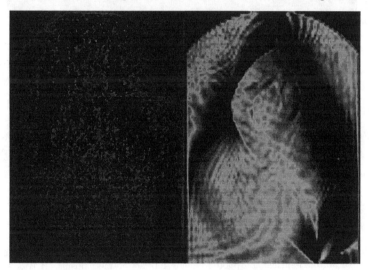

Figure 2: Overlayed thermopaper measurements and MiRa simulation without mode stirrer

the energy density $\sim \vec{E}(\vec{x})^2$, thermopaper was exposed to the millimeter waves and darkened at the more intense regions. Due to nonlinear effects, reliable results have to be obtained by considering a set of thermopapers. For this reason, Fig.2 shows in the applicator's vertical plane with the Gaussian beam input at the bottom several computationally overlaid photos of thermopaper measurements with removed mode stirrer. Good agreement with the simulation (right) can be seen: the beam is reflected at the cylindrical applicator surface and focused by the spherical top leading to caustics of high field intensity and an inhomogeneous distribution. Fig.3 shows results for the operating stirrer. A much higher degree of homogeneity can be realized in the photograph and simulation. The corresponding mean field obtained by the MiRa-Code was calculated from 20 different positions of the mode stirrer. A significant phase pattern in the right applicator region can be seen very well in the simulation and experiment. The intense caustic due to the spherical top has vanished (due to modeling the stirrer simply as a disc a small field enhancement remains in the simulation). An analysis of the single positions with

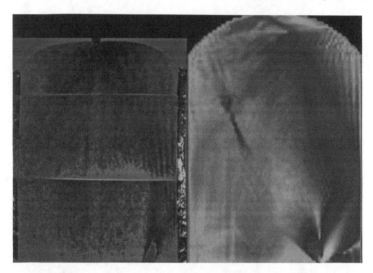

Figure 3: Single thermopaper measurement and MiRa simulation with mode stirrer

the MiRa Code showed the beam being reflected at the bottom and cylindrical surface leading to moving field enhancements in the defined material processing area. This behavior has been verified by temperature measurements of processed ceramics with temperature peaks following the mode stirrer period.

As a conclusion from experiment and simulation, the current applicator configuration is not able to provide uniform conditions on a larger scale for millimeter-wave processing of several samples.

Ceramics sintering simulation by the THESIS3D-Code

Dissipation of millimeter waves in nonpolar, dielectric materials represents a combination of dipole relaxation mechanisms and photon-phonon interaction mechanisms. It is described by the dielectric loss factor $\epsilon_r^{''}$. A thermodynamical approach gives a possibility to solve numerically the equations for such a collective system. In this case, relevant material properties and their dependence on temperature have to be obtained from experiments. Energy conservation and thermal transfer are the basic laws for the temperature development $T(\vec{x}, t)$ of a material represented by the inhomogeneous diffusion equation

$$c_V(\vec{x}, T)\rho(\vec{x}, T)\frac{\partial}{\partial t}T(\vec{x}, t) - \sigma_T(\vec{x}, T)\Delta T(\vec{x}, t) - \mathrm{grad}\sigma_T(\vec{x}, T) \cdot \mathrm{grad}T(\vec{x}, t) = p_{eff}(\vec{x}, t) \qquad (9)$$

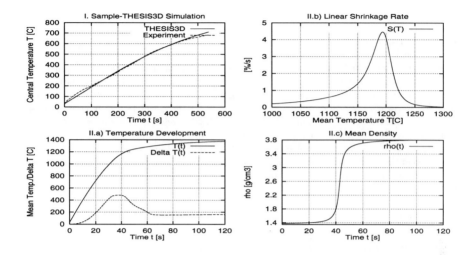

Figure 4: I.Comparison THESIS3D/Experiment - II.Calculations for small sintered sample

$\sigma_T(\vec{x}, T)$ is the thermal conductivity, $\rho(\vec{x}, T)$ the density and $c_V(\vec{x}, T)$ is heat capacity of the material. The effective power density $p_{eff}(\vec{x}, t)$ results in contributions of microwave heating $p_{elec} = \frac{dP_{elec}}{dV}$, eq.(8) and heating losses by radiation $p_{rad} \sim T^4$, conduction p_{cond} and convection $p_{conv} \sim T$. Due to the dependence of the material parameters on the temperature T, the thermal transfer process is a highly nonlinear problem. The THESIS3D-Code (THErmodynamical SIntering Simulation), a staggered grid FDTD-Code has been developed for the time and space integration of (9). Dynamic contracting grids are implemented to simulate the temperature dynamics during the densification process and estimate the densification rate of a ceramic sample. Experimental data for e.g. Al_2O_3 have been taken for these calculations from the literature: $\sigma_T(T)$, thermal diffusivity $\alpha(T)$ [9], $c_V(T)$ [10], $\epsilon'(T), \epsilon''(T)$ [11], emissivity $\epsilon(T)$ [12]. In Fig.4I., the measured time development of the sample's central temperature in comparison to the simulation is depicted at an GOTS input power of P_{elec}=0.6 kW for an Al_2O_3 sample without thermal insulation (volume $\approx$ 30 cm^3,initial density $\approx 1.4 \frac{g}{cm^3}$). Only 12 % of the input power is initially dissipated in the green body. The calculation showed that the inner region of the sample was heated more strongly as a result of the volumetric heating by millimeter-waves and the thermal radiation at the sample's surface; this caused the densification and contraction to start at the centre. At this input power, no relevant densification has been achieved. Another sample (Fig.4II.a)-c)) of half the size ($\frac{1}{8}$ Volume) is predicted to be sintered at P=1.4 kW, heating up to a mean equlibrium temperature of $\approx$ 1390°C. The sintering process leads to an increased thermal conductivity during the contraction reducing the temperature difference center/surface $\Delta T(t)$. The calculations estimate to achieve $\approx$ 97% of the theoretical density for this sample.

Outlook and conclusions

From our point of view, two basic engineering problems for industrial applications remain:
The choice of resonator design:
In order to achieve field homogeneity over most of a large overmoded resonator, optimized geometries have to be developed and tested by simulation and experiment.

Low thermal gradients:
An effective reduction of thermal gradients within ceramic samples during sintering by thermal insulation or hybrid combination with external heating has to be achieved.
Theoretical modeling and simulation can give important contributions to these developments in micro- and millimeter-wave material processing technology. Electromagnetic wavefields can be represented by a ray expansion. The optical approach of the MiRa-Code leads to a dramatic reduction in memory allocation and computation time for the calculation of large overmoded applicators suitable for design studies. With the THESIS3D-Code, time dependent material dynamics are investigated and predictions for densification and temperature gradients are obtained. An implemented zip concept between the THESIS3D and MiRa-Code allows a self consistent simulation of the nonlinear field/thermal effects in dielectric ceramics during the heating. These simulations are now in progress for obtaining a better understanding of the generation and dynamics of hot spots.

Acknowledgement

We thank Prof. Edith Borie for critically reading the manuscript, Dr. G. Müller and P. Mehringer (University of Stuttgart,IPF) for fruitful discussions.

References

[1] A.C. Metaxas, R.J. Meredith, "Industrial Microwave Heating", Peregrinus Ltd., ISBN 0906048893

[2] Yu.V. Bykov,A.G. Eremeev, V.A. Flyagin et al., L.Feher, M. Kuntze, G. Link, M. Thumm, "Gyrotron System for Millimeter-Wave Processing of Materials", Microwave and Optronics 1995, May 30-June 1, Stuttgart Sindelfingen, Germany

[3] Yu.V. Bykov, V.E. Semenov, "Processing of Material Using Microwave Radiation" in "Application of High-Power Microwaves", Artech House Inc. Norwood MA, (1994) ISBN 0-89006-699-x,319-351

[4] Kurt L. Shlager, John B. Schneider, "A Selective Survey of the FDTD Literature", IEEE Ant. and Propagation, Vol.37 No.4, August 1995,39-57

[5] Zh. Huang,M.F. Iskander,J. Tucker, H.D. Kimrey, "FDTD Modeling of Realistic Microwave Sintering Experiments", MRS Symp. Proc., Vol. 347, 1994, 331-345

[6] M.F. Iskander,R.L. Smith,O. Andrate, H. Kimrey, L. Walsh, "FDTD simulation of Microwave Sintering of Ceramics in Multimode Cavities", IEEE Transactions on Microwave Theory and Techniques, Vol.42, May 1994, 167-184

[7] L.Feher, G.Link, M.Thumm, "Microwave Raytracing in Large Overmoded Industrial Resonators", Proc. Meeting ACERS'95, Microwaves: Theory and Application in Mat. Proc. III Symp., Cincinnati, 1995

[8] L.Feher, G.Link, M.Thumm, "Theoret. Aspects of Microwave Raytracing Calculations in Screened Structures", Proc. Latsis Symp. of Comp. Electromagn., Sept. 19-21 th, Zürich 1995, 236-241

[9] F.Raether, G. Müller, "New In Situ Measuring Methods For The Optimization Of Sintering Processes", Proc. Fourth Euro-Ceramics Vol.2, European Ceramic Society, Faenza 1995

[10] H.U. Nickel, "Hochfrequenztechn. Aspekte zur Entwicklung rückwirkungsarmer Ausgangsfenster für Millimeterwellengyrotrons hoher Leistung", Thesis 1995, Univ. of Karlsruhe

[11] "Millimeter-Wave Dielectric Property Measurement Of Gyrotron Window Materials", Rockwell International Science Center, April 1984

[12] R. Spiegel, "Wärmeübertragung durch Strahlung, Vol.1", Springer Verlag 1988

MICROWAVE-ASSISTED IGNITION

J.K. Bechtold, M.R. Booty, and G.A. Kriegsmann,
Department of Mathematics,
Center for Applied Mathematics and Statistics,
New Jersey Institute of Technology,
University Heights,
Newark, NJ 07102-1982

Abstract

The microwave heating and ignition of a combustible material is modeled and analyzed in the small Biot number and large activation energy regimes. Both the temporal and spatial evolution of the temperature within the material are described. The ignition characteristics are determined by a localized equation for the perturbation to the inert temperature, which is shown to exhibit thermal runaway behavior. Analysis of this local equation provides explicit ignition conditions in terms of the physical parameters in the problem.

Introduction

In recent years, microwave heating has been proposed as an alternative to ignite materials during the process of self-propagating high-temperature synthesis (SHS). It has been shown that microwaves may possess several advantages over more conventional heating techniques [1,2]. For example, they are more efficient since the internal chemical energy of the sample is used during the actual conversion process rather than externally applied energy provided by ovens. Microwaves also penetrate the material to produce volumetric heating, which is more uniform than that of a current-carrying coil, and thus can be used to increase control over the ignition process. Finally, the deposition of heat in the sample interior causes the combustion wave to originate there and propagate outward. This may lead to products of higher purity, since contaminants in the fresh mixture will be driven toward the surface of the product, rather than the interior.

In this paper we develop a simple one-dimensional model for the microwave heating and ignition of a solid combustible material. Our model accounts for the coupling between the electric field and the temperature inside the material. We perform an asymptotic analysis for both small Biot number and large activation energy. The former limit is appropriate in many microwave heating experiments when the convective heat loss at the boundaries is small compared to the heat conduction across the sample. The latter limit is typical of most combustion systems and results in chemical activity being confined to a narrow layer in the sample.

Our model gives a complete description of the heating process up to the point of thermal ignition. In addition, a local analysis of the onset of reaction results in explicit ignition conditions in terms of the parameters in the problem.

Formulation

We consider a plane, time-harmonic electromagnetic wave of frequency ω impinging normally upon an isotropic solid combustible material which fills the region $0 < x < d$. Our theory assumes that the time required for heat to diffuse through the distance of an electromagnetic wavelength is much larger than the period of the microwave. Thus, Maxwell's equations for the electromagnetic field can be averaged to yield the standard time-harmonic vector wave (Helmholtz) equation for the electric field inside the slab. This is coupled to the diffusion equation for the temperature distribution of the sample, in which the electromagnetic source term has been integrated over one microwave period [3].

The appropriate equations for the electric field, which is given by the real part of $\mathbf{E} = [U(x)\exp(-i\omega t)]\mathbf{k}$, and the temperature T are given by

$$\frac{d^2 U}{dx^2} + k_1^2[1 + i\frac{\sigma_0}{\omega\epsilon_1}e^{\chi(T-T_0)/T_0}]U = 0, \qquad 0 < x < d, \tag{1}$$

$$\frac{\partial}{\partial t}(\rho c_p T) = \frac{\partial}{\partial x}(K\frac{\partial T}{\partial x}) + \frac{\sigma_0}{2}e^{\chi(T-T_0)/T_0}|U|^2 + \hat{A}e^{-\frac{E_a}{T}+\frac{E_a}{T_c}}, \qquad 0 < x < d. \tag{2}$$

The parameters ϵ_1 and ϵ_0 are the electrical permittivities of the material and free space, respectively, $k_1 = (\omega/c)\sqrt{\epsilon_1/\epsilon_0}$ is the electromagnetic wavelength inside the sample, T_0 is the ambient temperature, σ_0 is the electrical conductivity of the slab at T_0, and $e^{\chi(T-T_0)/T_0}$ represents the temperature dependence of the effective conductivity. Implicit in the definition of k_1 is our assumption that the magnetic permeability of the material is the same as that of free space. Also, K is the thermal conductivity of the material, ρ is its density, c_p is its thermal capacity, $A = \hat{A}e^{E_a/T_c}$ is the reaction frequency factor of the material, T_c is the ignition temperature, and E_a is the activation temperature of the solid material.

These equations must be solved subject to appropriate boundary and initial conditions. Continuity of the tangential electric field and the magnetic field at $x = 0$ and $x = d$ imply that both U and its first derivative are continuous there. Also, we allow for heat loss at the surface due to both Newton's law of cooling and radiation. Thus we impose the following surface boundary conditions

$$\frac{dU}{dx} + ikU = 2ikE_0, \quad x = 0, \qquad \frac{dU}{dx} - ikU = 0, \quad x = d. \tag{3}$$

$$\pm K(T)\frac{\partial T}{\partial x} = h(T - T_0) + e_r(T^4 - T_0^4), \tag{4}$$

where $\pm$ are to be imposed at $x = 0$ and d, respectively. Here h is a convective heat constant, e_r is the radiative emissivity of the surface, and E_0 is the strength of the electric field. Implicit in this model of surface heat transfer is the assumption that the sample is situated in an unbounded environment, the temperature of which is maintained at the ambient value T_0. Finally, we assume that the sample is at the ambient temperature initially, i.e. $T(x,0) = T_0$.

The nonlinear character of the problem is now apparent. The electric field penetrates the reactive material and affects the temperature distribution through its presence in (2). This in turn causes changes in the temperature-dependent effective electrical conductivity of the material, which then modifies the electric field through its presence in (1). The presence of a temperature-dependent energy release term in (2) exacerbates this nonlinear character.

Dynamic Heating and Onset of Combustion

We first consider the chemically inert solution, which is appropriate when the temperature is too low for chemistry to be active. Explicit solutions for the temperature evolution can be constructed by considering the limit of small Biot number as has been discussed previously in the context of ceramic heating [3]. Basically, the diffusion term in the chemistry-free version of (2) is much smaller than all the other terms. Thus if we average this equation over the thickness of the slab we find that, to a first approximation, the temperature evolves according to the equation

$$\frac{\partial}{\partial t}(\rho c_p T) = -\frac{h}{d}[T - T_0 + \frac{e_r}{h}(T^4 - T_0^4)] + \frac{\sigma_0}{2d}e^{\frac{\chi}{T_0}(T-T_0)}\int_0^d |U|^2 dx \tag{5}$$

with initial condition given by $T(0) = T_0$. The temperature evolves to one of the steady-state solutions shown in Figure 1. We see that if the power input from the microwave source is sufficiently larger than the power lost at the surface, then the terminal value will lie on the upper branch of the S-shaped curve. On the other hand, if the ratio of these two competing effects is less than the critical value (p_c at the turning point) the final temperature will correspond to a solution on the lower branch. The middle branch of this curve is unstable as discussed in [3].

We can calculate corrections to the solution discussed above in the form $T_I = T^0(t) + BT^1(x,t)$, where $B = dh/K_0$ is the Biot number that is assumed small. The leading order term, T_0, is the spatially uniform solution of (5), and the correction, T^1, will possess spatial nonuniformities that are necessary to identify the hottest spots in the sample. Because of the temperature-sensitive nature of the reaction rate term, we anticipate that ignition will first occur near the location of these hot spots. We have calculated T^1 and found that there can exist more than one hot spot in the sample, and location of each hot spot varies in time. In Figure 2 we identify those regions below the S-shaped response curve, for $\chi = 2$, where multiple hot spots exist. Typical profiles of T_1, with $T_0 = 750^0 K$ are shown in Figure 3 for three different values of input power, p. We note that the hot spots usually reside between the center of the slab at $x = d/2$ and the surface on which the microwave impinges, $x = 0$.

We now consider the situation when the temperature at a single hot spot in the sample reaches the ignition temperature of the combustible material. Our analysis is similar to that done for problems concerning the radiant ignition of reactive solids [4]. Because the activation energy of these systems is typically quite large, reaction initiates in a narrow region about the point (x_c, t_c) at which the inert temperature, T_I, first reaches the critical ignition temperature, T_c. Here we sketch out the analysis of this region; details can be found in [5].

When reaction first starts to occur, it quickly generates more heat than the microwaves within a narrow region of the hot spot. This narrow region is characterized by rapid temperature evolution and steep spatial gradients. Consequently, the microwave source term in (2) is much smaller than the other terms. We can therefore ignore this term and analyze the dynamics of small perturbations from the inert solution. For this purpose it is convenient to introduce local variables for space, time and temperature in non-dimensional form, viz. $(T - T_I)/T_0 = \epsilon\theta$, $(x - x_c)/d = \sqrt{\epsilon\mu_1}\xi$, $(t - tc)/(\rho_0 c_{p0} d/h) = \epsilon\mu_2(\tau - \tau_c)$. Here

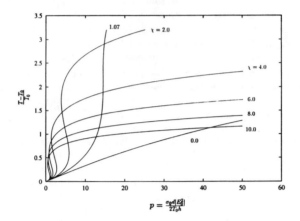

Figure 1. Steady state inert temperature versus applied power for different values of factor χ in electrical conductivity exponent.

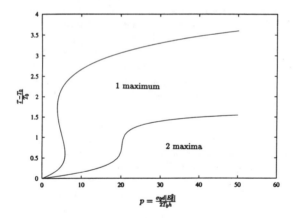

Figure 2. Regions in sample temperature - applied power plane where temperature has either one or two hot spots. Also shown is the steady state inert temperature for $\chi = 2$.

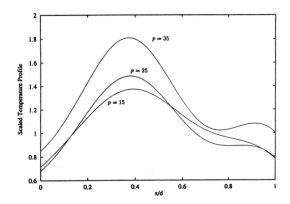

Figure 3. Scaled temperature profile T_1 versus distance x/d
for different values of p when $T_0 = 750^0 K$ and $\chi = 2$.

$\epsilon = T_c^2/(E_a T_0)$ is a small parameter, and we have defined $\mu_1 = (KT_0/d^2\partial_t(\rho c_p T))$, $\mu_2 = (hT_0/\rho_0 C_{p0} d\partial_t T)$, where $\partial_t(\cdot)$ is to be evaluated at the critical conditions (x_c, t_c). Ignition occurs at the critical time, t_c, which from our definition of the localized time variable τ corresponds to $\tau = \tau_c$, where we have defined

$$\tau_c = \ln[Ae^{-E_a/T_c}/\partial_t(\rho c_p T)_c]. \tag{6}$$

Substituting these expressions into (2) now leads to the following local equation for θ

$$\frac{\partial\theta}{\partial\tau} = \frac{\partial^2\theta}{\partial\xi^2} + e^{\theta+\tau-F\xi^2}, \tag{7}$$

where $F = \frac{B\mu_1}{2T_0}T_{1xx}(x_c, t_c)$. The requirement that this solution match to the inert solution outside of this narrow layer yields the conditions $\theta \to 0$ as $\xi \to \pm\infty$ and as $\tau \to -\infty$. The ignition characteristics of our system are governed entirely by this local equation. In Figure 4 we plot the spatial distribution of θ for several different times with $F = 1$. The temperature is seen to increase most rapidly near the hot spot, $\xi = 0$, and the solution blows up at the critical (local) time, τ_c. This singularity in the solution indicates that deviations from the inert state are much larger than the small perturbations we have considered, and we interpret this thermal runaway as the ignition point. This critical value, τ_c, can be evaluated numerically, as shown in Figure 5 with $F = 1$. Once the numerical value of this quantity is known equation (6) gives an implicit equation to determine the desired quantity t_c.

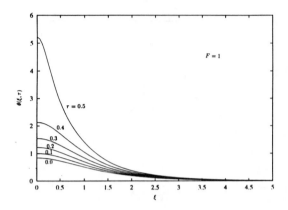

Figure 4. Local temperature $\theta(\xi, \tau)$ versus distance ξ at consecutive times τ.

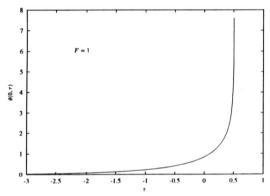

Figure 5. Maximum local temperature $\theta(0, \tau)$ versus time τ when $F = 1$.

Conclusions

We have derived a simple mathematical model which describes the nonuniform heating and local ignition of a solid combustible material due to microwave heating. The asymptotic theory employs the small Biot number assumption and exploits the fact that the activation energy is large for these systems. We show that the phenomena of ignition can be described by a single localized nonlinear differential equation. The solutions to this equation exhibit thermal runaway behavior, and explicit ignition times can be computed. These critical times are expressed in terms of various physical parameters in the problem.

References

1. Yiin, T. and Barmatz, M., "Microwave Induced Combustion Synthesis of Ceramic and Ceramic-Metal Composites," Microwaves: Theory and Application in Material Processing III, Ceramics Transactions, Vol. 59 pp. 541-547 (1995).

2. Sutton, W. H. in Microwave Processing of Materials III, MRS Symposium Proceedings, Vol. 269, pp. 3-20 (1992).

3. Kriegsmann, G. A., "Thermal Runaway in Microwave Heated Ceramics: A One-Dimensional Model," Journal of Applied Physics, Vol. 71 pp. 1960-1966 (1992).

4. Liñan, A. and Williams, F.W., "Radiant Ignition of a Reactive Solid with In-Depth Absorption," Combustion and Flame, **18** (1972), 85-97.

5. Bechtold, J.K., Booty, M.R. and Kriegsmann, G. A., "Microwave Induced Combustion: A One-Dimensional Model," in preparation (1996).

FDTD SIMULATION OF INDUCTION HEATING OF CONDUCTING CERAMIC WARE

Mikel J White, Magdy F. Iskander, and Shane Bringhurst
University of Utah, Electrical Engineering Department, Salt Lake City, UT 84112

ABSTRACT

Induction heating for the treatment of metals has been in commercial use since the mid 1960's. Traditional advantages of induction heating over the convection or radiation processes include speed of heating, possible energy savings, and the ability to customize the coil design to optimize the heating process. In this paper we used the Finite-Difference Time-Domain (FDTD) technique to simulate and analyze the induction heating process for highly conducting ceramics. In order to analyze frequency effects, simulations were performed at 300 kHz, 2 MHz, and 25 MHz. It is found that at higher frequencies coils with a pitch of 2" or greater became capacitive and generate a large, axial, electric-field component. This new axial electric field, in addition to the normally encountered azimuthal field, causes an improvement in the uniformity of the power deposition in the ceramic sample. If the sample occupies a large portion of the coil, uniformity may also be improved by using a variable-pitch coil, or by extending the length of the coil a few turns beyond the length of the sample. In a production-line arrangement, where multiple sample are place inside the coil, it is shown that maximum uniformity is achieved when the samples are placed coaxially.

INTRODUCTION

Induction heating is a cost-effective process and hence has been used for heating and forming of metals for many years. Energy losses due to resistive heating and losses associated with magnetic hysteresis are the two mechanisms of energy dissipation in induction heating. In most cases heating is dominated by resistive heating due to eddy-current losses while magnetic-hysteresis losses are negligible. In the case of induction heating of metal, the eddy currents flow on the surface of the ware, and the penetration depth is given by [1]:

$$\delta = (\frac{2}{\sigma \mu \omega})^{\frac{1}{2}} \tag{1}$$

σ = *Conductivity of the metal (S/m)*
μ = *Permeability of the metal (H/m)*
ω = *Radian frequency of the source (rad/s)*

Since penetration depth is inversely proportional to frequency, lower frequencies have most often been used in induction heating of steel.

Until recently, however, induction heating furnaces were designed without the use of sophisticated numerical techniques. Simple design ideas, coupled with experimental data, often guided designs of available induction furnaces. New applications, however, have prompted the development of design approaches that include computer modeling, simulation, and the use of numerical techniques. The expansion of induction heating into ceramic drying is one such application. New processes of ceramic drying by induction heating require analysis of operating frequency, determination of the coil pitch, relative size of sample and coil, and the placement of

Mat. Res. Soc. Symp. Proc. Vol. 430 © 1996 Materials Research Society

multiple samples inside the induction coil. In this paper we describe the use of the FDTD technique to evaluate design parameters of an induction coil that may be used for heating high-conductivity ceramic products.

FDTD SOLUTION PROCEDURE AND VALIDATION

Most of the simulations reported in this paper were performed using a 3D, 60 x 70 x 100 cell FDTD model. A 1.27 cm cell size was used in the simulations. Figure 1 shows a schematic of the FDTD model used in these simulations. The basic model consists of a 25.4 cm diameter coil, 60.96 cm in length and a pitch that varies from 2.54 to 10.16 cm. The sample has a diameter of 8.89 cm, a height of 12.7 cm, and dielectric properties of $\epsilon_r' = 255.0$ and $\sigma = 0.249$ S/m. Mür second-order absorbing-boundary conditions were used to limit the computation domain in the FDTD simulation. All simulations were performed on a RISC/6000 workstation.

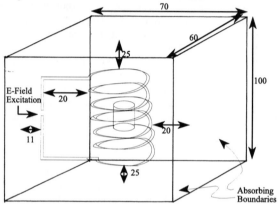

Figure 1, Drawing of 3D FDTD model used in the simulation. Dimensions given represent the number of cells, each with 1.27 cm width.

In order to validate the FDTD model, the inductance of an empty coil was calculated. The results were compared to available analytical values [2]. Table 1 shows a comparison of analytical inductance and the inductance calculated using the FDTD method. From these results, it may be observed that the FDTD calculations produce accurate inductance values and hence, the FDTD method is capable of accurately simulating an induction coil.

FDTD SIMULATION OF VARIOUS PARAMETERS INFLUENCING THE INDUCTION HEATING PROCESS

A. Frequency Effects

Operating frequency represents most important factor which influences the uniformity of power deposition in the sample. In all previous simulations of the induction heating process, quasi-static conditions were assumed and the possible existence of an axial, electric-field component was not

Table 1
Comparison of Analytical vs. FDTD Calculated Inductance

Pitch of coil	Analytical Inductance (μH)	FDTD inductance at 300 kHz (μH)	FDTD Inductance at 2 MHz (μH)
1 "	50.1	47.3	49.2
2 "	13.7	13.6	13.9
4 "	4.4	4.7	4.7

included in the formulation of the problem and the calculation of the results. Numerical simulation has been performed using the finite-element [3,4], finite-difference [5], and boundary-element techniques [6,7]. Since these simulations were performed on very high-conductivity samples, removal of the axial electric field was justified. However, based on our FDTD results, it will be shown that the role of the axial, electric-field component, created with an increase in coil pitch, cannot be ignored when samples of lower conductivity (such as ceramics) are considered.

Simulations were performed at 300 kHz, 2 MHz, and 25 MHz. Figure 2 shows the power deposition patterns in the sample at 2 MHz and 25 MHz. From Figure 2, it may be seen that the uniformity of the power deposition is greatly improved when the 25 MHz frequency is used. Simulation results at 300 kHz were similar to results at 2 MHz, and in both cases the heating pattern was predominantly at the surface. Figure 3 shows a line graph through a cross section of the sample as the simulation frequency was increased. From Figure 3 it may be observed that for the conductive ceramics used in our simulations, power penetration increases with frequency. This increase in penetration, and subsequently uniformity, results mainly from the increase in axial electric fields, which increase with frequency and pitch of the coil.

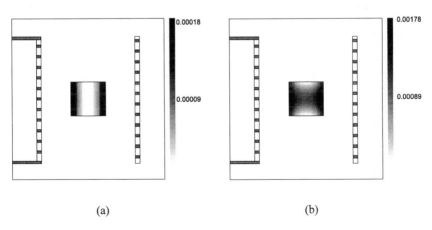

(a)   (b)

Figure 2, Power deposition patterns in a ceramic sample for (a) 2 MHz excitation and (b) 25 MHz excitation. Much improved uniformity may be observed in (b).

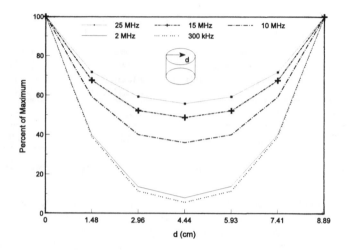

Figure 3, Line plot of power deposited on a cross-axis line through the sample center. Uniformity of power deposition increases with increasing frequency.

B. Coil Pitch and Length

If the sample occupies a small region of space at the center of an induction heating coil, the resulting heating pattern is expected to be uniform along the length of the sample. This is due to the uniformity of the electromagnetic fields in the interior of the coil. However, for large samples, and in particular as the sample length approaches that of the coil, the uniformity of the power deposition pattern will continue to deteriorate due to the diverging fields at the ends of the coil. In addition, fields are not uniform near the turns of the coil either, causing additional non-uniformity with any increase in the sample diameter. Uniformity of fields may, however, be improved by increasing the pitch of the coil, the length of the coil, and by using a variable pitch coil with a tighter pitch near its ends and a looser pitch in the middle. Figure 4 shows simulation results for a sample of height 30.5 cm and diameter of 15.2 cm placed in a coil of height 35.6 cm and 20.3 cm diameter. It may be observed that the fields are more uniform along the sides of the sample for the variable pitch coil simulation.

C. Placement of Multiple Samples

In a production line environment it is often necessary to heat multiple samples simultaneously so as to increase the throughput and the efficiency of the process. Therefore, several multiple sample arrangements were simulated, including side-by-side and co-axially stacked arrangements, to identify the most appropriate arrangement for providing adequate uniformity and efficiency of heating. Due to eddy current heating at frequencies below 10 MHz, it was observed that the placement of multiple samples in either arrangement does not affect the power deposition pattern. This observation was verified experimentally by placing two conductive samples in an induction coil available at the University of Utah. Qualitative temperature measurements were made by using

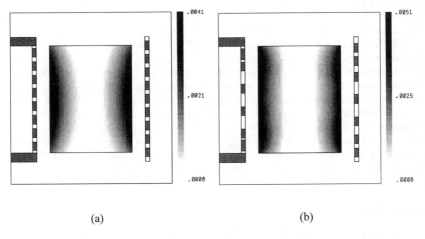

(a)   (b)

Figure 4, Simulated power deposition in induction heating experiments when the sample occupies a large portion of the coil. Results are shown for (a) uniform-pitch coil and (b) a variable-pitch coil

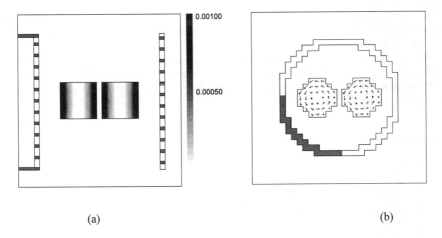

(a)   (b)

Figure 5, Simulation results of two side-by-side samples placed in a coil operating at 2 MHz: (a) Power absorption results and (b) vector field plot of the surface currents

381

temperature-sensitive liquid-crystal sheets, and by using a thermographic camera. Figure 5 shows the simulation results of two samples placed side-by-side in an induction coil. The vector, electric-field pattern shown in Figure 5b illustrates the circulation of eddy-currents and the dominance of surface heating, even when two samples were placed side-by-side.

With the increase in the operating frequency to 25 MHz, however, it is observed that the power deposition patterns change as additional samples are introduced into the coil. When samples are placed side-by side, it is observed that power deposition becomes less uniform near the axis of the coil as compared to power deposition near the coil turns. This is due to orientation of the axial electric field which is normal to the sample boundaries near the axis of the coil, and tangential to the sample boundary near the coil turns. The difference in dielectric constant from air (ϵ_r' = 1.0) to sample (ϵ_r' = 255.0) results in reduced penetration of fields near the axis of the coil, and stronger coupling of the tangential fields to the sample near the turns of the coil. When the samples are stacked along the axis of the coil, and placed close together, a more uniform power deposition pattern results. Therefore, an axial arrangement of stacked samples is preferred for multiple sample heating.

CONCLUSIONS

The FDTD code available at the University of Utah was used to obtain useful guidelines for the application of induction heating to conducting ceramics. Specifically, it is shown that it is possible to obtain more uniform power deposition by increasing the frequency above 10 MHz. Higher frequencies and increase in coil pitch creates axial electric fields which penetrate the sample, thus increasing the uniformity of absorbed power. When the samples occupy a large portion of the coil, the ends of the samples receive reduced heating as compared to the middle section of the sample. However, by increasing the length of the coil or by using a variable-pitch coil with a tighter pitch at the ends and a looser pitch in the middle, the power absorption at the ends of the sample may be increased and axial uniformity of heating, along the length of the sample, may be achieved.

Simulations were also made in which multiple samples were placed inside the coil. Simulation results and experimental results showed that the power deposition pattern is not affected by the introduction of a second sample when the operating frequency is lower than 2 MHz. At 25 MHz it is shown that the maximum uniformity is obtained by placing multiple samples in a stacked arrangement along the axis of the coil.

REFERENCES

[1] W.B. Kim and S.J. Na, *Surface and Coatings Technology*, Vol. 58, pp. 129-136, Feb. 1993.

[2] R. Lundin, *Proceedings of the IEEE*, Vol. 73, No. 9, pp. 1428-1429, Sept. 1985.

[3] K.F. Wang, S. Chandrasekar, and T.Y. Yang, *Journal of Mat. Eng. and Perf.*, Vol. 1(1), pp. 97-112, Feb 1992.

[4] J. Donea, S. Giuliani, and A. Philippe, *Int. Jnl. for Num. Meth. in Eng.*, Vol. 8, pp. 359-367, Sept. 1974.

[5] J.D. Lavers, *IEEE Trans. on Magnetics*, Vol. MAG-19, No. 6, pp. 2566-2572, Nov. 1983.

[6] T.H. Fawzi, K.F. Ali, and P.E. Burke, *IEEE Trans. on Mag.*, Vol. MAG-19, No.6, pp. 2401-2404, Nov. 1983.

[7] H. Tsuboi, M. Tanaka, F. Kobayashi, and T. Misaki, *IEEE Trans. on Mag.*, Vol. 29, No. 2, pp. 1574-1577, March 1993.

Part VII

Microwave Interactions and Mechanisms

INFLUENCE OF SPECIFIC ABSORBED MICROWAVE POWER ON ACTIVATION ENERGY OF DENSIFICATION IN CERAMIC MATERIALS

Y. BYKOV, A. EREMEEV, V. HOLOPTSEV
Institute of Applied Physics, Russian Academy of Sciences, 46 Ulyanov St., Nizhny Novgorod 603600 Russia.

ABSTRACT

Correlation between the rate of densification in powder ceramic materials and specific absorbed microwave power is determined by the experimental method. The approach is based on a comparison of the densification curves obtained at different rates of heating. The changes in the ramping rate are provided by varying the microwave power fed into the microwave furnace. Using the energy balance for the microwave heated samples, the correlation between the apparent energy of activation at the initial stage of densification and the value of the specific microwave power absorbed in heated materials are found. The experiments with silicon nitride-based ceramics allowed to determine the reduction in the value of the activation energy resulted from an increase in the specific absorbed microwave power.

INTRODUCTION

It is almost universally accepted opinion today that enhanced densification and decreased sintering temperature are inherent in microwave ceramics sintering. Numerous experimental results show that acceleration of sintering can not be attributed only to the specific distribution of thermal sources within the body being microwave heated, but has to do with more fundamental processes relevant to the physics of sintering [1, 2, 3]. Despite that much evidence points to the microwave accelerated ceramic sintering, the problem of the specific microwave influence on a sintering mechanism remains open. Up to date, there is no any reliable, verified by the direct experiment, theoretical explanation of a "microwave effect" in mass transport processes. On the other hand, the results of quite a few experiments testify against any difference between microwave and conventional sintering. Two factors are of primary importance from the experimental standpoint while comparing the results of microwave and conventional heating. The first of them concerns the reliability of correlation between the temperatures measured in the bodies which have quite opposite temperature profiles when heated conventionally and with microwaves. The importance of this factor is evident and it is usually taken into account in the experiments and analysis of the experimental results. However, as far as, the case in point is an influence of the microwaves on mass transport, the strength of the electromagnetic fields in processed materials is not less essential characteristic. The fact that material is heated in a microwave oven is not necessarily a reason for the presence of microwave fields in the materials themselves. An intricate scheme of thermal insulation, unknown microwave absorptivity of the processed material and insulation, especially at elevated temperatures, are frequent occasions. These factors make true estimates of the microwave fields penetrated into materials virtually impossible. Those experiments wherein the microwave power feeding an oven is measured, are not capable of giving a direct knowledge of the power absorbed in the heated material. In most cases, the relationship between feeding power and power absorbed in both the samples and in the thermal insulation remains unknown and varies during the heating. Hence, even a comparison of the results obtained in experiments run under a variety of feeding powers appears to be unhelpful. Yet, a direct measurement of the electric fields within a sample heated with microwaves to the temperatures well above 1500°C is the task of a great difficulty and the authors do not know any successful results of such measurements.

In the present report, a calorimetric method for a measurement of the specific microwave power absorbed in the sintered sample is described. The method was applied to measure the specific microwave power absorbed in the Si_3N_4 - based powder compacts at the initial stage of densification. Concurrent determination of the activation energies of densification and the α - β phase transformation allows to establish a correlation between a decrease in the activation energies and the specific absorbed power and define the extent to which the activation energies are affected by the latter. Separate experiments on conventional and microwave sintering of the ZrO_2 - powder compacts illustrate the idea that the results of microwave sintering can crucially depend on the value of the absorbed power.

EXPERIMENTAL PROCEDURES

The powder compacts of composition α - Si_3N_4 + 2% Al_2O_3 + 9% Y_2O_3 were used in the present study. The green samples were obtained from the company "Teknologia" (Obninsk, Russia). Processes for the green samples preparation is described elsewhere in this Proceedings [4]. Samples were placed in the center of 60mm dia × 60mm light fused quartz vessel and covered with Si_3N_4 powder mixed with Al_2O_3 - Y_2O_3 powder additives in the same proportions as athe sample itself. For microwave processing, the vessel with a sample was placed inside the multimode furnace of a 30GHz Gyrotron - based Heating System [5]. The furnace has been evacuated and then filled up with dry nitrogen up to 1atm pressure. The sample temperature was measured by (W + 5%Re - W + 20%Re) thermocouple, the head of which was inserted into a narrow bore drilled to the center of a sample. Microwave heating was done in a computer control mode. The constant heating rate in a range of 30-100°C per minute was maintained by controlling the output power of the gyrotron. Having reached the maximal prescribed temperature, microwave power was switched off and a sample cooled down together with the thermal insulation. All samples were weighed and their sizes were measured before and after heating to determine the weight loss during processing and the linear shrinkage. Densities were determined by the Archimedes method after one-hour boiling of samples in distilled water. The phase composition of the samples was identified by X-ray diffraction analysis.

ZrO_2 (8% Y_2O_3) powder was obtained by the conventional method of $ZrSiO_4$ thermal dissociation in the presence of ytria oxide. The product of solution was crushed and milled to the average grain size $\leq 1\mu m$ using ZrO_2 - balls. Green samples were uniaxially pressed to density of 2.7gr/cm^3, and conventionally prefired at the temperature 900°C for 2 hours. The procedure of microwave sintering of the ZrO_2 - compacts was similar to those Si_3N_4 - samples and differed in using Al_2O_3 - powder for thermal insulation and air as a gas environment. Samples temperature was measured by K-type thermocouple placed beneath the samples. To simulate the fast conventional sintering with the heating rate equal to the rate of microwave heating, the samples were placed within completely closed metal box filled with Al_2O_3 powder. The thermocouple wires passed through the small holes in the wall of the box and the sample lay on the thermocouple head.

RESULTS AND DISCUSSION

The method for finding the relationship between the appeared activation energy of densification Q_d and specific microwave power absorbed in ceramic samples is based on the comparative study of densification for the samples heated with the different heating rates $\dot{T}_+$.

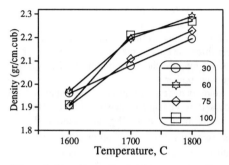

Figure 1. Density vs maximal temperature for different rates of heating.

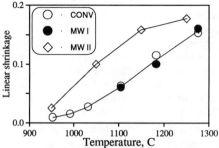

Figure 2. Linear shrinkage of ZrO$_2$ samples vs. temperature under different processing modes and constant rate of heating; P_{abs}(MW II) $\gg$ P_{abs}(MW I).

In Figure 1 are shown the densification curves corresponding to various heating rates - 30, 60, 75, and 100° C/min. All the curves start with approximately equal densities at the temperature T = 1600°C. But the densification proceeds in a different manner at elevated temperatures. Densification goes on faster as the heating rates are increased. An accelerated densification of the faster heated samples takes place despite a decrease in their time of being at the elevated temperature.

Another feature peculiar to the densification is the change in its rates. Despite the further increase of the temperature above 1700°C, the rate of densification becomes lower. Since the total densification of the samples is not large (density remains less than 65%ρ_{th}), this decrease of the densification rate is not associated with typical slowing down of sintering while approaching the theoretical density of the material, but connected with a change in the conditions of heating. At elevated temperatures, a relatively larger part of microwave power is absorbed in the thermal insulation. As a result, the temperature of the part of the thermal insulation adjacent to the sample grows faster than at lower temperatures, and this leads to a reduction in the thermal loss from the sample surface. This fact is supported by the concurrent variations of the sample cooling rates $\dot{T}_-$ after switching off the microwave power. As it is seen from Table I, the cooling rates fall when the maximal temperature of samples is exceeding T = 1700°C.

Table I. Specific microwave absorbed power P_{abs}, reduction in activation energies of densification ΔQ_d and $\alpha \rightarrow \beta$ phase transformation ΔQ_{tr}, and the cooling rates at various temperatures and rates of heating.

(dT/dt)$_+$,	P_{abs}, W/cm^3			ΔQ_d, kJ/mol		ΔQ_{tr}, kJ/mol
°C/min	1600°C	1700°C	1800°C	1600÷1700°C	1700÷1800°C	1600÷1800°C
30	5.5	6.0	3.5			
60	6.0	7.0	4.5	-21.0	-8.5	
75	6.5	7.5	4.5	-23.0	-16.0	-14.5
100	7.0	8.0	5.0	-34.0	-9.5	-32.0
(dT/dt)$_-$, °C/min	185	220	115			

In the conditions when the thermal insulation is invariable for all processed samples, an increase in the heating rate is provided by an increase in the specific microwave power P_{abs} absorbed in the samples. To determine qualitatively how the enhancement of densification is connected with the value of P_{abs}, we will use the densification curves, presented in Figure 1 for evaluation of the apparent activation energy of densification, and the equation for the energy conservation for finding the specific absorbed microwave power.

What we really need to know for the aim of this study is not the magnitude of the apparent activation energy of densification but its variation caused by a change in the value of P_{abs}. In these experiments, only the first stage of densification has been studied. The ratio of the total variation of density in all the experimental runs was less than $15\%\rho_{th}$. In addition, since the variation of $\Delta\rho/\rho$ among the samples heated with different heating rates to a given maximal temperature T_{max} was also small ($< 6\%$), we will use the straightforward equation for the dependence of the densification rate $d\rho/dt$ on the temperature

$$\frac{d\rho}{dt} \approx \frac{A}{T} \cdot \exp\left(-Q_d/R \cdot T\right) \qquad (1)$$

(here A is a constant coefficient, R is the gas constant) and omit all other factors connected with the dependence of the rate of densification on the grain size, grain and pore distribution, etc., assuming these parameters remain constant at this first stage of sintering.

The difference in microwave heating rates may result in the difference in the spatial distribution of the temperature across the samples. Since, the thermocouple head in these experiments was located at the center of the samples, the maximum temperature of them was the same for all the samples heated at various heating rates to a given value of the temperature. However, the different results of densification even at fixed maximal temperature of the samples may be caused by a non-uniformity of the temperature across the sample. The temperature non-uniformity, that is the temperature drop ΔT between the center of a sample and its surface is proportional to the rate of cooling $\Delta T \sim r^2 \dot{T}_- / k$, where r is the sample size, k is the coefficient of thermal conductivity. In these experiments the cooling rates depended on the maximal temperature to which samples were heated, but, within the limits of the experimental error, did not depend on the rates of heating. Therefore for present purposes the temperature may be considered to have same profile over the volume of the samples heated to a given maximal temperature. The results of the calculation of the difference in the values of Q_d for various heating rates (60, 75, 100°C/min) as compared with Q_d^{30} for the heating rate of 30 °C/min are given in Table I.

Given the experimental data on the samples cooling rates and the rates of their heating it is easy to estimate the specific microwave power absorbed in the samples

$$P_{abs} = \sigma_{eff} E^2 = c\rho(\dot{T}_+ + |\dot{T}_-|) \qquad (2)$$

(σ_{eff} is the effective microwave conductivity, E is the strength of the electromagnetic field in the sample) assuming that in the vicinity of $T = T_{max}$ the thermal loss is the same at the heating and cooling stages.

The calculated magnitudes of P_{abs} for the different rates of heating, with values of the density taken from the experiments and with $c = 0.7J/gr\cdot grad$ [6] are listed on Table I. The values of P_{abs} progressively increase with the increase in the heating rate for each value of T_{max}. Along with this, a reduction in P_{abs} is observed at the temperatures above $T_{max} > 1700°C$. This fact, as it was mentioned above, reflects the decrease in the fraction of the microwave power absorbed in the sample itself and an increase in the fraction absorbed in the thermal insulation. Dependence of the activation energy reduction ΔQ_d versus the specific microwave absorbed power can be followed from the data given in Table I. In each of the two temperature ranges, $T < 1700°C$ and $T > 1700°C$, the activation energy Q_d decreases (the value of $\Delta Q_d = Q_d - Q_d^{30}$ goes down) as the value of P_{abs} increases. At $T > 1700°C$, each of the parameters P_{abs} and ΔQ_d are less in magnitude than at $T < 1700°C$.

Similar relationship between P_{abs} and the activation energy can be obtained from analysis of the experimenal kinetics curves of the $\alpha \rightarrow \beta$ phase transformation occurring in the samples heated at different rates. The activation energy for $\alpha \rightarrow \beta$ phase transformation Q_{tr} was calculated using the following equation for the transformation rate:

$$\frac{d[N]}{dt} \sim \frac{B}{T} \cdot [N] \cdot \exp(-Q_{tr}/R \cdot T), \qquad (3)$$

here $[N]$ is the ratio of α and β phases content in samples.

The magnitude of Q_{tr} for the heating rate 30°/min has been found equal to 380kJ/mol, which agrees with the value $Q_{tr} = (300 \div 500)$kJ/mol given in [7] for the activation energy of phase transformation observed at conventional heating. The differences in the magnitudes of Q_{tr} corresponding to various heating rates are given in Table I. As is seen, an increase of the specific absorbed microwave power P_{abs} from 5 to 8W/cm^3 leads to a reduction of the order of 4÷8% in the magnitude of Q_{tr}.

The data presented in this paper allow to determine quantitatively to what extent the reductions in the apparent activation energy for various mass transport processes are connected with another energetic characteristic, namely, with specific absorbed microwave power. Notice that the data of the experiments permit also to make an even more essential conclusion. Assuming that those slight changes in the sample density, which correspond to heating of the samples at various heating rates, do not alter significantly the magnitude of σ_{eff} and that σ_{eff} is not affected by the value of the heating rate, there is a good reason to believe that this parameter σ_{eff} remains nearly constant for all the samples irrespective to their heating rates. Then, the mechanism which is really responsible for the decrease in the activation energies, that is for the enhancement of mass transport processes, is not of the energetic character, but is connected with square of the amplitude of electric field E^2. If this is the case, by using the experimental data for P_{abs} it is possible to estimate the amplitude of the electric field at which microwaves are capable of affecting mass transport at sintering of Si$_3$N$_4$ - ceramics. At $P_{abs} = 5$W/cm^3, and assuming for Si$_3$N$_4$ - ceramics $\varepsilon\sim 9$ and $tan\delta \sim 6 \cdot 10^{-2}$ at $T \approx 1700°C$ we obtain that the amplitude of the electromagnetic field E is of order of 100V/cm.

To check whether the enhancement of densification is really connected with a specific influence of the microwave electromagnetic field, and not caused by simple increase in the rates of heating, a special set of experiments with ZrO$_2$ - ceramic compacts has been done. A metal box was used in these experiments for shielding the samples from the microwaves. It was not possible to carry out such tests with the Si$_3$N$_4$ - ceramics needed to be heated to a temperature as high as 1800°C for sintering. ZrO$_2$ samples were sintered at the constant heating rate of 50°C/min. Sintering was done in three different modes:

a) a sample was placed in the metal box (imitation of conventional heating);

b) and c) - samples were heated with microwaves under condition of strong (case b) and weak (case c) microwave absorption in a thermal insulation, i.e. the cases of low and high values of P_{abs}, respectively. The results of those experiments are given in Figure 2. As is seen from a comparison of the kinetic curves, a crucial factor responsible for an enhancement of sintering is the value of the specific microwave absorbed power, but not the rate of heating.

CONCLUSION

A calorimetric method was applied to measure the specific microwave power absorbed in Si_3N_4 - based ceramic samples at the initial stage of densification. Concurrent determination of the activation energies of densification and the $\alpha \rightarrow \beta$ phase transformation allowed to define the extent to which the activation energies were affected by the specific absorbed microwave power. Cooling rate data comparisons argue that apparent activation energy differences are not due to different temperature spatial gradients. The results of a comparative study of conventional and microwave sintering of the ZrO_2 -powder compacts showed that the rate of sintering crucially depended on the value of the absorbed power.

ACKNOWLEDGMENT

This work was supported in part by the Russian Basic Science Foundation through grant No95-02-05000-0. A. Sorokin and T. Borodacheva contributed valuable assistance to the conduct of this research.

REFERENCES

1. M.A. Janney and H.D. Kimrey, in Microwave Processing of Materials II, edited by W.B. Snyder, W.H. Sutton, M.F. Iskander and D.L. Johnson (Mater. Res. Soc. Proc. **189**, Pittsburgh, PA 1991), pp. 215 - 227.

2. Yu. Bykov, A. Eremeev and V. Holoptsev, in Microwave Processing of Materials IV, edited by M.F. Iskander, R.J. Lauf and W.H. Sutton (Mater. Res. Soc. Proc. **347**, Pittsburgh, PA 1994), pp. 585 - 590.

3. M. Willert-Porada, T. Gerdes, S. Vodegel, in Microwave Processing of Materials III edited by R.L. Beaty, W.H. Sutton and M.F. Iskander (Mater. Res. Soc. Proc. **269**, Pittsburgh, PA 1992), pp. 205 - 210.

4. Yu. Bykov, A. Eremeev and V. Holoptsev, this Proceedings.

5. Yu. Bykov, A. Eremeev, V. Flyagin, V. Kaurov, A. Kuftin, A. Luchinin, O. Maligin, I. Plotnikov and V. Zapevalov, in Ceramic Transactions, Microwaves: Theory and Applications in Material Processing III, edited by D.E. Clark, D.C. Foltz, S.J. Oda and R. Silberglitt (Am. Cer. Soc., **59**, Westerville, Ohio, 1995) pp. 133 - 140.

6. G. Ziegler, J. Heinrich, G. Wotting, J. Mater. Sci., **22**, pp. 3041 - 3086 (1987).

7. F.K. VanDijen, A. Kerber, V. Vogt, W. Pfeiffer and M. Schulze, Key Engineering Materials, **89 - 91**, p. 24 (1994).

Kinetics of the Carbon Monoxide Oxidation Reaction Under Microwave Heating

W. Lee Perry, Joel D. Katz, Daniel Rees, Mark T. Paffett
Los Alamos National Laboratory, Los Alamos, NM 87545

Abhaya Datye,
University of New Mexico, Albuquerque, NM 87131

Abstract

915 MHz microwave heating has been used to drive the CO oxidation reaction over Pd/Al_2O_3 without significantly affecting the reaction kinetics. As compared to an identical conventionally heated system, the activation energy, pre-exponential factor, and reaction order with respect to CO were unchanged. Temperature was measured using a thermocouple extrapolation technique. Microwave-induced thermal gradients were found to play a significant role in kinetic observations.

Introduction

Microwave energy is an alternative to conventional heating in catalytic processes and has proved successful in driving the CO oxidation reaction over Pd and Pt on γ-Al_2O_3 without significantly affecting the reaction kinetics. There are many ways the microwave might interact with the catalyst to have an effect on chemical kinetics in this system:
1) space charge effects around the supported metal particles such that an induced surface charge has significant effect on the dipolar CO molecule either on/near the surface;.
2) differential heating of the metal versus the ceramic support; 3) the oxidation of the polar CO molecule could be affected both in the gas phase and near the catalyst surface by the electric field. If any of these effects were important, a significant alteration in the kinetics would be observed with respect to conventional heating.

In most of the literature reviewed [1-4], beneficial effects of microwave heating were shown. However, little has been done to understand the effect of microwaves on the underlying kinetic and mechanistic process due to the difficulties experienced in measuring temperature and the non-isothermal nature of a microwave heated packed-bed. Because of the temperature gradients and lack of temperature data, it is difficult to ascertain the root cause of the beneficial effects. In the research presented, we have taken steps to determine the temperature and temperature gradients so that an accurate comparison of chemical kinetics during microwave and conventional heating can be made.

Experimental

In this paper we will compare the kinetics of the CO oxidation reaction over Pt and Pd/γ-Al_2O_3 as observed in a specially constructed microwave heated differential [5] reactor. These kinetics were compared to a conventionally heated reactor. To eliminate the effects of temperature gradients, a differential mass of active supported catalyst was sandwiched between two masses of inactive support inside a quartz tube, which was radially insulated with microwave transparent sapphire wool. The active material was on the maxima of the temperature profile, and the short length of the active material approximated an isothermal bed. A schematic of a typical experimental reactor is shown in Fig. 1.

Mat. Res. Soc. Symp. Proc. Vol. 430 ©1996 Materials Research Society

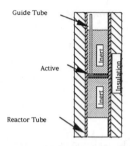

Guide Tube

Active

Reactor Tube

Figure 1. Schematic of reactor configuration used in both the microwave and conventional experiments. "Differential" active catalyst zonewas axially thin to minimize thermal gradients, and inert material was also added to minimize gradients. Insulation was provided to minimize radial gradients.

We chose the CO oxidation reaction over a supported metal catalyst because the reaction kinetics are well known, and because of the diverse dielectric properties of the various elements in the system: CO is a polar molecule, O_2 and CO_2 are non-polar, Al_2O_3 is a dielectric, and Pt and Pd are conductors. The literature [6,7] has revealed the conventionally-heated reaction kinetic and mechanistic details, with the desorption of CO being the rate limiting step.

The thermal reactor system was heated by a digitally controlled 1" tube furnace. The microwave reactor was heated by a magnetron based system operated on a continuous basis at 915 MHz. The power used for the experiment was 850W to 1000W and was delivered to a resonant applicator via rectangular waveguide in the TE_{010} mode. The reactor tube was inserted through the rectangular resonant applicator parallel to the direction of the electric field. Product analysis was performed using gas chromatography .Reactant ratios were fixed by mass flow controllers, and the pressure was regulated by a bypass valve and monitored with a capacitance manometer.

The catalyst used for the final experiments was 5% (by weight) Palladium supported on γ-Al_2O_3. Preliminary experiments used 1% Pt/γ-Al_2O_3. To determine the activation energy, E_a, and the pre-exponential, A, factor in the Arrhenius expression temperature versus reaction rates were recorded. The flow rate for CO was 2 cc/s and 1 cc/s for O_2, and the reactor pressure was fixed at 1000 Torr. The dependency on CO partial pressure was found by using an excess of oxygen and the CO partial pressure ranged from 35 to 95 Torr. The reactor pressure was held at 1000 Torr and the temperature was 135 C for these experiments.

Temperature was measured in the microwave system by inserting a thermocouple into the bed after the microwave power was turned off. For the final experiments, the thermocouple was inserted through a "guide tube" (see Figure 1) until it came to rest in the hottest (active) part of the bed. The temperature decay curve was recorded on a x-y recorder. The temperature versus time curve was linearized by taking the natural log of the data (the decay was assumed to be exponential) and coordinates were entered into a curvefitting routine. The routine provided an expression in the form y=mx+b where b was the initial temperature. The thermal mass of the thermocouple was not insignificant with respect to the mass of the bed and a corrective procedure was employed: The digitally controlled tube furnace was allowed to reach a known equilibrium temperature, furnace power was terminated, the furnace was opened, the thermocouple was inserted and the recorder was started. Several data points were recorded and a best fit line expression was then used to correct the temperature measurements.

Results and Discussion

Apparent kinetics. Early experiments were performed using an integral-type reactor [5]. A schematic of this reactor is shown in Figure 2 along with the comparative Arrhenius plot. The experiment used 2 grams of 1% Pt/Al_2O_3. The thermocouple extrapolation technique was employed and the probe was inserted such that it was in contact with the packed bed as shown in Fig 2. The microwave-heated packed bed was shown by thermal imaging in Figure 3 to be very non-isothermal. The non-isothermal nature of this packed bed precluded accurate temperature measurement and resulted in the apparent rate enhancement , activation energy and pre-exponential .

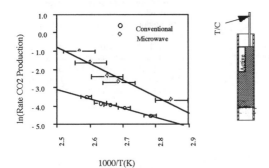

Figure 2. Comparative Arrhenius plot and Schematic of an integral-type packed-bed reactor. Error bars reflect maximum possible error.

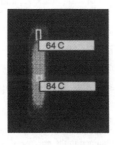

Figure 3. Thermal (infrared) image of microwave heated packed bed. Temperatures shown are the average temperatures of the areas outlined by the boxes.

A subsequent experimental design used α-Al_2O_3 for the inert insulating material and γ-Al_2O_3 was used to support the active metal (see Fig. 1). Figure 4 shows a comparative Arrhenius plot and the associated thermal profile obtained from this differential configuration. The plot clearly shows an apparent rate enhancement. The probe was inserted into the active portion of the bed as shown in Figure 1. A qualitative experiment was performed, and γ-Al_2O_3 was observed to absorb microwave energy more efficiently than α-Al_2O_3. Thus, the true reaction temperature was greater than the average, observable temperature as shown in Figure 4.

The true reaction temperature of the γ-Al_2O_3 was not observable with our technique for two reasons: 1) the mass of the active portion of the bed was much less than the inactive portion so the thermocouple would primarily "see" the inactive (cooler) mass; 2) the time required for the isolated γ-Al_2O_3 particles to cool to the temperature of the α-Al_2O_3 was much less than the time required to insert the thermocouple. This time for cooling can be estimated rigorously by performing a transient heat conduction analysis. Because of the rapid cooling and low mass fraction, the reaction was occurring at a hotter temperature than was observed by the thermocouple and yielded the apparent results shown in Figure 4.

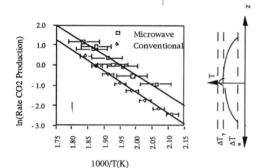

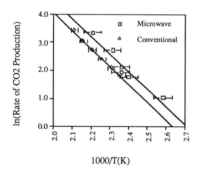

Figure 4. Result for differential type reactor where active catalyist was supportedon γ-alumina and insulating alumina was α-type. This material mismatch gave rise to the qualitative temperature profile shown here. The rise at z=0 results from more efficient absorption of microwave energy by the γ-alumina. Error bars reflect maximum error.

This experiment illustrates the difficulty in observing the proper temperature when one material absorbs more efficiently than another. The situation is analogous to the case where the supported metal particles absorb energy preferentially to the support. The metal particle temperature would not be observable due to the reasons discussed above, and a discrepancy similar to that shown in Figure 4 would result if preferential absorption were occurring.

True kinetics. A comparative Arrhenius plot is shown in Figure 5 for CO reaction over Pd/γ-Al$_2$O$_3$. The reaction rates were nearly identical within experimental error. The slightly higher rate observed in the microwave reactor resulted from thermal gradients which cannot be eliminated and the reaction surface was always hotter than the average temperature observable using a thermocouple.

Figure 5. Plot shows comparison of data obtained in the microwave-heated reactor vs. the conventionally heated reactor. The slope of both lines is nearly identical, yeilding an activation energy of approximately 13 kcal/mol. The proximity of the data indicates no significant microwave effect. Error bars represent maximum error.

The conductive palladium did not couple effectively with the microwave energy with respect to the alumina support. Had the metal coupled more strongly, the local metal surface temperature would be greater than the support, and the rate would appear faster in the microwave than the conventional furnace. The specific metal temperature was not observable for reasons described above.

The other possible effect considers a specific microwave effect on the desorption rate of CO, the most likely area for a specific microwave effect. A specific effect would appear as a change in CO reaction order. Figure 6 shows the rate dependence as a function of CO partial pressure, yielding reaction orders (slopes) which are nearly identical. These results support the conclusion that the process is unaffected by the heating mode.

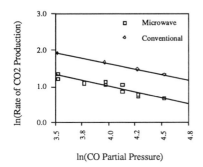

Figure 6. Plot indicates the dependence of the overall rate of reaction on CO partial pressure in both the microwave and conventional reactors. The slopes are -0.59 and -0.64, respectively. The near equality of the slopes indicates that no significant alterations of the reaction pathway occurs in the microwave reactor. The conventional temperature was 135 C; the microwave temperature was not measured. Pressure is in Torr.

Conclusion

The primary conclusion obtained from the data presented is that no alterations in the kinetic behavior of the reaction were observed using the Pd/Al_2O_3 catalyst. This was a significant result considering the potential for direct interaction of microwave energy with both the small Pd catalyst particles and the highly polar CO reactant.

We would like to stress the overriding theme of this research: The highly non-isothermal nature of the microwave heated packed bed can lead to improper conclusions about rate behavior. Taken at face value, early experimental results would have pointed to a significant rate enhancement which was clearly erroneous. Therefore, to correctly observe true chemical kinetics, the unique heat transfer effects that are present in a microwave heated experimental reactor must be carefully considered.

Bibliography

1. C-Y. Cha in Proceedings: Microwave-Induced Reactions Workshop, edited by M. Burka, R. D. Weaver, J. Higgins, (Electric Power Research Institute, Palo Alto CA, April 1993), p. A-2.
2. J. K. S. Wan and T. A. Koch in Proceedings: Microwave-Induced Reactions Workshop, edited by M. Burka, R. D. Weaver, J. Higgins, (Electric Power Research Institute, Palo Alto CA, April 1993), p. A-3.
3. C. Chen, P. Hong, S. Dai, and J. Kan, J. Chem. Soc. Faraday Trans., **91**(7), p. 1179. (1995)
4. M. S. Ioffe, S. D. Pollington, J. K. S. Wan, Journal of Catalysis, **151**, p. 349 (1995).
5. H. S. Fogler, Elements of Chemical Reactor Engineering, 2nd Edition, Prentice Hall, New Jersey, 1992, p. 217.
6. N. W. Cant, P. C. Hicks and B. S. Lennon, J. Catal. **54**, p. 372 (1978).
7. P. J. Berlowitz, C. H. F. Peden, and D. W. Goodman, J. Phys. Chem., **92**, p. 5213 (1988).

INVESTIGATIONS OF MICROWAVE ABSORPTION IN INSULATING DIELECTRIC IONIC CRYSTALS INCLUDING THE ROLE OF POINT DEFECTS AND DISLOCATIONS

Benjamin D.B. Klein[a], Binshen Meng[a], Samuel A. Freeman[c], John H. Booske[a,c], and Reid F. Cooper[b,c]

[a] Department of Electrical and Computer Engineering
[b] Department of Materials Science and Engineering
[c] Materials Science Program
University of Wisconsin, Madison, WI 53706

ABSTRACT

A theoretical model of microwave absorption in linear dielectric (non-ferroelectric) ionic crystals that takes into account the presence of point defects was synthesized and verified using $NaCl$ single crystals. In the next stage of this research, we will introduce a controlled density of dislocations into the single crystal $NaCl$ samples and study the effect on the microwave absorption mechanisms (ionic conduction, dielectric relaxation and multi-phonon processes) both theoretically and experimentally. Qualitative outlines of this modified theory are presented. The loss factor ϵ'' has been measured in the dislocation-free case by a cavity resonator insertion technique and the experimental results are in good agreement with the theoretical model. We describe the sample preparation technique that will be used to produce a controlled dislocation density in single crystal samples that will also be studied in our cavity resonator insertion system.

INTRODUCTION

A complete predictive theory of the absorption of microwave radiation by polycrystalline ionic solids is crucial to several materials applications of growing interest and importance, namely: (1) materials selection and fabrication for advanced radomes as well as window and insulator structures in high power coherent microwave sources, (2) microwave processing of ceramics, and (3) determining the complex dielectric properties of ceramic substrate materials for use in microwave and high speed digital circuits. The task of creating such a theoretical model is a formidable one, however, as there are many possible absorption mechanisms that must be considered. We have approached the problem incrementally. In our previous work, we developed and experimentally verified a complete theory of microwave absorption in single crystal ionic solids with extremely low dislocation densities, including the role of point defects[14]. In our current research, we are working to extend this theory to the case of single crystal ionic solids with low but nonnegligible concentrations of dislocations. By comparing experimental results with those previously obtained, we will isolate the contribution of the dislocations, and tailor our theory to match the results.

In this paper, we first briefly review the outlines of the previously verified theory of microwave absorption including point defects. We then lay the foundation for the extension of our theory to include dislocations by considering the possible modifications to our theory in a qualitative way. We then present the experimental data which verified our previous theory, and explain the sample preparation method which will allow us to use our resonator-based measurement system in our future work.

THEORETICAL MODEL OF ABSORPTION NEGLECTING DISLOCATIONS

In this section we will present the experimentally verified theory which was synthesized to predict microwave absorption in single crystal ionic solids with very low disloction densities. This theory incorporates three radiation absorption mechanisms that have been extensively studied in different frequency ranges, extrapolating them to the frequency range of interest and summing their contributions. The three mechanisms are: ohmic loss, dielectric loss, and photon-phonon interactions. We have found that in our frequency and temperature range none of these mechanisms can be safely ignored and ϵ'' must be written in the most general form:

$$\epsilon'' = \epsilon''_c + \epsilon''_d + \epsilon''_{mp}, \tag{1}$$

where $\epsilon''_c, \epsilon''_d$ and ϵ''_{mp} represent contributions from ionic conduction, dielectric relaxation and multi-phonon processes, respectively. The relevant expressions for each of these loss mechanisms are given below. We will be working specifically with $NaCl$, which is the material used in our experimental work; however, it is clear that this theory could easily be applied to other ionic crystalline materials as well.

Mat. Res. Soc. Symp. Proc. Vol. 430 © 1996 Materials Research Society

Ionic Conduction

In a Schottky-disorder compound such as $NaCl$, we treat the vacancies as the charge carriers in the bulk crystal. In our model, we choose to neglect the contribution to the ionic current due to the motion of anion vacancies[14]. Calculations based on these considerations lead to a thermally-activated expression for ϵ_c'', which is[6]

$$\epsilon_c'' = \frac{ne^2b^2\nu_0}{\omega\epsilon_0 kT} e^{-\frac{\Delta g_m}{kT}} \tag{2}$$

where n is the free cation vacancy number density, b is the vacancy jump distance ($b = 3.99\mathring{A}$), ν_0 is some characteristic lattice frequency[3], $\Delta g_m (= \Delta h_m - T\Delta s_m)$ is the Gibb's free energy for cation vacancy jumps.

To obtain the equilibrium number density n of free cation vacancies, we must simultaneously solve the mass-action-law equations for the concentration of Schottky vacancies, vacancy-vacancy pairs, and vacancy-impurity pairs, along with enforcing overall charge neutrality[6, 3]. The details of this are presented elsewhere[14]. It should be noted that this calculation will also yield the equilibrium densities of vacancy-vacancy and vacancy-impurity pairs, which are used later.

Dielectric Relaxation

Following Breckenridge[1, 2], we will consider the relaxations due to both vacancy-vacancy $(v - v)$ and vacancy-impurity $(v - i)$ pairs. We will treat each of these relaxations as single relaxation time processes, based on many researchers' work[9, 10, 11, 12]. If we assume that these relaxation times are due to jumps of one nn vacancy to other nn lattice sites (w_1 type jumps[8]), we can derive the following expression for ϵ_d''[1, 2, 7, 8]:

$$\epsilon_d'' = \frac{Nx_{vv}e^2a^2(\epsilon_\infty + 2)^2}{54\epsilon_0 kT} \frac{\omega\tau_{vv}}{1 + \omega^2\tau_{vv}^2} + \frac{Nx_{vi}e^2a^2(\epsilon_\infty + 2)^2}{28\epsilon_0 kT} \frac{\omega\tau_{vi}}{1 + \omega^2\tau_{vi}^2} \tag{3}$$

where ϵ_∞ is ϵ' at frequencies much greater than the inverse of the relaxation time, ϵ_s is ϵ' at frequencies much less than the inverse of the relaxation time, N is the number density of normal cation sites, x_{vv} is the mole fraction of vacancy-vacancy pairs, x_{vi} is the mole fraction of vacancy-impurity pairs, a is the lattice constant for $NaCl$, τ_{vv} is the relaxation time for $v - v$ pairs, and τ_{vi} is the relaxation time for $v - i$ pairs. The relaxation times are functions of temperature as follows:

$$\tau_{vv} = \frac{1}{4\nu_0} e^{\frac{\Delta g_{mvv}}{kT}} \tag{4}$$

and

$$\tau_{vi} = \frac{1}{2\nu_0} e^{\frac{\Delta g_{mvi}}{kT}} \tag{5}$$

where Δg_{mvv} and Δg_{mvi} are the Gibbs free energies for the w_1 type vacancy jumps around another vacancy and an impurity, respectively.

Similar to the contribution from ionic conduction discussed earlier, the contribution from the ion jump relaxation process is very temperature sensitive. However, the magnitude of ϵ_d'' is typically greater than ϵ_c'' by several orders of magnitude, except at high temperatures in relatively pure crystals.

Phonon-Photon Interactions

In crystalline materials with the rock-salt structure, multi-phonon absorption of microwave energy is dominated by a lifetime-broadened two-phonon difference process [4, 5]. Based on the work in [4, 5], ϵ_{mp}'' can be calculated as:

$$\epsilon_{mp}'' = \frac{(\epsilon_s - \epsilon_\infty)\omega_f^3\Gamma}{(\omega^2 - \omega_f^2)^2 + (\omega_f\Gamma)^2}, \tag{6}$$

where ω_f is the resonant frequency of the fundamental reststrahlen transverse-phonon mode. Note that while ϵ_s and ϵ_∞ are still interpreted as the low- and high-frequency dielectric constants, respectively, they will have different numerical values in this calculation than in the dielectric relaxation calculation. This is due to the fact that the two-phonon absorption peaks at a much higher frequency than the dielectric relaxation absorption. The relaxation rate constant Γ corresponding to the reststrahlen mode is given by

$$\Gamma = \int_0^{K_{BZ}} dk\, g(k)\, L(\mathcal{E}). \tag{7}$$

where $g(k)$ is determined by the geometry and structure of the crystal, and $L(\mathcal{E}) = \frac{\gamma/\pi\hbar}{(\omega_{ji}-\omega)^2+\gamma^2}$ is a normalized Lorentzian function with $\mathcal{E} = \hbar\omega - \hbar\omega_{ji}$, $\omega_{ji} = \omega_j - \omega_i$ [4]. However, Eq. (7) is not very useful for the actual calculation of the temperature dependence of Γ. In Ref. [4], a more useful approximate expression is derived from the above, which was used in our calculations[14].

Our calculations in the lower microwave region $(2 - 20\,GHz)$ show that ϵ''_{mp} varies much less sensitively with temperature than contributions from ionic conductivity or ion jump relaxation in this frequency range. In particular, ϵ''_{mp} varies approximately linearly with both temperature and frequency.

THEORY OF MODIFICATIONS TO ABSORPTION DUE TO DISLOCATIONS

In the next stage of our research, we will incorporate dislocations into our samples in a controlled fashion (as described in the next section) and experimentally measure the modified microwave absorption in our samples, attempting to isolate the absorption due to the presence of the dislocations. Therefore, in this section we will describe qualitatively an extension of the above theory which will incorporate the effect of dislocations on microwave absorption. This extension is being developed for comparison to our anticipated experimental results. A key part of our experimental technique is that we will plastically deform the samples at high temperatures in order to create the dislocations, and then allow the samples to cool slowly to room temperature (rather than quenching them) in order that the point defects are allowed to reach their thermal equilibrium concentrations. We will not allow the samples to anneal long enough to rid themselves of the newly created dislocations, however. In addition, the dislocation densities will be kept low enough that ideal solution theory still applies to the entire system.

Ionic Conduction

Dislocations would be expected to modify the ionic conductivity in two ways[15]. First, if we imagine that the point defect concentrations have reached thermal equilibrium values, they must be in equilibrium with respect to their interactions with dislocations. Thus the dislocations may alter the concentration of free vacancies which contribute to bulk conductivity. Second, the conductivity along the length of a dislocation will be very different than that of the bulk crystal, due to the altered atomic arrangement and hence activation energy for migration in a dislocation[6], and also due to the enhanced concentration of vacancies near the dislocation core[15]. We will now explore each of these effects further.

If we are to calculate the "thermal equilibrium" concentration of free vacancies in the presence of dislocations, we must account for all possible vacancy-dislocation interactions. If we ignore effects such as vacancy clusters caused by moving dislocations (which we can "anneal out" by cooling the sample slowly), the interactions of interest are (1) via dislocation jog creation, annihilation, and motion (climb) [15], and (2) in ionic crystals, via electrostatic interactions with (charged) edge dislocations. If (1) were the only interaction, one could use thermodynamic arguments to convince oneself that the equilibrium vacancy concentration will be unchanged[16]. In an ionic crystal, however, there will be a significantly higher equilibrium vacancy concentration in the vicinity of a dislocation due to (2)[15]. As long as the dislocation density is relatively low this effect will not alter the overall vacancy concentration greatly. Indeed, experimental observations of the change in conductivity due to this effect have described the conductivity returning to the original value after a short time has passed[15].

The effect of altered conductivity along the axis of a dislocation is a more significant effect, and will persist even after point defect equilibrium is achieved. The widely accepted model for this is the dislocation pipe model[15], in which the dislocation is treated as a high conductivity (low migration activation energy) path through the crystal. Some experimental evidence suggests that this effect may be unimportant in doped or otherwise impure crystals[17]. To obtain an unambiguous characterization, one can isolate the change in the conductivity experimentally by taking either dc conductivity measurements or ac impedance spectroscopy measurements (at frequencies far below those at which relaxation and multi-phonon effects are important).

399

Dielectric Relaxation

Dislocations may create electrostatic fields that result in a vacancy cloud surrounding the dislocation[15]. In particular, though screw dislocations are uncharged, an edge dislocation has alternating positive and negative charges along it's length, and vacancies can be electrostatically bound to these charges. We can imagine these as dipole pairs with the vacancy free to move about relatively easily, and the dislocation less so. Thus we may expect at least two extra relaxation terms in Eq. (3): one due to sodium vacancy-dislocation relaxations and the other due to chlorine vacancy-dislocation relaxations. The sodium vacancies are far more mobile than the chlorine vacancies, so we may expect the sodium vacancy term to dominate; it may even be the dominant overall relaxation term, as the number of dipoles may be large. After isolating the effect of the change in conductivity due to the dislocations, we will subtract this off of experimentally measured absorption curves to look at the dielectric relaxation effects. Since these processes are both exponentially dependent on temperature, their contributions to the absorption curve should be easily distinguishable from the contribution due to multi-phonon absorption.

Phonon-Photon Interactions

The phonon spectrum will be modified by the presence of dislocations in the crystal, but if we keep the dislocation density small, we would expect that the phonon spectrum would not be greatly affected. There has been some theoretical work done in this regard[18], but we primarily expect to get information on the changes in this mechanism experimentally. At this time it seems likely that the two-phonon absorption will still be the dominant phonon-photon interaction, with little change in magnitude. Since the temperature dependence of this absorption mechanism is quite different from that of the other two mechanisms, it will be easily isolated and studied.

EXPERIMENTAL CONSIDERATIONS

In this section we will set forth our most recent experimental results, which verify nicely our dislocation-free model, and we will describe our sample preparation method to create single crystal $NaCl$ samples with controlled dislocation densities that may be used in our present measurement system.

Verification of Dislocation-Free Theory

To experimentally evaluate our theoretical model of microwave absorption, we have measured ϵ'' in $NaCl$ samples with very low dislocation densities as a function of temperature, frequency, and Ca^{++} impurity dopant concentration. The measurements were performed with cylindrical single crystal specimens by measuring the change in the loaded Q of a temperature controlled cylindrical cavity resonator due to a sample inserted in a small hole in the top of the cavity. The experimental configuration is described in detail in Ref. [13]. The relative measurement error was established[13] to be ±1%.

In Fig. 1, the experimental and theoretical results are plotted together in order to allow direct comparison. The theoretical results were obtained using the dislocation-free theory equations given above, with the impurity concentration determined by inductively coupled plasma spectroscopy at the Wisconsin State Laboratory of Hygiene[14]. As can be seen, the agreement between the two is very satisfactory. In Fig. 2, two of the theoretical curves from Fig. 1 are broken down into their components, in order to illustrate the difference between the temperature dependence of the multi-phonon process and the relaxation and conduction processes.

Sample Preparation: Next Stage of Research

Dislocations are introduced in materials through plastic deformation. We have obtained a large "brick" of single-crystal sodium chloride from the Harshaw Company, which will be carved into smaller bricks. These small bricks will then be deformed at high temperatures and allowed to cool slowly to partially anneal the material and allow the point defects to reach equilibrium. Since our experimental system requires rod-shaped samples of small radii, we will then employ a coring drill obtain samples useful for our measurement system. This method allows the possibility of controlling the dislocation orientation in our sample relative to the (known) orientation of the electric field inside the resonant cavity. We may therefore detect variations in the absorption when the electric field is oriented either along the length of the dislocations or perpendicular to the dislocations. If modified conductivity along the length of the dislocations proves to be a dominant effect, this experimental approach will conveniently isolate that phenomenon.

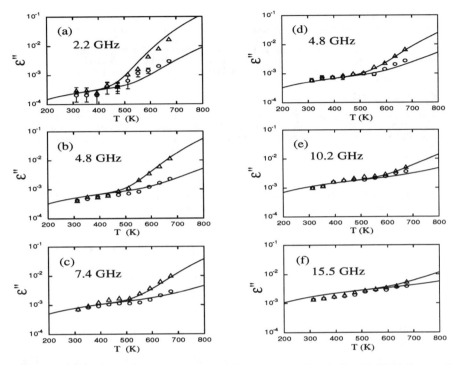

Figure 1: The experimentally measured ϵ'' and theoretically predicted ϵ'' in $NaCl$ single crystals at the frequencies of (a) $2.2\,GHz$, (b) $4.8\,GHz$, and (c) $7.4\,GHz$ versus temperature for the first cavity, and (d) $4.8\,GHz$, (e) $10.2\,GHz$, and (f) $15.5\,GHz$ versus temperature for the second cavity. In all the plots, o: undoped crystal (nearly-pure crystals), $\triangle$: doped crystals. —: theory with $40\,ppm\ Zn^{++}$ for the undoped sample and $405\,ppm\ Ca^{++}$ doping level for the doped sample, respectively, for the first cavity; theory with $40\,ppm\ Zn^{++}$ for the undoped sample and $190\,ppm$ Ca^{++} doping level for the doped sample, respectively, for the second cavity.

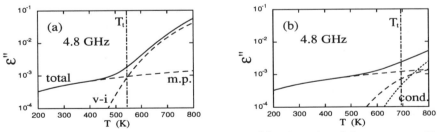

Figure 2: The theoretical curves at $4.8\,GHz$ from Fig. 1(b), split up into their components. The solid line is the total ϵ'' and the dashed lines are the contributions due to multi-phonon processes (m.p.) and $v-i$ pair dielectric relaxation (v-i), in both cases. (a) doped case, and (b) undoped case with dotted line representing the contribution due to ionic conductivity (cond.). Also indicated on the figure is the approximate value of T_t, the transition temperature discussed in the text.

CONCLUSION

From our above results, it is clear that microwave absorption in single crystal ionic solids with extremely low dislocation densities has been completely characterized. The first stage of our program to characterize microwave absorption in polycrystalline ionic solids has been a success.

We contend that the absorption effects due to the presence of dislocations in the solid will take the form of modifications to the three absorption processes detailed above. Therefore, although we do not know the parameters yet, we do know the functional form that the dislocation absorption effects must take.

ACKNOWLEDGEMENTS

The authors gratefully acknowledge the partial financial support of the Electric Power Research Institute, the Wisconsin Alumni Research Foundation, the NASA Graduate Student Researcher Program, and the National Science Foundation through the Presidential Young Investigator Award Program; and of Nicolet Corp. for the donation of the digital oscilloscope used in the experiments.

REFERENCES

[1] R.G. Breckenridge, *Imperfections in Nearly Perfect Crystals*, Ed. W. Shockley, J.H. Hollomon, R. Maurer, and F. Seitz (J. Wiley and Sons, N.Y., 1952), Chap. 8.

[2] R.G. Breckenridge, *J. Chem. Phys.* **16**(10), 959(1948).

[3] I.E. Hooton and P.W.M. Jacobs, *J. Phys. Chem. Solids* **51**(10), 1207(1990); Hooton and Jacobs, *Can. J. Chem.* **66**, 830 (1988).

[4] M. Sparks, D.F. King, and D.L. Mills, *Phys. Rev.* **B26**, 6987(1982), and references contained therein.

[5] K.R. Subbaswamy and D.L. Mills, *Phys. Rev.* **B33**, 4213(1986).

[6] W.D. Kingery, H.K. Bowen, and D.R. Uhlmann, *Introduction to Ceramics* (Wiley and Sons, NY, 1976).

[7] A.B. Lidiard, in *Handbuch der Physik*, Vol. 20, Ed. by S. Flugge (Springer-Verlag, Berlin, 1957).

[8] A.B. Lidiard, *Bristol Conference Report on Defects in Crystalline Solids, 1954* (The Physical Society, London, 1955).

[9] R.W. Dreyfus and R.B. Laibowitz, *Phys. Rev.* **A135**, 1413 (1964).

[10] C.H. Burton and J.S. Dryden, *J. Phys. C* **3**, 523 (1970).

[11] P. Varotsos and D. Miliotis, *J. Phys. Chem. Solids* **35**, 927(1974).

[12] J.S. Dryden and R.G. Heydon, *J. Phys. C* **11**, 393 (1978).

[13] B. Meng, J.H. Booske, and R.F. Cooper, *Rev. Sci. Instrum.* **66**, 1068 (1995).

[14] B. Meng, B.D.B. Klein, J.H. Booske, and R.F. Cooper, to be published in *Phys. Rev. B*.

[15] F.R.N. Nabarro, *Theory of Crystal Dislocations* (Oxford University Press, 1967).

[16] J.P. Hirth and J. Lothe, *Theory of Dislocations* (McGraw-Hill, 1968).

[17] H. Kanzaki, K. Kido, and T. Ninomiya, *J. Appl. Phys.* **33**, 482(1962).

[18] B. Y. Balagurov, V.G. Vaks, and B.I. Shklovskii, *Fiz. Tverd. Tela* **12**, 89(1970).

MICROWAVE EFFECTS ON SPINODAL DECOMPOSITION

M. WILLERT-PORADA
University of Dortmund, Department of Chemical Engineering, Division of Material Science, Dortmund, F.R. Germany

ABSTRACT

Microstructural differences between microwave and conventionally annealed solid solution TiO_2-SnO_2 and TiO_2-SnO_2-$(Al_2O_3)_{0,02}$-samples demixed by spinodal decomposition are reported. The decomposition is retarded upon microwave annealing as compared to conventional heat treatment. This result is in accordance with theoretical predictions, if a superposition of the microwave field with an internal periodic electrostatic potential resulting from kinetic effects is assumed.

INTRODUCTION

The origin of microstructural differences observed in microwave sintered ceramics in comparison to conventionally sintered ceramics still remains unclear. Recently it was suggested, that microwave effects leading to accelerated densification in the early stage of microwave sintering or accelerated grain growth upon aliovalent doping in the late stage of microwave sintering of oxide ceramics are related to the intrinsic space charge ubiquitous in compounds with a high degree of ionic bonding [1]. In experimental work the difference in shrinkage rates for alumina and alumina-zirconia ceramics sintered conventionally and in a microwave field was clearly documented [2]. Because of different diffusion paths operative at different stages of the sintering process, analysis of microwave effects based on sintering experiments is quite complicated. Furthermore, theoretical work on the role of space charge for sintering is sparse [3]. In contrast, space charge effects on spinodal decomposition of ionic solid solutions $AX_{(s)} + BX_{(s)} \leftrightarrow (A,B)X_{(ss)}$ and on diffusion in ionic solids were thoroughly investigated by different theoretical approaches [4,5], with evidence given theoretically and experimentally for a susceptibility of this process to a local electrostatic potential and to aliovalent dopants. In the following, the theoretical background of the spinodal decomposition for the well known TiO_2-SnO_2-system will be used in order to define parameters, which could influence the spinodal decomposition of this solid solution in a microwave field.

The TiO_2-SnO_2-system exhibits a miscibility gap at low temperature. Spinodal decomposition in this tetragonal system occurs in the temperature range between 800°C and approximately 1400 °C [6], as indicated in Figure 1. Doping with the aliovalent Al_2O_3 facilitates spinodal decomposition, in accordance with theoretical predictions given by Virkar and Plichta [4]. For the analysis of effects originating from electromagnetic fields some basic presumptions are made:

• Upon heat treatment in a microwave field an electromagnetic field is established in materials with a high penetration depth for microwaves

• In ionic solid solutions the fluxes of the cationic species can be coupled because of the different mobility of the cations A^+ and B^+ [5]

• Net flux of zero for electrical charge [5] is assumed; no long range electrical fields build up

• The chemical potential of the anion X^- is the same for both A^+ and B^+ and the electrostatic potential of the lattice is periodic [4]

Mat. Res. Soc. Symp. Proc. Vol. 430 © 1996 Materials Research Society

Only the first presumption is not established in literature, because the analysis of diffusion or phase changes is usually performed assuming the absence of an external electromagnetic field.

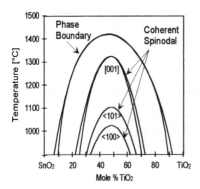
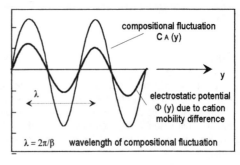

Figure 1: Low temperature section of the TiO_2-SnO_2 phase diagram (after [7]) and a schematic representation of the compositional and electrostatic fluctuation within the spinodally decomposed miscibility gap (after [4])

When an initially homogeneous solution of TiO_2 and SnO_2 is cooled to a temperature inside the spinodal, the composition will become periodic with an amplitude A and a wave number β, following Cahns theory of spinodal decomposition [8], such that

$$N_A - N_A^0 = A\cos\beta \qquad N_B = 1 - N_A = 1 - N_A^0 - A\cos\beta \qquad (1)$$

with N_A and N_B as the site fractions of A^+ and B^+, respectively and β the wave length of the fluctuation.

For the diffusion process yielding the compositional fluctuation a coupled flux of the cations in binary ionic solid solutions has to be considered [5]. The coupling occurs as follows: assuming the mobility of A^+ ($M_A=[V]M_{Va}$) to be larger than the mobility of B^+ ($M_B=[V]M_{vb}$), regions with excess cation vacancies will develop in the areas depleted in A^+ and excess anion vacancies will develop in the area enriched in A^+. The area with excess cation vacancies becomes negatively charged, enhancing flux of B^+ into this region. Consequently, flux of A^+ into the area already enriched in this cation will be retarded because of the relative positive charge developed from excess anion vacancies. At a phase boundary between the solid solution and the components AY and BY the coupling of fluxes would be visible by the formation of a concentration profile and the migration of the boundary. Inside the miscibility gap at conditions which fulfill the requirements for a spinodal decomposition in terms of elastic strains, however, for ionic solid solutions a periodic electrostatic potential is build up according to an analysis given by Virkar and Plichta [4]. Starting with the electrochemical potentials of the cations A^+ and B^+, η_A and η_B, and the flux-equations:

$$\eta_A = \mu_A + e\phi \qquad \eta_B = \mu_B + e\phi \qquad (2)$$
$$J_A = -C_A[V]M_{Va}\mathrm{grad}\eta_A \qquad J_B = -C_B[V]M_{Vab}\mathrm{grad}\eta_B \qquad (3)$$

and following the condition, that long range electrical fields are not build up, e.g. $|J_A| = |J_B|$, the gradient in local electrostatic potential is described by eq. (4), with α as the interaction parameter of mixing for regular solutions used in Cahns theory of spinodal decomposition [8]. From this equations it is evident, that without a difference in cationic vacancy mobility no electric field is developed concomitant with a compositional fluctuation. However, with such a difference a periodic electric field with the same wavelength as the compositional fluctuation but $90°$ out of phase with the composition modulation is developed. A sketch of this result for $\Omega < 0$ is shown in Figure 1.

$$\frac{d\phi}{dy} = \frac{(2kT - \alpha)[M_{Va} - M_{Vb}]A\beta \sin\beta y}{e(M_{Va} + M_{Vb})} \qquad \Omega = \frac{(2kT - \alpha)[M_{Va} - M_{Vb}]}{e(M_{Va} + M_{Vb})} \qquad \frac{d\phi}{dy} = \Omega A\beta \sin\beta y$$

(4)

The influence of this internal periodic electric field with respect to the free energy for a composition gradient and to diffusion is analyze in [4]. The important results of this analysis are the following: the location of the spinodal, i.e. the thermodynamic of the decomposition is unaffected by the electrostatic field, but the critical wavelength β_c is affected. Applied to diffusion, the electrostatic potential affects the interdiffusion coefficient which in turn is dependent on the wave length β. Furthermore the amplification factor $R(\beta)$ is depending on the electrostatic potential, by a term $\varepsilon \Omega^2 \beta^2$. *Because the term $\varepsilon \Omega^2 \beta^2$ is always positive, it subtract out of the free and elastic energy term, thereby reducing the amplification factor and retarding the kinetics of decomposition [4].* A small change in the mobility difference of the cations will through the amplification factor result in a large change in the kinetics of spinodal decomposition in a ionic solid solution.

Cation diffusion is strongly affected by aliovalent doping which increases the vacancy concentration. In this case the influence of the different cation mobilities is diminished by the total increase in cation diffusion, as could be deduced from the experimental finding of an accelerated spinodal decomposition in TiO_2-SnO_2-solutions by Al_2O_3-doping [4].

The main goal of the work presented here is the analysis of microwave effects. Therefore, in order to identify the effect of an alternating external electric field on spinodal decomposition, TiO_2-SnO_2 and TiO_2-SnO_2-Al_2O_3-solid solutions were subjected to heat treatments in a conventional furnace and in a microwave field. According to the theoretical background, the microwave field could affect the spinodal decomposition by superposition with the local electrostatic potential in analogy to material transport upon sintering, as suggested in [1]. With regard to sintering, microwaves were assumed to polarize periodically the space charge layer of ceramic crystallite grain boundaries, thus reducing the vacancy transport retarding effect of the space charge build up due to the different energy of formation for the anionic and cationic defects. Assuming such an polarization effect during spinodal decomposition in a microwave field, the intrinsic periodic local electric field build up by the kinetic effect of different A^+ and B^+-cation mobilities could superimpose with the externally applied field causing:

- phase changes in the periodic electrostatic fields given by eq.(4), altering the direction of spinodal decomposition (the decomposition follows particular crystallographic directions [6,7])
- an enhancement of the mobility difference between the A^+ and B^+-cations and therefore modulation of the wave length β
- local changes of the static dielectric constant, in analogy to observations for halides at lower frequency external fields [9], and from there changes in β

As stated in the previous analysis [1], the frequency of microwave radiation is to high to be followed directly by migrating ions, and to low to directly influence lattice vibrations. However, polarization of local electrostatic fluctuations might occur, generating additional gradients as driving force for material transport. Evidence for this is found in alkali halides subjected to microwave radiation [10].

EXPERIMENTAL

Commercial TiO_2, SnO_2 and Al_2O_3-powders with >99%purity (Hereaus, Germany, Alcoa A16)) and specific surface area of 10 m^2/g, 3 m^2/g and 12 m^2/g respectively, were employed. The powders were mixed by 15 min. milling in ethanol, in a planetary mill using Al_2O_3 milling media. Two 150 g batches of material were prepared: one with a $(TiO_2)_{0,5}(SnO_2)_{0,5}$ composition and one with a $(TiO_2)_{0,49}(SnO_2)_{0,49}(Al_2O_3)_{0,02}$ composition. Green parts with 10-15 g weight (ϕ 20 mm, h 10-20 mm) were made by isostatic pressing at 300 MPa. To obtain a solid solution the green parts were sintered at 1500°C for 2h in a conventional furnace in air followed by quenching from 1500°C to 500°C within 5 min. Compositional homogeneity was tested at the outer and inner part of the samples by XRD. As visible from the XRD-data (Figure 2.), in alumina containing green parts partial spinodal decomposition occurred during the cooling procedure.

Heat treatment was performed on air at 900°C for 1h and with a second and third batch of samples at 1100°C for 1h. The samples were heated with 1000°C/h heating rate in a conventional furnace and in a 2.5 kW multi-mode rectangular microwave cavity with 2.45 GHz frequency, at two different power levels. A pyrometer is used for temperature measurements in the microwave experiments, with emissivity adjusted to the thermocouple reading upon the conventional heating experiments (on cooling, to avoid contribution from heating elements radiation), using the same casket as in the microwave experiments. The TiO_2-SnO_2-sintered parts show very low loss at 20°C and also at 900°C. Therefore, for microwave experiments hybrid heating is used, by placing the samples on top of a preheated 80% Al_2O_3-20% SiO_2 susceptor or on a 80% Al_2O_3-20% TiC susceptor. The microwave power density necessary for heating at 1000°C/h is 0.02 W/cm^3 for the alumina-silica and 0.01 W/cm^3 for the alumina-TiC susceptor. To avoid further contamination, a powder bed composed of TiO_2 and SnO_2 is used between the sample and the susceptor. The penetration depth of microwaves into the casket is ascertained by microwave calorimetry, as described elsewhere [11].

The heat treated samples were investigated by SEM-EDX and XRD at the bottom, top and middle section to estimate the degree of compositional changes and the microstructural changes. Only samples with homogeneous elemental distribution were taken for closer analysis. To insure the origin of topographic contrast from compositional fluctuation, BSE (mass contrast) and SE-electrons were used for recording the SEM-micrographs with two separate detectors.

RESULTS AND DISCUSSION

As shown in Figure 2, decomposition of the $(TiO_2)_{0,49}(SnO_2)_{0,49}(Al_2O_3)_{0,02}$ can be directly detected by SEM (scanning electron microscopy) and ascertained by satellite peaks in the X-ray diffractogramms of the samples [4].

The alumina containing samples undergo at 1100°C further decomposition, which, based on the XRD-results, indicates a high content of nearly pure TiO_2-lamellae. For TiO_2 the (101)-peak is found at 2 Θ > 34° (Cu K_a-radiation [12]). The lattice constants of the oxides are very similar, with a= 4.594; c=2.958 Å for TiO_2 and a=4.7380, c=3.1867 Å for SnO_2, respectively [7]. The strong topography contrast in the SEM-micrograph of the alumina containing samples as well as the lack of satellite peaks in the samples after heat treatment at 1100°C indicates loss of coherency and demixing into the pure oxides. This happens upon microwave as well as conventional annealing, as shown in Figure 3 by the X-ray diffraction results.

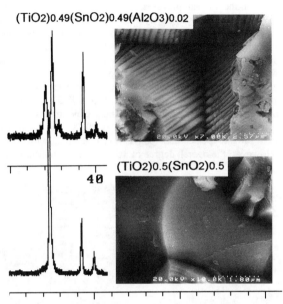

Figure 2. Microstructure of partially decomposed TiO_2-SnO_2 solid solutions; the satellite peaks in the alumina containing sample are typical for spinodal decomposition and correspond to a lamella microstructure of the crystallites along one direction

With respect to microstructural changes due to annealing at 1100°C for different times, the microwave annealed samples display a very unique microstructure, not found in conventionally annealed samples. Examples are shown in Figure 4.

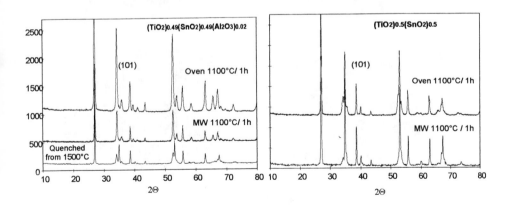

Figure 3. XRD-results of samples after 1100°C microwave or conventional annealing

This „tweed" structure is found for samples heat treated in a microwave field regardless the power density at 0.02 W/cm^3 for the 80% Al$_2$O$_3$-20% SiO$_2$ susceptor or 0.01 W/cm^3 the 80% Al$_2$O$_3$-20% TiC suszeptor. However, this structure is more pronounced for samples annealed at 1100°C for 1h than for samples after 0.5h hold time at 1100°C, as shown also in Figure 4. the wavelength of the „lamellae" in microwave annealed samples is approximately 200 nm as compared to 400 nm for the conventionally annealed samples.

For alumina free samples scanning electron microscopy is sufficient to visualize the lamellae structure, as indicated by the micrograph in Figure 2. From XRD-measurements a retarded spinodal decomposition is indicated for microwave annealing experiments at 1100°C/ 1h (Figure 3 an 4) and 0.5h as compared to conventional heating. At 900°C/ 1h the microwave annealing yields more pronounced satellite peaks than conventional heating, with approximately 30% higher satellite peaks. However, the overall degree of decomposition does not exceed 30 %.

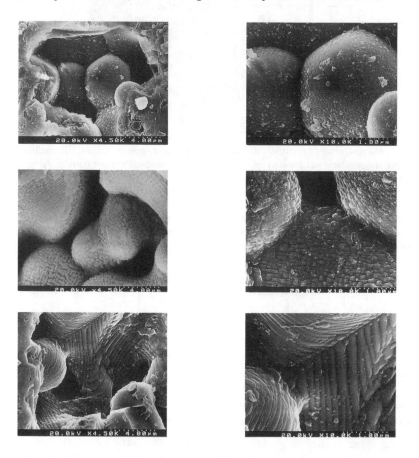

Figure 4: Microstructure of microwave annealed TiO$_2$-SnO$_2$-Al$_2$O$_3$-samples as compared to conventional annealing; from top to bottom MW, 1100°C/0.5 h; MW, 1100°C/1h; conventional 1100°C/1h

This preliminary results indicate, that the influence of an electromagnetic field on spinodal decomposition is only partially explained by the theory. The wave length β of the compositional fluctuation appears to be similar in microwave and conventional annealed samples containing an aliovalent additive. However, the different shape of the compositional fluctuation found under microwave conditions could be understood as a visible effect of an alternation of the phase of the electrostatic fluctuations connected with the compositional one, as schematically shown in Figure 1. In aliovalent additive free samples the retardation of decomposition upon microwave annealing at 1100°C fits the prediction of such an effect caused by an electrostatic fluctuation. However, at low decomposition temperatures this effect is reversed.

CONCLUSION

Spinodal decomposition of ionic solid solutions could be a useful tool to investigate microwave effects on diffusion in ionic solids, as indicated by preliminary results obtained for TiO_2-SnO_2-/+Al_2O_3-solid solutions. A retardation of the decomposition reaction upon microwave annealing is found for 1100°C, in accordance with theoretical analysis, whereas the opposite effect is indicated in experiments at 900°C. However, theoretical predictions are not naturally fulfilled by the experimental results, because the relative significance of aliovalent doping and internal electrostatic field fluctuations on the kinetics of the decomposition reaction is not known.

REFERENCES

[1]M. Willert-Porada, in Microwaves; Theory and Application in Materials Processing III, edited by D.E. Clark, D.C. Folz, S.J. Oda and R. Silberglitt, Ceram. Trans. Vol. 59 (1995) pp. 193-203
[2] F.C.R. Wroe, A.T. Rowley, Theory and Application in Materials Processing III, edited by D.E. Clark, D.C. Folz, S.J. Oda and R. Silberglitt, Ceram. Trans. Vol. 59 (1995) pp. 69-78
[3] S.J. Bennison, M.P. harmer, J. Am. Ceram. Soc. , 68 [1] C-22-C-24 (1985)
[4] A. V. Virkar, M. R. Plichta, in J. Am.. Ceram. Soc., 66 [6],451 (1983)
[5] A. R. Cooper,Jr., J. H. Heasley, J. Am. Ceram. Soc., 49 [5], 280 (1966)
[6] V.S.Stubican, A.H. Schultz, J. Am.. Ceram. Soc., 4, 211 (1970)
[7] M. Park, T.E. Mitchell, A.H. Heuer, J. Am. Ceram. Soc., 58 [1-2],43 (1975)
[8] J. W. Cahn, „On Spinodal decomposition", Acta Met., 9, 795 (1961)
[9] E. Fatuzzo, S. Coppo, J. Appl. Phys., 43 [4], 1467 (1972)
[10] S.A. Freeman, J.H. Booske, and R.F. Cooper in Theory and Application in Materials Processing III, Ceram. Trans. Vol. 59 (1995) pp.185-192
[11] T. Gerdes, M. Willert-Porada, K. Rödiger, and H. Kolaska, ibid., pp. 423-432
[12] G. Croft, M.J. Fuller, Br. Ceram. Soc. Trans. J., 78 [3] 52 (1979)

ACKNOWLEDGMENT

Financial support of DFG, Wi-854/4-1 is gratefully acknowledged.

THERMAL PHENOMENA UNDER MICROWAVE FIELD IN THE ORGANIC SYNTHESIS PROCESSES: APPLICATION TO THE DIELS ALDER REACTION

R.SAILLARD*, M.POUX**, M.AUDHUY-PEAUDECERF***
* 19 rue Guy de Maupassant 92500 RUEIL MALMAISON, FRANCE
**ENSIGC, 18 chemin de la Loge 31078 TOULOUSE CEDEX, FRANCE
***ENSEEIHT, 2 rue Camichel 31071 TOULOUSE CEDEX, FRANCE

ABSTRACT

The influence of the microwave heating on chemical reactions were investigated. The kinetic of the Diels Alder reaction were studied under Microwave irradiation at a frequency of 2.45 GHz in a single mode cavity and were compared to the kinetic obtained by a conventional heating. Experiments were carried out in a liquid solvent in order to have a better control of the medium temperature measurement. In a second part, the presence of a catalytic solid phase was introduced. Some thermal fluctuations which are due to an heterogeneity of the electric field were detected in the medium. They reduce the precision of the results and cause problems of experimental reproducibility. A thermoluminescent material allow a good visualisation of these phenomena.

In addition, the profiles of the electric field intensity were modelled by a 2D finite elements method in our reactor in the presence of a solvent. Despite the small size of the sample and the use of a monomode cavity which both limited the heterogeneities of the medium temperature, we showed a great heterogeneity of the electric field intensity and as a result the heterogeneity of the temperature in our sample.

In order to avoid these phenomena which induce a lack of reproducibility, a stirring device was developed.

The values of the kinetics obtained under the 2 heating modes with the introduction of the stirring device. So, it induces a good control of the medium temperature.

All those investigations prompted us to the conclusion that there is no difference between microwave heating and a classical heating in the studied reaction.

INTRODUCTION

The microwave technology induces a selective and rapid heating. So, the introduction of this technology in the organic synthesis processings would be interesting. This potential economical interest has induced a lot of investigations for several years. Chemical conversion rates under microwave heating reported in the litterature generally present a great kinetic acceleration. The main field of investigations under microwave heating is carried out in a classical microwave oven and in a dry medium, without any solvent [1-6]. Conversion rates obtained are generally very high and change with the support and the studied chemical reactions. It seems that a local overheating at the surface of the solid particles could be responsible for the observed effect. Unfortunatly, the absence of medium temperature measurement does not allow the comparison of the results with a classical reaction process. Therefore we decided to investigate the effects of microwave field on the chemical reactions [7]. These investigations were made to conclude on the existence or not of a specific effect of the microwave on the chemical reaction.

The aim of this work is to compare the kinetics obtained in a chemical reaction carried out under microwave irradiation with the kinetics obtained under a conventional heating.

To compare the kinetic of the reaction, the control of the medium temperature was crucial. It was necessary to work with an identical and constant temperature in the two modes of heating. Therefore, we decided to investigate a chemical reaction which can be carried out in presence of liquid solvent.Thus the temperature measurement of the solution was easier.

We worked with and then without catalysis solid phase of the reaction. The introduction of the solid phase allowed the introduction of a heterogeneous medium.

The chemical reacion studied is a Diels Alder reaction : the addition of the anthracen on the maleic anhydrid. We compared the chemical conversion rate.

EXPERIMENT

Microwave device : The microwave device is presented in fig.1. It consists in two parts:
(i) the magnetron at a frequency of 2.45 gHz and the waveguide giving a TE10 monomode propagation. The waveguide was equiped with an HI tuner for the minimization of the reflected power, a circulator and a water load, to absorb the reflected power and to protect the magnetron.
(ii) The reactor was located at a distance of λ/4 cm away from the short circuit, and was protected by a brass tubing in order to avoid a leakage of waves. The reactor is dipped into the waveguide through a Teflon guide which ensured the correct location.

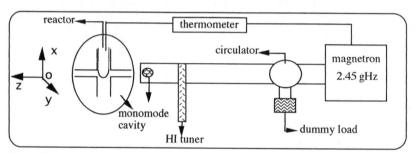

fig.1 : microwave device

Running conditions : The microwave reaction was carried out in a test tube (internal diameter 20 mm), fitted by a condenser. the small size of the reactor was choosen to limit the electric field heterogeneities. Experiments ran under atmospheric pressure. The temperature was measured with a calibrated fiber optic probes (luxtron) connected to a bravo 286 computer. The regulation of the temperature was performed by means of a system acting directly on the power of the magnetron. A stirring system has been introduced into the reactor to have a homogeneous temperature of the solution and to generate a solid-liquid suspension.

Reaction and analytical conditions : In a typical run, Reactants (maleic anhydrid 99% purity from Janssen and anthracen 98+% are in stoechiometric quantities with a concentration of $2.247 \ 10^{-2}$ mol l^{-1} and are completed at 10 ml by dioxan (98% purity from Aldrich). If the heterogeneous solid phas is introduced into the solution, 0.25g of solid particles are introduced in the solution. These particles are either graphite particles or 2 different batchs of dissipatives ceramics, one with dielectric properties permitting a good heating under microwave and the other with the dielectric properties and with a catalytic effect on the diels alder reaction due to the presence of lewis acids sites.
The reaction was carried out at 100°C during 8 hours; at this time, the chemical conversion rate of the reaction was 52%. To confirm the presence of side-products, the mixture was analysed by HPLC (UV detector diod array : Merck L-3000; pump Merck L-6200A, Spherisorb column ICS plus C18 inversed phase (250mm, internal diameter : 4.6 mm; particles diameter : 8μm); eluent : water-acetonitrile(80-20v/v); flow : 0.4ml-8mn, 1ml/min). the reaction conversion rate was determined by analysing anthracene by UV spectrophotopetry (Hitachi U200) (λ : 377,5nm)

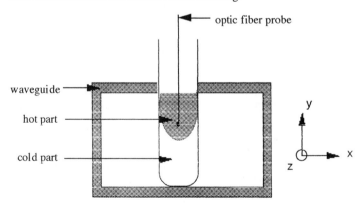

RESULTS

Our first investigations rested on the reactions in a homogeneous phase, without any solid particles and without any stirring system. The results were different in the two modes of heating with a very bad reproducibility between each experiments carried out with the same apparent parameters. Sometimes results obtained under microwave heating showed a slight acceleration of the chemical kinetic of the reaction. Generally, they showed a decrease of the kinetic. These results which contradicted all the initial results found in the literature prompted us to a better knowing of our global system : medium temperature, electric field... Clearly, it was necessary to discover the no undercontrol parameters.

Visualization of the temperature heterogeneities with thermoluminescent material : The introduction of a thermoluninescent material in the reactor showed us two areas with different temperatures (fig.2). This material were firstly tested in a classical microwave oven : it did not heat. The material discoloration was only due to the enhancement of the liquid medium temperature. The microwave theory indicates a variation of the electric field only along the Ox axis and Oz axis (see fig.1 for the axes). The hypothesis for the electric field modelling consideres an infinite dimension along the Oy axis. This induces a constant electric field along this axis. So, the temperature, proportional to the electric field, should be homogeneous along this axis.The position of the reactor in the cavity induces an assymetry and a preferential way of the waves. The temperature is higher in the area with an intense electric field.

The fiber optic probe was located in the hot temperature area. So the average temperature of the solution was lower than the measured temperature. The kinetic of the reaction was proportional to the temperature. Therefore the chemical conversion rate found under microwave oven was lower than the one found in a conventional heating.

Figure 2 : heterogeneities of temperature

413

Field modelling and temperature heterogeneities : The modelling[8] of the electric field intensity absorbed by a solvent carried out using a finite elements method[9] established the non homogeneity of the electric field in the reactor which induced non homogeneous temperatures in the solution. Dielectric values of the studied solvent necessary for the calculations were already determined[10]. First, the electric field was calculated in the empty waveguide, then in the waveguide containing the quartz reactor and lastly the quartz reactor with the solvent. For the solvent, ethyl alcohol, these experiments were carried out at two different temperatures, 20°C and 78°C. the presence of the reactor did not significantly change the field distribution in the guide. Figure 3 shows the electric field distributions along the Oz axis of the waveguide E(z). With a small reactor placed in a monomode waveguide the field intensity was far from being homogeneous in the liquid.

Those two observations prompted us to introduce a stirring system in the reactor in order to avoid the temperature heterogeneities in the solution.

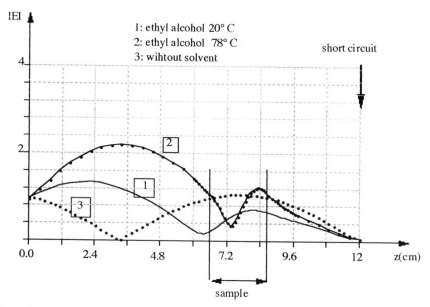

figure 3 : Field modelling with ethyl alcohol along the oz axis of the waveguide

Introduction of a stirring system : The use of the stirring system in the reactor submitted to the microwave heating permitted to obtain a good reproducibility of the kinetic constants for the reaction carried out in a homogenous phase, as seen in the table 1. The chemical conversion rates obtained were identical for both microwave and conventional heating. In this case, there was no specific effect of the microwave heating.

We introduced a heterogeneous phase and saw if this phase could generate local overheating and increase the kinetic of the reaction[11]. The solid particles incorporated called "Heat Captors" were dissipative ceramics and graphite. They can play a double part : (i) they can heat under microwave because of their dielectric properties. We noted that when we carried out experiments by heating these solids under microwave and without solvent, the temperature could increase from 20°C to about 400°C in 30 seconds. The control of the medium temperature induces the use of particles having very small dimensions and of course the presence of a stirring device.

An increase in the particle size would reduce the contact area with the liquid. The heat transfer would decrease and local hot zone could develop at the surface of the particles and could induce some specific effects. (ii) Some "heat captors" may also have a catalytic effect because of the presence of lewis acid sites. Runs were carried out under microwave or conventional heating at 100°C.

solid particles	heating mode CH : oil batch heating MW : microwave	kinetic constant $(10^{-3}\ mol\ l^{-1}\ s^{-1})$
none	CH MW	0.93(± 0.7%)
graphite	CH MW	1.15 (±1%)
céramiques with lewis acid sites	CH MW	1.24 (±0.4%)
céramiques no lewis acid sites	CH MW	0.97 (±2.5%)

Table 1 : Reaction kinetic constants at 100°C

The results are reported in table 1. We can see a good reproduciblity of the experiment results and that microwave and classical heating do not induce a difference in the kinetic constant values.

CONCLUSIONS

The kinetics of the addition reaction of anthracen on maleic anhydrid with the presence or not of a solid phase determined under microwave or conventional heating are compared. The use of a stirring device and the choice of the solid size particles make possible the strict control of the temperature. The results showed the same kinetic values for the two heating modes with the presence or not of a solid phase in the liquid. The introduction of solid particles in order to create an heterogeneous media does not induce any specific effect under microwave heating when the temperature is supervised. In fact, it was clearly established for this reaction that the modification of the kinetics obtained in heterogeneous media under microwave heating is the result of a thermal effect. It is necessary to take care of the measurement method to have a good knowledge of all the parameters acting on the results as temperature heterogeneities, heat transfer....

ACKNOWLEDGMENTS

We thank EDF for the financial support of this work. We thank M.Delmotte and H.Jullien, CNRS LM3, ENSAM, 151 Bd de L'Hopital 75013 PARIS and the group of M.Lallemant, laboratoire de Recherches sur la Réactivité des solides - B.P. 138 - 21004 DIJON Cedex France, for tg δ determinations.

REFERENCES

1.   R.S. Varma, A.K. Chatterjee, M. Varma, Tetrahedron Lett. 1993, 34, pp 4603-4606.
2.   A.Molina, J.J; Vaquero, J.L. Garcia-Navio, J. Alvarez-Builla, Tetrahedron Lett. 1993, 34, pp 2673-2676.

3.   L.A. Martinez, O. Garcia, F.Delgado, C. Alvarez, R. Patino; Tetrahedron Lett. 1993, 34, pp 5293-5294.

4.   M. Csiba, J. Cleophax, A. Loupy; J. Malthête, S.D. Gero, Tetrahedron Lett. 1993, 34, pp 1787-1790.

5.   A. Loupy, A. Petit, M. Ramdani, C. Yvaneff, m. Majdoub, B. Labiad, D. Villemin, Can. J. Chem. 1993, 71, pp 90-95 .

6.   B. Rechsteiner, F. Texier-Boullet, J. Hamelin, Tetrahedron Lett. 1993, 34, pp 5071-5074.

7.   R.Saillard, Phénomènes thermiques sous champ micro-ondes dans les procédés de synthèse organique : Application à la réaction de Diels Alder, Doctoral thesis, Institut national Polytechnique toulouse France, June 1995.

8.   R. Saillard, M. Poux, J. Berlan, M. Audhuy-Peaudecerf, Tetrahedron, 1995, 51, pp 4033-4042

9.   M. Audhuy-Peaudecerf, M. Barbosa, H. Majdabadino, 11ème Colloque Optique Hertzienne et Diélectrique, Hammamet (Tunisie), sept. 91.

10.   H.Jullien, M. Delmotte, Private communication.

11.   R. Saillard, M.Poux, Tetrahedron Lett. accepted, to be printed

HIGH-FREQUENCY FIELD EFFECTS ON IONIC DEFECT CONCENTRATIONS AND SOLID-STATE DIFFUSION PROCESSES

S.A. FREEMAN*, J.H. BOOSKE**, R.F. COOPER***
*Materials Science Program, Univ. of Wisconsin, Madision, WI 53706, freeman@cae.wisc.edu
**ECE Department, Univ. of Wisconsin, Madision, WI 53706, booske@ece.wisc.edu
***MS&E Department, Univ. of Wisconsin, Madision, WI 53706, cooper@engr.wisc.edu

ABSTRACT

We describe computer simulations of a microwave-induced driving force for ionic transport. The simulations are based on a model which predicts rectification of ionic fluxes at interfaces and a resulting depletion or accumulation of defects near the interface. Some details of the model are discussed, results of the simulations are presented, and the impact of these effects on sintering and diffusion is discussed.

INTRODUCTION

Our early investigations [1-3] of the "microwave effect" focused on possible mechanisms of enhanced diffusion rates due to microwave field effects on ionic lattices. Our theoretical and experimental work showed conclusively that microwave field strengths are not high enough to effect solid-state diffusion events. With our ionic current measurements [4,5], we did however find experimental evidence of microwave-induced ionic transport in NaCl. Concurrently, a theoretical model [6] was put forth that suggested that defect concentrations would be perturbed and rectified at interfaces/boundaries. This "ponderomotive effect" would manifest itself as a driving force for charge and mass transport. The general agreement between the theory and the measurements of microwave-induced currents has led us to further investigate the role of this effect in other transport processes such as diffusion and sintering. We describe here our efforts at modeling this effect on the computer, the results of the computer simulations, and some speculation of how this effect would impact diffusion and sintering rates.

COMPUTER MODEL

The basis of the "ponderomotive force" theory derives from applying perturbation techniques to a coupled set of non-linear partial-differential continuum equations [6] that describe charge flow in ionic materials (and also describe physical problems in plasma physics, semiconductor physics, electrochemistry, ceramics science, and other fields). The first equation describes the particle flux (J) of a charged species (i) that can result from either a concentration gradient (∇C) or a electrical potential gradient (E):

$$\overline{J_i} = -D_i \overline{\nabla} C_i + \frac{D_i C_i}{kT} q_i \overline{E} \tag{1}$$

where D is the diffusion coefficient, q is the electric charge, and kT is the thermal energy. The two components of flux are sometimes called *diffusion* and *drift*, respectively. This expression

417

of equation (1) assumes (in the *diffusion* term) ideal solution behaviour of the charged species and (in the *drift* term) Ohmic conduction. TheE-field in equation (1) is a linear superposition of the microwave field, the intrinsic space-charge field (near surfaces/interfaces), and an internal field resulting from the unequal perturbations of differently charged species.

The continuity equation can also be applied such that the concentration time-rate-of-change depends on the divergence of the flux and on the presence of source/sink terms (S):

$$\frac{\partial C_i}{\partial t} = -\overline{\nabla} \cdot \overline{J}_i + S_i \qquad (2)$$

Since the defects are charged, Poisson's equation also applies:

$$\overline{\nabla} \cdot \overline{E} = \frac{1}{\varepsilon_r \varepsilon_0} \sum_i q_i C_i \qquad (3)$$

where ε_0 is the permittivity of free space, and ε_r is the relative dielectric constant of the solid. It is through equation (3) that the electric field and defect concentrations become coupled.

The problem is simplified with a few key assumptions. Since we are modeling a very small crystal/grain with dimensions smaller than the microwave wavelength, we assume that the electric field strength is uniform across the dimensions of the crystal. A one dimensional model is used with the assumption that the electric field is aligned in that direction, perpendicular to the crystal surface. Only half of a single crystal is considered such that one boundary is the surface and the other edge is the "bulk". Furthermore, it is known for NaCl that vacancies of Na and Cl are the only mobile species. NaCl was chosen because of its ideal properties and because it has been very carefully studied and has well-characterized defect parameters. However, the generality of the effects studied here is not lost by the choice of NaCl, since the continuum equations could be derived for any charged mobile species in other solid materials. Thus, the model is very general, and the results can be expected to apply to many materials.

Perturbations from the initial, equilibrium values of *vacancy* concentrations (C), electrostatic potential (V), and electric field (E) are allowed, with the resulting equations:

$$C_{Na}(x,t) = C_{Na}^i(x) + C_{Na}^p(x,t) \quad and \quad C_{Cl}(x,t) = C_{Cl}^i(x) + C_{Cl}^p(x,t) \qquad (4)$$

$$E(x,t) = -\frac{\partial V(x,t)}{\partial x} = E^i(x) + E^p(x,t) + E^{mw}(t) \qquad (5)$$

Here the subscripts refer to the VACANCY species, superscript *i* refers to the initial value (unchanging in time), superscript *mw* denotes the microwave field, and superscript *p* is the perturbation (varying spatially and temporally) from the initial value due to the microwaves.

We apply these perturbations to equations (1) - (3) for the 1-D case and with the other assumptions previously mentioned. For initial conditions, we use the boundary space-charge theory of Kliewer and Koehler [7]. For the boundary conditions on the "bulk" side, we assume symmetry such that the divergences of both the fluxes and the E-field are zero. On the surface boundary, we assign the reference potential (V) to be zero, and we do not allow for fluxes of vacancies to or from the surface/interface. Instead, exchange (or creation/annihilation) of vacancies with the interface is handled by the source/sink (S) term. We use the Simulation Generator (SimGen) software package [8] to numerically solve the continuum equations. The program employs a finite-difference calculation based on the Newton method. Newton damping, Gaussian elimination, and Scharfetter-Gummel discretization techniques are also included.

RESULTS

The results of the simulations can be naturally classified into three categories -- static, high-frequency, and quasi-static (varying slowly in time compared to the microwave period). The *static* case is determined by the initial conditions and is uninteresting except in relation to how the spatial variation of the initial vacancy concentrations interacts with the spatial variation of the high-frequency results. The perturbations in equations (4) and (5) have both a high-frequency (same frequency as the microwave field) oscillatory nature and a slowly varying component. The perturbations (C = [Vacancy]) are shown in Figures 1 for Cl vacancies during two full microwave cycles. The high-frequency oscillations are clearly seen. In equation (1) these sinusoidal oscillations are multiplied by the sinusoidal microwave field, yielding a sine^2 component of the flux. It is through this interaction that the fluxes and concentrations acquire a quasi-static, time-averaged-non-zero character. Two distance scalings are apparent in Figure 1; the sharp peak near the surface is from the ponderomotive effect, while the longer-scale hump is characteristic of the Debye-Huckel distance associated with the boundary space-charge.

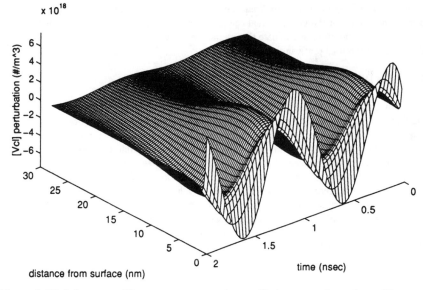

Figure 1: High-frequency Cl vacancy concentration oscillations near the surface. The simulations are performed at 1 GHz, 800 K, 10 ppm doping, and 10^6 V/m field strength.

The flux rectification or "ponderomotive" action is expressed in terms of a transport coefficient (C_T) and a driving force, which we call the "ponderomotive force" or pmf:

$$J_{Na}^{pmf} = C_T \times pmf \qquad (6)$$

$$C_T = \left(\frac{D_{Na} C_{Na}^i}{kT} \right), \quad pmf = \left\langle \frac{C_{Na}^p}{C_{Na}^i} q_{Na} E^{mw} \right\rangle \qquad (7)$$

The angle brackets here indicate time-averaging over one cycle. The magnitude of the pmf depends on the magnitude of the defect concentration perturbations. These perturbations are in turn dependent on the microwave field strength, how quickly the defects can move to the surface region (e.g. diffusion coefficients and therefore temperature), how much time the defects have to move (e.g. microwave frequency), and on the rate of vacancy creation or annihilation at the surface (which we model as the source/sink term (S) in equation (2)).

We have chosen to investigate the magnitude of the pmf over a range of field strengths, impurity doping levels, frequencies, and temperatures. Figure 2 displays the pmf versus each variable independently. The calculated pmf varies linearly with the square of the E-field ($\propto$ microwave power), which is consistent with the theoretical expectations and with our ionic current measurements. It is somewhat surprising, on the other hand, that doping the crystal with divalent impurities yields only a small enhancement of the driving force. The characterizations of the pmf versus the frequency and the temperature show peaks that suggest there are some values for the variables that can lead to optimum microwave-enhancements. Further efforts are underway to evaluate the interactions between the variables and to search the variable space for optimal pmf values.

EFFECTS ON DIFFUSION AND SINTERING

Having now found a range of values for the magnitude of the microwave-induced driving force (Fig. 2), we wish to compare that to typical values for the driving force in other processes. In our ionic current experiments, for example, the external bias on the NaCl crystal corresponded to about 10^{-16} N, so it is reasonable that we measured microwave-induced currents of about the same scale as that driven by the external bias. The driving force for sintering derives from the reduction of surface area of the pores. This driving force changes drastically during the sintering process, but an average value for the driving force can be determined from the surface energy of NaCl [9]. We calculate this to be ~ 10^{-17} - 10^{-16} N for micron size grains of NaCl. Driving forces during chemical diffusion experiments are also time dependent, starting out very large (~ infinite) and diminishing over time. For a concentration profile dropping from 100% to 0.1% over a distance of 10 - 100 microns, the driving force is about 10^{-15} - 10^{-14} N. On the other hand, the driving force for radioactive isotope tracer diffusion is virtually nil, and such processes proceed mostly by "random-walk" mixing.

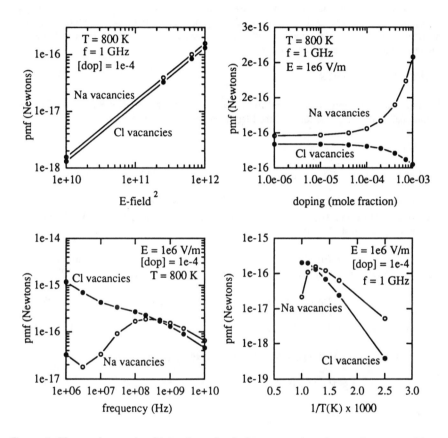

Figure 2: The ponderomotive driving force (pmf) characterized vs. the simulation variables.

Comparison of these driving forces with the pmf numbers displayed in Figure 2 suggests that certain processes might see greater microwave-enhancements than others. Conventional sintering, for example, has a naturally small driving force, and microwave-sintering almost always shows a dramatic rate enhancement over conventional sintering. On the other hand, the diffusion literature shows more variation for the reported microwave enhancements, and this could be explained by the difference in driving forces between chemical and tracer diffusion. For common experimental conditions, for example, the ponderomotive driving force may be much larger than the tracer diffusion driving force and much smaller than the chemical diffusion driving force.

CONCLUSIONS

A computer model has been developed to simulate microwave-induced ponderomotive driving forces in ceramic materials. With these simulations, we have been able to characterize the magnitude of this driving force versus field strength, doping, frequency, and temperature. The magnitude appears large enough to affect such processes as sintering and tracer diffusion, which have inherently small driving forces.

ACKNOWLEDGEMENTS

We would like to acknowledge support from the NASA Graduate Student Researcher Program, the National Science Foundation, and the Electric Power Research Institute.

REFERENCES

1. J.H. Booske, R.F. Cooper, and I. Dobson, *J. Mater. Res.*, 7[2], p. 495 (1992).

2. J.H. Booske, R.F. Cooper, L. McCaughan, S.A. Freeman, and B. Meng, **Microwave Processing of Materials III**, Matls. Res. Soc. Proc., **269**, ed. by R.L. Beattty, W.H. Sutton, and M.F. Iskander, p. 137 (1992).

3. S.A. Freeman, J.H. Booske, R.F. Cooper, B. Meng, J. Kieffer and B.J. Reardon, **Proc. Microwave Absorbing Materials for Accelerators Conf.**, Newport News, VA, 1993.

4. S.A. Freeman, J.H. Booske, and R.F. Cooper, *Rev. Sci. Inst.*, 66[6], p. 3606 (1995).

5. S.A. Freeman, J.H. Booske, and R.F. Cooper, *Phys. Rev. Lett.*, 74[11], p. 2042 (1995).

6. K.I. Rybakov and V.E. Semenov, *Phys. Rev. B*, **49**[1], p. 64 (1994).

7. K.L. Kliewer, J.S. Koehler, *Phys. Rev.*, **140**[4A], p. A1226 (1965).

8. K.M. Kramer, W.N.G. Hitchon, *Computer Physics Comm.*, **85**, p. 167 (1995).

9. W.D. Kingery, H.K. Bowen, D.R. Uhlmann, **Introduction to Ceramics**, John Wiley and Sons, NY (1976).

KINETICS OF REDUCTION OF IRON OXIDES USING MICROWAVES AS POWER SOURCE

I. GOMEZ, J. AGUILAR, M. GONZALEZ AND J. MORALES
Universidad Autónoma de Nuevo León, Facultad de Ingeniería Mecánica y Eléctrica, Apartado Postal 076 F, Cd. Universitaria, San Nicolás de los Garza, N.L. CP. 66450, México.

ABSTRACT

This work deals with kinetic description of carbothermic reduction of iron oxides using microwaves as power source. Previous researches shown that it is possible to conduct this kind of processes successfully, but real kinetic comparisons between conventional and microwaves procedure have been presented partially. The aim of this work is to describe reduction kinetics, taking into account how the iron oxide is reduced by microwaves compared with conventional energy supply. In this study we used iron ore in pellet shape and dust. We found that both, pellet and dust reduction stops when it reaches approximately 40%, even at whole power

INTRODUCTION

In recent years, there has been considerable growth of interest in applying microwave energy for materials processing. The mechanism of heating is not understood completely, but there are some theories that explain the possible behavior of the materials exposed to a microwave field [1]. Recently, it has been shown that the use of microwaves for the process of carbothermic reduction is possible [2]. Now we present the results obtained from the kinetic comparisons between microwave and conventional reduction of iron oxides.

Iron oxide reduction is described by two simultaneous reactions:

$$Fe_xO + CO \rightleftharpoons xFe + CO_2 \qquad (1)$$
$$CO_2 + C \rightleftharpoons 2\,CO \qquad (2)$$

To perform kinetic studies, it is necessary to identify the processes involved inside the global process. In the case of reduction kinetics of metal oxides, the reduction rate is governed by the interplay rate at which CO (the reducing agent) is consumed by the reaction with Fe_xO itself (1) and the rate at which it is produced by Boudouard equilibrium reaction (2).

EXPERIMENTAL PROCEDURE

Two kinds of experiments were conducted, the first one was over the pellet shape (spheres) and the second experiment was on hematite dust (Fe_2O_3). The materials were mixed in a ratio $Fe_2O_3 / C = 1/3.5$.

Iron ore for tests was provided in pellet from the mexican state of Colima. This material is called "Alzada" and its reducibility properties are well known. Main characteristics of this samples are: pellet radius 0.6 cm, density 4.22 gr./cm^3, weight 3.8 gr., total iron 66.5 wt%, Fe^{+2} 0.64 wt%, gangue 5.3 wt% (Gangue composition; CaO:37 wt%, MgO:11 wt%, SiO$_2$:38 wt%, Al$_2$O$_3$:14 wt%). Carbothermic reduction tests with microwaves were conducted in a conventional microwave oven (2450 MHz, 800 Watts). The mixture was placed inside a crucible made of pure aluminum oxide and thermally isolated with ceramic fiber. Then, the oven was programmed for different times of microwave exposure. When each time was reached, the specimen was removed and chemically analyzed.

Temperature measurements are difficult in a microwave field, but they are necessary for controlling the process and for having kinetic comparisons, e.g. to ensure the same temperature or the same energy input rate for conventional as well as microwave reduction. A grounded and shielded thermocouple (3.175 mm diameter, including the stainless steel shield) was placed into the system, in such a way that part of it is inside of the specimen area and the rest is almost out of the microwaves field. The measurement system consist then in connecting a thermocouple to a controlling device.

In the second kind of experiment, a mixture of hematite dust and carbon, was placed inside the crucible and then placed at the same position in the cavity as the pellet samples. Reduction tests in the microwave field were divided in two series. In the first one there was no control of temperature. In the second serie, tests were carried out with temperature control in two ways, either keeping the temperature constant over the whole reaction period or stopping it at the moment that desired temperature was reached.

Conventional reduction tests were conducted in an electric resistance furnace at 1100°C. In this case the furnace was heated until desired temperature was reached and stabilized. Crucibles with specimens were placed inside the furnace, and removed one by one and chemically analyzed.

RESULTS AND DISCUSSION

Results of microwave tests without temperature control are presented in Figure 1. This figure shows a solid curve which was generated with a computer model [3] simulating the reduction of the iron ore with CO at 1100°C.

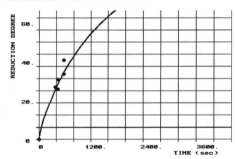

This simulated curve is based upon a topochemical reduction model that accepts that temperature inside the pellet is almost constant. According to Okura [4] differences between surface and center of the pellet are 25°C maximum when reduction was carried out with hydrogen. Considering that reduction with hydrogen is strongly endothermic, then in this case differences will be smaller. This situation is in agreement with results reported by other researchers [5]. Dots over the curve correspond to results of microwave experiments, the data show that reduction with microwaves has the same behavior than conventional one, up to 40%. Higher degrees of reduction are not achievable using microwave radiation for heating. Kinetic constant within this range is 787 x 10^{-3} seg^{-1} for first order reaction.

Figure 1. Pellet reduction degree against time. Continuos line: calculation for conventional heating at 1100°C. Dots: reduction in a microwave field at 800 watts

To make a complete thermal analysis, besides temperature measurement, the phase composition was studied. One way to estimate temperatures is to identify and analyze the different phases that are present in the pellet. Figures 2 and 3 show two SEM photos taken on specimens exposed to microwave radiation without temperature control. In Figure 2 the only

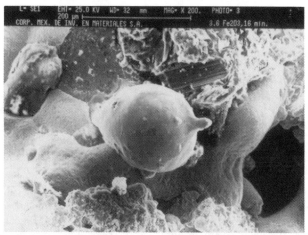

Figure 2. Specimen exposed to microwaves for 16 minutes (the large drop is melted iron)

found element in the large drop and the general background was metallic iron, there is carbon too, but separate of the iron. Figure 3 shows, besides melted iron, the bright drops contain just iron, silicon and oxygen, and the only possible compound with these elements that is thermodynamically feasible is fayalite (Fe_2SiO_4). Melting points are $1537°C$ and $1217°C$ for iron and fayalite respectively, which shows that minimum temperature was over $1200°C$. This information allows us to plot microwave results over conventional heating as shown in Figure 1.

Taking the reaction with CO it is possible to find out that Bouduoard equilibrium makes easier to conduct reduction at low temperatures rather than high, this explains why simulated curves between $1100°C$ and $1500°C$ with CO are almost the same. Therefore this same curve can be taken for any temperature between $1100°C$ and $1500°C$.

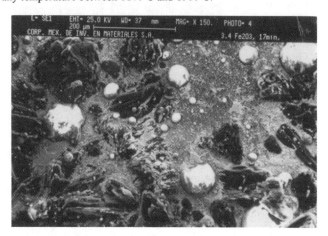

Figure 3. Specimen exposed to microwaves for 17 minutes, bright drops are melted fayalite, black zones are graphite and background gray is iron

This condition makes possible to compare tests conducted with microwaves at certain power against tests conducted conventionally at certain temperature. Accepting that diffusion of the reducing agent is the controlling step on reduction rate, when metallic iron appears on the surface, it melts and covers the pellet. At this stage of the reaction the reduction stops. Furthermore, melted fayalite closes pores, thus porosity decreases, and this also stops reduction.

Dust processing and temperature measurement

Even when in this case there is a better contact (more intimate) between reagents, final achieved reduction is not higher than pellet reduction. In the case of reactions where the test was stopped when the desired temperature was reached, no reduction took place. Reduction process stops because each grain or dust particle is covered by melted iron, as the pellet was, but in this case a metallic grid is formed on the inner wall of the crucible and this grid avoid the microwaves arriving to the center of the material and this produces the same results found in pellet tests.

From a graphic of temperature against time (Figure 4) we can see that temperature is risen until the set point is achieved (this is an example at 800°C), then the on/off controlling device shut down the microwave field and temperature decreases until it is too low that microwave field is connected again (the on/off control returns to "on"). The most important part in this graphic is related with the zone where microwaves are disconnected, here the measured temperature is as accurate as the thermocouple is capable. Notice that the heating part is within the same range and the same behavior, If the thermocouple were affected by the microwave field, then the temperature behavior would be very different from the cooling part. Looking at the ripples, the frequency is the one that the on/off control gives according to the heating rate, very different that the rippling that is caused by the pulses of the magnetron or insufficient shielding of the thermocouple.

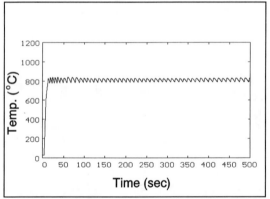

Figure 4. Graphic of temperature against time of a microwave test conducted at 800°C.

Thermocouple is not affected by the microwave field, but the microwave field is affected by the thermocouple shield. In other words, the thermocouple is at the center of the system surrounded by absorbent material (the one that we want to process) and protected from the microwave field by the shield and the material surrounding the thermocouple. Then microwaves pass through the material and they are partially absorbed, while the unabsorbed portion reaches the thermocouple shield and they are reflected through the material again, in such a way that part of this reflected wave is absorbed by the material. So the material that is just surrounding the thermocouple (the closest one) absorbs energy twice, one from the microwave beam and one from the reflected wave from the thermocouple shield.

Chemical analysis of the tests conducted on dust with temperature control show a similar behavior to those carried out conventionally at the same temperature, but reduction also stops when the reduction degree reaches about 40%. On the other hand the tests that where stopped when the desired temperature was achieved did not shown any appreciable reduction.

Unreacted centers versus cold centers

It was exposed above one reason to explain how reduction stops due to melted iron covering the porous. Other reason for this could be that film of melted iron (or the non-stoichometric iron oxides) shields the pellet against microwaves, so energy is not arriving to the center of the pellet. This argument is very important, because in this condition microwaves could not avoid the "cold centers". This also true if the process is diffusion controlled even when the microwaves could heat volumetrically. In the case of reduction with temperature control, iron did not melt, but there is a metallic grid made of the reduced iron with open porous.

As was described above in the case of hematite dust, final achieved reduction was same than pellet reduction. For having a comparison between the effect of the shield against the reducing agent we took the dust reduction in a crucible and thought that dust that is close and over the inner wall of the crucible might form a metallic net that shields the material at the center of the crucible.

One conclusion from this is that even when it is supposed that microwave processing supplies energy to the whole volume, this assumption comes from processes were the system is small enough (specially the thickness) to have a difference between the surface an the inner part.

Due to process is depending on how intimate is the contact between the reagents (iron oxide and reducing agent), iron oxide that is far from the reducing agent will not be reduced, even when it has enough energy to react. This place will appear as a "unreacted center", but it is not a "cold center". In this way, microwave heating can avoid "cold centers" if the system is not too thick, but it will not avoid "unreacted centers", because these ones come from a different phenomena.

Factors affecting reduction rate

From above explanation it is clear that diffusion of CO through reduced layer plays a very notable role. Accepting this statement, the description of the reduction process in the pellet is as follows: Reduction begins outside the pellet were there is not any barrier to the reducing agent. Then metallic iron covers the pellet, and reducing agent has to pass through it to reach the interface reaction zone.

The process is slightly different for conventional and microwave reduction. In the first case reduction rate decreases because the mass of iron oxide to reduce inside the pellet is decreasing with cube of radius. In the microwave case, besides this situation, external layer melts and covers the pellet with liquid iron. This layer does not allow to pass reducing agent as easy as it is just with reduced oxide (without melt) or iron oxide and it reflects the microwave radiation.

When temperature was controlled, the metallic film does not appear as painting way, but even so the reduction is not faster than conventional one, and reduction in this case does not go beyond 40% consistently. This means that this process is controlled by diffusion of the reducing agent mainly, besides means that even when nature of microwaves is oriented to heat the whole volume metallic grid would interfere with wave propagation.

427

CONCLUSIONS

The analysis of our results let us conclude the following:

- Iron oxides absorb microwaves, therefore it is possible to use them as a power source for conducting a reduction process.
- It is true that microwave heating is fast, but this reason not is enough because the control for conducting the process is by means of diffusion of carbon atoms.
- Reduction rate is about the same for the two heating schemes.
- In our case, microwave heating just avoids "cold centers", but it does not avoid "unreacted centers".

ACKNOWLEDGMENTS

We express our gratitude to CONACYT (National Science and Technology Council) for its financial support.

REFERENCES

1. A. Metaxas and R. Meredith, Industrial Microwave Heating, (Peter Peregrinus Press, London, 1983)
2. I. Gómez and J. Aguilar in Dynamics in Small Confining Systems II, edited by J.M. Drake, J. Klafter, R. Kopelman and S.M. Troian (Mater. Res. Soc. Proc. 366, Pittsburgh, PA 1994), p. 347-352
3. J. Aguilar, Kinetic constants for description of iron ore reduction, (7th International Symposium on Transport Phenomena in Manufacturing Processes, Acapulco, México 1994, 126)
4. A. Okura, R. Maddox, Tetsu-to Hogane, **61** (9), 2151 (1975)
5. Von Bougdandy and H.J. Engell, The reduction of iron ores, (Springer-Verlag, Berlin, 1971)

MODIFICATIONS ON BULK CRYSTALLIZATION OF GLASSES BELONGING TO M$_2$O-CaO-SiO$_2$-ZrO$_2$ SYSTEM IN A 2.45 GHz MICROWAVE FIELD

C. Siligardi*, C. Leonelli*, Y. Fang**, D. Agrawal**
*Department of Chemistry, University of Modena, Via Campi 183, 41100, Modena, Italy, Manfredini@imoax1.unimo.it
**Intercollege Material Research Laboratory, The Pennsylvania State University, University Park, PA, 16802, USA

ABSTRACT

The potential for microwave processing of a single phase material is often limited due to the dependence of dielectric losses upon the chemical bonding and temperature of the material. We will present results showing the effect of microwave absorption on bulk crystallization of glasses belonging to M$_2$O-CaO-SiO$_2$-ZrO$_2$ system (where M= Li, Na, K). The glass samples were devitrified using both microwave and conventional heating. The effect of Li$^+$, Na$^+$, K$^+$ on crystallization is quite remarkable and is a function of ion size. This is true especially in the microwave heating where the important dielectric losses observed in silicate glasses are related to the motion of alkali cations throughout the glass matrix. X-ray diffraction analysis was performed on the powdered samples to determine crystalline phases. Results of microstructure and microanalysis on these glass-ceramic samples will also be presented.

INTRODUCTION

Glasses have some significant advantages over many crystalline materials seeking certain properties required for a specific application. Firstly, when properly made, they are isotropic. Secondly, due to absence of a microstructure, their properties depend almost entirely on composition, although there is a slight effect of thermal history. Further more, many properties are simply a function of the composition so that any value of a particular property, within the limits of the possible range, can be obtained by composition adjustment.

ZrO$_2$ is well known as a common component of industrial glasses and enamels because of its ability to increase chemical resistance and improve mechanical and thermal properties of the glasses. The high melting point, high refractive index, low coefficient of thermal expansion and high electrical resistivity have made ZrO$_2$ an important refractory material as well as a ceramic opacifier and ceramic insulator. These properties of ZrO$_2$ make ZrO$_2$-containing glasses and glass-ceramics very attractive, especially when they are used as matte glaze components or monolithic floor tile. Moreover, many researchers have demonstrated that ZrO$_2$ is a good absorber in a microwave field after reaching a critical temperature [1]. In this work we have investigated the effects of heat treatment using both microwave and conventional methods on the devitrification tendency of glasses belonging to the M$_2$O-CaO-SiO$_2$-ZrO$_2$ (where M=Li, Na, K) system.

EXPERIMENT

The glass compositions studied are shown in Table I. Mixtures of each composition containing the raw materials of calcite, quartz, zirconium silicate, Li$_2$CO$_3$, Na$_2$CO$_3$ and K$_2$CO$_3$ were melted in alumina crucibles in a electric melting muffle furnace at about 1500°C. The melts were quenched in a graphite crucible to obtain small pieces of glass.

Mat. Res. Soc. Symp. Proc. Vol. 430 ©1996 Materials Research Society

Table I. Compositions (wt%) of various glasses

Name	CaO	SiO$_2$	ZrO$_2$	Li$_2$O	Na$_2$O	K$_2$O
Li-glass	31.02	51.7	11.28	6		
Na-glass	31.02	51.7	11.28		6	
K-glass	31.02	51.7	11.28			6

The devitrification behavior of various compositions was studied by performing a series of heat treatments, both conventional and in a microwave field. The samples were heated for 30 minutes at 850°C; 900°C; 950°C; 1000°; 1050°C; 1100°C; 1150°C, respectively.

A 900W, 2.45 GHz commercial microwave oven (Panasonic mod. 6371NN) was used in this study [2]. An electric furnace (Remet mod. Melt-two) was used for conventional heat treatments.

The glass transition and crystallization temperatures were determined with a differential thermal analyzer (Netzsch mod. STA 409), DTA, on samples ground to a grain size of less than 20 μm. The DTA measurements were carried out on about 30 mg of sample in a Pt crucible. Data for each run were automatically collected from the DTA apparatus.

The measurements of dielectric properties at elevated temperature were made by using the resonant cavity technique [3,4] from the measured change in the resonant frequency and Q-factor of the cavity [5].

The influence of temperature on the crystallization of the glasses was studied by x-ray diffraction, XRD, on powdered samples, using a powder diffractometer (Philips mod. PW 3710), and CuKα-Ni filtered radiation. The bulk heated samples were ground to less than 20 μm and the XRD patterns were collected in the 10-50° 2θ range at room temperature. The scanning rate was 0.04°/s, step size 0.02°.

Polished glass-ceramic samples (using abrasive paper < 0.5 μm) were coated with a thin film of Au/Pd for scanning electron microscopy (SEM, Philips mod. XL 40) observations. Energy dispersion x-ray spectroscopy (EDS, EDAX PV9900) technique mounted on the SEM, was used to identify the chemical composition of the different phases in the samples.

RESULTS

The DTA curves for the different glasses show very similar features. A change in the specific heat due to the glass transition and an exothermic peak corresponding to crystallization, have been recorded. The crystallization ability of a glass can be measured either by the difference between the glass transition temperature, Tg, and the maximum crystallization peak temperature, Tc, or by the sharpness of the peak (half width, HW). The Tg and Tc values are listed in Table II.

Table II. Values obtained by DTA analysis

Glass	Tg (°C)	Tc (°C)	Δ(Tc-Tg) (°C)	HW (°C)
Li-glass	621	781	160	20
Na- glass	739	930	191	23
K- glass	799	999	200	27

The Li$^+$ has high field strength (Z/a^2= 0,23), so it influences the bonds with neighboring atoms to a higher degree than ions of low field strength, as Na$^+$ and K$^+$, (Z/a^2=0,19 for Na$^+$ and Z/a^2=0,13 for K$^+$). Therefore, Li$^+$ is making the glass structure stiffer, hence it is easier to rearrange into a stiff structure such as the crystalline type. The K-glass bulk crystallization temperature is probably higher than the temperature given by the thermal analysis result, consequently the sample melts before the completing of the crystallization. Na-glass has an intermediate behavior, since the samples devitrified at higher temperature than Li-glasses.

Figure 1 shows the composite ε' and ε'' vs. temperature plot for the compositions featuring the three different modifier cations. The six different patterns are qualitatively quite similar. From this figure the temperatures of the dielectric loss onset can be determined and they are Li-glass, 730°C; Na-glass, 860°C; K-glass, 960°C.

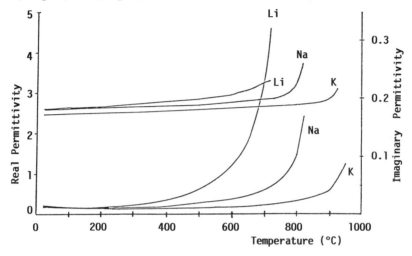

Figure 1. ε' and ε'' vs temperature plot of all the glasses at 2.21 GHz.

A satisfactory correlation was observed between the onset temperatures of dielectric losses and the Van der Walls radii of three cations (r_K=1.38 Å, r_{Na}=1.02 Å, r_{Li}= 0.74 Å) [6]. The ionic mobility of the modifier cations is mainly responsible for the interaction between the microwave field and the material. Moreover, the glass transition temperature, an indirect measure of the glass viscosity (corresponding to $10^{13.3}$ poise), decreased with the ionic volume of the alkaline ions, as shown in Table II. Note that the present measurements were performed on powdered glass samples, higher values are reasonably expected in the case of bulk samples.

XRD and SEM observations of K-glass, show a very slight surface devitrification in both conventional and microwave treatments.

X-ray diffraction analysis showed that wollastonite-2M (JCDPS files no. 10-489, no. 27-88), wollastonite-1A (JCDPS files no. 27-1064, no.29-372) and baddeleyite (JCDPS file no. 36-420) are the main crystalline phases in Li- and Na-glass-ceramics. The relative proportions of these phases vary depending upon the composition of the parent glass, the heat treatment temperature and the type of thermal treatment. A summary of the crystallization products of these two glasses is given in Table III. In the microwave specimens, at low temperatures, the major crystalline phase is wollastonite-2M. As the temperature is increased, the amount of wollastonite-1A also increased, while both crystalline phases were present in conventional samples at all temperatures. The enhancement or inhibition of certain crystalline phase was mainly judged from the presence or absence of its specific diffraction lines and the increase or decrease in their relative intensities. The samples heat treated by a conventional method devitrified at a temperature 50°C lower than in the presence of a microwave field, however, crystallization took place only on the surface. From macroscopic observation by optical microscope, on the contrary, the microwave treated samples show both surface and bulk crystallization.

Table III. Phase developed after heat treatments

T (°C)	Li-GLASS phases developed		Na-GLASS phases developed	
	Microwave	Conventional	Microwave	Conventional
850	Partially devitrified Woll.2M, $Li_2ZrSi_6O_{15}$ (Woll.1A)	Partially devitrified Woll.2M, Woll.1A $Li_2ZrSi_6O_{15}$	Glassy	Glassy
900	Partially devitrified Woll.2M, $Li_2ZrSi_6O_{15}$ (Woll.1A, ZrO_2)	Woll.2M, Woll.1A (ZrO_2, $Li_2ZrSi_6O_{15}$)	Glassy	Glassy
950	Woll.2M, Woll.1A ZrO_2	Woll.1A, Woll.2M ZrO_2, (glassy phase)	Glassy	Partially devitrified Woll.2M, (Woll.1A)
1000	Woll.1A, Woll.2M ZrO_2, (glassy phase)	Woll.1A, Woll.2M ZrO_2, (glassy phase)	Partially devitrified Woll.2M, (Woll.1A)	Woll.2M, Woll.1A
1050	Woll.1A, Woll.2M ZrO_2 (glassy phase)	Woll.1A, Woll.2M ZrO_2, (glassy phase)	Woll.2M, (Woll.1A, ZrO_2)	Woll.2M, Woll.1A ZrO_2
1100	Woll.1A, Woll.2M ZrO_2, glassy phase	Woll.1A, Woll.2M ZrO_2, glassy phase	Woll.1A, Woll.2M ZrO_2 (glassy phase)	Woll.2M, Woll.1A ZrO_2 (glassy phase)
1150			Woll.1A, Woll.2M ZrO2, glassy phase	Woll.2M, Woll.1A ZrO_2, glassy phase

( ) = incipient formation

Moreover, we note that only Li-glass, heated with both treatments, segregates a small amount of a new phase, (detected also by XRD), $Li_2ZrSi_6O_{15}$ at low temperature (850 and 900°C). At higher temperature this phase disappears, concomitant with an increasing amount of an amorphous phase, as shown in Figures 2 (a, b).

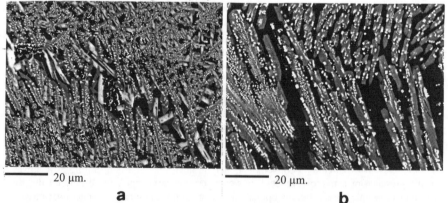

20 µm.   20 µm.

a   **b**

Figure 2. SEM micrographs of Li-glass heated at (a) 850 and (b) 1000°C in a conventional oven.

At temperature higher than 950°C, the following reactions take place:

$$Li_2ZrSi_6O_{15} \rightarrow ZrSiO_4 + liquid1 \qquad (1)$$
$$ZrSiO_4 + liquid1 \rightarrow ZrO_2 + liquid2 \qquad (2)$$

Under thermodynamic equilibrium conditions [7], pure $Li_2ZrSi_6O_{15}$ decomposes at 1078°C in $ZrSiO_4$ and liquid. Moreover, the presence of the zircon diffraction peaks in the Li-glass-ceramics only at 950°C may suggest the mechanism (2). By EDS analysis it was possible to confirm both mechanism because the amorphous phase consists of high amount of SiO_2 at high temperature. SEM examination reveals very tiny white crystals of ZrO_2, present at low temperature in all Li-glass-ceramics samples even if their presence can not be revealed by XRD until 950°C. The main crystalline phase was wollastonite occurring in two distinct crystalline forms, the 1A-form as elongated crystals of about 20 μm, and the 2M-form in smaller round shaped crystals (< 10 μm). The microstructure is such that the zirconium oxide crystals are randomly dispersed in the wollastonite matrix indicating the simultaneous precipitation of the two phases. On the contrary, in the case of Na-glass-ceramics, ZrO_2 grows on the outer surface of wollastonite crystals upon both heat treatments (Figures 3).

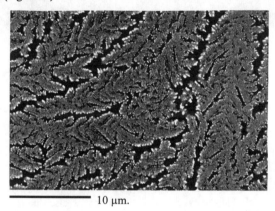

10 μm.

Figure 3. SEM micrograph of the Na-glass heated at 1050°C in microwave oven for 30 min.

In the last composition the samples treated by conventional and microwave processing show different microstructures: at lower temperature the conventional sample has two different crystalline forms of wollastonite, while in the microwave sample only the elongated type of crystal was observed. At higher temperatures in the conventionally treated material roundish wollastonite-2M crystals immersed in a amorphous matrix were observed upon microstructural examination, while in the microwave heated glass-ceramic oriented long-shaped wollastonite-1A crystals (Figure 4 a, b) were seen.

CONCLUSIONS

The samples treated by conventional method showed only surface crystallization while the glasses treated by microwave energy showed a feeble bulk crystallization. The composition containing lithium oxide devitrifies at the lowest temperature, 850°C, while K-glass presents only a surface opalescence; Na-glass has an intermediate behavior and devitrifies at high temperatures

1000-1050°C. Such a compositional effect on crystallization temperatures was observed in microwave treated samples, at a temperature approximately 50°C lower than the conventional treatment. The more significant effect of microwave energy on such glass-ceramics was on the stability of the different crystalline phases: wollastonite-1A was favored at high temperature, wollastonite-2M was favored at low temperature; both crystalline phases were present in conventional samples at all temperatures.

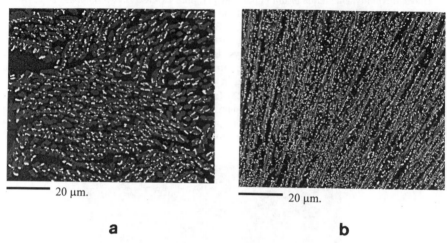

20 μm.   20 μm.

a   b

Figure 4. SEM micrographs of Na-glass treated at 1150°C in for 30 min: (a) conventional and (b) microwave oven.

ACKNOWLEDGMENTS

Cristina Siligardi would like to thank Prof. R. Roy for the opportunity of working at the IMRL, The Pennsylvania State University, PA, USA. Thanks are also due to Dr. Neil Greenacre for technical and scientific support and Dr. Tom Cross for the opportunity of working at the Department of Electrical and Electronic Engineering, University of Nottingham, UK.

REFERENCES

1. W.H. Sutton, Mat. Res. Sc. Bull., 11, 22, (1993).
2. Y. Fang , Ph.D. thesis, The Pennsylvania State University, 1994.
3. M. Arai, J.G.P. Binner, G.E. Carr, T. E. Cross in Ceramic Transactions, Microwaves: Theory and Application in Materials Processing II, (The American Ceramic Society, Westerville, Ohio, 1993).
4. H.A. Bethe, J. Schwinger, NRDC Report D1-117, 1943.
5. F. Horner, T. A. Taylor, R. Dunsmuir, J. Lamb, W. Jackson, J.Inst. El. Eng. 93, 53, (1946).
6. F.A. Fusco, H.L. Tuller in Proceeding Elettro Ceramics and Solid State Ionics 88-3, 167 (The Electrochemical Society, Inc., 10 South Main St. Pennington, NJ, 1988).
7. P. Quintana and A. R. West, Trans. J. Br. Ceram. Soc., 80, 91 (1981).

POSSIBILITY OF MICROWAVE-CONTROLLED SURFACE MODIFICATION

K. I. RYBAKOV[1,2] AND V. E. SEMENOV[1]

[1]Institute of Applied Physics, Russian Academy of Sciences, Nizhny Novgorod, Russia
[2]Institute for Plasma Research, University of Maryland, College Park, MD

ABSTRACT

Results of the theoretical study of surface effects in ionic crystalline solids under the action of high-frequency electric fields of moderate intensity are presented. The averaged ponderomotive action of the electric field on the charged vacancies within the crystal causes directional mass transport that leads to development of a surface instability. The analysis shows that the proposed effect can result in the formation of a periodic profile on the surface.

Experiments on microwave processing of materials, including annealing of semiconductors [1], sintering of ceramics [2], *etc.*, have revealed substantial reduction in the activation energies and other specific features in the course of mass transport processes in solids under the action of the high-frequency (HF) electromagnetic field. Estimates [3] and purposely designed experiments [4] show that this specificity cannot be attributed to nonuniformity of temperature distribution which exists during the microwave exposure, and thus suggest existence of an effect of a nonthermal nature. A model of nonthermal microwave influence on mass transport processes in crystalline solids has been proposed by the authors [5,6]. The averaged ponderomotive action of the microwave electric field on charged vacancies in ionic crystalline solids was shown to create driving forces for mass transport and cause reduction in its activation energy. This model has received an experimental proof in the study of microwave-induced DC currents in ionic crystals [7].

In the previous paper [6] it was shown that the ponderomotive-driven mass transport can cause enhancement of plastic deformation of grains in ceramics undergoing microwave sintering. In this presentation, an effect that can probably make possible the use of microwave processing for surface modification and related purposes is discussed. It is shown that a planar surface of an ionic crystalline solid can become unstable in a tangential HF electric field, which means that a periodic corrugated profile will be developing on it. A similar instability is shown to be possible in a doped layer with higher vacancy concentration.

The analysis that is the subject of this paper is carried out on the basis of the model that is basically similar to that used in [6]. An ionic crystalline solid with the vacancy mechanism of diffusion predominating is considered within the continuum approximation. The flux of vacancies of each sort is determined by diffusion and drift:

$$\mathbf{J}_\alpha = -D_\alpha \nabla N_\alpha + D_\alpha N_\alpha \frac{e_\alpha}{kT} \mathbf{E}, \tag{1}$$

435

where the subscripts are introduced to distinguish vacancies of different sorts, $\mathbf{J}_\alpha$ is the flux density, D_α is diffusivity, e_α is the electric charge, and N_α is dimensionless (normalized to the density of sites in the crystalline lattice) concentration of vacancies of sort α, $\mathbf{E}$ is the vector of electric field, and the vacancy mobility is expressed with the help of the Nernst-Einstein relation.

In the absence of sources and sinks of vacancies, concentration and flux density are linked by the continuity equation:

$$\frac{\partial N_\alpha}{\partial t} + (\nabla \cdot \mathbf{J}_\alpha) = 0. \tag{2}$$

The electric field is influenced by the distribution of charged vacancies. For the case when the characteristic dimensions of the problem are much less than the electromagnetic wavelength within the solid the equations for the field may be written in the quasielectrostatic limit:

$$\nabla \times \mathbf{E} = 0,$$
$$\nabla \cdot \mathbf{E} = 4\pi \Sigma\, e_\alpha N_\alpha / \Omega, \tag{3}$$

where Ω is vacancy volume, summation is over sorts of particles, and the lattice dielectric constant is assumed to be unity for simplicity.

The boundary conditions for Eqs. (1)-(3) on the surface of the solid are the following. The total vacancy concentration at the surface is equal to the total equilibrium vacancy concentration:

$$\Sigma\, N_\alpha|_S = \Sigma\, N_{0\alpha}. \tag{4}$$

Also, there is the restriction on the vacancy fluxes meaning that there is no electric current through the surface:

$$\Sigma\, e_\alpha \mathbf{J}_{\alpha n}|_S = 0. \tag{5}$$

For the electric field, there are boundary conditions of continuity of both its tangential and normal components at the surface. The external electric field is specified by the vector $\mathbf{E}^0$.

Eqs. (1)-(3) with the boundary conditions described above comprise the mathematical model of the problem. The system of equations can be solved using a perturbation technique similar to that developed in [5]-[6]. As a zeroth approximation, a planar surface of the solid in a tangential HF electric field is considered. Small excursions from planarity result in small perturbations in the electric field (in particular, its normal component becomes non-zero) and vacancy concentration, which are the first-order quantities. The time-averaged action of the electric field on the charged vacancies produces quasistationary flows.

The solving procedure for the equations of the model uses the following simplifying assumptions in addition to the described above. The solid is assumed to have two sorts of vacancies with equal diffusivities D and equilibrium concentrations N_0 and with opposite electric charges e and $-e$. (It can be shown that in the case of different diffusivities the ponderomotive mass transport will be controlled by the smaller diffusivity [8].) In such solid the HF perturbations of concentration of vacancies of the two sorts ν_1, ν_2 have opposite signs: $\nu_1 = -\nu_2 \equiv \Omega \rho / 2e$. The first-order HF charge density

ρ is obtained, to a certain accuracy, from the 2-dimensional quasielectrostatic problem with the complex dielectric permittivity

$$\epsilon = 1 - i4\pi G/\omega \tag{6}$$

that takes only the vacancy conductivity $G = 2N_0 e^2 D/\Omega kT$ into account. The linear phase of the surface instability is considered, which provides a small geometric parameter for the quasielectrostatic problem. With ρ obtained, the quasistationary surface displacement velocity is obtained by averaging Eqs. (1)-(2) in the first order of the perturbation theory, without specifically solving them:

$$V = -J_n = \int_{\rho \neq 0} \mathrm{div}\left[D(y)\langle \rho \mathbf{E}^0 \rangle\right] dy. \tag{7}$$

Here the integration is performed over the thickness of the space charge layer, in the direction normal to the surface, and J_n denotes the normal component of the time-averaged vacancy flux.

The simplified description of the development of the surface instability can be given as follows. Consider the Cartesian coordinate system with x being a direction along the planar surface $y = 0$ of the solid. The external tangential HF electric field is $\mathbf{E}^0 = \mathbf{x}_0 \operatorname{Re} E_0 \exp(i\omega t)$. Assume that the surface gets slightly perturbed, so that it is described by the equation $y_s = h \sin \kappa x$, $\kappa h \ll 1$ (Fig. 1). From the solution of the quasielectrostatic problem with ϵ determined by Eq. (6) below this surface and $\epsilon = 1$ above it, the following expression for the HF density of vacancy charge can be obtained:

$$\rho = \frac{E_0}{2\pi} \frac{1-\epsilon}{1+\epsilon} \kappa h \cos \kappa x \, \delta(y - y_s) \exp(i\omega t). \tag{8}$$

The vacancy charge, of course, is induced within the crystal, close to its surface. The actual thickness of the charge layer is the smaller of the Debye-Huckel radius, $\sqrt{D/4\pi G}$,

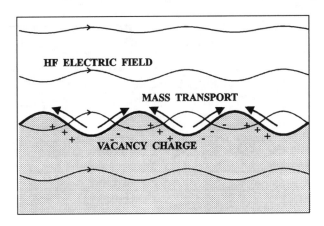

Fig. 1. Surface instability caused by tangential HF electric field

437

and the characteristic diffusion length, $\sqrt{D/\omega}$ [5].

Charged vacancies are driven by the electric field. The oppositely directed atomic flows cause plastic deformation. The velocity of surface displacement can be obtained by inserting Eq. (8) into Eq. (7) and performing the averaging. The result is:

$$V = \frac{E_0^2}{4\pi} \frac{\Omega D}{kT} \frac{\epsilon''^2}{4 + \epsilon''^2} \kappa^2 h \sin \kappa x, \tag{9}$$

where $\epsilon'' = \mathrm{Im}\,(\epsilon) = 4\pi G/\omega$. Note that within our approximation Eq. (9) can be rewritten as

$$\partial y_s/\partial t = \Gamma y_s, \tag{10}$$

where

$$\Gamma = \frac{E_0^2}{4\pi} \frac{\Omega D}{kT} \frac{\epsilon''^2}{4 + \epsilon''^2} \kappa^2 \tag{11}$$

is the instability growth rate.

It may be added that a similar instability is possible in a doped layer with higher vacancy concentration (Fig. 2). Calculation of the growth rate for it gives

$$\Gamma = \frac{E_0^2}{4\pi} \frac{\Omega D}{kT} \frac{\epsilon''^2 e^{\kappa d} \left(e^{\kappa d} - 1\right)}{\epsilon''^2 \left(e^{\kappa d} - 1\right)^2 + 4e^{2\kappa d}} \kappa^2, \tag{12}$$

where d is the unperturbed width of the layer.

The development of the above described surface instability in a real solid will be suppressed by the stresses of surface tension $\sigma = 2\gamma/r$ (γ is the coefficient of surface tension, r is the surface curvature radius). These stresses affect vacancy concentration at the surface,

$$N = N_0 \exp(\sigma \Omega/kT), \tag{13}$$

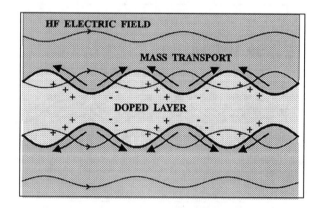

Fig.2. Instability of the doped layer

which causes diffusion vacancy transport leading to surface smoothening. The corresponding surface displacement velocity is

$$V = -\frac{2\gamma\Omega N_0}{kT}(D + \kappa a D_a)\kappa^3 h \sin \kappa x, \qquad (14)$$

Thus, there will be competition between "ponderomotive" instability and the surface tension.

In Eq. (14) both bulk and surface diffusion are accounted for by considering the near-surface amorphized layer (Fig. 3) of thickness a in which vacancy diffusivity is D_a ($D_a \gg D$). Existence of the amorphized layer can cause enhancement of the "ponderomotive" mass transport [6]. For the model with

$$\epsilon = \begin{cases} 1, & y > y_s; \\ \epsilon(y - y_s), & y_s - a < y \le y_s; \\ \epsilon_b, & y \le y_s - a \end{cases} \qquad (15)$$

calculations similar to the described above give (under the assumptions $\kappa a \ll 1$ and $\epsilon''_{max} \gg \epsilon''_b$, $\epsilon''_{max} \gg 1$, where $\epsilon_{max} \equiv \epsilon(y = y_s)$) the following values for the induced charge density and the surface displacement velocity:

$$\rho = \frac{E_0}{2\pi} \frac{\epsilon_b \kappa h e^{i\omega t} \cos \kappa x}{1 + \epsilon_b + \kappa \epsilon_b \int\limits_{(a)} dy/\epsilon} \left[\left(\frac{1}{\epsilon_{max}} - 1\right)\delta(y - y_s) - \frac{\partial}{\partial y}\left(\frac{1}{\epsilon}\right) \right]; \qquad (16)$$

$$V \simeq \frac{E_0^2}{4\pi} \frac{\Omega D_a}{kT} \kappa^2 h \sin \kappa x. \qquad (17)$$

The velocity (17) is at least D_a/D times greater than (9).

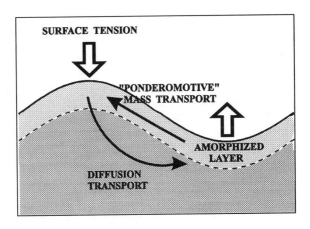

Fig. 3. Mass transport in the amorphized layer

The net growth rate of the surface instability that accounts for surface tension and for the enhancement of ponderomotive mass transport due to existence of the amorphized layer can be derived from Eqs. (14) and (17):

$$\Gamma = \frac{\Omega}{kT}\kappa^2 \left[\frac{E_0^2}{4\pi}D_a - 2\gamma\kappa N_0(D + \kappa a D_a)\right]. \tag{18}$$

By equating $\partial\Gamma/\partial\kappa$ to zero, the prevailing period of the corrugated profile (for which Γ is maximum) can be estimated. In particular, if bulk diffusion is predominating, i.e., $D \gg \kappa a D_a$, the period $(2\pi/\kappa)$ can be expressed from the following relation:

$$2\gamma\kappa = \frac{2}{3}\frac{E_0^2}{4\pi}\frac{1}{N_0}\frac{D_a}{D}. \tag{19}$$

It can be seen from Eq. (19) that the period of the profile formed on the surface is such that the stresses of surface tension ($\sim 2\gamma\kappa$) and the equivalent ponderomotive stress [6]

$$\sigma_E = \frac{E_0^2}{4\pi}\frac{1}{N_0}\frac{D_a}{D} \tag{20}$$

are of the same order of magnitude. Note that σ_E can be great in a very moderate electric field. For example, for $E_0 \simeq 1$ kV/cm (which can be considered a typical value for technological systems of RF and microwave processing (see, e.g., [4])), $N_0 \simeq 10^{-5}$ and $D_a/D \simeq 10^3$ the equivalent ponderomotive stress reaches 10^7 Pa. The period of the relief formed in these conditions lies in the micrometer range, and, as seen from Eq. (19), it can be controlled by adjusting the microwave power.

ACKNOWLEDGEMENT

Partial support of this work by Russian Basic Science Foundation (grant # 95-02-05000-a) and NATO (Linkage grant # HTECH LG 940364) is acknowledged.

REFERENCES

1. P. Chenevier, J. Cohen, and G. Kamarinos, *J. Physique – Lettres*, **43**, 291 (1982).
2. S.J. Rothman, in *Microwave Processing of Materials IV*, edited by M.F. Iskander, R.J. Lauf, and W.H. Sutton (Mat. Res. Soc. Proc. **347**, Pittsburgh, PA 1994), p.9-18.
3. K.I. Rybakov and V.E. Semenov, *Phil. Mag.* (1996), accepted for publication.
4. Y. Bykov, A. Eremeev, and V. Holoptsev, in *Microwave Processing of Materials IV*, edited by M.F. Iskander, R.J. Lauf, and W.H. Sutton (Mat. Res. Soc. Proc. **347**, Pittsburgh, PA 1994), p. 585 - 590.
5. K.I. Rybakov and V.E. Semenov, *Phys. Rev. B*, **49**, 64 (1994).
6. K.I. Rybakov and V.E. Semenov, *Phys. Rev. B*, **52**, 3030 (1995).
7. S.A. Freeman, J.H. Booske, and R.F. Cooper, *Phys. Rev. Lett.*, **74**, 2042 (1995).
8. K.I. Rybakov and V.E. Semenov, in *Solid State Ionics IV*, edited by G.-A. Nazri, J.-M. Tarascon, and M. Schreiber (Mat. Res. Soc. Proc. **369**, Pittsburgh, PA 1995), p. 263 - 268.

GRAIN GROWTH BEHAVIOR DURING MICROWAVE ANNEALING
OF SILICON NITRIDE

M. HIROTA, M.E. BRITO*, K. HIRAO*, K. WATARI*, M. TORIYAMA* and T. NAGAOKA*
Proposal-based Advanced Industrial Technology R & D Program Office, Industrial Technology
Department, New Energy and Industrial Technology Development Organization (NEDO), 28th
Floor, Sunshine 60 Bldg., 1-1 Higashi-Ikebukuro 3-chome, Toshima-ku, Tokyo 170, JAPAN
*National Industrial Research Institute of Nagoya, 1-1 Hirate-cho, Kita-ku Nagoya, Aichi 462,
JAPAN

ABSTRACT

A comparative study of grain growth behavior in silicon nitride under conventional and
microwave annealing is presented. Microwave annealed specimens showed a faster growth rate
as indicated by the quantitative microstructural analysis. The phenomenon was used in combination
with seeding techniques to develop a silicon nitride exhibiting a bi-modal microstructure.
Microwave annealing was carried out using a microwave radiation frequency of 28 GHz.

1. INTRODUCTION

Silicon nitride exhibits excellent properties at high temperature, such as good mechanical
properties, high resistance to corrosion and high creep resistance. For these reasons, silicon nitride
is being used for applications where metals can not be used, for example in high-temperature gas
turbines, and heat exchangers. Nevertheless, the wide use of the silicon nitride is restricted due to
its brittle fracture behavior. Typical silicon nitride obtained by pressureless sintering consists of
relatively small grains, exhibiting a high fracture strength but a low fracture toughness. On the other
hand, silicon nitride sintered at high temperature under a high nitrogen pressure ($\approx$ 10 MPa),
exhibits a bi-modal microstructure, where some large grains are developed within a fine grained
matrix. This material exhibits a high fracture toughness but the strength is lowered owing to the
existence of large grains. To overcome this problem, Hirao *et al.* [1] demonstrated that addition
of a few vol% single crystal β-Si_3N_4 as seeds made it possible to control the size and amount of large
elongated grains. They have obtained a silicon nitride having a well controlled bi-modal
microstructure after sintering under 1 MPa of nitrogen pressure. This material has both high fracture
toughness and high fracture strength.

In recent years, microwave heating technology has been used for ceramic processing.
However, microwave studies in this field are still at a developmental stage. This heating technology
is very different from the conventional heating process, and provides some advantages: (1) the use
of microwave radiation makes the heating period shorter and (2) densification seems to be completed
at relatively low sintering temperatures. Furthermore, Cheng *et al.* [2] point out that properties of
alumina-mullite ceramics sintered by microwave radiation are often superior to that of the
conventional products. Tiegs *et al.* [3] also indicated that the mechanical properties of microwave
sintered silicon nitride, especially its fracture toughness are significantly improved over conventionally
sintered material. They concluded that this was attributed to enhanced growth of elongated grains
in the microwave annealed sample. Such tendency in grain growth can be also seen in microwave
annealed samples of silicon nitride [4].

In this report, bi-modal microstructure were obtained by the combination of seeding and
microwave heating techniques under nitrogen pressure of 0.1 MPa. The grain growth behavior during

microwave annealing of the silicon nitride was investigated, and a comparison of grain growth between conventional and microwave annealing was carried out.

2. EXPERIMENTAL

2.1. Preparation of specimen

Two series of specimen with and without seed particles were prepared for the annealing test. The β-Si_3N_4 single crystal particles were synthesized as seed particles by the method reported previously [5]. The mean diameter of a β-Si_3N_4 seed particle is 1.4 μm with a mean length of 5.6 μm. The specimens mainly consist of α-Si_3N_4 (E-10 grade, Ube Industries Ltd., Japan), 5 wt% Y_2O_3 (purity > 99.9 %, Hokko Chemicals, Ltd., Japan), 2 wt% Al_2O_3 (purity > 99.9 %, Hokko Chemicals, Ltd., Japan). They were mixed up by planetary milling using methanol as the mixing medium. In the case of preparing specimen with seed particles, 2 vol% β-Si_3N_4 seed crystals were ultrasonically dispersed in the slurry to avoid fracture of the particle. After drying the slurry, the powders were hot pressed at 1700 °C with 40 MPa pressure in BN-coated graphite dies under 0.1 MPa nitrogen for 1 hour. All samples for annealing test were cut from the same hot pressed billet. The density of the material used is above 99 % of the theoretical density.

2.2. Annealing

Although 2.45 GHz is the most common microwave processing frequency, in this work, microwave annealing was carried out using a 28 GHz gyrotron source. The use of microwaves at 28 GHz frequency makes it possible to produce ceramics that are not easily heated at lower frequencies such as 2.45GHz [6]. The samples were placed top of a boron nitride plate in a silicon nitride crucible, which was further surrounded by alumina heat insulating material (Fibermax 18R, Toshiba Monofrax Co., Ltd., Japan). Temperature was measured by a W-Re thermocouple. The samples were annealed at 1750 °C and kept for 4 or 8 hours under a nitrogen pressure of 0.1 MPa. Conventional annealing was done in a BN-coated graphite crucible filled with a powder mixture of 40 vol% boron nitride and 60 vol% silicon nitride powders. The annealing conditions were the same as in the microwave annealing.

2.3. Microstructural Analysis

Annealed samples were polished and plasma-etched (Plasma Reactor Model PR-41, Yamato Science Co. Ltd., Japan) for microstructure observation using scanning electron microscopy (SEM; JSM-T330AS, JEOL Ltd., Japan). In order to quantitatively characterize the microstructure of specimens, diameter, length and area fraction of large elongated grains were measured by using an image analyzer.

3. RESULTS AND DISCUSSION

Microstructures of specimen before and after annealing are shown in Fig. 1. The as hot-pressed specimen without seed particles consist of only small grains (Fig. 1a). The as hot-pressed specimen with 2 vol% seed particles (Fig. 1b) exhibits a bi-modal microstructure composed of small matrix grains and large rodlike grains. Comparing diameters and lengths of these rodlike grain with those of the original seed particles (diameter: 1.4 μm, length: 5.6 μm), no remarkable grain growth was observed. The sample without seed particles was annealed by conventional heating for 8 hours (Fig. 1c), and the grain size became slightly larger as compared to the as hot-pressed microstructure

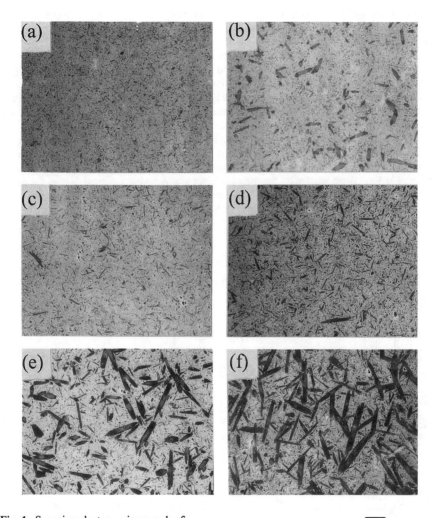

Fig. 1 Scanning electron micrograph of
  (a) as hot-pressed specimen without seed particles
  (b) as hot-pressed specimen with 2 vol% seed particles
  (c) conventionally annealed specimen without seed particles
      at 1750 °C under 0.1 MPa N_2 for 8 hours
  (d) microwave annealed specimen without seed particles
      at 1750 °C under 0.1 MPa N_2 for 8 hours
  (e) conventionally annealed specimen with 2 vol% seed particles
      at 1750 °C under 0.1 MPa N_2 for 8 hours
  (f) microwave annealed specimen with 2 vol% seed particles
      at 1750 °C under 0.1 MPa N_2 for 8 hours.

10μm

of Fig. 1a. Remarkably differences were observed in grain growth behavior for similar annealing times in the microwave furnace as revealed by Fig. 1d. This seems to indicate that the use of microwaves for annealing enhances grain growth.

On the other hand, when the specimens with seed particles are annealed by conventional method, considerably grain growth is obtained (Fig. 1e). Judging from the fact that the grain growth of the specimens without seed particles is less remarkable (Fig. 1c and d) at the same conditions, it can be concluded that larger grains grew from seed particles. In particular, when the specimen with seed particles is microwave annealed, grain growth and changes in microstructure are very drastic, as revealed in Fig. 1.

In order to discuss quantitatively grain size distribution of specimen with seed particles, measurement of mean diameter, mean aspect ratio and area fraction was conducted using an image analyzer. The large elongated grains which have a diameter larger than the original seed particle diameter, 1.4 μm, were selected for the analysis. Since grains having a diameter above 1.4 μm are not observed in the microstructure of the specimen without seed particles, this value was chosen as the onset for growing from seeds. Since elongated grains have a hexagonal prism shape, the

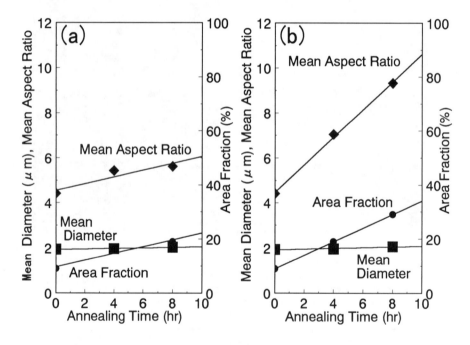

Fig. 2 Graph of the mean diameter, the 10% mean aspect ratio and the area fraction for the elongated seed particles as a function of annealing time.
(a) Conventionally annealed specimen with seed particles
(b) Microwave annealed specimen with seed particles.

diameters of each grains were directly adopted from the measured values. Because all grains are oriented randomly, the apparent lengths of each grains are smaller than the real one except when the c axis is parallel to the observation surface. It has been proposed that the real average aspect ratio should be the mean value of the 10 % highest observed ratios (length/diameter), because the shape of the grains, *i.e.*, the aspect ratio of the grains, is nearly constant [7-8]. Then, the mean aspect ratio was used for a quantitative characterization of the grain growth of the annealed specimen.

Graph of the mean diameter, the 10 % mean aspect ratio obtained from above mentioned method and the area fraction for the elongated seed particles as a function of annealing time is shown in Fig. 2. Both mean diameters, of conventionally annealed samples and microwave annealed samples, are almost constant. However the mean aspect ratio is increased with the annealing time. Furthermore, the mean aspect ratio of the microwave annealing is increasing with a higher rate than for conventional annealing. The area fraction of both microwave and conventional annealing is also increased, showing for microwave annealing the higher value. From the comparison between these two curves, it may be concluded that the use of microwave heating should be more effective to promote the formation of a well controlled bi-modal microstructure than conventional heating.

4. CONCLUSIONS

From the present experimental results, the following conclusions can be drawn:
(1) Grain growth of silicon nitride seems to be enhanced by microwave annealing.
(2) Considering the relatively mild annealing conditions (nitrogen pressure of 0.1 MPa at 1750 °C), the use of microwave is very effective to obtain well controlled bi-modal microstructures, when coupled to seeding techniques.

REFERENCES

[1] K. Hirao, T. Nagaoka, M. E. Brito and S. Kanzaki, *J. Am. Ceram. Soc.*, 77[7], 1857-1862(1994).
[2] J. Cheng, Y. Fang, D. K. Agrawal, Z. Wan, L. Chen, Y. Zhang, J. Zhou, X. Dong and J. Qui, in Microwave Processing of Materials IV, edited by M. F. Iskander, R. J. Lauf and W. H. Sutton(*Mater. Res. Soc. Proc.* 347, Pittsburgh, PA, 1994)pp. 557-562.
[3] T. N. Tiegs, J. O. Kiggans, Jr., H. T. Lin and C. A. Willkens, *ibid.*, 347, pp.501-506.
[4] T. N. Tiegs, J. O. Kiggans, Jr. and H. D. Kimrey Jr. in Microwave Processing of Materials II(*Mater. Res. Soc. Proc.* 189, Pittsburgh, PA, 1991)pp. 267-272.
[5] K. Hirao, A. Tsuge, M. E. Brito and S. Kanzaki, *J. Ceram. Soc. Jpn.*, 101[9], 1071-1073(1993).
[6] T. Saji, *New Ceramics*, 8, No.5, 21-30(1995), in Japanese.
[7] G.Wötting, B. Kanka and G. Ziegler, Non-Oxide Technical and Engineering Ceramics, edited by S. Hampshire(Elsevier, London, 1986), pp.83-96.
[8] M. Mitomo, M. Tsutsumi, H. Tanaka, S. Uenosono and F. Saito, *J. Am. Ceram. Soc.*, 73[8], 2441-2445(1990).

KINETICS AND MECHANISM OF THE MICROWAVE SYNTHESIS
OF BARIUM TITANATE

Hanlin Zhang, Shixi Ouyang, Hanxing Liu and Yongwei Li
National Key Laboratory for Synthesis and Processing of Advanced Materials,
Wuhan University of Technology, Wuhan 430070, P. R. China

ABSTRACT

The formation kinetics of $BaTiO_3$ from the solid-state $BaCO_3$ and TiO_2 powder in a microwave field was investigated. The quantitative XRD analysis and the model considered the volume change between reactant and product were used in this experiment. Results show that the formation rate of $BaTiO_3$ in a microwave field is much faster than upon conventional heating. The activation energy of the solid state reaction for $BaTiO_3$ was measured as 58 kJ/mol. This indicates the enhancement of diffusion by the microwave heating process.

INTRODUCTION

The number of applications of microwave radiation for the synthesis of materials is increasing because of the rapid synthesis rate, low energy consumption and high product quality observed upon microwave processing. The applications in synthesis of many inorganic compound materials have been reviewed in several articles[1-3]. With the development of electronic industry, the need for $BaTiO_3$, as an important electronic material, is growing. Therefore, the study on a new method of synthesizing this compound is very interesting. The authors[4] have investigated the microwave synthesis of $BaTiO_3$ in the solid state. The results show that the synthesis time is much shorter, the particle size smaller, and the distribution of the particle size narrower, in comparison with a conventional synthesis at a similar temperature. To study the kinetics of the $BaTiO_3$ formation process in the solid state in the microwave field is useful, in order to obtain fundamental information about factors controlling the reaction.

Little information exists about the synthesis of $BaTiO_3$ in a microwave field. However, the reaction kinetics and mechanism of the conventional formation for barium titanate in the solid state have been investigated[5-11]. Generally, the overall net reaction can be expressed by the following equation:

$$BaCO_3(s) + TiO_2(s) = BaTiO_3(s) + CO_2(g) \qquad (1)$$

However, on the way to pure barium titanate, intermediate steps are involved in reaction (1). They have been examined by several authors[5-11]. First, a small amount of $BaTiO_3$ forms at the interface between $BaCO_3$ and TiO_2. Second, Ba_2TiO_4 is produced by the

447

reaction of $BaTiO_3$ with $BaCO_3$ during exposure in hot air. Then Ba_2TiO_4 continues to form until no $BaCO_3$ is left. Finally, the Ba_2TiO_4 phase is gradually consumed by reaction with the remaining TiO_2 to form $BaTiO_3$. Small amounts of other intermediate titanates ($BaTi_3O_7$ and $BaTi_4O_9$) have also been reported to form during the course of the reactions [7].

The main purpose of the present work was to study the kinetics of the $BaTiO_3$ formation process and to measure the activation energy of the diffusion, to elucidate the overall reaction mechanism.

EXPERIMENTAL PROCEDURE

The experimental materials were the commercial chemicals $BaCO_3$ and TiO_2 powders with a chemical purity grade. Equimolar amounts of $BaCO_3$ and TiO_2 powders were first mixed in a ball-mill jar filled with ethanol for 10 hours, then dried in an infrared dryer, and compacted inside a round steel die at a uniaxial pressure. Green compact disks with a volume of ϕ13x3mm^3 were formed. The green compacts were placed in a fibrous thermal insulation in a 2.45GHz single mode, TE_{103}, microwave resonant cavity and rapidly heated to 950oC,1000oC,and 1050oC, respectively. Holding times were varied at each temperature. The samples were quenched in air to room temperature after switching off the microwave power. The amounts of $BaTiO_3$ in specimens as-received were determined by a quantitative XRD analysis.

RESULTS AND DISCUSSION

Fig. 1 shows the dependence of the formed $BaTiO_3$-fraction, a, on the reaction time, t, at temperatures between 950oC to 1050oC in a microwave field. As can be seen, the formation rate of $BaTiO_3$ is quite fast.

The time for complete reaction at 1050oC was less than ten minutes, while at 1100oC the reaction takes less than three minutes(not shown in the figure). It can also be seen from Fig. 1 that $BaTiO_3$ formed during the period of the temperature increase for the 1000°C and 1050°C experiment. As well-known, the $BaCO_3$ reaction with TiO_2 to form $BaTiO_3$ starts above a critical reaction temperature in a certain condition. The higher the temperature increases over the critical

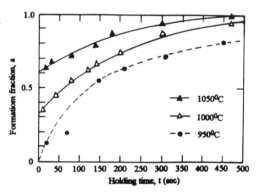

Figure 1. Kinetic growth curves for $BaTiO_3$ formation from the solid state $BaCO_3$ and TiO_2 powders in a microwave field.

temperature, the faster the formation rate of $BaTiO_3$ and the larger the $BaTiO_3$ amounts formed within the same time. In the present experiment, even though the samples were heated from room temperature to the holding temperature in a microwave cavity as fast as possible, to reach higher temperature more time is necessary. In this case, $BaTiO_3$ formed during the heating process and the amount increased with the time and temperature increase. Therefore, the initial formation fractions were not zero at $t=0$ in Fig.1. This result also reflected the characteristics of the microwave synthesis in the enhancement of the reaction.

The rapid formation of $BaTiO_3$ in a microwave field would require a reaction mechanism different from conventional synthesis. Actually, in the present work, the formation amounts of intermediate phase, Ba_2TiO_4, were not sufficient to be detected by XRD analysis of all samples used for the kinetic study. This suggests that $BaTiO_3$ could form through a direct reaction between $BaCO_3$ and TiO_2 while the intermediate steps are thought existing in the reaction process of the conventional formation for $BaTiO_3$[5-10]. It could be related to the external electric field applied to reactants and products in the microwave synthesis process. The external electric field was useful to cation's diffusing in the direction of the field[12-14]. Comparing with a conventional synthesis, the external electric field in a microwave synthesis would be the main cause of the rapid synthesis.

On the basis of the model[15] for a diffusion-controlled reaction with a volume change:

$$
\left(\frac{V+1}{V}\right)^{\frac{1}{3}} \left[3(1+Va)^{\frac{1}{3}} + \sum_{n=1}^{\infty} \frac{\frac{1}{3}(\frac{1}{3}-1)\cdots(\frac{1}{3}-n+1)}{(n-\frac{1}{3})\,n!\,(-1)^n(1+V)^n} (1+Va)^{n+\frac{1}{3}} \right]
$$

$$
-\left(\frac{V+1}{V}\right)^{\frac{2}{3}} \left[\ln(1+Va) + \sum_{n=1}^{\infty} \frac{\frac{2}{3}(\frac{2}{3}-1)\cdots(\frac{2}{3}-n+1)}{n\,n!\,(-1)^n(1+V)^n}(1+Va)^n \right]
$$

$$
-\frac{3}{2}(1-a)^{\frac{4}{3}} - \frac{3}{2}V(1+Va)^{\frac{4}{3}} = KDt - \frac{3}{5} - \frac{3}{2}V
$$

$$
+\left(\frac{V+1}{V}\right)^{\frac{1}{3}} \left[3 + \sum_{n=1}^{\infty} \frac{\frac{1}{3}(\frac{1}{3}-1)\cdots(\frac{1}{3}-n+1)}{(n-\frac{1}{3})\,n!\,(-1)^n(1+V)^n} \right]
$$

$$
-\left(\frac{V+1}{V}\right)^{\frac{2}{3}} \sum_{n=1}^{\infty} \frac{\frac{2}{3}(\frac{2}{3}-1)\cdots(\frac{2}{3}-n+1)}{n\,n!\,(-1)^n(1+V)^n} \tag{2}
$$

where V is the ratio of the volume difference between product formed and reactant consumed to the reactant volume, a, the mole fraction reacted, K, a constant, and D, the diffusion coefficient. If $y(a)$ represents the left terms of Eq. (2), B_0 does the right constant terms, we can obtain:

$$
y(a) = KDt + B_0 \tag{3}
$$

Substituting $D=K_1\exp(-Q/RT)$ into Eq. (3), transferring terms, and taking the nature logarithm of the equation obtained, we get:

$$
R\ln[y(a)-B_0] = -Q/T + R\ln(KK_1H) \tag{4}
$$

where H is a constant holding time. In this study, $V = 0.933$, substituting the V value, and the t and a value at each temperature into Eq.(2), taking $n < 10$, the calculation values are illustrated in Fig. 2. The straight lines in Fig. 2 were drawn from regression equations corresponding to Eq. (3). As can be seen, the experiment data agreed with the model selected.

Substituting the B_0 value and $y(a)$ value at each temperature at $t = 140$ second into Eq. (4), we can obtain the dependence of $f(a)$ ($= Rln[y(a)-B_0]$ on the reciprocal of temperature as shown in Fig. 3. The slope of the curve was the diffusion activation energy, Q, which calculated from regression equation by $f(a) = Q/T + A$ ($A = Rln(KK_1H)$) was about 58 kJ/mol. The Q value is much lower than that in conventional reactions[11]. The explanation for the above description could be that, in the case of microwave heating, since the dipole moment inside dielectrics would align in an electric field, the interaction among the dipole moments in the vertical direction of the electric field would be reduced, which could make the resistance of moving ions along the electric field reduce, resulting in the reduction of the diffusion activation energy and the increase of the diffusivity.

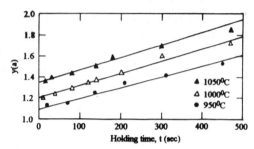

Figure 2. Dependence of y(a) on holding time t at different temperature in a microwave synthesis.

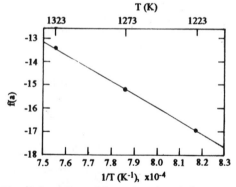

Figure 3. Dependence of f(a) on the reciprocal of temperature in fixing holding time t at 140 second.

CONCLUSIONS

The microwave synthesis of $BaTiO_3$ from the solid state $BaCO_3$ and TiO_2 is much faster than the conventional synthesis at a similar temperature. It can be concluded that the enhancement of diffusion reaction by microwave heating is obvious. The activation energy of diffusion reaction formation of $BaTiO_3$ in a microwave field was measured as about 58 kJ/mol, which was a quite lower value, comparing with that in conventional diffusion reactions.

REFERENCES

1. D. M. P. Mingos and D. R. Baghurst, Br. Ceram. Trans. J., **91**, P.124 (1992).

2. D. M. P. Mingos, Adv. Mater., **5**, P. 857 (1993).

3. H. Zhang, H. Liu and S. Ouyang, Mater. Rev., (to be published, in Chinese).

4. H. Zhang, Y. Li, H. Liu and S. Ouyang, J. Mater. Sci. Lett., (accepted).

5. T. Kubo and K. Shinriki, J. Chem. Soc. Jpn., Ind. Chem. Sect., **57**, p.621 (1954).

6. W. Trzebiatowski, J. Wojciechowska and J. Damm, Rocz. Chem., **26**, p.12 (1952).

7. L. K. Templeton and J. A. Pask, J. Amer. Ceram. Soc., **42**, p.216 (1959).

8. M. Cournil, M. Soustelle and G. Thomas, Oxid. Met., **13**, p. 89 (1979).

9. A. Beauger, J. C. Mutin and J. C. Niepce, J. Mater. Sci., **18**, p. 3041 (1983).

10. Idem, ibid,, 18, p. 3542 (1983).

11. A. Amin, M. A. Spears and B. M. Kulwicki, J. Amer. Ceram. Soc., **66**, p. 733 (1983).

12. J. D. Katz, R. D. Blake and V. M. Kenkre, in Microwaves: Theory and Applications in Materials Precessing, edited by D. E. Clark, F. D. Gac and W. H. Sutton, (Ceram. Trans. 21, ACS, Westerville, Ohio,1991), p. 95.

13. O. Stasiw and J. Teltow, Ann. der Phys., **1**, p. 26 (1947).

14. K. I. Rybakov and V. E. Semenov, Phys. Rev. B, **49**, p. 64 (1994).

15. C. Zhang, Y. Yang and G. Zhang, Acta Phys. Chem. Sinica, **4**, p. 539 (1988).

STATISTICAL COMPARATIVE ANALYSES OF ENGINEERING PROPERTIES OF MICROWAVE AND CONVENTIONALLY SINTERED ALUMINA

K.R. Binger[1], S.A. Freeman[2], D.J. Grellinger[1], R.F. Cooper[3], and J.H. Booske[1]
University of Wisconsin - Madison, Madison, Wisconsin 53706
1 Department of Electrical and Computer Engineering
2 Material Science Program
3 Department of Material Science and Engineering

ABSTRACT

Processing conditions such as temperature, soak time, and heating rate affect the final density of conventionally-sintered and microwave-sintered ceramics. Of additional importance is the question of whether microwave-sintered ceramics display *intrinsically* superior macroscopic engineering properties compared with conventionally-sintered control specimens. An analysis using the Yates algorithm indicates that the processing condition which has the largest impact on the density of the specimen is the heating method (microwave vs. conventional). The microwave-sintered specimens resulted in higher densities and higher fracture strengths. However, it was determined that the higher fracture strengths were due to the higher sintered densities rather than a significantly different microstructure.

INTRODUCTION

Engineering properties of microwave and conventionally-sintered alumina were compared using statistical experimental design methods. Factorial experiments allowed us to examine effects of processing variables such as temperature, soak time, and heating rate in both microwave and conventional furnaces on properties such as density and fracture strength of sintered alumina. Using correct experimental design methods [1] allows one to comprehensively survey the large parameter space of a multivariable process with a minimum number of runs, yet still generate statistically meaningful data leading to rigorously valid conclusions.

EXPERIMENTAL METHOD

A 2^N full factorial statistical design was formulated with a total of four factors (N=4). The four experimental variables were processing temperature, soak time, heating rate, and heating method (microwave vs. conventional). Each processing parameter was assigned either "+" or "-." For example, the heating method was either microwave (+) or conventional (-) (figure 1). Each possible combination of the four processing conditions was desired yielding 16 different runs.

FURNACE PROCESSING CONDITIONS		
level / variable	+	-
temperature	1550 C	1475 C
soak time	120 min.	30 min.
heating rate	8 C/min.	4 C/min.
heating method	microwave	conventional

Figure 1. Processing conditions.

Measurements of density and fracture strength were performed on samples from each run. This data was evaluated using the Yates algorithm, which required the runs to be put in standard order with all factors low continuing down until all factors were high (figure 2) [1]. The measured response, such as density, corresponding to each run was also listed in the same order and incorporated into the Yates algorithm. The algorithm was used to extract from the data which parameters had a significant effect on the result. For example, there were eight pairwise comparisons of the rate effect . Run asc013 and asc012 (figure 2) compare "+" and "-" rates

453

while holding all other factors the same. Similarly, the next seven pairs compare the rates ("+" and "-") while keeping the other factors the same. The average of these eight comparisons was the "main effect of the rate," which measured the average effect of rate over all conditions of the other variables. If the "main effect of rate" was 10, for example, then the effect of rate was to increase the density by about 10 units (i.e. 85% dense vs. 95% dense). For the density results in figure 2, the Yates algorithm gives the effects in figure 3.

Run Number	- = 4 + = 8 Rate	- = 1475 + = 1550 Temperature	-=30 +=120 Soak Time	- = conv. + = micr. Heating Method	Density
asc013	-	-	-	-	93.6
asc012	+	-	-	-	93
asc016	-	+	-	-	97.9
asc015	+	+	-	-	97.9
asc017	-	-	+	-	96.6
asc011	+	-	+	-	96.6
asc014	-	+	+	-	97.8
asc010	+	+	+	-	98.8
asm022	-	-	-	+	97.7
asm024	+	-	-	+	98
asm023	-	+	-	+	98.9
asm020	+	+	-	+	98.9
asm026	-	-	+	+	98.3
asm028	+	-	+	+	99
asm025	-	+	+	+	99.3
asm021	+	+	+	+	99.5

Figure 2. Table for a 2^4 factorial experiment to be used in the Yates algorithm.

rate=r, temperature=T, time=t, heating method=h

variable	r	T	r/T	t	r/t	T/t	r/T/t	h	r/h	T/h	r/T/h	t/h	r/t/h	T/t/h	r/T/t/h
effect	0.2	2	0.1	1.3	0.3	-0.8	0.03	2.2	0.1	-1.1	-0.3	-0.6	-0.13	0.7	-0.08

Figure 3. The effects on density determined by the Yates algorithm.

There are many effects close to zero and some substantially larger (both positive and negative). To separate significant effects from experimental error, the row of numbers (figure 3) was sorted and then plotted on a normal probability scale. The horizontal axis of the probability plot is the effect of the processing parameter on the result (density). Since there are 15 effects to plot, the probability (vertical axis) assigned to each effect is the cumulative fraction (e.g. 1/15 for the first effect). For points that lie in the central linear region (figure 4), it is not possible to separate these small effects from the measurement error. Hence, they are statistically insignificant. The outlying points correspond to experimental variables that significantly contribute to the measured result (density or strength). This method allowed us to evaluate whether each factor had a significant impact on a particular response [1].

In order to obtain fracture strengths of the samples from various runs, biaxial flexure tests were conducted. We used the ball-on-ring test where a disk-shaped sample of sintered alumina was supported by a ring and centrally loaded with a ball [2]. The tests were set up on an Instron testing machine. The apparatus which included the ball was lowered, the displacement was induced, and the load was measured by the machine [3]. This situation for a concentric point load is approximated by a central zone of uniform stress with finite dimensions. The maximum radial and tangential stresses at the center are equal and given by

$$S = \left[\frac{3P(1+v)}{4\pi t^2}\right] \times \left[1 + 2\ln\left(\frac{a}{b}\right)\right] + \left[\frac{(1-v)a^2}{(1-v)R^2}\right] \times \left[1 - \left(\frac{a^2}{b^2}\right)\right] \quad (1)$$

where P is the load, t the thickness of the disk, a the radius of the circle of support, b the radius of the region of uniform loading, R the radius of the disk, and v is Poisson's ratio[3]. From the measured loads at fracture for each disk and the dimensions of the configuration and specimen, the fracture stresses could be calculated.

Ceramics have an inherent dispersion of strength. Two seemingly identical samples may not both be capable of surviving the same load. Porous ceramics contain microscopic cracks and angular holes, and if one type of ceramic material is cut into pieces, each piece contains a different distribution of crack lengths. Thus, two pieces which appear identical may not both endure the same load because one piece's distribution may contain longer flaws and would break under a smaller load.

Weibull, a Swedish engineer, developed a way to mathematically characterize strength statistics. He defined the fraction of identical samples which survive loading to a stress (σ) to be the survival probability, P_s.

$$P_s = \exp\left[-\left(\frac{\sigma}{\sigma_0}\right)^m\right] \qquad (2)$$

where σ_0 and m are constants. As σ increases, the number of failing samples increases, and P_s decreases. The tensile stress that allows 37% (1/e) of the samples to survive is σ_0. The Weibull modulus, m, indicates how quickly the strength decreases as σ_0 is approached. The higher the value of m, the less the variation in fracture strength between pieces of one sample[4].

Taking the natural log of the previous equation twice yields the expression

$$\ln\left[\ln\left(\frac{1}{P_s}\right)\right] = m \times \ln\left(\frac{\sigma}{\sigma_0}\right). \qquad (3)$$

Plotting $\ln(\ln(1/P_s))$ versus $m \ln(\sigma/\sigma_0)$ should yield a straight line of slope m. Common practice is to fracture a large number (>20) samples, arrange fracture strengths in ascending order, assign survival probabilities (the fraction of samples that survived that fracture strength), and then plot $\ln(\ln(1/P_s))$ versus $m \ln(\sigma/\sigma_0)$. By performing a linear regression, the slope (m) and the intercept (m ln (σ_0)) can be used to determine m and σ_0. Through this method, we developed values for the Weibull modulus, m, from which one can infer whether different processing conditions yield different microstructure or defect distributions over macroscopically large volumes.

RESULTS AND DISCUSSION

The 16 measured density values, one for each combination of processing conditions of the sintered alumina samples, were incorporated into the Yates algorithm. The resulting numbers were plotted on a normal probability scale (figure 4). One set of points is grouped together on a line at zero on the horizontal axis. Significant, outlying values correspond to processing parameters of soak time, processing temperature, and heating method, as well as combinations of these. All values for single parameter effects were positive. This says that a high condition of that processing parameter results in greater density than the low condition. The condition which has the largest impact on the density is the method of heating - microwave versus conventional.

In addition to the alumina powder which was pressed into the samples used for the data in figure 2, a similar set of experiments was run on another type of alumina powder. The second powder, Al_2O_3II, was Sylvania CR-30 spray dried alumina. The powder used for the data in figures 2 and 3 (Al_2O_3I) was Reynolds Chemical RCHPDBM alumina. Al_2O_3II had a particle size ranging from 10-75 microns, while Al_2O_3I had a larger particle size ranging from 25-480 microns. The green density of Al_2O_3I was approximately twice that of Al_2O_3II after similar pressing conditions, most likely due to the fact that Al_2O_3I had a larger range between the small and large sizes of the particles which allowed for higher initial packing. Al_2O_3I also had a larger variation in particle morphology, and the nonspherical shapes may have packed more densely than spheres (the approximate shape of the Al_2O_3II particles).

Densities of Al_2O_3II were put into the Yates algorithm. The resulting probability plot (figure 5) was in good agreement with the Al_2O_3I plot. Statistically significant processing conditions for Al_2O_3I were significant for Al_2O_3II as well. According to the algorithm, the heating method had the greatest effect on the density, while the temperature, soak time, and combinations of these also played significant roles in the density. In both cases, varying the ramp-up rate (4°/min. vs. 8°/min.) did not have a statistically significant effect on the density.

Values of fracture stress determined by the flexure tests were also evaluated by the Yates algorithm. From the ball-on-ring tests the fracture load (P) was determined and used to evaluate the fracture stress (S) by equation (1). The stress was calculated by dividing the load (P) by the square of the thickness of the disk (t). The other conditions of the experiment contribute a factor of 3 to the values of the stresses given (figure 6). The results of the statistically significant parameters were less clear than the two density plots. However, the most influential parameter was again the heating method, followed by temperature and soak time. Stress versus density for Al_2O_3I was also plotted (figure 7). The samples able to withstand the higher loads were those with the higher densities (microwave-sintered).

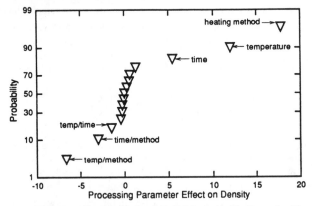

Figure 4. Parameter effects on density of Al_2O_3I determined by Yates algorithm.

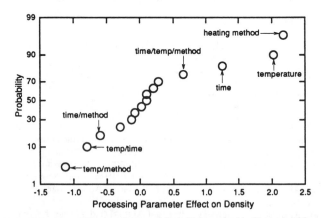

Figure 5. Parameter effects on density of Al_2O_3II determined by Yates algorithm.

Weibull statistics were used on the stress values of the sintered alumina Al_2O_3I. Stresses of conventionally-sintered samples were grouped together, put in ascending order and analyzed according to Weibull's method. This was repeated for stress values of the microwave-sintered alumina. The $\ln(\ln(1/P_s))$ versus the $\ln(\sigma/\sigma_0)$ of both conventionally-sintered and microwave-sintered results were plotted (figure 8). The Weibull modulus was ~13 for the conventionally-sintered samples, and ~12 for the microwave-sintered samples. The similar slopes (Weibull modulus) of the two lines indicate that the rate at which the strength decreases as σ_0 is approached is the same for both types of sintering.

If any apparent enhancements associated with the microwave sintering are due solely to inverted temperature profiles (i.e. interior temperature higher than the surface temperature), then the data establish that *most* of the interior of the microwave-sintered specimen internal temperature must be approximately 100°C hotter than its surface temperature. This is determined by observing how much the temperature affected the final density of the conventionally-sintered and microwave-sintered samples. Figure 3 shows that the effect (on Al_2O_3I) of changing the temperature from low to high was to increase the final density by 2, and the effect of the method of heating (changing from conventional to microwave) was to increase the density by 2.2.

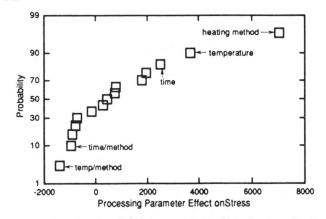

Figure 6. Parameter effects on fracture stress of Al_2O_3I determined by Yates algorithm.

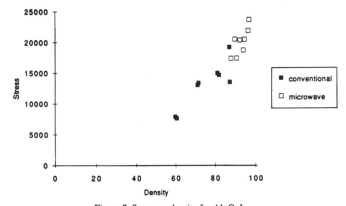

Figure 7. Stress vs. density for Al_2O_3I.

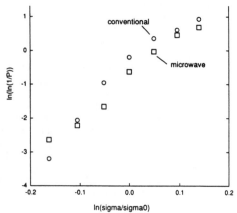

Figure 8. Plots to determine the Weibull modulus for conventional and microwave sintering.

The difference between the low and high levels of temperature was 75°. Dividing this number by 2 (the effect of temperature), and then multiplying this result by 2.2 (the effect of heating method), gives the apparent internal temperature enhancement of about 85°. Performing these same manipulations on the results from the second powder (Al_2O_3II) gives an inverted temperature profile of about 105°. Future experiments are under design to evaluate if such inverted temperature profiles are present in these specimens.

CONCLUSION

Processing conditions such as heating method, temperature, and soak time contribute to the densities and fracture strengths of sintered alumina. A Yates algorithm analysis indicates that the method of heating (microwave versus conventional) has the largest impact on the result, be it density or strength of the sample. The densities of microwave-sintered alumina were higher than those of the conventionally-sintered samples for two different types of alumina powders. Microwave-sintering also resulted in higher fracture strengths of alumina. However, the Weibull modulus appears to be insensitive to the choice of heating method. Hence, the higher fracture strengths of microwave-heated specimens are interpreted to result from the higher sintered densities rather than a significantly different microstructure.

ACKNOWLEDGMENTS

We would like to acknowledge NSF, EPRI, and NASA through the graduate student researchers program.

REFERENCES

1. George E.P. Box, William G. Hunter, and J. Stuart Hunter, *Statistics for Experimenters*. New York: John Wiley & Sons (1978).

2. G. DeWith and H.M. Wagemans, "Ball-on-Ring Test Revisited," in *J.Am.Cer.Soc.,* Vol.. 72, pp. 1538-1541 (1989).

3. D.K. Shetty, A.R. Rosenfield, P. McGuire, G.K. Bansal, and W.H. Duckworth, "Biaxial Flexure Tests for Ceramics," in *Ceramic Bulletin,* pp. 1193-1197 (1980).

4. M.F. Ashby and D.R.H. Jones, *Engineering Materials,* pp. 169-177, Pergamon (1986).

STUDY OF MICROWAVE-DRIVEN CURRENTS IN IONIC CRYSTALS

K.I. RYBAKOV*, V.E. SEMENOV*, S.A. FREEMAN**, J.H. BOOSKE**, and
R.F. COOPER**
* Institute of Applied Physics, Russian Academy of Sciences, Nizhny Novgorod, Russia
** University of Wisconsin, Madison, Wisconsin 53706

ABSTRACT

The recent theory of averaged ponderomotive action of microwave E-field in solids
is expanded to describe quasistationary ionic currents driven by that action. New
experimental results on the dynamics of the microwave-induced currents in AgCl and
NaCl are presented. Agreement between experiment and theory provides further and
stronger evidence for general validity of the theoretical model.

INTRODUCTION

Over the past decade, results from experimental investigations of microwave processing of materials have repeatedly suggested the existence of an unknown, non-thermal
interaction between high-strength microwave fields and ionic materials. This "microwave effect" became a separate subject of experimental and theoretical studies [1].

Freeman, Booske, and Cooper performed ionic conduction experiments [2] that were
specifically designed [3] to test for solid-state diffusion enhancements from intense
microwave irradiation. The results clearly showed no increase in the bulk diffusion
coefficient. Instead, it was found that microwave fields could by themselves induce
quasistationary ionic currents to flow in the crystal.

Concurrent to these experiments, Rybakov and Semenov [4] showed theoretically
that in ionic materials the oscillatory fluxes of charged point defects, driven by the
microwave E-field, are rectified near crystal surfaces and boundaries, resulting in quasistationary near-surface depletion or accumulation of defects. This constitutes a driving force for diffusional flow of defects within the material, which may lead to creep
deformation [5] and formation of quasistationary distribution of electric potential [6].

This paper describes results of a collaborative effort to verify the validity of the
theoretical model by directly applying it to analysis of more recent experimental results.

EXPERIMENTAL SYSTEM

A full description of the experimental system appears elsewhere [3]. In essence, AgCl
and NaCl single crystals are placed into a standard Ku-band rectangular waveguide.
The microwave power is fed into the waveguide as the TE_{10} transmission mode at 14
GHz in short single pulses. The ends of the crystals are coated with metallic electrodes.
In the case of NaCl, plasma sputtered platinum is used. For AgCl, conductive silver
paint is employed. The current is measured by a current amplifier and displayed
on an oscilloscope. The closed electrical circuit in which the microwave-induced dc

459

currents flow thus consists of the ionic crystal with the metallic electrodes and the input impedance of the measuring device. The resistance of the ionic crystal is very high compared to the input resistance of the measuring device.

THEORY

The analytical model is generally similar to that used for description of mass transport phenomena related to the averaged ponderomotive action of microwave field in an ionic crystal [4,5]. The model accounts for diffusion of mobile charge carriers in the solid and their drift in the electric field. The flux of charge carriers is expressed as

$$J = -D\nabla N + \frac{qND}{kT}\mathbf{E},\tag{1}$$

where N is the concentration, D is the diffusivity, q is the electric charge of the carriers, and $\mathbf{E}$ is the electric field vector. As long as action of sources and sinks of charge carriers can be neglected, N and $\mathbf{J}$ are linked by the continuity equation,

$$\frac{\partial N}{\partial t} + (\nabla \cdot \mathbf{J}) = 0.\tag{2}$$

The electric field in the solid is assumed to be quasistatic:

$$\nabla \times \mathbf{E} = 0;$$

$$(\nabla \cdot \epsilon'\mathbf{E}) = \rho/\epsilon_0,\tag{3}$$

where ϵ_0 is dielectric permittivity of free space, ϵ' is relative dielectric constant of the solid, and the density of free electric charge ρ is determined by the deviations of charge carrier concentrations N from their equilibrium value in the solid N^0.

The analysis of Eqs. (1)–(3) is carried out in one dimension with all vectors directed along the x axis. We assume that there is only one sort of mobile charge carrier in each solid: vacancies or interstitials in the ionic crystal, electrons in the metal. We also consider the contact region between the bulk of the ionic crystal and the electrode as a separate part of the circuit. This contact region has the same crystalline structure as the bulk of the ionic crystal; however, it may have different conductivity due to relatively high concentration of defects (introduced when forming electrodes) which can act as electron donors or acceptors. The subscripts i, m, and c refer to the bulk of ionic crystal, metal, and the contact region, respectively. Diffusivities and equilibrium concentrations are assumed to be constant within each of the described parts of the circuit, with the following relationships between them: $D_m \gg D_c$, D_i; $N_m^0 \gg N_c^0$, N_i^0.

The boundary conditions at the interfaces include continuity of the conduction current density $j \equiv qJ$ and the displacement current density $\epsilon_0\epsilon'\partial E/\partial t$. Also, permeability of the metal-ionic crystal interface S for carrier fluxes is taken into account by introducing kinetic coefficients β_c, β_m [7]:

$$j|_S = q_c\beta_c[N_c|_S - N_c^0] - q_m\beta_m[N_m|_S - N_m^0].\tag{4}$$

The boundary condition (4) arises due to the difference in crystalline structure of the solids, which "slows" carrier fluxes.

The analysis of the described model [8] shows that the microwave field induces a quasistationary electromotive force (emf) which is localized at each of the interfaces between the crystal and electrodes. The emf density distribution is given by

$$f = \frac{\epsilon_0 \epsilon'}{4qN^0} \frac{\partial |\mathcal{E}|^2}{\partial x},$$ (5)

where $\mathcal{E}$ is the local amplitude of the microwave field obtained from the solution of the electrodynamic part of the problem. The emf induced on the metal side of the contact is negligible due to high concentration N^0 of free electrons.

Qualitatively, the dynamics of the quasistationary currents can be illustrated in terms of the equivalent circuit shown in Fig. 1. The expression for the emf F is [8]

$$F = \int_{(c)} f(x)dx = \frac{\epsilon_0 E_0^2}{4\epsilon_c' q_c N_c^0}(|A|^2 + 2\mathrm{Re}A),$$ (6)

where the integral is taken over the contact region,

$$A = \frac{i\beta_m/\omega\lambda_m}{(\beta_c\epsilon_c'/\epsilon_m'\omega l_c) + (1/\epsilon_m') - i[(\beta_m/\omega\lambda_m) + (\beta_c\epsilon_c'/\epsilon_m'\omega l_c)]},$$ (7)

$\lambda_m = \sqrt{D_m\epsilon_0\epsilon_m'/\sigma_m}$ is the Debye-Huckel radius in the metal, and $l_c = \sqrt{2D_c/\omega}$ is the characteristic diffusion length in the contact region of the ionic crystal. The amperemeter in the circuit is ideal, and other parameters are defined as usual:

$$R_i = d_i/\sigma_i S, \quad C_i = \epsilon_0\epsilon_i' S/d_i, \quad R_c = d_c/\sigma_c S, \quad C_c = \epsilon_0\epsilon_c' S/d_c,$$ (8)

where d_c, d_i are the lengths of the contact region and of the ionic crystal, respectively ($d_c \ll d_i$), and S is the cross-section area of the crystal. The contributions of the electromotive forces generated at the two contacts are additive due to the superposition principle, therefore we will consider only one of them. According to Eq. (7), the absolute value of the emf F reaches its maximum when $\beta_m/\omega\lambda_m \gg 1$, which yields $A \approx -1$. For simplicity, the results will imply that this condition is fulfilled.

From the analysis of the circuit in Fig. 1 it follows that the dynamics of the current is strongly dependent upon the properties of the contact region. In particular, when $\sigma_c \gg \sigma_i$ (i.e., $R_cC_c \ll R_iC_i$) the measured current decreases with time:

$$I = \frac{F}{R_i} + \frac{F}{R_c}\frac{C_i}{C_c}\exp\left(-\frac{t}{R_cC_c}\right) = \frac{\epsilon_0 SE_0^2}{4\epsilon_c' q_c N_c^0 d_i}\left[\sigma_i + \sigma_c\frac{\epsilon_i'}{\epsilon_c'}\exp\left(-\frac{\sigma_c t}{\epsilon_0\epsilon_c'}\right)\right].$$ (9)

Description in terms of the equivalent circuit is inaccurate in the sense that it does not account for the diffusion term present in Eq. (1). From a more general solution [8] it follows that the current within the contact region exhibits diffusional behavior governed by the equation

$$\frac{\partial j}{\partial t} = D_c\frac{\partial^2 j}{\partial x^2}$$ (10)

before it relaxes exponentially (cf. Eq. (9)). The relationship between the current within the contact region and the measured current, and hence the time dependence of the latter vary with parameters σ_i/σ_c (or R_cC_c/R_iC_i, which is the same) and d_c/d_i. In the limiting case $\sigma_i/\sigma_c \ll 1$, for $t \ll \epsilon_0\epsilon_c'/\sigma_c$ we obtain:

$$I = \frac{\epsilon_0 S E_0^2 \sigma_c}{4\epsilon_c' q_c N_c^0 d_i} \frac{1}{\sqrt{2\pi\omega t}}. \tag{11}$$

At $t \sim \epsilon_0\epsilon_c'/\sigma_c$ the relaxation becomes exponential, described by Eq. (9).

In the case $d_c/d_i \ll \sigma_c/\sigma_i \ll 1$ Eq. (11) is valid only for $t \ll \epsilon_0\epsilon_i'/\sigma_i$, after which the current increases as $\sqrt{t}$ until it reaches the steady-state value at $t \sim \epsilon_0\epsilon_c'/\sigma_c$.

Finally, in the case $\sigma_c/\sigma_i \ll d_c/d_i$ the observed current displays the same time dependence as in the previous case until $t \sim (\epsilon_0\epsilon_i'/\sigma_i)(d_i/d_c)$, but instead of stabilizing at a steady-state value, it consequently experiences another decrease described by

$$I = \frac{\epsilon_0 S E_0^2 \sigma_c}{4\epsilon_c' q_c N_c^0 d_c} \frac{1}{\sqrt{2\pi\omega t}}, \tag{12}$$

$(\epsilon_0\epsilon_i'/\sigma_i)(d_i/d_c) \ll t \ll \epsilon_0\epsilon_c'/\sigma_c$, and reaches its steady-state value at $t \sim \epsilon_0\epsilon_c'/\sigma_c$.

RESULTS AND DISCUSSION

The most important theoretical results can be summarized as follows. When a contact of an ionic crystal with metal is subjected to the microwave field, a quasistationary (dc) emf is induced. It is proportional to the microwave power, i.e. to the square of the E-field strength. This result is consistent with an earlier experimental observation [2].

The magnitude of the emf depends significantly upon the kinetic properties of the crystal-metal interface and frequency. As discussed in [2], the two contacts of the same crystal with the electrodes generally are not identical, therefore the net emf is non-zero. However, at low frequencies, when $\beta_m/\omega\lambda_m \gg 1$ for both contacts (cf. Eq. (7)), the contacts become equivalent, and the net emf disappears, as it was observed in [2].

When the microwave power is pulsed on abruptly, the dc current exhibits a spike with the peak value significantly exceeding the steady-state. Then relaxation of the current occurs until steady state is established. When the power is turned off, there is a similar spike of the oppositely-directed current that is also followed by a relaxation.

The time dependence of the current during relaxation depends upon the properties of the contacts. At times less than $\epsilon_0\epsilon_c'/\sigma_c$ the dynamics of the observed current is determined by diffusional processes within the contact region and may be complicated and not even monotonic. In fact, this sensitivity of the current dynamics to the physical parameters may be exploited to obtain certain information about the structure of the contact region from data on the current dynamics. On longer time scales, the current may stabilize (if $\sigma_c \ll \sigma_i$) or exhibit further decrease governed by space charge relaxation processes within the contact region. In the simplest case (when the dipolar relaxation processes are not taken into account) the current decreases exponentially.

These theoretical predictions are in general agreement with the experimental results. A typical oscilloscope trace of the current (upper trace) and microwave power (lower trace) from the experiments with single crystals of NaCl and AgCl is shown in Fig. 2. The oscilloscope is used in "Average" mode so that the trace in Fig. 2 is actually an average of 16 separate pulsed experiments. This effects some smoothing of the traces and enhances the signal-to-noise ratio.

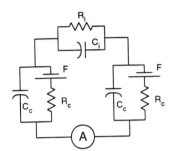

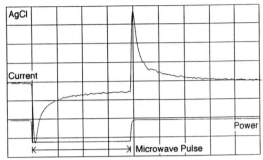

Figure 1: Equivalent circuit used to model the ionic crystal (i) and the contact region (c). F is the emf from the ponderomotive action, and A is the current measurement instrumentation.

Figure 2: Oscilloscope traces of the ionic current (upper trace) in a AgCl crystal at 180 C in response to a pulse of microwave radiation (lower trace). The horizontal scale is 0.5 msec/div. The vertical scale is 500 pA/div for the current trace.

When the microwave pulse is applied, a large spike of negative current results. This is followed by a relaxation of the current during the pulse. Following the pulse, a spike of positive current occurs and then relaxes back to zero current. The full magnitude of the spikes is likely truncated by the slower rise time of the current amplifier electronics. It is not until about 100 μsec after the spike that the data analysis actually begins. The current in NaCl appears to respond slower to the microwave-induced emf.

In Fig. 3, the data on the relaxation of the current in NaCl following the microwave pulse are plotted on a logarithmic scale versus time. It can be seen that the relaxation in NaCl following the microwave pulse is exponential to very good accuracy. This is consistent with the theoretical model that does not account for diffusion effects on the current (cf. Eq. (9)). According to the theory, the time constant of the relaxation should be $\epsilon_0\epsilon_c/\sigma_c$. Calculation of the conductivity in the contact region, σ_c, using the time constant from the experiments ($\sim$ 10 msec) gives a value $2 \cdot 10^{-10}$ (Ohm $\cdot$ cm)$^{-1}$. This is larger than the bulk conductivity of "pure" NaCl at this temperature, which suggests that a higher level of impurities exists in the contact region. Rutherford Backscattering Spectroscopy (RBS) study of crystal ends showed presence of Pt ions within the contact regions. It also confirmed asymmetry of the contacts which is essential for existence of non-zero net current.

The data from the experiments with AgCl are best fit by plotting the current versus $1/\sqrt{t}$ (Fig. 4). Initially, the time dependence of the current demonstrates excellent agreement with Eq. (12), which results from the theoretical model that covers diffusion effects on the current. After 1 msec, the current dynamics begin to deviate from this behavior and asymptote towards a steady state current ($\sim$ 200 pA during the pulse or zero following the pulse), in accordance with the theoretical expectations. The fact that this transition is observed at the millisecond time scale suggests that the conductivity in the contact region is much lower than the bulk conductivity of AgCl crystals. A possible reason for such a decrease in the conductivity may be due to Ag$^+$ interstitial defects recombining with electrons (from semiconduction-type defects) to form neutral species. In the same vein, the Frenkel defect formation equilibrium could be perturbed near the Ag-painted electrode (due to a lack of free surface), resulting in a lower concentration of Ag$^+$ defects in the contact region.

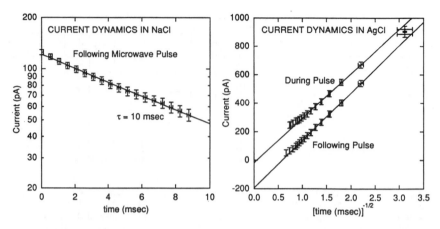

Figure 3: The current dynamics in NaCl following the microwave pulse are best fit by an exponential decay.

Figure 4: The current dynamics in AgCl during and after the microwave pulse show a $t^{-1/2}$ dependence for the first msec before asymptoting towards a steady state current.

REFERENCES

1. S.J. Rothman, in *Microwave Processing of Materials IV*, edited by M.F. Iskander, R.J. Lauf, and W.H. Sutton (Mat. Res. Soc. Proc. **347**, Pittsburgh, PA 1994), p.9-18.
2. S.A. Freeman, J.H. Booske, and R.F. Cooper, Phys. Rev. Lett., **74**, 2042 (1995).
3. S.A. Freeman, J. Booske, and R. Cooper, Rev. Sci. Inst. **66**, 3606 (1995).
4. K.I. Rybakov and V.E. Semenov, Phys. Rev. B, **49**, 64 (1994).
5. K.I. Rybakov and V.E. Semenov, Phys. Rev. B, **52**, 3030 (1995).
6. K.I. Rybakov and V.E. Semenov, in *Solid State Ionics IV*, edited by G.-A. Nazri, J.-M. Tarascon, and M. Schreiber (Mat. Res. Soc. Proc. **369**, Pittsburgh, PA 1995), p. 263 - 268.
7. Y.E. Geguzin, *Fizika spekanija* (Nauka, Moscow, 1984) (in Russian).
8. K.I.Rybakov, V.E.Semenov, S.A.Freeman, J.H.Booske, R.F.Cooper, *to be published*.

ACKNOWLEDGEMENTS

We would like to acknowledge the financial support of the Russian Basic Science Foundation (grant 95-02-05000-A), NASA, the National Science Foundation, and the Electric Power Research Institute.

A Comparison Study on the Densification Behavior and Mechanical Properties of Gelcast Versus Conventionally Formed B₄C Sintered Conventionally and by Microwaves

P.A. Menchhofer*, M.S. Morrow**, J.O. Kiggans*, D.E. Schechter**

* Oak Ridge National Laboratory, Oak Ridge TN., 37831-6087
** Y-12 Development Division, Oak Ridge, TN., 37831-8096

ABSTRACT

This investigation compares microstructures and mechanical properties of both Gelcast B₄C and "conventionally" die-pressed B₄C, using both microwave and conventional sintering methods. The microstructures and final mechanical properties of B₄C specimens are discussed.

INTRODUCTION

The utilization of microwave energy for reaching high temperatures necessary to densify B₄C powder is compared with conventional means of sintering by evaluating the mechanical properties after densification. Microwave energy has been shown to be an effective means for achieving high sintered densities, even though temperatures of ~2250°C are required. In this study, green preforms of B₄C specimens were sintered by both conventional and microwave heating. This study also utilized an advanced forming method called "Gelcasting" developed at ORNL[1]. Gelcasting is a fluid forming process whereby high solids suspensions of powders containing dissolved monomers are cast into a mold, then polymerized or "gelled" in situ.

EXPERIMENTAL

Appropriate amounts of powders were weighed to yield a 50:50 blend of B₄C (#3000 and #1500 powders)[†] + 5wt% (Thermax) carbon powder as a sintering aid[2]. This B₄C mixture was first "washed" in methanol, which removes B₂O₃ as trimethyl borate [3]which significantly improves the powder dispersion. The methanol treatment of boron carbide powder consisted of the placement of powders in a plastic (HDPE) beaker, then filling remaining volume with methanol to fully cover all powder. This was followed by drying in an oven at 60-70° C to evaporate the methanol. This was repeated 3 times for efficacy. Samples of mixed powder were uniaxially cold pressed in stainless steel dies to 15KSI using a light coating of stearic acid as a die lubricant, followed by isostatic pressing at 50KSI.

Gelcasting slurries were prepared using a 15% aqueous solution of methacrylamide[††] : methylbisacrylamide[††] (MAM:MBAM at 2.5:1 respectively). The pH was adjusted to >11 using small additives of tetramethyl ammonium hydroxide (TMAH) [††]. Darvan #7[†††] and PVP K-15[††††] each at 0.5% by weight of the B₄C powder were added as dispersants.

[†] B₄C from E.S.K. Kempten, Germany (#3000 and #1500 powders)
[††] Sigma Chemical Co., St. Louis, MO
[†††] R.T. Vanderbilt Company, Inc., Norwalk, CT.
[††††] GAF Chemical Corp., Wayne, NJ: Polyvinylpyrrolidone K-15

Mat. Res. Soc. Symp. Proc. Vol. 430 © 1996 Materials Research Society

The B_4C powder was then added to the monomer premix (to yield a solids loading of 45% by volume) while stirring on a laboratory mixer until fluid. The production of samples was completed as follows: The plastic (HDPE) mold was prepared by spraying with Polyester Parafilm. The B_4C slurry was degassed on ro-vap laboratory evaporator for ~ 30 minutes at approximately 0°C. (to prevent air entrapment in the finished samples) prior to addition of the catalyst tetramethylethylenediamine[†] (TEMED) first at the rate 0.1 microliter / gram (of slurry) followed by the ammonium persulfate[†] (APS). Note, the pH of the A.P.S. had to be adjusted as follows by: addition of 500 microlitres TMAH/ 25cc of an aqueous 10% APS solution. (which increased pH to ~8.5). This mixture was then stirred for 1-3 minutes on the Ro-Vap while continuing to degas for ~ 2 minutes. The APS was then added at the rate 4 microliters / gram of slurry, with an additional 2 minutes for stirring, before removal and casting. Note, the order of addition was found to be essential to achieve optimal dispersion.

The casting of samples was accomplished simply by pouring the slurry into molds to yield specimens (approximately .5" thick by 2.5"dia), which were then covered and placed in an oven at 70°C. Samples were allowed ~ 60 mins. for gelation. Specimens were removed from the molds and dried at room temperature. They were debindered in argon (to prevent oxidation and the formation of B_2O_3) to 600° C and were sintered by either conventional or microwave heating. Half of the green preforms were sintered conventionally at (2250°C) and half by microwave heating (estimated at 2250°C).

After sintering, the densities of samples were measured by the Archimedes method. The sintered samples were then diamond machined and polished. Hardness and toughness measurements were calculated from micro-indents made on the polished samples. Polished samples were then etched using an alkaline electrolytic process and photographed on a Leica Reichert MEF4-A microscope at 1000X using differential interference contrast to enhance the microstructural features.

RESULTS

For the B_4C materials sintered by microwave heating, the die-pressed and the Gelcast materials had comparable densities, with the Gelcast material being slightly higher at 94% of theoretical. The hardness and toughness values were also comparable. Further sintering studies are being conducted and improved methods for temperature measurement in the microwave cavity are being explored. For specimens sintered conventionally, those produced by gelcasting exhibit improvements in the final properties when compared to samples formed by "conventional" methods. For the B4C samples processed by conventional sintering, the densities were ~8% higher for those formed by Gelcasting (95% of theoretical) as compared to die-pressed-isopressed samples, (only 87 % of theoretical density). The higher densities of the sintered gelcast materials resulted in improved mechanical properties.

Densities of the sintered materials are compared in figure 1. Note for all graphs in figures1-4: the specimens A2-A4 were sintered by microwaves, the specimens B1-B2 were sintered conventionally). Hardness and toughness measurements are compared in figure 2. Correlations of density, hardness and toughness are shown in figure 3.

Typical microstructures are presented in figure 4. In the microwave sintering experiments, it was apparent that the expected sintering temperature of 2250°C was exceeded, as evidenced by excessive grain growth and crystallographic twinning of some grains[4] (specimens A2,A3,A4 in figure 4). After sintering, the average grain size of samples fabricated by Gelcasting was ~ 5 μm compared to ~10 μm for the microwave sintered material. For the specimens sintered conventionally, the Gelcast specimen (B1 in figure 4) exhibited finer grain size, a more uniform microstructure, and improved mechanical properties.

[†] Sigma Chemical Co., St. Louis, MO.

Fig. 1

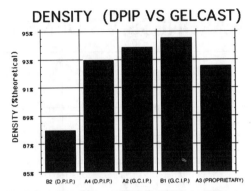

Fig. 1. Density values of die-pressed vs. Gelcast specimens are compared.

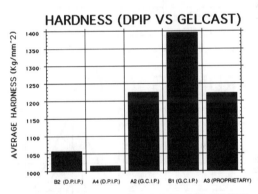

Fig. 2

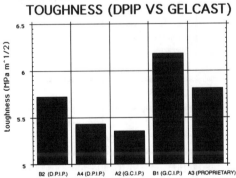

Fig. 2. Hardness and toughness measurements of die-pressed vs. Gelcast specimens are compared.

HARDNESS VS DENSITY

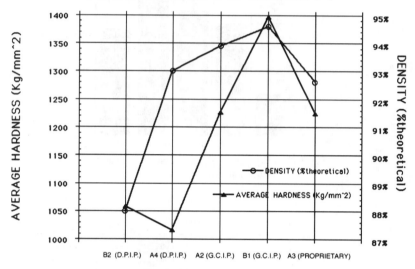

HARDNESS VS TOUGHNESS

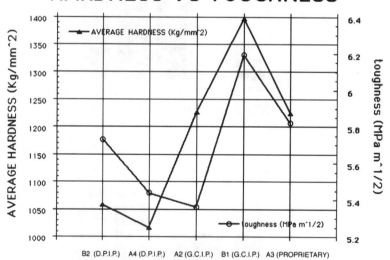

Fig. 3. Correlations of density, hardness and toughness values of die-pressed vs. Gelcast specimens are compared.

Fig. 3

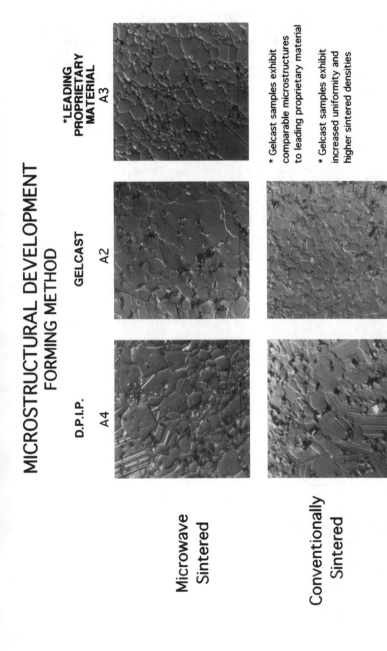

Fig. 4. Microstructures of die-pressed vs. Gelcast specimens are compared.

Fig. 4

469

CONCLUSIONS

Although the expected sintering temperature of 2250°C was beleived to be exceeded, the densification of B4C by microwave energy is an effective method for densification. As further studies are conducted and improved methods for temperature measurement in the microwave cavity are resolved, it is expected that improvements in mechanical properties will follow. For the comparison of Gelcast vs die-pressing forming processes, specimen formed by Gelcasting exhibited finer grain size, more uniform microstructures, and improved mechanical properties.

REFERENCES

1.   Young, A. C., O.O. Omatete, M. A. Janney, and P.A. Menchhofer "Gelcasting of Alumina" J. Am. Ceram. Soc., 74[3] 612-18(1991)

2.   S.L. Dole and S. Prochazka "Densification and Microstructure Development in Boron Carbide" Ceram. Eng. Sci. Proc., 6 [7-8] (1985)

3.   P.D.Williams and D.D. Hawn, "Aqueous Dispersion and Slip Casting of Boron Carbide Powder: Effect of ph and Oxygen Content", J. Am. Ceram. Soc.,74 [7] 1614-18 (1991)

4.   S.L. Dole and S. Prochazka and C. I. Hejna "Abnormal Grain Growth and Microcracking in Boron Carbide" J. Am. Ceram. Soc., 68 [9] C-235-C-236 (1985)

ACKNOWLEDGMENTS

Work supported by the U.S. Dept. of Energy under contract DE-ACO5-96OR22464 with Lockheed Martin Energy Systems, Inc.

EFFECT OF MICROWAVE TREATMENT ON THE BULK AND NEAR-CONTACT PROPERTIES OF SEMICONDUCTORS OF TECHNICAL IMPORTANCE

A.A. BELYAEV*, A.E. BELYAEV*, I.B. ERMOLOVICH*,
S.M.KOMIRENKO*, R.V. KONAKOVA*, V.G. LYAPIN*, V.V. MILENIN*,
Yu.A. TKHORIK*, M.V. SHEVELEV**
*Institute of Semiconductor Physics, National Academy of Sciences of Ukraine,
45 prospect Nauki, 252650 Kiev-28, Ukraine
**E.O. Paton Electric Welding Institute, National Academy of Sciences of
Ukraine, 11 Bozhenko St., 252650 Kiev-5, Ukraine

ABSTRACT

We investigated the effect of microwave radiation on the physico-chemical properties of some semiconductor materials ($Cd_xHg_{1-x}Te$, GaAs) and metal (Al, Pt, Cr, Mo, W) - GaAs Schottky barrier diode structures. From comparison with the results of the other treatments (heat annealing, g-irradiation) some suggestions are made concerning possible mechanisms of the interaction between microwave radiation and the objects studied.

INTRODUCTION

Methods of processing of semiconductor materials and device structures by high-power (non)coherent radiation are among the promising lines of microelectronic technologies. They enable one to purposefully change the composition of the materials processed, as well as their structural and phase states. Among the merits of these methods are also the possibilities of (i) selective action on multiphase structures, (ii) realization of short-term treatments, (iii) control over the active areas and dosing energy supply to these areas. All the above features arouse interest in studying interaction between the microwave radiation in the frequency range 10^9 to 10^{11} Hz (which has not been adequately investigated) and various semiconductor materials and heterostructures.

Up to now only the destructive aspects of high-power irradiation of semiconductor devices have received primary emphasis. A deep insight into the nature of the physico-chemical processes occurring in irradiated objects, as well as relating the physical and chemical properties of semiconductor materials and heterostructures to the parameters of microwave processing, seem to be of use for (i) improving the resistance of microelectronic devices to high-power microwave irradiation and (ii) technological applications of the microwave radiation in the above range.

Some positive potentialities of microwave processing of semiconductors are discussed below.

Mat. Res. Soc. Symp. Proc. Vol. 430 ©1996 Materials Research Society

EXPERIMENT

We took, as the objects of investigation, (i) $Cd_xHg_{1-x}Te$ ($x \sim 0.2$) single crystals, single crystal wafers of GaAs doped with Sn or Te (the electron concentration n_e changing from 10^{16} to 2×10^{17} cm^{-3}) and (ii) metal - GaAs Schottky barrier diode structures fabricated using layer-by-layer electron beam sputtering of metals (Al, Pt, Cr, Mo or W) onto the chemically cleansed (100) face of the n - n^+ - GaAs structures. The metal layers (80 to 100 nm thick) were deposited in a unified processing circuit (pressure $p = 10^{-4}$ Pa).

The samples studied were exposed to microwave processing (in a cm wavelength range). The irradiation condition corresponded to that in a free space. The magnetron output power W was 5 kW, and the intensity of microwave irradiation was varied by changing the distance between the samples studied and the waveguide window. (It should be noted that greater irradiances were needed to change the near-contact characteristics, as compared with those needed to change the bulk properties.)

In our investigations we have used a complex of experimental methods, including Auger electron spectroscopy, measurements of photoluminescence, photoconductivity kinetics, I - V and C - V curves, galvanomagnetic measurements. This enabled us to get information on the composition and electrophysical properties of the samples studied.

RESULTS AND DISCUSSION

Bulk Properties

Microwave processing resulted in changing the defect structure of both semiconductor materials - $Cd_xHg_{1-x}Te$ and GaAs. The corresponding changes depended on the doping level and the concentration of intrinsic defects. A pronounced dependence of these changes on the absorbed dose of microwave energy was also found.

Given in Table I are some parameters of $Cd_xHg_{1-x}Te$ samples measured before and after microwave processing and plastic deformation. One can see that for the sample 1 which contained a large amount of extended defects (such as small angle boundaries, dislocations, etc) the parameters measured changed drastically due to microwave processing. On the other hand, for the more perfect sample 2 the parameters changed rather slightly. (Note that for this sample the trends of both electron concentration n_e and minority carrier lifetime t_p changes were opposite to those in the previous case.) From the temperature dependence of the minority carrier lifetime t_p before and after microwave processing it was found that the concentration of recombination centers, N_r, has dropped to one third of its initial value. One should also note that microwave processing has made the samples more resistant to plastic deformation - compare the parameter changes for the sample 3 (plastic deformation after microwave processing) with those for the sample 4 (plastic deformation of the initial sample).

Let us consider possible mechanisms of the effect of microwave irradiation on the defect structure of the crystals studied. The sample heating during the microwave processing stimulates the diffusion of recombination centers (both

Table I. Some parameters (electron concentration n_e and mobility m, minority carrier lifetime t_p, energy position of recombination centers E_t below the bottom of the conduction band and their concentration N_t) measured for the initial $Cd_xHg_{1-x}Te$ samples (IS) at 77 K and after microwave processing (MP) or plastic deformation (PD).

Samp-le, No	Composi-tion, x	Type of treatment	$n_e \times 10^{-14}$, cm^{-3}	$\mu \times 10^{-4}$, $cm^2/V \times s$	$\tau_p \times 10^6$, s	E_t, eV	$N_t \times 10^{-13}$, cm^{-3}
1	0.24	IS	69	2	0.61		
		MP	254	0.6	0.25		
2	0.215	IS	5.1	20	1.4	0.1	11
		MP	4.9	15	2.4	0.1	3.6
3	0.22	IS	4.8	14	2.1	0.07	6.25
		MP	4.6	9.6	3.0	0.07	2.5
		MP+PD	4.6	8.1	2.0	0.07	7.2
4	0.22	IS	1.68	3.84	1.57		
		PD	26	0.11	0.33		

intrinsic point defects and impurities). The stable extended defects (big clusters of point defects, small angle boundaries, dislocations) can serve as the drains for the diffusing recombination centers. (Some unstable defects which were "frozen" at the room temperature can dissociate during microwave heating, and their components may also diffuse to the above drains.) Such a gettering can result in t_p increase.

Rather slight changes in the sample 3 parameters after plastic deformation (which was preceded by microwave processing) can be also understood within such a notion. On the one hand, the creation of dislocations in this case occurs in a material containing lesser amounts of various point defects than before the microwave processing. This fact reduces the probability of point defect clustering around the newly formed dislocations. But, on the other hand, presence of large amounts of various extended impurities with gettered point defects around them prevents the motion of the newly formed dislocations.

We have studied also the photoluminescence spectra of GaAs samples at 77 K. For the initial GaAs : Sn samples two overlapping bands were observed (see Fig.1): the high-energy (maximum at $hn = 1.15 - 1.20$ eV for different samples) and the low-energy (maximum at $hn = 0.993 - 1.01$ eV) ones. The high-energy peak was 2 to 5 times higher than the low-energy one. After microwave processing (irradiance 90 W/cm^2, duration $t = 6$ s) the spreads of peak positions disappeared. The new high- and low-energy band peak positions were 1.185 and 1.01 eV, respectively. By this a predominant growth of the high-energy band peak occurred. The narrowing of both bands indicates that the sample structure became more ordered. This may result from the mobility growth for the line, as well as point, defects followed by gettering of the last ones.

The long-term ($t \geq 60$ s) microwave processing led to the intensity drop for both high- and low-energy bands of photoluminescence. This fact may be due to enhancement of the degradation processes in the GaAs material.

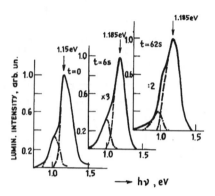

Fig.1. GaAs : Sn photoluminescence spectra: before ($t = 0$) and after microwave processing ($t = 6$ and $t = 62$ s).

For GaAs : Te samples we observed a single band of photoluminescence with the peak at $hn = 1.20$ eV. Its parameters did not change even under a rather long-term (up to $t = 60$ s) microwave processing. So this material (GaAs : Te) is more resistant to the microwave radiation than GaAs : Sn.

Near-Contact Properties

Given in Table II are some electrophysical parameters of the Schottky barrier diode structures exposed to various treatments - microwave processing, heat annealing and g-irradiation. These treatments acted in different manners on both atomic and electronic subsystems of the heterostructures studied.

From the data given in Table II one can conclude that microwave treatment resulted in both structural and chemical transformations of the interfaces. These transformations gave rise to the changes in such parameters as ideality factor n and minority carrier diffusion length L_p. Following is brief discussion concerning possible mechanisms that may be responsible for the changes observed.

The thermal mechanism is connected with contact heating due to the absorption of microwave energy. Analysis of the concentration depth profiles for the heterostructure components taken before and after microwave processing (see Fig.2), as well as their correlation with the corresponding profiles in the case of heat annealing, indicate that the role of heating up is not important.

The electrostatic mechanism is due to the voltage drop across the barrier junction. This voltage drop may lead to electric fields which (even being below the critical values that determine avalanche or tunnel breakdowns) can substantially affect the diffusive redistribution of the contact components - see [1]. An intense metal - GaAs interdiffusion starts at the absorbed energy values about 2/3 the critical value needed for breakdown to occur.

The electrodynamic mechanism stems from the electronic subsystem departure from equilibrium. In this case (see [2]) even a slight electron temperature rise can

Table II. Some parameters (barrier height j_B, ideality factor n, reverse current density j_R at bias $V = 5$ V, minority carrier diffusion length L_p) measured for the metal - GaAs Schottky barrier structures before (IS) and after different treatments. MP - microwave processing (frequency $f = 10^{10}$ Hz, magnetron output power $W = 5$ kW, duration $t = 1$ s); HA - heat annealing at 573 K during $t = 10$ h (for Mo - GaAs and W - GaAs heterostructures $t = 8$ h); GI - g-irradiation (^{60}Co), dose absorbed $D = 10^5$ Gy.

Heterost-ructures	Type of treatment	φ_B, eV	n	$j_R \times 10^6$, A/cm²	L_P, mcm
Al - GaAs	IS	0.55 - 0.58	1.68 - 2.20	1.2 - 3.0	1.6 - 1.9
	MP	0.57 - 0.58	1.30 - 1.40	0.1	2.0
	HA	0.51 - 0.55	1.90 - 2.30	20 - 50	0.7 - 0.8
Pt - GaAs	IS	0.88 - 0.95	1.12 - 1.37	0.01 - 0.06	2.1 - 2.2
	MP	0.88 - 0.89	1.18 - 1.24	0.05 - 0.06	2.1 - 2.2
	HA	0.53 - 0.57	2.37 - 2.80	6000	0.5 - 0.7
	GI	0.92 - 0.95	1.08 - 1.10	0.01 - 0.015	2.5 - 2.8
Au - Pt - GaAs	IS	0.88 - 0.95	1.17 - 1.30	0.09	2.0 - 2.2
	MP	0.90 - 0.92	1.12 - 1.15	0.01	2.3 - 2.5
	HA	0.51 - 0.60	2.20 - 2.70	8000 - 9000	0.55
	GI	0.93 - 0.96	1.10 - 1.20	0.01	2.4 - 2.6
Cr - GaAs	IS	0.73 - 0.75	1.17 - 1.24	0.04 - 0.06	0.5 - 0.7
	MP	0.76 - 0.77	1.08 - 1.09	0.01	1.2 - 1.4
	HA	0.73 - 0.75	1.20 - 1.24	20 - 25	0.5 - 0.6
	GI	0.75 - 0.77	1.08 - 1.10	0.01 - 0.012	1.2 - 1.6
Au - Cr - GaAs	IS	0.70 - 0.76	1.12 - 1.17	0.1 - 0.6	0.9 - 1.1
	MP	0.75 - 0.77	1.04 - 1.08	0.12 - 0.14	1.0 - 1.3
	HA	0.56 - 0.62	1.80 - 2.10	5000 - 7000	0.12
	GI	0.74 - 0.76	1.07 - 1.10	0.05 - 0.1	1.2 - 1.5
Mo - GaAs	IS	0.68 - 0.69	1.16 - 1.23	2 - 2.5	2.3 - 2.8
	MP	0.68 - 0.69	1.09 - 1.14	0.2 - 0.4	2.5 - 2.7
	HA	0.68 - 0.69	1.10 - 1.12	10	0.2 - 0.4
	GI	0.68 - 0.71	1.10 - 1.12	0.1	2.6 - 2.8
W - GaAs	IS	0.65 - 0.66	1.20 - 1.40	1 - 2	1.7 - 2.0
	MP	0.69 - 0.70	1.09 - 1.12	0.1 - 0.2	2.1 - 2.2
	HA	0.65 - 0.66	1.10 - 1.17	10 - 20	0.6 - 0.9
	GI	0.69 - 0.72	1.08 - 1.17	0.1 - 0.35	2.1 - 2.4

drastically affect the impurity-defect composition of the semiconductor near-surface layer. For g-radiation treatment this mechanism is decisive. For some metal - GaAs structures the electrophysical parameters after microwave processing are very close to those after g-irradiation. So one may conclude that in this case the electrodynamic mechanism is decisive in structural relaxation under microwave processing.

One more factor may lead to structural ordering of the interfaces. It is connected with appearance of the nonstationary elastic strain gradients due to the practically instantaneous heating up of the disordered semiconductor near-boundary regions. (Such a disordering occurs during the contact formation.) In this case both sets (before and after microwave processing) of the concentration depth profiles of the contact components may be practically the same. Taking into account the interactions in the rather small regions of defect clustering (where the elastic strain fields exist), one can substantially reduce the potential barriers and thus achieve transformation of such regions [3].

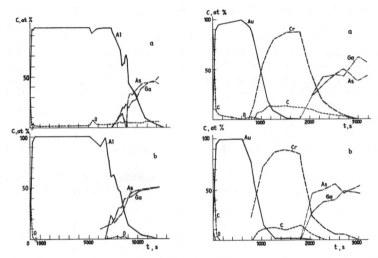

Fig.2. Concentration depth profiles for different components in the Al - GaAs and Au - Cr - GaAs structures before (a) and after (b) microwave processing (t is time of ion etching).

CONCLUSIONS

We have found a number of changes in the physico-chemical properties of various semiconductor materials and metal - GaAs heterostructures, resulting from microwave processing of these objects. This indicates at the definite potentialities of using such a processing in the microelectronic technology. Still further investigations are needed to elucidate the mechanisms of the physical and chemical processes occurring in the above objects under the action of microwave irradiation.

ACKNOWLEDGMENT

This work was partially supported by the Ukrainian State Committee for Science and Technologies.

REFERENCES

1. V.V. Antipin, V.A. Godovitsyn, D.V. Gromov, A.S. Kozhevnikov and A.A. Ravaev, Zarubezhnaya Radioelektronika 1995 (1), 37.

2. I.K. Sinishchuk, G.E. Chaika and F.S. Shishiyanu, Fiz. Tekh. Poluprovodn. 19, 674 (1985).

3. V.D. Skupov and D.I. Tetelbaum, Fiz. Tekh. Poluprovodn. 21, 1495 (1987).

THE MAGNETIC FIELD SENSOR ON THE BASE OF YBaCuO CERAMICS WITH SHARP JUMP OF MAGNETIC REPLY

Kh.R. ROSTAMI , A.A. SUKHANOV, and V.V. MANTOROV.
Institute of Radioengineering and Electronics RAS. Mokhovaya st. 11, Moscow, 103907, Russia, e-mail: aas195@ire216.msk.su.

ABSTRACT

The magnetic field sensor with sensitivity 10^{-3} Oe and work temperature 77^0K is created on the base of YBaCuO ceramics having a sharp jump of the diamagnetic moment in the region of the first Josephson critical magnetic field. The magnetic field measurement methods and the sensor principle are described. Ways to improve the obtained results are indicated.

INTRODUCTION

The use of high-temperature superconductor (HTSC) for applied purposes represents great interest in connection with the superiority of numerous parameters (sensibility, precision, rapidity, little volumes, economy, etc.) of the devices created on their base over parameters of devices created on the base of other known materials [1]. In this work, the possibility of HTSC use for creation of a simple sensitive magnetic field sensor is demonstrated.

EXPERIMENTAL PROCEDURE AND RESULTS.

The work principle of the proposed magnetic field sensor and the magnetometer created on its base consists in registration of diamagnetic moment M response of a HTSC cell on a measured continuous magnetic field or use of a magnetic flux trapping effect for registration of peak value of the measured magnetic field. As the sensible element, we used YBaCuO ceramics synthesized by us that have a sharp increase in magnetic reply in the region of the first Josephson critical magnetic field H_{c1}^J (Fig. 1). The technology of such ceramics synthesis is described in detail earlier [2]. In Fig. 1, the dependence of the diamagnetic moment M of $YBa_2Cu_3O_{7-x}$ ceramics on the external magnetic field H_0 is presented.

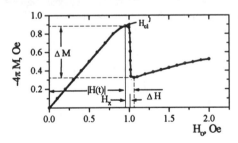

Fig. 1. The diagram of sensor work. M - diamagnetic moment, H_0 - external magnetic field, H_{c1}^J - the first Josephson critical magnetic field, H(t) - supporting magnetic field, H_x - measured magnetic field, ΔH - transition width, ΔM - the diamagnetic moment jump.

Mat. Res. Soc. Symp. Proc. Vol. 430 © 1996 Materials Research Society

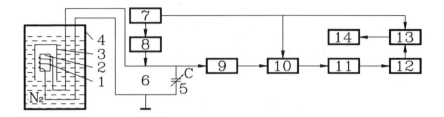

Fig. 2. Block-scheme of magnetometer. 1 - induction coil, 2 - HTSC core, 3 - HTSC shield, 4 - cryogenic block, 5 - capacitor, 6 - oscillatory contour, 7 - regulated current generator, 8, 9 - high-harmonics filters, 10 - synchronic detector, 11 - voltage amplifier, 12 - voltage to frequency transformer, 13 - subtractor of frequency, 14 - frequency measurer.

The magnetometer block-scheme is presented in Fig. 2. A detailed description of its work can be found in [3]. The main part of magnetometer - the sensor consists of an induction coil (1) within which the core of the HTSC (2) is placed. The coil and core are placed inside the HTSC magnetic shield (3) which thermally contacts with the copper rod. The heater is wound bifilarly on the other end of the rod (this part of the scheme is not shown in Fig. 2). With the help of a cryogenic block (4), the temperature around the HTSC core is regulated in the range 4.2 - 300⁰K.

The method of magnetic field registration is as follows: The HTSC core is brought into superconducting (SC) state at zero magnetic field. Then, with the help of the coil around the HTSC core, the support-measuring field H(t) is created and the diamagnetic moment of the HTSC core (2) is measured. The magnitude of the measuring field H(t) is chosen from the condition $(H_{c1}{}^J - |H(t)|)/ H_{c1}{}^J \ll 1$. Later on, with the help of the heater, the magnetic shield (3) escapes from the SC state (by this, the access of the measured magnetic field to the HTSC core (2) is ensured). Next, the diamagnetic moment of the core (2) is measured once more. The magnitude of the measured magnetic field is determined from the difference between the magnitudes of the diamagnetic moments δM. A HTSC rod of coarse-grained $YBa_2Cu_3O_{7-x}$ ceramics was used as the sensor core. It had the critical temperature of SC transition $T_c = 92$-93.5⁰K. For 77⁰K, the HTSC core had the sharp jump of the diamagnetic moment $\Delta M \approx 0.56$ Oe per unite of volume at $H_{c1}{}^J \approx 1.1$ Oe and $\Delta H \approx 10^{-2}$ Oe. The magnitude and frequency of the sinusoidal support-measuring field were $|H(t)| = 1.099$ Oe and f = 10 kHz. With the help of this sensor, the sensibility to the measured magnetic field was not worse than 10^{-3} Oe and a precision of measurement of up to 0.01% was reached. The use of the HTSC core with a sharper jump of ΔM, a lower magnitude of $H_{c1}{}^J$, and a maintenance of work temperature near T_c can increase the magnetometer sensibility by at least one order of magnitude.

REFERENCES

1. J.E.C.Williams. Superconductivity and its application. Pion Ltd. London. 1970.

2. Kh.R.Rostami, A.A.Sukhanov, V.V.Mantorov. Low Temp. Phys. 22, 42, (1996).

3. Kh.R.Rostami. Patent Application No 95106569/07(011699) from 19.04.95.

Part VIII

Microwave Processing Using Variable Frequency Sources

SPREAD SPECTRUM MICROWAVE POWER GENERATION
FOR IMPROVING EFFICIENCY

W.M. VAN LOOCK, University of Ghent, B-9000 Ghent, Belgium, vanloock@intec.rug.ac.be

ABSTRACT

Microwave power for heating applications is normally generated in the designated ISM frequency bands which occupy a band of 4%. Actual microwave generators, such as are used in domestic ovens utilise only a small fraction of this bandwidth. It is being demonstrated that spreading the power uniformly over the full ISM band by controlled frequency modulation dramatically reduces all levels of potential electromagnetic interference. With such controlled modulation telecommunication channels can operate within the ISM bands without serious problems because the leakage levels are reduced by 20 to 30 dB with no additional shielding costs. One simple (though not optimum) modulating waveform is a large ripple voltage on the magnetron power supply. Frequency modulation that spreads the energy over the full ISM band also improves the overall energy efficiency in multimode heating applications.

INTRODUCTION

Telecommunication and information technology is conquering the whole spectrum of electromagnetic energy. This is creating the need to control all radiated energy emissions in order to prevent interferences between systems operating close to one another. Hence, microwave energy for heating applications such as for the processing of materials has to be generated in shielded systems and the energy spilled into space must be limited to the emission levels set and enforced by the Authorities. At 400 MHz a typical limit may be 40 dBmV/m at 10 metres distance [1]. This level is equivalent to a total leakage of 33 nW. To illustrate this, suppose we have a housing of a 1 kW system for ISM (industrial, scientific, medical and domestic applications of electromagnetic energy) at that frequency. It must have a shielding efficiency of roughly 105 dB to comply with the regulation. This creates a severe problem, both economically and technically. Therefore, it is better to generate microwave energy for heating within the designated ISM frequency bands, avoiding all other frequencies, even as by-products. Unlimited spilling of energy into space is normally allowed in these bands but some countries have already set limits on the emissions out of concern for human safety.

Actual microwave generation within the designated ISM bands is normally performed in small bands which are typically 3 MHz wide and are mostly located around the centre frequency. Some frequency bands are unused, in particular near the microwave ISM band edges, and telecommunication channels are in fact already operating in the ISM bands.

Some countries are using the 915 +/-13 MHz ISM band for telecommunication services (TACS, GSM, etc.) and these countries do not allow 915 MHz for power applications. The 2.45 +/- .05 GHz is designated world-wide for ISM but the World Administrative Radio Conference (1992) has allocated bands within this 2.4 - 2.5 GHz band for telecommunication purposes (RLAN, LPD, etc.), and many other services are planned.

We propose microwave power generation with tightly controlled frequency modulation (FM) so that its resulting frequency spectrum spreads the energy evenly over the full designated ISM band, avoiding a peak in a narrow band. Spreading the power uniformly over the ISM band when generating microwaves has many advantages. It drastically reduces all radiated emission levels and telecommunications equipment can operate within the ISM bands. Controlled

481

spreading of the energy over the designated ISM band instead of small band operation improves heating uniformity and increases overall energy efficiency.

THEORY

Frequency modulation is well known in telecommunications. It is defined as that type of modulation in which the instantaneous frequency of the wave to be modulated is equal to the constant frequency of the wave plus a time varying component that is proportional to the magnitude of the modulating waveform [2]. A simple example applied to microwaves at an operating frequency f_0 is frequency modulation of the sinusoidal electric field E(t) expressed as

$$E(t) = E \sin \tau(t) \tag{1}$$

where the phase angle $\tau(t)$ is varied by frequency modulation. The instantaneous frequency of this electric field is given by the time derivative of the phase angle, which is f_0 for the non-modulated microwaves. A frequency modulating waveform V(t) will give:

$$2p\, f_0 + k\, V(t) \tag{2}$$

with k being a proportionality constant. The peak frequency deviation Δf_0 represents the plus or minus peak variation from the frequency f_0 and, when V(t) is symmetrical, is given by

$$\Delta f_0 = \frac{1}{2\pi} k[V(t)]_{max} \tag{3}$$

A simple modulating waveform would be

$$V \sin 2\pi f_m t \tag{4}$$

The peak frequency deviation given by (3) can be rewritten as

$$\Delta f_0 = m.f_m; \qquad m = \frac{kV}{2\pi.f_m} \tag{5}$$

with m representing the modulation index.
As a result of frequency modulation with a wave of frequency f_m, many side frequencies, each with a spacing of f_m, are generated above and below the operating frequency f_0 which is being modulated. The frequency spectrum produced by FM with a sinusoidal waveform is easily obtained with standard Fourier analysis. The amplitudes of the side frequencies can be expressed in terms of Bessel functions $J_n(m)$, of the first kind and order n. The details can be found in most textbooks. However, determining analytical expressions for FM with high values of m and with more complex waveforms is more difficult. The side frequencies in the resulting spectrum are always infinite in number. The minimum and maximum instantaneous frequencies, f_{min} and f_{max} respectively, can be determined from (2) and (5):

$$f_{min} = f_0 - mf_m, \qquad f_{max} = f_0 + mf_m \tag{6}$$

It has been demonstrated [2] that most of the power is concentrated within a bandwidth BW which is somewhat more than twice the maximum frequency deviation m:

$$BW \cong 2(m + 1)fm \cong 2\Delta f_0 \qquad (7)$$

provided the modulation index m is large.

APPLICATIONS

The key for spreading the energy uniformly over the ISM band is the modulating waveform. This waveform should be derived cost-effectively from the line frequency (50 or 60 Hz) and its harmonics.
The bandwidth for 915 MHz is 26 MHz. For safety reasons we will use only 20 MHz. The modulation index for a 50 Hz modulating waveform is then roughly m = 200,000. This is an enormous index value compared with the indexes used for FM telecommunications. The resulting spectrum, which consists of lines every 50 Hz, is difficult to calculate; special techniques have to be used.

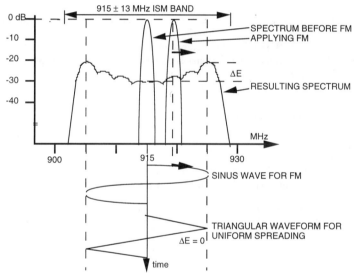

Fig. 1: FM with a sine wave for spreading the energy over a large part of the ISM microwave band. A triangular waveform will give a uniform spreading. Spreading of the energy over the band considerably reduces the peak level when generating microwaves for heating applications.

Fig. 1 shows the spectrum before and after applying frequency modulation with a 50 Hz sine wave.
The energy in the spectrum of the modulated waveform will tend to concentrate at those frequencies corresponding to points where the time derivative of the modulating waveform is small. This leads to a non uniformity. An optimum modulating waveform would give a flat spectrum. A triangular waveform, for example, will uniformly distribute all energy over the largest part of the band and ΔE is zero. In practice, other optimum waveforms are needed to counteract problems such as amplitude modulation due to the magnetron characteristics.

In the time domain, FM means that the microwave frequency is simply swept back and forth at the rate of the line frequency, as is shown in Fig. 1. The sweeping should be limited within the ISM band. The resulting spectrum must stay within the band limits.

A simple form of FM for generating microwaves can be obtained by allowing a sufficient ripple. The voltage supply for high power 915 MHz magnetrons comes normally from bridge rectifiers, and typically from a 3-phase bridge rectifier that produces a ripple of 4% at 6 times the line frequency. A ripple on the voltage supply will frequency modulate the microwave power. A ripple on the current of the electromagnet of the magnetron, reflections from the load and other phenomena will also FM the microwave power. In a first approximation, these phenomena are considered to be a resulting ripple which frequency modulates the output power between f_{min} and f_{max}. It can be shown that the power P_x which is then spilled in a frequency channel Δf_x with centre frequency f_x is given by

$$P_x = \frac{0.26\Delta f_x}{f_{max} - f_{min}}[1 - (0.87 + 0.13\frac{f_x - f_{min}}{f_{max} - f_{min}})^2]^{-1/2}P \qquad (8)$$

with P being the total microwave power. The telecommunication channels around 915 MHz usually have channels of 30 kHz. Equation (8) shows that the leakage levels into these channels are reduced by 20 to 30 dB, provided the ripple allows a sweeping over 20 MHz.

The corresponding interference reduction has been verified for cordless telephones operating at 915 MHz [3], [4]. The design of a power supply for microwave generation must therefore aim to produce a large ripple, which produces FM, so that the resulting spectrum covers practically the whole ISM band. A ripple waveform generated by diode rectification is not the optimum waveform but it is more easy to implement.

2.45 GHz

Most magnetrons at 2.45 GHz, the frequency of the domestic oven, operate pulse-wise. The magnetron is then switched on and off at the rate of the line frequency, which results in a dirty spectrum.

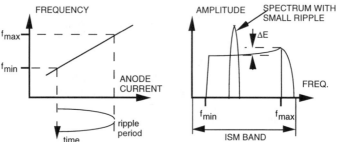

Fig. 2: FM with a sufficient ripple amplitude and the resulting spectrum.

Telecommunication channels above 2 GHz typically use a bandwidth of 100 kHz. A 20 to 30 dB reduction of the potential interfering levels can be obtained with an equation similar to (8) for a ripple resulting from a two-bridge rectification [5]. See Figure 2 for a typical result. In order to counteracting amplitude modulation, it may be necessary, for example, to have an inverted ripple to achieve a 3/4 coverage of the ISM band. Otherwise the power concentrates in the upper end of the ISM band.

Spreading the energy over the designated band also makes it possible to use telecommunication channels within the ISM band without expensive shielding of the microwave power system.

ENERGY EFFICIENCY

Improvement of the overall heating efficiency is possible in multimode microwave power systems. This can be demonstrated in its application to the domestic oven cavity with dimensions a, b and c. The choice of a, b and c should be such that there are as many resonances as possible within the band 2.4 - 2.5 GHz. However, if the magnetron generator does not provide the frequencies of these resonances then the modes will not be excited, except when the quality factor of the mode is low and there is coupling.

The apparent quality factor Q at 2.45 GHz centre frequency and 100 MHz bandwidth can be calculated as $Q = 2,450/100 = 24.5$.

At optimum loading Q must be lower than 24.5. However at low loading, for example with 100 ml water instead 1 litre, Q increases and the efficiency decreases.

The overall efficiency of the oven with a conventional pulsed power supply decreases by 10 to 20% when the load decreases. Full band ISM operation will then improve the overall efficiency for smaller loads [6].

CONCLUSIONS

Microwave energy for heating applications such as for the processing of materials should be generated with proper frequency modulation for using the broader part of the designated ISM band.

This FM is easy to implement, as for example by allowing a sufficient ripple voltage on the anode supply of the magnetron. The resulting spreading of energy reduces the levels of any out of band emissions and telecommunication channels can be used within the ISM band without increasing the costs for shielding.

Full band operation also improves heating uniformity and heating efficiency, in particular with low loads in multimode applications.

Other problems which are solved straightforward manner at the expense of more complicated power supply circuits include magnetron pulling and pushing

REFERENCES

1   European Standard EN 55011, Limits and methods of measurement of radio interference characteristics of industrial, scientific and medical (ISM) radio-frequency equipment, 1991.

2   George E. Happell and Wilgred M. Hesselberth, Engineering electronics, Mc Graw-Hill Book Company, Inc. New York, 1953, p 345-384.

3   W. Van Loock, "EMC of cordless telephones and 900 MHz ISM sources", EMC Conference, Zurich 1989, paper 112 R4.

4   W. Van Loock, "Microwaves at 896/915 MHz as a valuable ISM tool and how to avoid interference", Australia's Electronics Convention Proc. 1991, p 501-504.

5   W. Van Loock, "Broadband ISM operation for the reduction of radiated emissions", Int. Wroclaw Symposium on EMC, June 1996, paper 69/13.

6   A. Mackay B., W.R. Tinga and W.A.G. Voss, "Frequency agile sources for microwave ovens", J. Microwave power, 14 (1), 1979, p. 63-76.

FINITE-DIFFERENCE TIME-DOMAIN SIMULATION OF MICROWAVE SINTERING IN A VARIABLE-FREQUENCY MULTIMODE CAVITY

Mikel J White[1], Steven F. Dillon[1], Magdy F. Iskander[1], and Hal D. Kimrey[2]
[1]Electrical Engineering Department, University of Utah, Salt Lake City, UT 84112
[2]Oak Ridge National Laboratories, Oak Ridge, TN 37831

ABSTRACT

There have been recent indications that variable-frequency microwave sintering of ceramics provides several advantages over single-frequency sintering, including more uniform heating, particularly for larger samples. The Finite-Difference Time-Domain (FDTD) code at the University of Utah was modified and used to simulate microwave sintering using variable frequencies and was coupled with a heat-transfer code to provide a dynamic simulation of this new microwave sintering process. This paper summarizes results from the FDTD simulations of sintering in a variable-frequency cavity. FDTD simulations were run in 100-MHz steps to account for the frequency variation in the electromagnetic fields in the multimode cavity. It is shown that a variable-frequency system does improve the heating uniformity when the proper frequency range is chosen. Specifically, for a single ceramic sample (4 x 4 x 6 cm^3), and for a variable-frequency range from $f = 2.5$ GHz to $f = 3.2$ GHz, the temperature distribution pattern was much more uniform than the heating pattern achieved when using a single-frequency sintering system at $f = 2.45$ GHz.

INTRODUCTION

Numerical modeling has been used by our group to model microwave sintering in multimode cavities operating at a single-frequency [1-3]. It is observed that the resulting power deposition patterns are often non-uniform. This leads to large temperature gradients, cracking, and the inability to sinter large samples. Recent research has suggested, however, that variable-frequency multimode cavities may be used to improve the uniformity of heating [4]. The FDTD method provides an efficient and cost-effective tool for modeling microwave sintering using variable frequencies, and for examining the reported advantages of this new technique.

In this paper, we report the results of using the FDTD method to simulate sintering in variable-frequency multimode cavities, and compare these results with those obtained when a single-frequency multimode cavity of the same size is used for sintering. The simulation model and results from both the microwave power deposition and heating patterns will be presented.

MODEL DESCRIPTION

A model of a microwave cavity, including the feed waveguide, Alumina shelf, and a ceramic sample, was created using the FDTD code. This was done using a cell size of 0.5 cm. The dimensions of the multimode cavity are 40 x 40 x 30 cm^3, and the resulting FDTD model was 82 x 82 x 80 cells. The dimensions of the sample were 4 x 4 x 6 cm^3. The feed waveguide was positioned to launch a TE$_{10}$ mode in the vertical direction as shown in Figure 1.

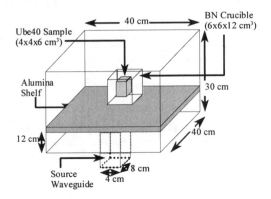

Figure 1, Schematic of FDTD model for microwave sintering experiments

The sample support shelf was Alumina, with material properties $\varepsilon_r' = 8.809$ and $\sigma = 6.7$ x 10^{-3} S/m. The crucible was Boron Nitride with material properties $\varepsilon_r' = 3.459$ and $\sigma = .0406$ x 10^{-3} S/m. Inside the crucible, and surrounding the sample, was foam insulation with material properties $\varepsilon_r' = 1.655$ and $\sigma = .0176$ x 10^{-3} S/m. The sample used in these simulations was modeled as Ube40, which is Si_3N_4 with 40 wt. % SiC sintering aids. The material properties of the Ube40 are $\varepsilon_r' = 6.497$ and $\sigma = 8.25$ x 10^{-3} S/m.

INTEGRATION OF FDTD AND HEAT-TRANSFER CODES

In order to create a dynamic model of heating using microwaves, the FDTD code and the finite-difference heat-transfer code available at the University of Utah were integrated. Approximate mathematical functions were fit to the measured data for ε_r^* of the various materials as a function of temperature. The FDTD code uses the temperature calculated from the heat-transfer code and the fitted approximate mathematical functions to calculate the dielectric properties of each cell as the temperature changes. A flow diagram illustrating the calculation process is shown in Figure 2.

SIMULATION PARAMETERS OF VARIABLE-FREQUENCY MICROWAVE SINTERING

To determine the frequency step size that may be used in simulating variable-frequency sintering, two primary tradeoffs were considered. First, sufficiently large frequency steps are necessary to minimize the number of computational runs. Second, the frequency step size should be sufficiently small to observe and account for variation in field patterns. An initial frequency step of 50 MHz was used to observe field patterns in an empty cavity. After numerous simulations in which the movement of the field patterns in the cavity was observed, it was determined that a step size of 100 MHz was the largest possible frequency step size in order to adequately track the movement of fields throughout the cavity.

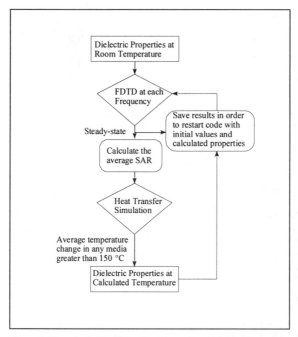

Figure 2, Flow chart of FDTD and heat-transfer code integration

After determining the frequency step size, it was necessary to determine the minimum frequency band required to achieve uniform power deposition in a typical sample. A 4 x 4 x 6 cm^3 sample of Ube40 (Si$_3$N$_4$ + 40 wt. % SiC sintering aid) was modeled and used in the simulations. In previous experiments, samples of this size have proven to be difficult to heat uniformly using a single-frequency source. Simulations were performed from 2.3 GHz to 4 GHz, and the power deposition patterns were observed and averaged over the band. The most uniform power deposition pattern was obtained by averaging (assuming a fast sweep time) microwave power absorption values from 2.5 GHz to 3.2 GHz. This average microwave power was then used in the heat-transfer calculation. Based on this calculation procedure, it is shown that the percent uniformity (average/maximum %) of power deposition at room temperature for the variable-frequency system from 2.5 GHz to 3.2 GHz is 69%. For a typical single-frequency experiment at 2.45 GHz, the percent uniformity is 18%, which is a much lower value. Room-temperature, microwave power deposition patterns for the single-frequency and variable-frequency simulations are shown in Figure 3. The reported improvement in the microwave power deposition pattern may be observed by comparing Figure 3b (variable-frequency simulation) with Figure 3a (single-frequency simulation).

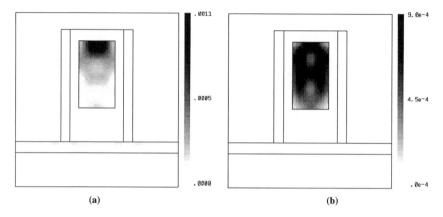

Figure 3, Power deposition patterns for (a) single-frequency (f = 2.5 GHz) and (b) a variable-frequency experiment obtained by averaging the power deposition over the frequency range from f = 2.5 GHz to f = 3.2 GHz

RESULTS

After determining the frequency step size (100 MHz) and the frequency range (2.5 GHz to 3.2 GHz) for a given sample size (4 x 4 x 6 cm^3), the combined FDTD/heat-transfer code was used to simulate dynamic heating of this sample up to the sintering temperature. For purposes of comparison, a single-frequency simulation was performed for the same sample at 2.45 GHz. By observing the simulation results, it was determined that using a variable-frequency system, uniform power deposition may be obtained, and a uniform temperature distribution in the sample may be achieved during the microwave sintering process. It was also determined that uniform temperature distribution was not possible for single-frequency microwave sintering of the 4 x 4 x 6 cm^3 sample. Temperature-distribution pattern results at intermediate and sintering temperatures are shown in Figure 4 for the single-frequency simulation and in Figure 5 for the variable-frequency system.

Simulations were also performed using other sintering arrangements including: samples of different sizes, different starting frequencies, and different overall frequency bands. As may be expected, it is observed that the frequency range needs to be adjusted in order to achieve uniform heating when the sample size is changed. For example, when the same frequency range that was successfully used in the previous arrangement was used in an arrangement in which the sample length was increased from 6 cm to 8 cm, the resulting heating pattern in the longer sample was not as uniform as in the shorter sample case. Also, if multiple samples are included, a large frequency band is necessary to obtain uniform heating.

CONCLUSIONS

The FDTD and Heat-Transfer codes were integrated to provide a dynamic model for simulating microwave sintering using both single- and multiple-frequency systems. For a microwave cavity of dimensions 40 x 40 x 30 cm^3, a frequency step size of 100 MHz was found to be adequate for simulating a variable-frequency sintering system. For the specific sample size

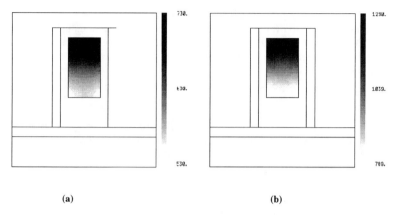

(a)   (b)

Figure 4, Temperature distribution at two different points in the single-frequency microwave sintering simulation, (a) at medium temperature (730 °C) and (b) at the sintering temperature (1300 °C)

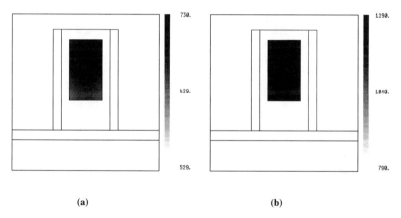

(a)   (b)

Figure 5, Temperature distribution at two different points in the variable-frequency microwave-sintering simulation, (a) at medium temperature (730 °C) and (b) at the sintering temperature (1300 °C)

($4 \times 4 \times 6$ cm^3) and the experimental arrangement used in these simulations, a frequency range from 2.5 GHz to 3.2 GHz provided uniform power deposition and temperature distribution patterns. Although uniform field patterns may be obtained in an empty cavity over a somewhat narrower frequency sweep, the introduction of ceramic samples into the cavity distorts the fields and makes it necessary to increase the frequency range to obtain the desired uniformity. Based on these simulation efforts, it is also shown that sintering arrangements need to be modeled in order to determine the frequency range necessary to obtain uniform heating. Different sample materials and sizes may require different experimental arrangements, including the frequency band and the sweep speed. For example, the successful arrangement for the $4 \times 4 \times 6$ cm^3 sample did not result in uniform heating of a $4 \times 4 \times 8$ cm^3 sample of the same material.

General guidelines for variable-frequency microwave sintering were difficult to obtain at this time and are expected to evolve as this research continues. Modeling using the developed coupled FDTD Heat-Transfer code provides a cost effective way of determining the most efficient sintering arrangement, including the frequency range necessary to obtain uniform heating for a given sample size.

ACKNOWLEDGMENTS

Research sponsored (in part, if applicable) by the US. Department of Energy, Assistant Secretary for Energy Efficiency and Renewable Energy, Office of Industrial Technologies, Advanced Industrial Concepts (AIC) Materials Program, under contract   DE-AC05-84OR21400 with Martin Marietta Energy Systems, Inc.

REFERENCES

[1]   M.F. Iskander, R.L. Smith, O. Andrade, H. Kimrey, and L. Walsh, IEEE Tran. On Microwave Theory and Tech., Vol. 42, pp. 793-800, May 1994

[2]   M.F. Iskander in Microwave Processing of Materials II, edited by W.B. Snyder, Jr., W.H. Sutton, M.F. Iskander, and D.L. Johnson (Mater. Res. Soc. Proc. 189, San Francisco, CA, 1990) pp.149-171.

[3]   M. Subirats, M.F. Iskander, M.J White, and J. Kiggans, presented ant the 1996 Spring Meeting, San Francisco, CA, 1996 (unpublished).

[4]   Richard S. Garard, J. Billy Wei, and Zak Fathi in 30th Microwave Power Symposium, (IMPI Proc. 30, Denver CO, 1995) pp. 22-24.

ADHESIVE BONDING VIA EXPOSURE TO VARIABLE FREQUENCY MICROWAVE RADIATION

FELIX L. PAULAUSKAS, APRIL D. McMILLAN and C. DAVID WARREN
Oak Ridge National Laboratory, P.O. Box 2009, Oak Ridge, Tennessee 37831-8048

ABSTRACT

Adhesive bonding through the application of variable frequency microwave (VFM) radiation has been evaluated as an alternative curing method for joining composite materials. The studies showed that the required cure time of a thermosetting epoxy adhesive is substantially reduced by the use of VFM when compared to conventional (thermal) curing methods. Variable frequency microwave processing appeared to yield a slight reduction in the required adhesive cure time when compared to processing by the application of single frequency microwave radiation. In contrast to the single frequency processing, the variable frequency methodology does not readily produce localized overheating (burnt or brown spots) in the adhesive or the composite. This makes handling and location of the sample in the microwave oven less critical for producing high quality bonds and allows for a more homogeneous distribution of the cure energy. Variable frequency microwave processing is a valuable alternative method for rapidly curing thermoset adhesives at low input power levels.

INTRODUCTION

In the past few years, Oak Ridge National Laboratory (ORNL) has been involved in a national initiative to conduct research aimed at promoting the use of lighter weight materials in automotive structures for the purpose of increasing fuel efficiency and reducing environmental pollutant emissions [1]. Key to this initiative is the joining of non-ferrous materials. The most promising method for joining the wide spectrum of alternate materials that may see automotive implementation is adhesive bonding. Unfortunately, the implementation of adhesives on the manufacturing plant floor suffers from a significant economical obstacle. Structural adhesives often require long cure times which translate into elevated per part costs. In order to make the use of adhesives for primary structures a reality in high production rate consumer goods industries, technologies for reducing the cure time of the adhesive must be discovered, developed and deployed. Variable frequency microwave radiation may be one solution to this problem.

This investigation pertains to the effect of variable frequency microwave radiation on the cure time and mechanical properties of an epoxy adhesive. Mechanical properties were evaluated by single lap-shear testing. The VFM samples were compared with data obtained via conventional (thermal) processing and single (fixed) frequency microwave curing of similar samples. It was found that cure time is substantially reduced by VFM radiation, when compared to conventional curing techniques. It was also found that VFM radiation curing yielded only a slight reduction in cure time when compared to single frequency microwave radiation processing but did so without producing localized overheating.

The substrate materials evaluated in this work are glass and a urethane-based composite with glass fiber reinforcement (SRIM-part). Glass substrates (annealed soda lime glass microscope slides) were bonded via microwave radiation curing of an epoxy adhesive to eliminate the variable represented by the absorption of microwave energy by substrates. The SRIM composite is a glass reinforced isocyanurate, DOW MM364, which is currently being investigated for use by the domestic automotive industry. The glass mat is a random swirled glass with a U750 binder and comprises 55 wt% of the composite. The adhesive tested in this work is the BF Goodrich 582E which is a toughened epoxy currently being investigated for structural automotive applications.

The power distribution within a microwave cavity resulting from a fixed frequency source is extremely heterogeneous. As is shown in Figure 1A, a single frequency source yields regions of very high energy and other regions of very low energy within any given cavity plane. If a different plane is chosen within the microwave cavity, a totally different energy profile will exist. By changing any one of many variables including the frequency, cavity size, input power, etc., totally different energy profiles may be achieved but each would have the same characteristic non-uniformity. Adhesives being cured by microwave energy that are subjected to these heterogeneous power distributions would have regions that are well cured, areas that are barely cross-linked and other sectors which are over cured (burnt). One method to overcome this problem when using fixed frequency sources is by placing the specimens on a rotating plate during irradiation or to have deflectors in the cavity rotating during the curing process.

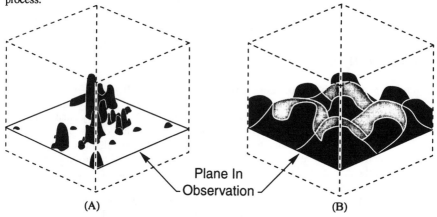

Plane In Observation

(A) (B)

Figure 1. (A) Typical energy distribution in a single frequency microwave oven. (B) Typical energy distribution in a variable frequency microwave cavity. Note that the cumulative effect of many energy profiles in the VFM results in a homogenization of the total energy distribution.

Variable frequency microwave techniques are based on the concept of identifying a central frequency and rapidly sweeping the microwave energy over a range of frequencies greater than and less than the central frequency creating a range of effective frequencies. Since each instantaneous frequency has a different electric field and subsequently a different energy deposition profile, the sweeping of frequencies overlays the energy distribution profiles of many frequencies. The cumulative effect of all these instantaneous but different modes will result in a homogenization of the total energy deposition in the material being processed as is schematically indicated in Figure 1B. Theoretical models and experimental results [2,3,4,5] corroborate this effect. The uniform energy distribution allows for even curing within materials in the cavity. The result is that adhesive samples can be cured in a more homogenous manner, without burnt spots and without uncured regions. Generally, a variable frequency microwave can be made to function as a single frequency source by choosing a center frequency and making the frequency range equal to zero. By doing this the center frequency becomes the fixed frequency.

EXPERIMENTAL PROCEDURE

To determine the correlation between cure times and input power for single frequency microwave irradiation, glass slides were joined using the epoxy adhesive and four to five 30 mil (0.030 inch) glass beads strategically dispersed in each uncured joint to accurately control the bondline thickness. Subsequently, all samples were exposed to varying power levels of microwave radiation using a Cober SF6 power supply which provided up to 5.5 kw of fixed 2.45 GHz radiation in a 61 x 61 x 61 cm multi-mode cavity. Samples were visually inspected to determine the degree of polymerization (crosslinking) by noting the color changes and also hardness of the adhesive. This adhesive changes from olive to bright green when crosslinking is well advanced. Results were verified by differential scanning calorimeter (DSC) testing of two or three samples for proper cure. Curves were then constructed for each power level relating input power to required curing time. This process was then repeated for the isocyanurate composite using the same processing conditions.

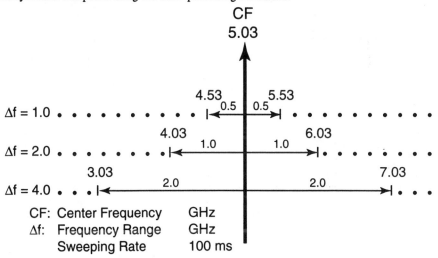

Figure 2. Variable Frequency Microwave Processing Spectrum

To determine the relationship between the required cure time and the input power for variable frequency microwave processing, a center frequency for the VFM of 5.03 GHz was selected. Frequency ranges of 1.0, 2.0, and 4.0 GHz were chosen and a sweep rate of 100ms was employed as is shown in Figure 2. For each frequency range, identical glass slide adherents were joined using the same adhesive. Four to five 30 mil (0.030 inch) glass beads selectively placed in the joint to accurately control the bondline thickness. By a trial and error approach, the cure time required for each frequency sweep range was identified. Once this was accomplished for the glass slides with the absence of lossy effects, it was repeated for the composite material under the same conditions (center frequency, frequency ranges, sweep rate, adhesive, bondline thickness and sample preparation procedures).

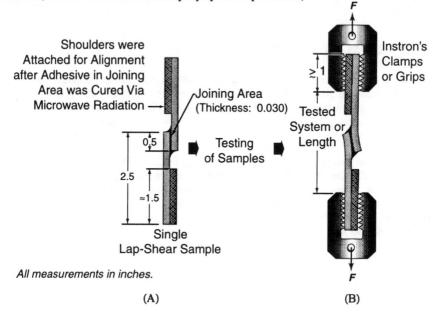

Figure 3. (A) As-Processed Specimen and Joint Configuration for the Single Lap Shear Tests. (B) System Deformation and Displacement in the Single Lap Shear Test

The characterization of the bond strength in the processed samples was determined using single lap shear samples. Composite specimens 2.5 inches long and 1.0 inch wide were bonded to form single lap shear joints with a 0.5 inch overlap as is shown in Figure 3. Glass beads were also used in the lap shear samples to hold a consistent bondline thickness. For single frequency microwave processing four distinct power level and curing time combinations were selected by preliminary trials and multiple specimens fabricated. The maximum power selected was 1200 Watts which corresponds to the shortest required curing time. Higher input powers caused the adhesive to be expelled from the joint after only a short irradiation. This may be related to a rapid exotherm or localized overheating in the adhesive due to a too fast deposition of energy into the epoxy prior to completion of the crosslinking reaction. In an attempt to further reduce the curing time while avoiding the exotherm, a two step, two energy level curing method was employed. The curing process was started with a high level of input power for a very short processing time and subsequently the power was reduced to a lower level and an appropriate processing time was given to complete the crosslinking process. This process was repeated for several high and low power level combinations and is presented in other work [6].

For the variable frequency microwave processing trials, a 5.03 GHz center frequency was selected with frequency sweeps of 1.0, 2.0 and 4.0 GHz employed. Because of the intrinsic nature of the VFM oven, the maximum deliverable forward power is decreased with an increase in the size of the frequency range. Average forward power was about 190W for $\Delta f = 1$ GHz, 160 W for $\Delta f = 2$ GHz and about 130 W for $\Delta f = 4$ GHz. The VFM equipment utilized in this project was Lambda Technologies Inc., Model T-4001E, a multimode cavity which is capable of delivering forward power at ranges over 2.5 to 7.5 GHz or 7.5 to 17.5 GHz and has a nominal maximum forward power of 200 watts. Equal numbers of samples were processed for each processing condition for both single and variable frequency microwave studies.

A control group of conventionally (thermally) cured samples was fabricated where the samples were cured for 45 minutes at 150°C. For all specimens (conventional and microwave processed), the urethane substrate was subjected to a drying cycle of 101°C for 48 hours prior to bonding. This was done to eliminate the interaction of residual water on the substrate with the incident microwave energy. The drying operation [7] is also a secondary post-cure operation for this material which completes the cure of any non-crosslinked material. This post cure could likely be accomplished directly with the microwave if processing conditions are properly designed.

To aid in better alignment and to minimize bending, 1.5 inch shoulders were bonded on each specimen as is depicted in Figure 3. Lap shear tests were then conducted on all samples using a Model 1125 Instron tensile testing machine and a crosshead displacement rate of 0.05 inches per minute. The complex test geometry of the single lap shear specimen imparts stresses to the substrates and adhesives which are difficult to define. Failure normally occurs due to peel (Mode I) stresses rather than shear (Mode II) stresses as the test name implies. Similarly, strain is also difficult to relate to machine displacement for the single lap shear test due to complex test geometry, specimen geometry and off-set load path [8].

SINGLE FREQUENCY MICROWAVE RESULTS

Microwave Processing

The first system evaluated was glass slides adhesively bonded using single frequency microwave (2.45 GHz) radiation over a range of input powers. An analysis of the data reveals that the epoxy cure time approximately followed the power law theory in that as electric field intensity increased, cure time decreased [6]. The empirical relationship indicated by the experimental data is that the power (P) varies as the inverse power of the cure time of the adhesive for this material combination. In the high power region, the experimental data indicates that the power is the inverse square root of the cure time, while for the lower input power region, the input power is inversely proportional to the exposure or cure time. Through analyzing the experimental data presented in Figure 4, an appropriate empirical function was determined which describes this data to a high degree of accuracy. The empirical relationship which describes the observed data is a power law function with an exponent of -0.612 and a constant term of 1464.

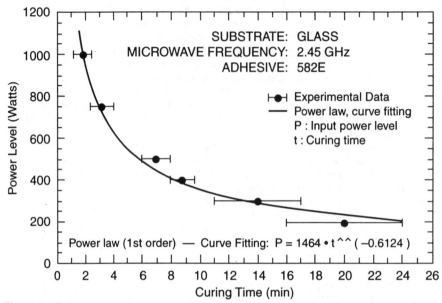

Figure 4. Cure Time vs. Power Level for the Adhesive on the Glass Substrates. Power Law Approximation.

Cure times below six minutes, which correspond to input powers about 500 watts, produced samples with a significant volume fraction of bubbles in the as-cured adhesive. It is likely that the energy deposition into these samples was too rapid resulting in excessive heating of the adhesive prior to sufficient crosslinking to prevent the flow of the adhesive. As a result, there exists an upper threshold energy for curing this adhesive. As the input energy was decreased, the time required for complete crosslinking was increased. At sufficiently low power inputs crosslinking would not be promoted. For this material system the lower threshold is below 150 watts. Therefore for this material and microwave system the effective curing range of the adhesives is between 150 and 500 watts and the complete cure can be achieved in as little as six minutes.

The second system evaluated was the glass fiber reinforced isocyanurate composite previously described. The composite was adhesively bonded using single frequency microwave (2.45 GHz) radiation over a range of input powers. An analysis of the data show that this material system also approximately follows the power law theory which is shown in Figure 5. The empirical function which best describes this data is a power law function with an exponent of -0.503 and a constant term of 1410.

Input power of about 550 watts, which corresponds to a cure time of about eight minutes, is the maximum input power limit due to the formation of bubbles in the adhesive. The lower limit for complete crosslinking for this material system is below approximately 300 watts. Therefore for this material and microwave system the effective curing range of the adhesive is between 300 and 550 watts and the complete cure can be achieved in as little as eight minutes. It is not surprising that the urethane substrates had higher input power limits and required

substrates bonded with the same adhesive and irradiated at the same frequency. The difference between the lossy characteristics of the two substrates would require more energy and longer cure times due to absorption of a portion of the input energy by the composite.

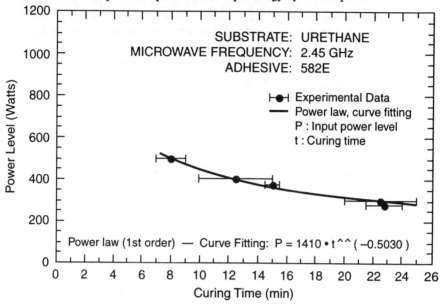

Figure 5. Cure Time vs. Power Level for the Adhesive on the Composite Substrates. Power Law Approximation.

Mechanical Evaluation

The characterization of the mechanical bond strength in the processed samples was determined using single lap-shear samples. In this evaluation, only urethane-glass substrates were studied. Table I gives the results of the single lap shear tests. Data is reported as maximum load carried and maximum total crosshead displacement (TCHD) rather than as stress and strain since for the single lap shear specimen configuration stress and strain cannot be directly correlated to machine displacements or load cell signals due to the complex test geometry, specimen geometry and off-set load path [8].

As seen in Table I, the ultimate tensile strength of the microwave processed samples cured for relatively short times was much lower than the conventionally processed samples. When the input power was reduced and the cure time lengthened to within the "moderate" range, the ultimate strengths of the microwave processed adhesives was nearly identical to that for the thermally cured adhesives. The microwave processed samples also had an increased TCHD and a decreased slope to the elastic loading curve. This indicates that the microwave processed samples may be less stiff than the conventionally cured samples and have a corresponding greater ductility. Mechanisms for this occurring are unknown, however it may be related to the difference in the thermal stressing of the substrate and adhesive by the two curing methods.

Table I

Single Lap Shear Data for Single Frequency (2.45 GHz) Microwave Processed Samples. Substrates are Urethane/Glass Composite. Adhesive is Goodrich 582E. Ultimate Load is that required for failure. TCHD is the machine displacement at failure.

	Ultimate Load (lbf)	TCHD (inches)
Conventional Cure 45 min, 150°C	1420	0.065
MICROWAVE PROCESSED		
Short Cure 12-14 min	959	0.085
Moderate Cure 22-30 min	1405	0.065
Long Cure >30 min	1373	0.073

It was noted that for fully cured joints, failure occurred at very high applied loads and by fiber tear of the composite substrate and seldom at the interface or within the adhesive. Bubble formation within the joints of the samples cured by microwave radiation for short times was also very prevalent indicating that the energy deposition rate was too high. This resulted in much weaker joints.

VARIABLE FREQUENCY MICROWAVE RESULTS

Microwave Processing

The first system evaluated was comprised of glass slides adhesively bonded using variable frequency microwave (Center frequency: 5.03 GHz) radiation over a range of input powers and frequency sweep ranges. An analysis of the data reveals that the epoxy cure time approximately followed the power law theory. In the high power region the experimental data indicates that the power varies as the negative power of the cure time, while for the lower input power region, the input power is inversely proportional to the exposure or cure time [6]. Through analyzing the experimental data presented in Figure 6, an appropriate empirical function was determined which describes this data to a high degree of accuracy. The empirical relationship which describes the observed data is a power law function with an exponent essentially of -0.21 and a constant term of approximately 308. With the variable frequency microwave, much less input power was required to achieve curing than with the single frequency microwave. This is due to an improved coupling efficiency obtained by frequency sweeping.

Cure times below fifteen minutes, which correspond to input powers in the region of 200 watts produced samples with a significant volume fraction of bubbles in the as-cured adhesive. It should be noted that 200 watts is the maximum input power available in this specific VFM system. It is likely that the density of energy deposition into these samples occurred too rapidly resulting in excessive heating of the adhesive prior to sufficient crosslinking to prevent the out flow of the adhesive or generation of exotherm bubbles. As a result, there exists an upper threshold energy for curing this adhesive with this methodology. As the input power was decreased the time required for complete crosslinking was increased. At sufficiently low power inputs crosslinking would not be promoted. For this material system the lower threshold is below 125 watts. Therefore for this adhesive/substrate material combination, VFM system and processing conditions the effective curing range of the adhesives is between 125 and 200 watts and the complete cure can be achieved in as little as fifteen minutes. It should be noted that the upper and lower power threshold for the variable frequency processing were much lower than for the single frequency processing. This indicates that from an energy standpoint, variable frequency processing may be much more efficient for this type of operation.

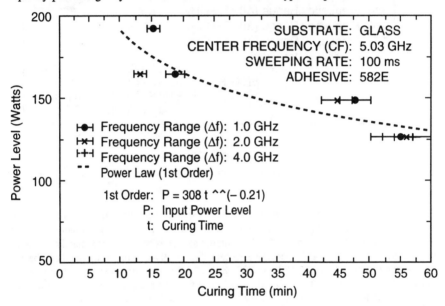

Figure 6. Cure Time vs. Power Level for the Glass Substrates. Power Law Approximation.

The second system evaluated was the glass fiber reinforced isocyanurate composite previously described. The composite was adhesively bonded using variable frequency microwave (Center Frequency: 5.03 GHz) radiation over the range of input powers and frequency sweep ranges previously indicated. An analysis of the data shows that this material system approximately follows the power law theory which is shown in Figure 7. The empirical function which best describes this data is a power law function with an exponent of -0.49 and a constant term of 604. This function easily fit the data from all three frequency ranges. With this exponent (approximately -0.5) a doubling of the input power will result in a four fold reduction in the curing time. A tripling of the input power will result in a nine fold reduction in the curing time. Any increase in input power will result in a much more drastic reduction in the required curing time for this adhesive/substrate material system. The lowest cure time possible, however, is always limited by the maximum input energy threshold which is a function of the specific material systems being studied and the frequency being utilized.

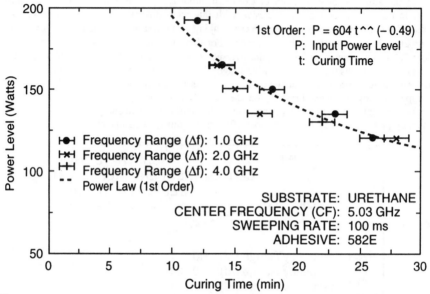

Figure 7. Cure Time vs. Power Level for the Composite Substrates. Power Law Approximation.

Input power of about 200 watts, which corresponds a to cure time below twenty-two minutes, is the maximum power limit due to the formation of bubbles in the adhesive. The lower limit for complete crosslinking for this material system is below approximately 150 watts. Therefore for this material and microwave system the effective curing range of the adhesives for these conditions is between 150 and 200 watts and the complete cure can be achieved in as little as fifteen minutes. This is significantly longer than the eight to twelve minutes required by the single frequency microwave but may be related to the lower power level available in the VFM used in this work. The time difference between the two methods may also be attributable to frequency being applied. (2.45 GHz for single frequency and a center frequency of 5.03 GHz for variable frequency)

No acceptable comparison can be established between the 5.03 GHz center frequency in the VFM processing and the 2.45 GHz frequency used in the fixed frequency processing. Even if the VFM center frequency were the same as the fixed frequency, a direct quantitative correlation could not be established between the two processing methods due to differences in the physics of the two processes. For fixed frequency microwave processing, the sample has to be located at the same place inside the microwave cavity. The determination of the optimum specimen placement in the oven has to be conducted by trial and error preliminary screening trials. Once a "good" location in the oven is determined, each batch of processed samples must be properly placed in that selected location. In this way, samples processed by fixed frequency microwaves can be compared to each other. For variable frequency microwave processing, the location of the samples inside of the cavity is not important which makes this technology more robust for manufacturing environments. Therefore, a direct correlation between these two microwave processing methods is not possible due to the location sensitivity of the single frequency processed samples. We will, however, draw a loose comparison between these two methods. This is done because neither of these technologies will be transferable to the manufacturing world unless comparisons are made which have understandable significance.

Mechanical Evaluation

The characterization of the bond strength in the processed samples was determined using the same test configuration and equipment as was used in evaluating the single frequency microwave cured samples. In this evaluation, only composite substrates were studied. Table II gives the results of the single lap shear tests. Data is reported as ultimate load (UL) carried and maximum total crosshead displacement (TCHD) rather than as stress and strain for reasons previous explained [8]. A detailed description of the mechanical evaluation results from fixed frequency (2.45 GHz) processing can be found in previous work [9, 10].

Table II

Single Lap Shear Data for Single and Variable Frequency Microwave Processed Samples
Substrates are Urethane/Glass Composite. Adhesive is Goodrich 582E.
Ultimate Load is that required for failure and is expressed in pounds.
TCHD is the machine displacement at failure expressed in inches.

	Single Frequency SF = 2.45 GHz		Variable Frequency CF = 5.03 GHz FR = 1.00 GHz		FR = 2.00 GHz	
	UL	TCHD	UL	TCHD	UL	TCHD
Short Cure 12-14 min	950	0.085	1005	0.040	960	0.041
Moderate Cure 22-30 min	1405	0.065	1390	0.065	1400	0.072
Long Cure >30 min	1373	0.073				

Conventional Cure: UL = 1420 lbf, TCHD = 0.065 inches

As seen in Table II and Figure 8, the samples processed by variable frequency microwave irradiation for 12 to 14 minutes had strengths and total elongations far below those of conventionally processed samples or the microwave processed samples cured for longer periods of time at lower energies. Post failure analysis revealed that these samples contained bubbles characteristic of a material that underwent too high of an energy deposition during curing. This same phenomenon has been previously noted [7] where the same adhesive was thermally cured at too high of a temperature. In that work, the exotherm of the epoxy produced the bubbles.

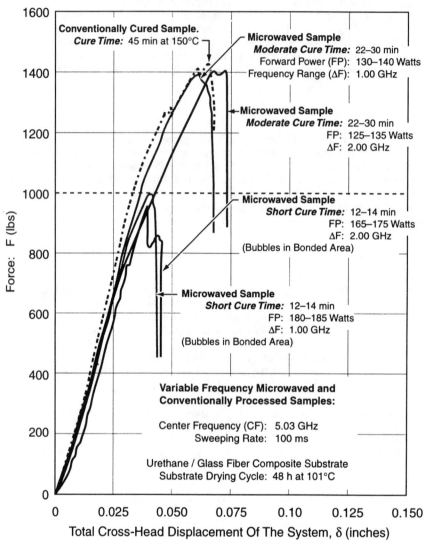

Figure 8. Typical Single Lap Shear Curves for Variable Frequency Microwave Processed Samples. Comparison to Thermally Cured Specimens is made.

The samples processed by lower input energies for longer periods of time had ultimate strengths almost exactly identical to those of the thermally cured samples. The total elongation (i.e. ductility, lack of stiffness) of the samples microwaved between 22 and 30 minutes was slightly greater than that of thermally cured samples. Thus equivalent failure strengths and increased ductility was noted at cure times that were between 1/2 and 2/3 those required for conventional curing. The increased ductility for variable frequency processed samples was not as extensive as that noted for single frequency processed samples. Further studies are required in this area to find an explanation for this observation.

CONCLUSIONS

1. The resultant energy deposition profile is more uniform for variable frequency microwave systems than for fixed frequency microwave systems. As a result, VFM curing is independent of sample placement in the microwave oven while fixed frequency microwave curing is highly sample position sensitive. Contrary to fixed frequency systems, samples cured in VFM ovens are processed with less generation of localized hot or burn spots which result in better quality bonds.

2. Variable frequency microwave technology represents a valuable alternative method for rapidly curing thermoset adhesives at low input power levels. For the substrates and adhesive used in this project, surprisingly low levels of forward (input) power were required for processing of the samples.

3. The application of microwave technology for joining of substrates using epoxy based adhesives significantly reduces the curing time to only a third to a quarter of the conventional cure time. This is accomplished while maintaining equivalent values of the ultimate tensile strength measured through the single lap shear test. This was observed for both fixed and variable frequency microwave technologies.

4. Microwave processed samples (fixed and variable frequency), when tested as single lap shear specimens exhibit less rigidity and more plasticity than conventionally processed samples.

5. Coupling of the Goodrich EXP 582E epoxy based adhesive to the fixed and variable frequency microwave radiation is extremely efficient for all the frequencies applied in this study.

6. Variable frequency microwave processing appears to yield a slight reduction in the required adhesive cure time when compared to processing by the application of a fixed frequency microwave source.

7. This technology may be extended to multiple-layered panels or components.

ACKNOWLEDGEMENTS

Research was performed at Oak Ridge National Laboratory and sponsored by the Office of Transportation Materials, U.S. Department of Energy, under contract No. DE-AC05-96OR22464 with Lockheed Martin Energy Research Corporation.

REFERENCES

1. Office of Transportation Materials, Materials for Lightweight Vehicles Program Plan, United States Department of Energy, (1992).
2. A.D. Surrett, R.J. Lauf, F.L. Paulauskas and A.C. Johnson, MRS Symposium Proceedings, 347, 691 (1994).
3. A.C. Johnson, R.J. Lauf and A.D. Surrett, MRS Symposium Proceedings, 347, 453 (1994).
4. R.J. Lauf, F.L. Paulauskas and A.C. Johnson, 28th Microwave Symposium Proceedings - International Microwave Power Institute, 150 (1993).
5. R.J. Espinosa, A.C. Johnson, L.T. Thigpen, W.A. Lewis, C.A. Everleigh and R.S. Garard, 28th Microwave Symposium Proceedings - International Microwave Power Institute, 26 (1993).
6. F.L. Paulauskas, T.T. Meek and C.D. Warren, SAMPE International Technical Conference Proceedings, 27, 114 (1995).
7. C.D. Warren, R.G. Boeman and F.L. Paulauskas, 1994 DOE Contractors Coordination Meeting Conference Proceedings, 2, 1 (1994).
8. R.G. Boeman and C.D. Warren, 10th Annual ASM/ESD Advanced Composites Conference Proceedings, 473 (1995).
9. F.L. Paulauskas, T.T. Meek and C.D. Warren, MRS Symposium Proceedings, 430 (1996) In press in this proceeding.
10. F.L. Paulauskas and T.T. Meek, MRS Symposium Proceedings, 347, 743 (1994).

OBSERVATION OF AN ELECTROMAGNETICALLY DRIVEN TEMPERATURE WAVE IN POROUS ZINC OXIDE DURING MICROWAVE HEATING

D. Dadon[1], D. Gershon[2], Y. Carmel[2], K.I. Rybakov[2,4], R. Hutcheon[3], A. Birman[2], L.P. Martin[1], J. Calame[2], B. Levush[2], M. Rosen[1]

[1] Department of Material Science and Engineering, Johns Hopkins University, Baltimore, MD.
[2] Institute for Plasma Research, University of Maryland, College Park, MD.
[3] Chalk River Laboratories, Chalk River, Ontario K0J1J0, Canada.
[4] Permanent address: Institute of Applied Physics, Nizhny Novgorod, Russia

ABSTRACT

Propagation of a sharp temperature wave was observed during microwave heating of porous zinc oxide in nitrogen and argon atmospheres. This wave initiated from the center of the sample and traveled at an average velocity of 0.2 cm/min towards its surface. This temperature wave was attributed to an anomalous peak in the imaginary part of the complex permittivity possibly caused by desorption of chemisorbed oxygen from the surfaces of ZnO crystallites.

INTRODUCTION

The recent growth in research on microwave processing of ceramic materials was motivated by the hope of achieving superior properties unattainable by conventional processing methods. This interest is related to the prospect of achieving uniform bulk heating as a result of the volumetric character of microwave absorption in nonmetallic materials. However, uniform power absorption does not necessarily guarantee a homogeneous temperature distribution. The balance between the absorbed microwave energy and the surface losses via radiation and thermal conduction can result in a temperature gradient within the sample. For example, a temperature gradients have been measured during microwave processing of porous ZnO [1]. The absorbed microwave power is proportional to the imaginary part of the complex permittivity, ε", and the square of local electric field. The former parameter is temperature dependent. The electric field depends on the distribution of the complex permittivity, $\varepsilon = \varepsilon' - i\varepsilon"$, within the material. Thus, a temperature gradient within the sample can lead to nonuniform power adsorption. This may further enhance the temperature gradient. In this paper, we describe experimental results showing how an anomalous peak in the temperature dependence of the complex permittivity of porous ZnO leads to an electromagnetically driven temperature wave propagating from the sample's core to its surface.

EXPERIMENTAL PROCEDURE AND RESULTS

Zinc oxide is a semiconductor material used extensively in varistors. A commercial zinc oxide powder with a -200 mesh particle size was uniaxially pressed without binder to approximately 54% of the crystal theoretical density ($5.61 g/cm^3$). This porous, polycrystalline material with a surface area of $3.1 \ m^2/g$ was microwave heated in a highly overmoded 2.45 GHz applicator. The samples were insulated in a porous alumina enclosure and loosely packed in ZnO powder, which served as a thermal insulator and buffer powder. As shown in Figure 1, two type K thermocouples measured the sample's core and surface temperatures.

Mat. Res. Soc. Symp. Proc. Vol. 430 © 1996 Materials Research Society

Heating of the ZnO samples was performed by applying microwave power in two 40W steps as shown in Fig. 2. This figure also shows the typical core temperatures of two ZnO samples, which were heated in two different gases at 1.1 atm. The core temperature of the sample heated in air increased each time power was increased and approached an asymtptotic temperature with a thermal time constant of about 15 min. In contrast, the core temperature of the sample heated in nitrogen showed a significantly different temperature evolution. At 100^0C, the core temperature started to rise at a notably higher rate than the sample heated in air. Between 230 and 290^0C, the core

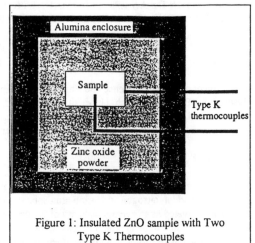

Figure 1: Insulated ZnO sample with Two Type K Thermocouples

temperature rose dramatically at a rate $> 200^0C/min$. The core temperature continued to rise at a lower rate to 580^0C. A similar temperature profile was observed during heating in argon. As shown in Fig. 3, the measured core and surface temperatures of the sample heated in nitrogen indicated the local nature of the heating. The sharp temperature increase took seven minutes to propagate from the core to the surface. Its average velocity was 0.2 cm/min.

Under certain conditions, this phenomenon was reproducible. After microwave heating a sample in a nitrogen atmosphere, the sample was then cycled up to 400^0C in a thermal oven with an oxygen atmosphere. When this sample was reheated in the microwave oven with a nitrogen atmosphere, the sharp temperature front reappeared. However, this phenomenon did not appear during a second microwave heating in a nitrogen environment if the sample was not thermally cycled in an oxygen environment.

An anomalous peak of the dielectric properties of porous ZnO has been observed and may be responsible for the observed phenomenon [2]. Therefore, the complex permittivity of a thermally heated ZnO was measured as a function of temperature in different atmospheres using the cavity perturbation technique [2]. As shown in Fig. 4, ε" at a frequency of 2.46 GHz (which is very close to the 2.45 GHz operating frequency of the microwave oven) of the for samples heated in nitrogen exhibited a significant peak from 200 to 500^0C. Comparison of Fig. 4a & b indicates that the peak increased in magnitude and possibly shifted to a higher temperature as the heating rate increased. The sample heated in air exhibited an insignificant temperature dependence below 600^0C. Fig. 4b demonstrates that this low temperature dielectric loss peak occurs during heating the sample in nitrogen, but not during cooling. Thus, this anomalous behaviour in ε" was not reversible.

Observations of microwave ignited combustion reactions or exothermic self-propagating high temperature synthesis (SHS) with similar propagation velocities have been reported in the literature [3,4]. A reduction in the specific heat of a sample due to a phase change could also result in a substantial temperature increase. During differential thermal analysis measurements, ZnO powder was thermally heated from room temperature to 800^0C in both air and argon atmospheres. No temperature difference was measured. X-ray diffraction analysis of the samples before and after the heating indicated no phase change. Thus, exothermic reactions or phase changes can not account for this temperature wave.

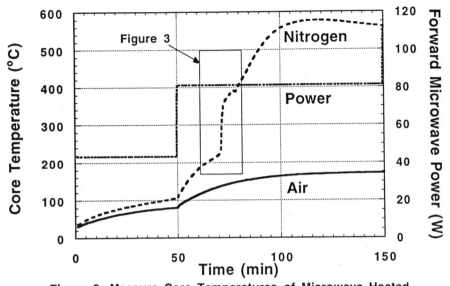

Figure 2: Measure Core Temperatures of Microwave Heated Porous Zinc Oxide in Nitrogen and Air Atmospheres

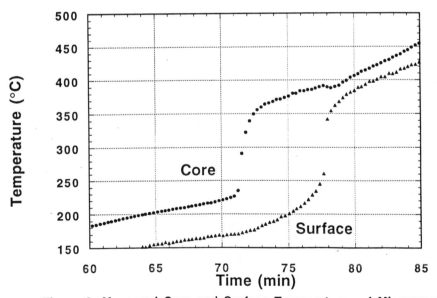

Figure 3: Measured Core and Surface Temperatures of Microwave Heated Porous Zinc Oxide in a Nitrogen Atmosphere

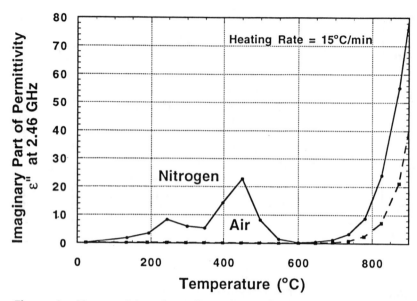

Figure 4a: Measured Imaginary Part of Permittivity of Porous Zinc Oxide during Thermal Heating in Nitrogen and Air Atmospheres

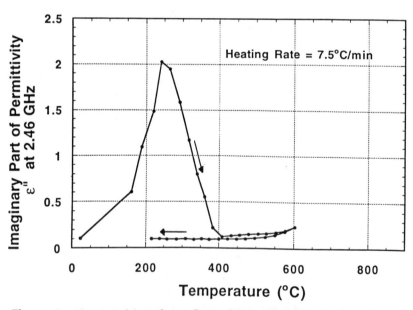

Figure 4b: Measured Imaginary Part of Permittivity of Porous Zinc Oxide during Heating and Cooling in a Nitrogen Atmosphere

DISCUSSION

The essential features of the observed temperature temporal profile can be seen in Fig. 3. The core and surface profiles are similar and appear shifted by about seven minutes. Initially, the temperature rise was comparatively slow. When it reached some critical value (around 240^0C), a very sharp temperature increase occurred. It is reasonable to assume that during this seven minute interval the temperature time dependence is similar for points between the core and the surface, ie. the temperature as a function of time and position can be represented by: $T = f(r - vt)$, where f is a step-like function and v is velocity.

The sharp increase in ε'' with respect to temperature, as shown in Fig. 4, improved the absorption of microwave energy and dramatically increased the local temperature. With an inverted temperature gradient typical to microwave heating, the core temperature was the first region to experience such an increase in ε''. When thermal conduction brought the temperature of the surrounding region to a critical value, this region also sharply increased its power absorption and temperature. Thus, this temperature front propagated from the sample's core towards its surface. The outer region of the sample shielded the inner regions from microwave power as its ε'' increased.

The temperature distribution across the sample is governed by the thermal conduction equation,

$$c\rho \frac{\partial T}{\partial t} - \nabla \cdot (\kappa \nabla T) = \varepsilon_0 \varepsilon'' \omega \langle \mathbf{E}^2 \rangle \qquad (1)$$

where c is thermal capacity, ρ is density, κ is thermal conductivity of the material, ε_0 is dielectric permittivity of free space, ω is the angular frequency of microwaves, $\mathbf{E}$ is the local electric field, and the angular brackets denote time average. Since ε'' peak depends on heating rate, an estimate of the maximum ε'' during microwave heating can be obtained from the core and surface temperature profiles. In fact, these profiles allow us to estimate both terms on the left side of Eq. (1). The value of $\partial T/\partial t$ is at least 3 ^0C/sec for $250 < T < 350^0C$. The maximum value of the second term on the left of Eq. (1) can be estimated as κ times the difference in temperature before and after the wave front (about 100 ^0C) divided by the square of the front length. Given the sharp character of the temperature rise, we can estimate the front length as the velocity of the thermal wave multiplied by the time of temperature increase on the front of the wave, which is on the order of 1 mm. Assuming that the field distribution in the multimode cavity is not significantly affected by the changes in the dielectric loss of the sample and knowing the total absorbed microwave power and the volume of the sample, we find that $\varepsilon_0 \omega \langle \mathbf{E}^2 \rangle \cong 50$ W/cm^3. Using the known values for $c \cong 0.5$ J/g$\cdot^0C$, $\rho \cong 3$ g/cm^3, and $\kappa \cong 0.1$ J/(cm$\cdot$sec$\cdot^0C$), an estimate of the peak value of ε'' in this process is about 20. This estimate correlates with the ε'' measurements performed at different heating rates, in the sense that the maximum value of ε'' is higher for faster heating rates.

Since this temperature wave and the ε'' peak do not occur in an air environment, this phenomenon could depend on the oxygen partial pressure. This partial pressure could result from chemical desorption of O_2 from the ZnO surface or depletion of intrinsic oxygen from the ZnO molecule. Göpel found that O_2 chemisorbs to ZnO surface as a singly charged ion and its desorption occurs from room temperature to about 380^0C. As oxygen desorbs as a neutral molecule, an electron is excited into the conduction band of the ZnO crystal [5]. An increase in the number of conducting electrons increases the material's ε''. Initially, the desorption rate of oxygen, which directly relates to ε'', has an exponential increase with temperature [5]. The desorption rate of oxygen eventually decreases as the finite amount of chemisorbed oxygen is depleted. The number of conducting electrons and the ε'' associated with the chemisorption mechanism should eventually decrease with time and temperature. This could explain why the heating rate in the sample is reduced at higher temperatures. Göpel also noted that reversible chemisorption of oxygen occurs for temperatures below 380^0C. This point is consistent with experimental observation that reheating samples only generated a temperature front if the samples were previously exposed to oxygen.

Several previous experiments measured the temperature dependence of ZnO conductivity (dc [6], 1.5kHz [7], and 10 GHz [8]) during cycling in various gas environments. The following common features were observed when the atmosphere lacked oxygen. (i)The conductivity exhibits a sharp maximum during initial heating. (ii) This temperature dependence is irreversible. No maximum and lower conductivity were observed during cooling and reheating. (iii) The results are very sensitive to the specific surface area of the material. During thermal cycling in air or oxygen, these features were not present. These previous works are in general agreement with our observations. Thus, the observations of ref. 6-8 and our own measurements can be interpreted in terms of the chemisorption mechanism [5], but can be also understood as intrinsic oxygen depletion of ZnO molecule, which makes it nonstoichiometric and consequently changes ε''.

CONCLUSIONS

1) Measurements of the complex permittivity in a thermal oven found a significant peak in ε'' with respect to temperature in a nitrogen atmosphere. This low temperature anomaly in ε'' can be explained by variations of oxygen content in porous ZnO.
2) Propagation of a temperature wave has been observed during microwave heating of a ZnO powder compact in a nitrogen atmosphere. The temperature front could be attributed to the measured peak in ε''. This phenomenon does not occur in subsequent heatings unless ZnO has been exposed to oxygen in the interim.

ACKNOWLEDGMENTS

This work was supported by the Division of Advanced Energy Projects, DOE, under contract No. DE-FG02-94ER 12140. Partial support from Army Research Office and a NATO linkage grant #HTECH LG 940364 is acknowledged. Technical support of K. Diller, V. Yun, J. Rodgers, and J. Pyle is greatly appreciated. The authors would like to thank I. Lloyd for helpful discussions.

REFERENCES

1. L.P. Martin, A. Birman, D. Dadon, Y. Carmel, D. Gershon, J. P. Calame, B. Levush and M. Rosen. Accepted for publication by J. Mat. Synth. and Proc.
2. R. Hutcheon, M. de Jong, F. Adams, G, Wood, J. McGregor and B. Smith., J. of Microwave Power and Electromagnetic Energy, 27 No. 2 pp. 93-99 1992.
3. D.E. Clark, I Ahmad and R.C. Dalton, Mater. Sci. and Eng., Vol. A144, pp. 91-7 (1991).
4. M. Willert-Porada, B. Fisher and T. Gerdes, Edited by D.E. Clark, W.R. Tinga and J.R. Saia (Am. Cer. Soc., Ceramic Transactions, Vol. 36, Micr. Theory and Application in Mater. Proc. II, Westerville, OH, 1993) pp. 365-75.
5. W. Göpel, Surf. Sci., Vol. 62, pp. 165 - 182 (1977).
6. P.Chandra, V.B.Tare, and A.P.B.Sinha, Indian J. Pure Appl. Phys., Vol. 5, pp. 313 - 7 (1967).
7. M. Takata, D. Tsubone, and H. Yanagida, J. Amer. Ceram. Soc., Vol. 59, pp. 4 - 8 (1976).
8. B.-K. Na, M. A. Vannice, and A.B. Walters, Phys. Rev. B, Vol 46, pp. 12266-77 (1992).

MICROWAVE PROCESS CONTROL THROUGH A TRAVELING WAVE TUBE SOURCE

G.J. VOGT, A. REGAN, A. ROHLEV, and M. CURTIN
Los Alamos National Laboratory, Los Alamos, New Mexico, 87545.

ABSTRACT

A rapid feedback control system was designed to operate with a traveling wave tube amplifier for regulating the sintering temperature of tows and tubes in a single mode microwave cavity. The control system regulated the microwave frequency and power absorbed by the sample in order to maintain the sample temperature. Testing with NICALON tows and mullite tubes demonstrated that the control scheme worked well for stationary and slowly moving (<10mm/min) samples, but failed for fast moving samples. Difficulty with measuring sample temperatures was resolved by using a light sensor to measure the emitted light intensity and to gauge the relative degree of heating. This work was supported by the Department of Energy, Office of Industrial Technologies, Advanced Industrial Materials Program.

INTRODUCTION

Rapid feedback control with a variable frequency microwave source can provide practical process control over the heating of continuous ceramic filament tows, rods, and tubes in a single mode cavity. Process temperatures can be regulated despite large variations in dielectric loss and heating as a filament tow or tube moves through the microwave cavity. The rapid and independent control over microwave frequency and power level needed for process control was obtained from a traveling wave tube (TWT) amplifier. Phase information from the microwave cavity can be used in a feedback loop to track the cavity frequency as it shifts due to changing dielectric properties of the load during heating [1-2]. By sampling the power in the cavity, process temperatures should be controlled by regulating the microwave power absorbed in the heated tow or tube [1-2]. Absorbed power by the sample was chosen to be the controlled variable, rather than temperature, because of the problems in measuring the sample temperature [1-2]. This work examined the assumption that constant absorbed power under steady state heating will produce a constant temperature in the tow or tube. Examples of applied process control are described for microwave heating of NICALON tows and mullite tubes. The challenging tasks of process temperature measurement and verification of heating control in tows and tubes are discussed.

EXPERIMENTAL

A block diagram of the microwave system with the frequency and power controllers is shown in Figure 1. Diagrams of the controllers were described in earlier work [1-2]. The TWT amplifier had an output frequency range of 2.5 to 8.0 GHz and an output power of 0-300 Watts. The cavity RF signal and power absorbed by the cavity were measured by an electric field probe positioned at an electric field maximum. The forward power (RF_f) and cavity (RF_c) signals are RF signals used by the frequency controller to modulate the TWT output frequency. An independent frequency counter monitored the output frequency.

The power controller monitored four power levels in the system. The forward power (P_f) is

Mat. Res. Soc. Symp. Proc. Vol. 430 © 1996 Materials Research Society

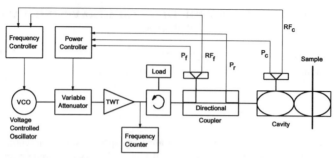

Figure 1. Schematic of microwave system with frequency and power controllers.

produced by the TWT and transmitted towards the cavity. The reflected power (P_r) is returned from the cavity due to the non-ideal matching of the iris, cavity, and load. The cavity power (P_c) is lost in the walls of the cavity due to its non-zero resistance. The absorbed power (P_a) is absorbed by the load in the cavity. The forward, reflected, and total cavity powers were measured directly by the controller, while the absorbed power was numerically derived. The difference between the forward and reflected power (P_f- P_r) was the total power in the cavity (P_a+P_c), either absorbed by the load or lost in the cavity walls. If P_c was subtracted from the difference P_f- P_r, the remainder was the absorbed power P_a given by

$$P_a = P_f - P_r - P_c \qquad (1)$$

The control system measured P_f, P_r, and P_c, performed the above arithmetic, displayed P_a, and controlled it by modulating P_f. Independent power meters measured and recorded the three power levels for the tests. For the P_a reading to be accurate, P_f, P_r, and P_c must be calibrated.

A network analyzer was used to measure and calibrate the attenuation between the TWT input and the P_f, P_r, and P_c ports. By replacing the cavity with a well-matched load, P_c and P_r were nearly zero. Hence, P_a equaled P_f. The TWT power was swept, and the gain and offset of the control system P_f port were adjusted so that the absorbed power corresponded to the forward power reading. To calibrate P_r, the load was replaced with a short which ensured that P_f equaled P_r and P_c was zero. The TWT power was again swept, and gain and offset of the control system P_r port were adjusted so that the absorbed power remained zero. The final step was to calibrate P_c. By placing the cavity back into the system with no load, P_a should be zero. The cavity was excited at its resonant frequency and again the TWT power was swept while the gain and offset of the control system P_c port were adjusted so that the absorbed power reading was zero. Thus with all three powers calibrated, the controller and the absorbed power display were calibrated to read an accurate P_a.

The TE_{10n} single-mode cavity in Figure 1 was described in earlier work [1-2]. The test cavity was tuned to operate near 2.95 GHz. The cavity had two pairs of opposing circular ports, shown in Figure 2, positioned at an electric field maximum. Test samples were loaded vertically through the ports aligned with the transverse electric field in the cavity. A second pair of ports, perpendicular to the transverse electric field in Figure 2a, provided for insertion of an Accufiber optical fiber thermometer (OFT) and for video recording of the heated sample. This local OFT lightpipe measured a mean temperature over a 4-5 mm sample length by line of sight. The angled port next to the local OFT port in Figure 2b held a global OFT sensor that viewed the entire 34-mm sample length in the cavity. The local OFT unit was a dual-wavelength sensor, measuring

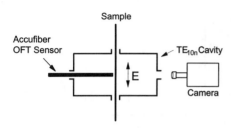

Figure 2a. Schematic cross-sectional view of the TE$_{10n}$ cavity, showing sample position with OFT sensor and video camera.

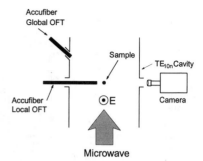

Figure 2b. Schematic top view of the TE$_{10n}$ cavity, showing positions of local and global OFT sensors.

temperature at 800 nm and 950 nm [1-2]. The global OFT was a sensitive single-wavelength sensor, operating as a light meter over a 1-2 μm band to gauge the relative degree of heating.

The feedback controllers were tested by microwave heating of stationary and moving test samples to examine the ability of the controllers to regulate the sample temperature and absorbed power. The sample length in the cavity was 34 mm long. The test samples included continuous NICALON tows (PVA sizing and 0.2 gm/m density) and thin-walled mullite tubes of two different diameters. The outer diameter and wall thickness for each tube were 1.6 mm x 0.2 mm and 4.7 mm x 0.8 mm.

RESULTS

Figure 3a shows the independently measured P_f, P_r, and P_c powers for a NICALON tow pulled at different speeds. The absorbed power P_a derived from Equation 1 is given in Figure 3b. While pulling at 8 mm/min, P_a was increased three times. Fluctuations in P_a occurred at faster pulling speeds in Figure 3b because the heated fraction of the tow length in the cavity was rapidly changing. The power controller also performed excellently for the mullite tubes at slow pull speeds (<10 mm/min). Our tests clearly demonstrated that the power controller operated as designed by maintaining P_a at a constant preset value. While pulling the 1.6 mm mullite tube at 40 mm/min, the controller automatically held P_a at a constant value, although greater than the preset one.

Figure 4 gives the response of the TWT output frequency and the frequency controller to stairstep changes in P_a for a stationary sample of 4.7 mm mullite tube. Decreases in P_a and sample temperature (not shown) produced increases in the cavity resonant frequency on the order of 0.2-0.5 MHz with probable decreases in the sample permittivity. Conversely, an increase in P_a and temperature yielded a decrease in the cavity resonant frequency. In general, the cavity resonant frequency with a stationary sample was constant under constant P_a for the test samples. However, pulling the sample or changing the pull speed produced a change in the resonant frequency. The frequency response in Figure 5 to a rapidly moving NICALON sample (also shown in Figure 3) was quite variable. The rapid frequency change reflected the large fluctuations in local temperature, also shown in Figure 5. From the video record of this sample, rapid changes in frequency occurred while the heated fraction of total tow length in the cavity was rapidly changing. In effect, the amount of tow sample heating was changing with time, although the amount of sample in the cavity remained constant.

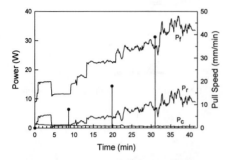

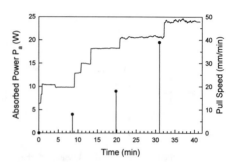

Figure 3a. Forward P_f, reflected P_r, and cavity P_c power levels used to heat a NICALON tow pulled up through the cavity at four different rates. Changes in pull speed are marked by vertical lines from a filled circle denoting the new speed. The tow was initially at rest.

Figure 3b. Absorbed power P_a used to heat a NICALON tow pulled up through the cavity at four different rates. Changes in pull speed are marked by vertical lines from a filled circle denoting the new speed. The tow was initially at rest.

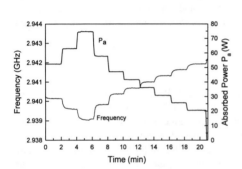

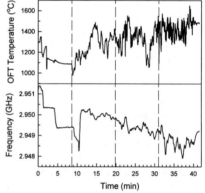

Figure 4. Microwave frequency and absorbed power P_a for a stationary 4.7 mm mullite tube heated at different P_a values.

Figure 5. Sample temperature and microwave frequency for a NICALON tow pulled up through the cavity at three different speeds (8, 18, and 39 mm/min at the first, second, and third dashed lines, respectively.) The sample temperature was measured by the local OFT.

The temperature of a stationary sample can be regulated by the controllers for a well-behaved sample, as described in earlier work [1-2]. The temperature history of a ill-behaved stationary sample is shown in Figures 6 and 7. The two-color temperature from the local OFT for the 1.6 mm mullite tube in Figure 6 steadily approached a constant value near 960°C, while decreasing by ~100°C over 11 minutes under a constant P_a. The signal from the global light sensor in Figure 6 remained quite constant, indicating that the mean sample temperature was also constant. The video record for the sample showed that the sample heating was limited to a hot spot approximately 3-4 mm long. Also, the hot spot slowly drifted up to a position only partially viewed by the local OFT, causing the measured drop in temperature.

An extreme example of the same phenomenon is given in Figure 7 for a 1.6 mm mullite tube.

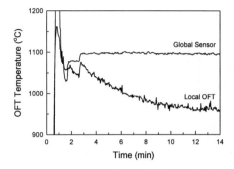

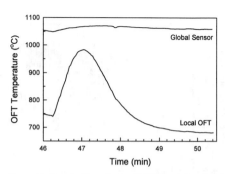

Figure 6. Signals from the local OFT and the global light sensor for a stationary 1.6 mm mullite tube. The global signal is plotted in arbitrary units.

Figure 7. Signals from the local OFT and the global light sensor for a stationary 1.6 mm mullite tube. The global signal is plotted in arbitrary units. A hot spot migrated from bottom to top of the tube sample.

Prior to the temperature record shown, the hot spot on the tube was positioned below the field-of-view of the local OFT by pulling the tube through the cavity in a downward direction. The temperature record in Figure 7 is for the sample after stopping the tube motion. The local OFT indicated a large, rapid increase and then a large fall in the sample temperature. The signal from the global light sensor was relatively constant with a small initial increase leading to a small gradual drop. The video record showed that the hot spot on the tube rapidly drifted upward past the local OFT and settled at a position above the field-of-view of the local OFT. The temperature maximum corresponds to the hot spot passing directly in front of the local OFT.

The presence of hot spot heating and hot spot drifting was observed for all mullite samples tested in this work and in earlier work [2]. The NICALON tows did not exhibit hot spot heating, but did clearly prefer to heat only on the upper half of the sample in the cavity.

This example of ill-behaved heating illustrates the problem of temperature measurement with a sensor, such as the Accufiber OFT, that viewed only a fraction of the sample in the microwave cavity. A global light sensor that viewed and measured the emitted light from the entire sample provided a better indicator of temperature control. The absorbed power P_a was selected as the potential control variable in this work precisely due to the extreme difficulty in measuring the sample temperature. The power controller tests indicated that controlling P_a can provide temperature regulation for stationary samples with different types of heating behavior.

The temperature record for a NICALON tow sample, shown previously in Figures 3 and 5, is plotted in Figure 8 for the local OFT and the global light sensor. The local OFT temperature varied significantly at each of three different pull speeds (8, 18, and 39 mm/min). Because temperature from the local OFT was negatively effected by rapid fluctuations in the heated tow fraction in the field-of-view, this measurement is not conclusive evidence against temperature control. The signal from the global light sensor in Figure 8 also varied significantly at the two greater pull speeds, indicating a lack of temperature control. At the slowest speed, the global light sensor signal suggests that mean tow temperature was under control, recalling from Figure 3b that P_a was increased three times in the first half of the test at this speed.

The temperature record for a 4.7 mm mullite tube is shown in Figure 9 at a slow pull speed of 8 mm/min. At the start of the motion, the hot spot was located above the local OFT and then was rapidly pulled past the local OFT, coming to a steady position below the local OFT. This motion relative to the local OFT is consistent with the temperature curve. The oscillation in the temperature curve after 35 minutes was due to an oscillation in the position of the hot spot. The

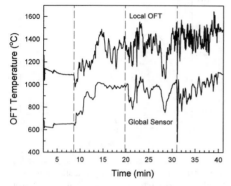

Figure 8. Signals from the local OFT and the global light sensor for a NICALON tow pulled up through the cavity at three different speeds (8, 18, and 39 mm/min at the first, second, and third dashed lines, respectively). The global signal was plotted in arbitrary units.

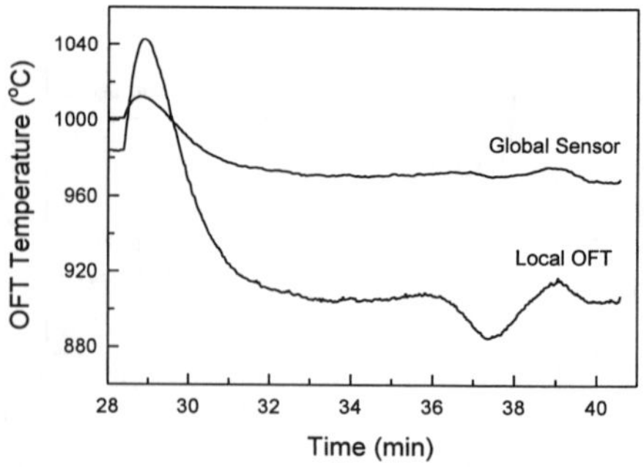

Figure 9. Signals from local OFT and global light sensor for a 4.7 mm mullite tube pulled down through the cavity at 8 mm/min. The global signal is plotted in arbitrary units.

signal from the global light sensor in Figure 9 suggests that the mean sample temperature was well controlled once the hot spot had reached its steady position within the cavity.

CONCLUSIONS

The power controller performed, as designed, by holding the derived absorbed power P_a constant for stationary samples, slowly moving mullite tubes, and rapidly moving NICALON tows. The frequency controller rapidly adjusted the TWT output to track the cavity resonance frequency as the sample temperature changed. The mean sample temperature was regulated for stationary and slow moving (<10 mm/min) samples by controlling the absorbed power P_a. For faster moving samples, the mean temperature was found to fluctuate by 200° to 400°C over several seconds despite overt control of P_a.

Temperature measurements by an Accufiber OFT sighting only a fraction of the tow or tube in the cavity were strongly hindered by non-symmetric heating in a TE_{10n} cavity and by the difficulty to target the hottest portion of the sample. Samples frequently heated at hot spots (mullite tubes) or only on the upper half of the vertical sample (NICALON). To assist the OFT, a second OFT (or a photodiode) was used to view the entire sample and to gauge the relative degree of heating by measuring the emitted light intensity.

REFERENCES

1. G.J. Vogt, A.H. Regan, A.S. Rohlev, and M.T. Curtin in <u>Microwaves: Theory and Application in Materials Processing III</u>, edited by D.E. Clark, D.C. Folz, S.J. Oda, and R. Silberglitt (Ceram. Trans. **59**, Westerville, OH, 1995) pp. 125-132.
2. G.J. Vogt, A.H. Regan, A.S. Rohlev, and M.T. Curtin in <u>Conference Proceedings for Microwave and High Frequency Heating 1995</u> (Cambridge University, Cambridge, U.K., 1995) pp. F3.1-F3.4.

Part IX

Alternate Microwave Sources

MILLIMETER WAVE MATERIAL SEPARATION SYSTEM

S. BIRKEN*, K. BIRKEN*, K. CONNOR**, G. SCHEITRUM***
* Energy Separation Systems, Clifton Park, NY
** ECSE Department, Rensselaer Polytechnic Institute, Troy, NY
*** Litton Systems EDD, San Carlos, CA

ABSTRACT

A new process is in development that uses high power millimeter wave energy to separate specific components from a composite material stream. The component to be separated from the material stream must have a combination of high loss tangent and low permittivity with respect to the other materials involved. In this case, when the material is subjected to a high power, high frequency RF source, the desired component can be vaporized and extracted without transferring significant heat to the remaining material. Since absorption increases significantly with frequency, it is important to use a millimeter wave source. A vacuum pump with sufficient pumping speed can transport the vaporized component for collection. The bulk material can either be collected or recycled through the system for further extraction.

INTRODUCTION

Innovative technologies are necessary to effectively separate specific components from a variety of composite materials in an energy efficient manner if, for example, impurities are to be removed from materials to increase their commercial value or sites contaminated with toxic and/or radioactive wastes are to be cleaned up. Rather than relying on chemical processes, the technology described here is based on the use of RF energy alone and therefore involves only the materials to be processed and not solvents or other chemicals that only add to disposal problems. Other materials processing techniques that use microwave or other RF heating share at least some secondary characteristics with the process to be described here [1, 2, 3]. Radio frequency heating of liquid wastes for volume reduction and stabilization in solid form, based on slow application of moderate levels of RF power, permits melting and stimulates out gassing from liquids and solids. Some constituent separation occurs, but is not the main purpose of the technology. Microwave accelerated organic synthesis has also been used as a source of selective heat in chemical processing, because of its potential for greatly reducing reaction times. However, the chemistry involved is still conventional and involves the use of solvents [4]. The technology described here makes use of the same tendency of some materials to preferentially absorb microwave energy that is exploited in the synthesis process, but has the advantage of not adding any additional materials, except possibly benign materials to help direct the microwaves to the optimum location. To take maximum advantage of selective heating, without adding significant energy to the bulk of materials involved, a process has been conceived that uses high power, high frequency radio frequency energy to separate the constitutive components of materials [5]. For conventional application of radio frequency energy for commercial purposes, the total applied power is relatively low compared to what is now becoming available from source manufacturers. It is the existence of this high power capability that makes possible this new approach.

DESCRIPTION OF PROCESS

RF power delivered per unit volume of material is given by the expression

Mat. Res. Soc. Symp. Proc. Vol. 430 © 1996 Materials Research Society

$$P_\mu = \frac{1}{2}\omega \, \epsilon_r\epsilon_o \, \tan\delta \, E^2 = 5.6x10^{-11} \, \tan\delta \, \epsilon_r \, fE^2 \qquad (1)$$

where ϵ_r is the dielectric constant, f is the frequency, E is the intensity of the electric field of the wave (volt/m), and $\tan\delta$ is the loss tangent. The electrical properties of materials vary a great deal with material composition, temperature, frequency of RF wave, time (as material is modified by heating), etc. It is possible, therefore, to selectively heat the component of a material that most readily absorbs RF energy. Note that the higher frequencies are better as is high power as expressed in the electric field intensity.

As an example of the application of this power deposition process, consider a sample surrogate soil with contaminants and waste as follows (at f=3 GHz, room temperature, and E=2x10^6 Volts/meter):

TABLE 1 Surrogate Material

Material	ϵ_r	$\tan\delta$	Power Density (Watts/cm^3)
Sandy Soil	2.55	.0062	10^4
UO$_2$	4.27	.00115	$3.3x10^3$
Cellulose	1.82	.028	$3.4x10^4$
Oil	2.16	.0043	$6.2x10^3$
Water	52	2.44	$8.5x10^7$

For a discussion of material data used in this paper, see [6]. The rate at which power is delivered to each material is seen to be very different. Note that, since water has a permanent electric dipole moment, it is the most effective absorber of RF energy and organic matter is usually much more absorbent than inorganic.

The rate of temperature rise is determined by the equation

$$Power = \frac{Heat}{Time} = \frac{M}{t}c_p(T-T_o) \qquad (2)$$

where M is mass (kgm), c_p is specific heat (joule/kgm·°K), (T-T$_o$) is temperature change (°K) and t is time (sec). The power delivered to the material by an RF source causes the temperature to rise if there are no significant mechanisms to convect or conduct the energy away. Also, if the heating can be achieved very rapidly, then only the very absorptive material component will increase in temperature because conduction and convection occur very slowly. High power sources permit such rapid heating. The differing specific heats and heats of vaporization of the materials then permit separation. For example, the power density necessary to vaporize sand is more than 50 times larger than that required for UO$_2$.

Large objects are not as uniformly heated as are small objects. Thus penetration is better for powdered materials than for solids. This phenomenon of nonuniform heating is exacerbated by thermal runaway in which the ability of the material to absorb energy goes up dramatically with temperature [7]. The microwaves should penetrate the particles and heat uniformly if the wavelength is somewhat larger than the particle. The free space wavelength is $\lambda = c/f$ where c is the speed of light. At 60 GHz, for example, the wavelength is 5 mm. The wavelength is about half this big inside most materials. The tendency of heat to be deposited in already hot regions can make microwave heating similar to that due to more conventional heat sources for which heat is applied externally and must propagate to the interior of an object by diffusion. A primary advantage in efficiency of RF heating -- that heat is deposited throughout the material rather than just at the surface -- is thus lost in large objects. To obtain uniform heating, much smaller objects (i.e. powder) are required. To both better localize the heating of the desired material component and to efficiently remove the vaporized material, the process will likely work best in vacuum.

A very important aspect of the proposed process involves timing. Conventional microwave heating is applied steady state so materials are heated slowly over a period of minutes or, occasionally, seconds. For such slow heating it is possible for thermal conduction and convection to become significant while the power is being applied. For this process, power is applied at such high levels that the most absorbent materials are heated before significant energy is transferred to the surrounding media (heating in less than one second), taking maximum advantage of thermal runaway.

DESIGN ISSUES FOR MICROWAVE SEPARATION

Ideally, in the proposed process, a continuous stream of small microwave absorbent particles are dropped through an evacuated microwave cavity. (See [3] for a general diagram of the system.) Since effective separation with RF energy is based on selective heating, it will be necessary to determine whether or not it is possible to bring some minority constituent of a complex mixture up to the necessary temperature for sublimation, without imparting much energy to the bulk material. Because microwave power is expensive, any process that uses it must either provide an outcome that is not available by any other means and/or must efficiently apply power to just the part of the material that needs it. To assess the general nature of this issue, we considered three kinds of material streams. (1) Most of the particles are some base material that does not readily absorb microwave power, while a small fraction of the particles are some absorbent impurity. The particles are either all impurity or all base, as would be the case when some waste material is mixed with soil. (2) Each particle is roughly the same consisting of mostly base material and with a small amount of impurity concentrated at one location, as would be the case when some impurity crystal is located inside a base mineral. (3) Amorphous material with small amounts of impurity randomly located throughout the base. Only the first two cases appear to offer attractive opportunities for application of RF energy.

Case (1) Here the heat goes to the impurity preferentially because the loss tangent is higher. The power in from the microwaves heats up the particle to a temparature T ($^\circ$K) which then radiates the heat away. Assuming black body radiation, *Stefan's Law* predicts that the output power is proportional to $T^4 - T_o^4$, where T_o is the ambient temperature. The final particle temperature is determined by equating the input and output powers. For a spherical particle with properties similar to those of uranium dioxide, f = 60 GHz, a field intensity of about 8×10^5 v/m, and particle radius about 2 mm, temperatures in the tens of thousands of degrees are possible. Clearly, the particle will vaporize at a much lower temperature; this calculation is only meant to show that the heating method proposed is realistic at moderate power levels.

Case (2) Consider a simplified problem of a spherical impurity lodged in a spherical bulk. The small sphere of radius r_1 is heated with microwaves. The heat is conducted into the bulk and finally radiated away from the surface of radius r_2. The goal of this application is to preferentially heat the impurity. However, some of the heat is conducted into the bulk while the impurity is heating up. In case (1), the heat left the impurity solely by radiation since the particles were separated by vacuum. Here, there will be some heating of the bulk. Heating of the impurity must occur over a much shorter time than the thermal time constant of the bulk, where R is the thermal resistance and C is the thermal capacity.

$$\tau = RC \approx \frac{\rho c_v}{k} \frac{r_2^3}{3r_1} \qquad (3)$$

If we choose typical numbers for materials found in soil of conductivity $k = 1$ (watts/m·°K), mass density $\rho = 2000$ (kgm/m^3), and specific heat $c_v = 1000$ (joule/kgm· K) and also assume 1% impurity, then τ is about 3 seconds. Thus, as long as heating occurs over a shorter time than this, heat transfer to the bulk will be given by the resistive term only. To determine the necessary conditions to raise material temperature from T_1 to T_2, set the power input equal to output power conducted from the impurity to the bulk:

$$P_{con} = \frac{T_1 - T_2}{R} \approx 4\pi r_1 k T_1 \qquad (4)$$

The geometric advantage of microwave heating for case (1) is clear because the heat from the impurity particles can pass to the bulk only through radiation. (The particles are insulated by vacuum.) For case (2), the bulk will be heated up to some extent. However, if the heating occurs over a short time compared to the thermal time constant, then the bulk heating will be small. Assuming that the heating occurs, for example, in milliseconds, then the average temperature of the bulk will be less than 10% of the impurity. (τ is proportional to the outer radius squared.)

Heating Rate

To determine the time it takes to heat up a quantity of material to some specified temperature requires the solution of a driven heat equation. Using a time-domain-finite-difference method for this analysis, we found that the results agree quite well with the application of equation (2). This expression implies that there is no heat transfer away from the material, that all power goes to raising the temperature. The TDFD model for the heating, assuming either black body radiation for particles in vacuum or conduction from a material of interest to a bulk material, showed that this expression gives a good ballpark number as long as the heating rate is high. The use of this simple expression is based on the assumption that material properties do not change with temperature or power level. Since microwave absorption tends to increase rather dramatically with temperature and, many times, falls off a bit with increasing power levels, this simple approach is not unreasonable if all we need is a very conservative estimate of the heating time.

The following table contains the information necessary to determine the time required to raise the temperature of a material sample a specified amount. Thermal conductivity is not used in equation (2), but it is significant for both the more general analysis of the TDFD method and to argue that the simple model is reasonable. The materials listed include *Test*, a fictitious material; uranium dioxide, a typical unwanted soil contaminant; and rutile, a mineral that has recently been found to have some interesting properties at microwave frequencies [5].

TABLE 2 Physical Material Properties

Material	Conductivity	c_p	ρ	ρc_p
Test	1	1000	2000	2×10^6
UO$_2$	5	235	10000	2.4×10^6
Rutile	6	680	4800	3.25×10^6

Two different power levels were assumed (800 kV/m and 1500 kV/m) which span a practical range for high powers and five different frequencies from for which commercial sources are available.

The electrical properties used hold for room temperature or a little above [5]. Finally, the required temperature rise for each material to reach a vapor state can be quite different. For *Test*, UO_2, and rutile these temperatures were 500°, 2700°, and 3000°, respectively.

For the three materials, the times to reach vapor state are shown below for the two power levels at each of the frequencies.

TABLE 3 Time to Reach Vapor State (Upper Bound)

Material	Freq.	17.43 GHz	28 GHz	40 GHz	110 GHz	140 GHz
Test (Low E)		1.8 sec	1.1 sec	.8 sec	.3 sec	.2 sec
Test (High E)		.5 sec	.3 sec	.2 sec	.08 sec	.06 sec
UO_2 (Low E)		8 sec	5.8 sec	3.8 sec	1.3 sec	1.1 sec
UO_2 (High E)		2.4 sec	1.6 sec	1.2 sec	.4 sec	.3 sec
Rutile (Low E)		.2 sec	.2 sec	9 sec	.12 sec	.08 sec
Rutile (High E)		.06 sec	.05 sec	2.5 sec	.03 sec	.03 sec

Different electrical properties are used for rutile at each frequency. One can readily conclude that higher power levels also give higher heating rates, but the numbers should not be treated as anything but rough approximations. This result can be misleading. The power density may be higher for the higher electric field, but less material will be heated per microwave tube. Since the separation process depends on high power to heat selected materials to vaporization before any other processes can occur at significant levels, exactly what is meant by high power must be determined for each application. Most radio frequency (RF) implementation problems will be reduced at lower electric fields, so an attempt must be made to find the lowest possible operating field. The time to heat to vaporization is not the quantity of greatest practical interest. Rather, it is the mass of material heated to vapor state per second per unit RF source, which is shown below for the same frequencies. Note that the power levels indicated here are determined by tube size and not by peak electric field, as was the case for the table above.

TABLE 4 Mass Per Second Per Unit RF Source

Material	Power Level	10 kw	100 kw	1 Mw
Test		.02 kg/sec	.2 kg/sec	2 kg/sec
UO_2		.016 kg/sec	.16 kg/sec	1.6 kg/sec
Rutile		.005 kg/sec	.05 kg/sec	.5 kg/sec

EXPERIMENTAL SYSTEM PLANS

We have specified a simple system (shown in Figure 1) for Proof-of-Principle studies using conventional gyrotrons and found that electric field strengths of the order of 10^6 V/m can be achieved with a fast focusing mirror for the range of relative permittivities expected for the materials of interest. We have done likewise for a pilot scale test facility.

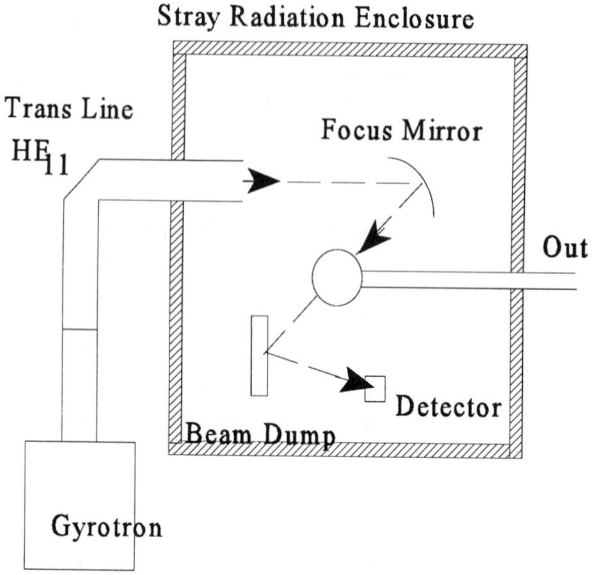

FIGURE 1 Proof - of Principle Experiment. Material Processed in Center Region With Gases Extracted Through Pipe Marked *Out*.

REFERENCES

1. NATIONAL RESEARCH COUNCIL, Microwave Processing of Materials, Publication NMAB-473, National Academy Press, Washington, D.C. 1994.

2. A. C. Metaxas, R. J. Meredith, Industrial Microwave Heating, Peter Peregrinus Ltd, London (1983).

3. R. D. Smith, Microwave Power in Industry, PERI Report EM-3645 (1984).

4. G. Majetich and R. Hicks, *Applications of Microwave-Accelerated Organic Synthesis*, Radiat. Phys. Chem **45** (1995) 567-579.

5. S. Birken and K. Birken, Patents 4,894,134 and 5,024,740

6. J. M. Borrego, K. A. Connor, and J. Braunstein, Measurements of Dielectric Properties for Intense Heating Applications, this symposium.

7. V. M. Kenkre, L. Skala, M. W. Weiser, J. D. Katz, Theory of microwave interactions I ceramic materials: the phenomenon of thermal runaway, J. Materials Science **26** (1991) 2483-2489.

PULSED 35 GHz GYROTRON WITH OVERMODED APPLICATOR FOR SINTERING EXPERIMENTS

A. W. FLIFLET, R. P. FISCHER, A. K. KINKEAD,* AND R. W. BRUCE*
Plasma Physics Division, Naval Research Laboratory, Washington, DC 20375, U.S.A.,
fliflet@ppd.nrl.navy.mil

ABSTRACT

The microwave sintering of nanocrystalline alumina compacts is currently under investigation at NRL. This paper will discuss an overmoded microwave furnace based on a 35 GHz pulsed gyrotron which is currently being set up to extend ongoing microwave sintering experiments at 2.45 GHz to 35 GHz. The gyrotron operates at 70 kV and currents up to 10 A. It is driven by a hard tube, variable pulse length (1–15 μs) modulator at repetition rates up to 1 kHz. The gyrotron can produce peak powers up to 100 kW at an efficiency of 20%, and average powers up to 200 W. The gyrotron output is transported via pressurized K_a-Band waveguide to an overmoded resonator containing the workpiece. In initial experiments, the resonator will consist of a piece of WR-284 waveguide. The operation of the system in preliminary sintering experiments is described.

INTRODUCTION

Microwave heating is emerging as an effective method of processing high performance ceramic materials. Advantages include the ability to deposit energy volumetrically in the sample, and the possibility of rapid heating and cooling profiles. To date microwave processing research is based mainly on conventional low-frequency (2.45 GHz) microwave ovens which do not couple microwave power efficiently to low-loss ceramics and often have large heating gradients. The development of powerful gyrotrons has opened up the millimeter-wave regime (>28 GHz) for processing ceramic materials. Millimeter-waves couple more strongly than conventional microwaves to low-absorption ceramics such as pure oxides, and highly uniform fields can be achieved in a compact overmoded cavity applicators. In addition, a directional beam can be formed to provide localized heating for applying coatings, joining, or brazing. A ceramic processing research facility based on continuous wave (CW) and pulsed gyrotron sources has recently been established at the Naval Research Laboratory (NRL). This paper will discuss an overmoded microwave furnace based on a 35 GHz pulsed gyrotron which has been set up to conduct microwave sintering experiments on small compacts of ultrafine grained ceramic materials. Microwave processing of ultrafine grain ceramics is of interest because the capability for rapid processing should retard grain growth, allowing fully densified materials to be achieved with grain size in the 10–100 nm range.

An excellent review of the field of microwave sintering of ceramic materials has been written by Sutton [1]. Sintering experiments using 2.45 GHz cavities have been reported by Katz et al. [2] and Blake and Katz [3]. Guidelines developed at the Oak Ridge National Laboratory for using multimode 2.45 GHz applicators have been given by Janney, Kimrey and Kiggans [4], and the use of 28 GHz radiation for microwave sintering has been discussed by Janney and Kimrey [5] and by Tiegs et al.[6]. There have been a number of recent studies of microwave sintering of nanocrystalline ceramics [7]–[12].

35 GHZ GYROTRON POWERED APPLICATOR

A schematic of a pulsed 35 GHz gyrotron powered microwave heating system for sintering experiments is shown in Fig. 1. The gyrotron operates in the TE_{01} circular waveguide mode at a voltage of 70 kV and currents up to 10 A. It is driven by a hard-tube, variable pulse length (1–15 μs) Rockwell Power Systems™ modulator which has a maximum pulse repetition frequency (PRF) of 1000 Hz. The gyrotron is based on a Varian VUW-8110 MIG-type electron gun and is placed in the

---

* SFA, Inc., 1401 McCormick Dr., Largo, MD 20771-5322

Mat. Res. Soc. Symp. Proc. Vol. 430 © 1996 Materials Research Society

6.35 cm diameter bore of a superconducting magnet built by American Magnetics, Inc. The magnet provides a 13.5 kG field at the gyrotron cavity. The gyrotron system was originally developed for microwave effects testing at NRL. The gyrotron can currently deliver up to 210 W average power to the applicator. The peak power is about 50 kW and the maximum duty factor is 0.5%. The gyrotron output efficiency is ~20% and the average power is controlled by varying the PRF. The gyrotron output waveguide, which also serves as the electron beam collector, has an OD of 1.59 cm. It is surrounded by a water jacket to provide cooling, as is the interaction cavity. After exiting an alumina output window, the microwave power passes through a mode converter which transforms the TE_{01} circular waveguide mode into the TE_{10} mode in rectangular Ka-Band waveguide. The power is then transported to the cavity applicator and monitored using standard Ka-band waveguide components. The waveguide and applicator are pressured with nitrogen to avoid microwave breakdown. The forward and reflected peak powers are sampled using crystal detectors and the average forward power is monitored using an Hewlett-Packard 432B Power Meter. The cavity applicator consists of a 17 cm length of WR-284 waveguide (width = 7.21 cm, height = 3.40 cm). The upstream end has a 1 cm diameter aperture for injecting the microwave power. The edges of the aperture are rounded to minimize microwave field enhancements. The downstream end consists of a 9.5 mm thick brass plate with an array of 2 mm diameter holes with centers separated by 2.54 mm. The brass plate serves as a viewing port and is covered by a Lucite plate to allow pressurization of the applicator. As shown in Fig. 1, the system is operated without a circulator to absorb power reflected from the applicator, however, this power does not significantly impact the operation of the gyrotron at the present power levels. In any case, the reflected power is quite low at sintering temperatures. Typical measured power reflection coefficients include: $R \approx 0.13$ for the empty cavity, $R \approx 0.05$ for cavity containing casket and compact at room temperature, and $R \leq 0.01$ when compact is hot.

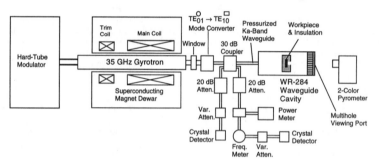

Fig. 1: Schematic of 35 GHz pulsed gyrotron with multimode resonator furnace.

The ceramic compact is placed in a Zircar SALI™ or ZYZ3™ fiberboard casket to provide thermal insulation from the cold-walled cavity. In previous experiments at 2.45 GHz, the ZYZ3™ fiberboard was found to be sufficiently lossy that it can be readily heated by microwaves. Thus, a casket made from this material provides a simple method of hybrid heating. In contrast, the less lossy SALI™ fiberboard provides thermal insulation without hybrid heating. A typical casket configuration is shown in Fig. 2. The casket has a small hole to allow the surface temperature of the compact to be measured using a Mikron™ M190 two-color infrared pyrometer. The pyrometer was calibrated by heating samples in a conventional furnace up to 1500°C and comparing the pyrometer reading obtained through the viewing port with that of the thermocouple temperature sensor installed in the furnace. The pyrometer was found to over estimate the sample temperature by about 50°C. This discrepancy is attributed mainly to the plastic plate used with the viewing port.

SINTERING EXPERIMENTS

To date the furnace has been used to heat nanocrystalline alumina compacts made with powders produced at NRL, alumina compacts made from commercial micron-grain-sized powders (Alcoa™

A16 and Baikolox™ Cr-10), and nanocrystalline titania compacts also produced at NRL. The nanocrystalline powders were prepared using a modified sol-gel technique [13]. The precursor alkoxide was suspended in absolute ethanol while stirring. A solution of distilled deionized water (DDW)–ethanol (excess water) was made into an aerosol with nitrogen and sprayed into the alkoxide suspension. The reactants were stirred for an additional 30 min. after the addition of the water solution and then filtered and washed with water. The precursor powders were calcined in air at 700°C for 2 hours to remove the organic moieties. The calcined powders were then uniaxially pressed to 14 MPa and cold isostatically pressed (CIP'ed) to 420 MPa.

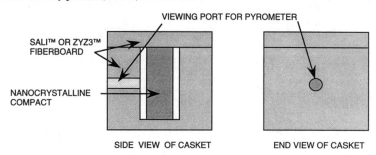

Figure 2: Schematic of casket used to provide thermal insulation from cold-walled cavity furnace.

All of the compacts could be heated to a temperature of ~ 1400°C within a few minutes as summarized in Table I. The nanocrystalline alumina compacts were heated using a SALI™ casket indicating that at 35 GHz these compacts can be heated from ambient temperature by direct deposition of microwave energy. On the other hand, the Alcoa™ A16 and Baikolox™ CR-10 compacts could be heated in the ZYZ3™ casket, but not in the SALI casket, indicating that even at 35 GHz, some hybrid heating is necessary for these coarser grained materials. The titania compacts could be readily heated using a SALI™ casket. This was expected since titania is significantly more lossy than alumina. The alumina results appear to be consistent with room temperature loss tangent measurements which indicate that the loss tangent of polycrystalline ceramics increases with decreasing grain size. A typical furnace temperature profile for hybrid heating of an Alcoa™ A16 compact is shown in Figure 3. The presence of hybrid heating is indicated by the fact that the workpiece does not become hot until after the casket is hot. The accuracy of the temperature measurement is estimated to be ±20°C. In addition, the internal temperature of the workpiece is expected to be somewhat higher than the surface temperature.

Table I. Processing Temperatures Achieved in 35 GHz Furnace

Material	Power (W)	Temperature (°C)	Hybrid Heating?
Nanocrystalline alumina	200	1429	No
Alcoa™ A16 alumina	200	1395	Yes
Baikolox™ Cr-10 alumina	208	1392	Yes
Nanocrystalline titania	100	1358	No

Preliminary sintering experiments have been carried out for several doped nanocrystalline alumina compacts, and densification estimates have been made. These results, shown in Table II, indicate that partial densification (≈ 70%TD [theoretical density]) is obtained at a temperature of ~ 1300°C. While there was no evidence of thermal runaway, these compacts had a tendency to crack during the sintering process, complicating the density measurement. A scanning electron micrograph (SEM) of the compact heated to 1429°C (alumina doped with 1% by weight MgO) is shown in Fig. 4. The initial grain size of the powder is estimated based on X-ray diffraction measurements to be ~ 10 nm. Fig.4 indicates a grain size of ~ 75 nm after partial sintering, and the presence of a micron-sized agglommerate.

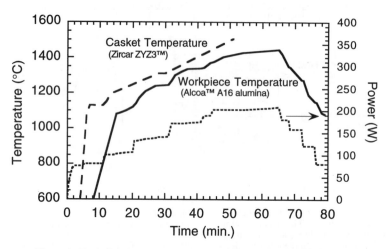

Figure 3: Typical hybrid temperature profile for Alcoa™ A16 compact.

Table II: Preliminary sintering data for nanocrystalline alumina compacts in 35 GHz furnace

Material	Peak Temp. (°C)	Hold Time (min.)	Initial Dens. (%TD)	Final Dens. (%TD)
Al_2O_3 + 1% MgO	1090	0	43%	52%
Al_2O_3 + 10% Y_2O_3	1205	0	42%	52%
Al_2O_3 + 1% MgO	1260	30 min.	43%	74%
Al_2O_3 + 1% MgO	1320	25 min.	43%	72%
Al_2O_3 + 1% Y_2O_3	1313	0	42%	72%
Al_2O_3 + 1% MgO	1429	0	43%	>70%

DISCUSSION AND CONCLUSIONS

A 35 GHz multimode microwave cavity furnace has been constructed, instrumented, and integrated into a microwave system that has been used to perform preliminary sintering experiments. The applicator power was easily controlled by varying the pulsed gyrotron PRF. Experiments using the 35 GHz furnace are continuing with the objective of achieving higher final densities and further characterizing the dependence of density and grain size on sintering temperature and hold time. The effect of additives on densification and microstructure of microwave sintered nanocrystalline materials is also being investigated.

The 35 GHz multimode cavity furnace has been shown to be an effective means of heating small ceramic compacts. The present sintering results indicate that nanocrystalline ceramics can be sintered with 35 GHz radiation without using hybrid heating techniques. High purity commercial fine-grained alumina could also be readily sintered with a simple form of hybrid heating. This data suggests that the nanocrystalline alumina is more lossy than the commercial alumina at room temperature for high frequency microwaves. Work on improving the furnace cavity and insulating casket design is in progress with the aim of improving heating uniformity and achieving higher workpiece temperatures.

Figure 4: SEM of nanocrystalline alumina compact sintered in 35 GHz
furnace using SALI™ fiberboard casket.

ACKNOWLEDGMENT

This work is supported by the Office of Naval Research.

REFERENCES

1. W. H. Sutton, "Microwave Processing of Ceramic Materials," *Ceramic Bulletin* **68**, 376 (1989)
2. J. D. Katz, R. D. Blake, J. J. Petrovic, and H. Sheinberg, "Microwave sintering of boron carbide," *Mat. Res. Soc. Sym. Proc.* **124**, 219 (1988).
3. R. D. Blake and J. D. Katz, "Microwave sintering of large ceramic bodies," *Ceramic Transactions* **36**, 459 (1993).
4. M. A. Janney, H. D. Kimrey, and J. O. Kiggans, "Microwave processing of ceramics: guidelines used at the Oak Ridge National Laboratory," *Mat. Res. Soc. Sym. Proc.* **269**, 173 (1992).
5. M. A. Janney and H. D. Kimrey, "Microwave sintering of alumina at 28 GHz," *Ceramic Powder Science II*, p. 919, G. L. Messing, E. R. Fuller, and H. Hausner, Editors, American Ceramic Society, Westville, Ohio (1988).
6. T. N. Tiegs, J. O. Kiggans, and H. D. Kimrey, "Microwave sintering of silicon nitride," *Mat. Res. Soc. Sym. Proc.* **189**, 267 (1990).
7. J. A. Eastman, K. E. Sickafus, J. D. Katz, S. G. Boeke, R. D. Blake, C. R. Evans, R. B. Schwarz and Y. X. Liao, "Microwave sintering of nanocrystalline TiO_2", *Mat. Res. Soc. Sym. Proc.* **189**, 273 (1990).
8. S. N. Kumar, A. Pant, R. R. Sood, J. Ng-Yelim and R. T. Holt, "Production of ultra-fine silicon carbide by fast firing in microwave and resistance furnaces", *Ceramic Transactions,* **21**, pp. 395-402, (1991).

9. D. Vollath, R. Varma and K. E. Sickafus, "Synthesis of nanocrystalline powders for oxide ceramics by microwave plasma pyrolysis", *Mat. Res. Soc. Sym. Proc.* **269**, 379 (1992).
10. D. Vollath, "Some activities of microwave processing of ceramics in Germany", Ceramic Transactions **36**, pp. 147-156 (1993).
11. J. Freim, J. McKittrick, J. Katz and K. Sickafus, "Phase transformation and densification behavior of microwave sintered γ-Al_2O_3", *Mat. Res. Soc. Sym. Proc.* **347**, 525 (1994).
12. J. Zhang, Y. Yang, L. Cao, S. Chen, X. Shong, and F. Xia, "Microwave sintering on nanocrystalline ZrO_2 powders", *Mat. Res. Soc. Sym. Proc.* **347**, 591 (1994).
13. L. K. Kurihara, G. M. Chow, and P. E. Schoen, "Low temperature processing of nanoscale ceramic nitride particles using molecular precursors," Patent Disclosure, U.S. Navy Case 82,737 (1995), application pending.

ADVANCED CERAMICS SINTERING
USING HIGH-POWER MILLIMETER-WAVE RADIATION

Y. Setsuhara *, M. Kamai *, S. Kinoshita *, N. Abe *, S. Miyake * and T. Saji **
* Welding Research Institute, Osaka University, Osaka, Japan, setuhara@jwri.osaka-u.ac.jp
** Fujidempa Kogyo Co., Ltd., Ibaraki, Japan

ABSTRACT

The results of ceramics sintering experiments using high-power millimeter-wave radiation are reported. Sintering of silicon nitride with 5%Al_2O_3 and 5%Y_2O_3 was performed in a multi-mode applicator using a 10-kW 28-GHz gyrotron in CW operation. It was found that the silicon nitride samples sintered with 28 GHz radiation at 1650°C for 30 min reached to as high as theoretical density (TD), while the conventionally sintered samples at 1700°C for 60 min resulted in the density as low as 90% TD. Focusing experiments of millimeter-wave radiation from the high-power pulsed 60-GHz gyrotron have been performed using a quasi-optical antenna system (two-dimensional ellipso-parabolic focusing antenna system) to demonstrate the feasibility of the power density of as high as 100 kW/cm². Typical heating characteristics using the focused beam were made clear for this system. It was found that the densification of yttria-stabilized zirconia (ZrO_2-8mol%Y_2O_3) samples to as high as 97% TD was obtained from the sintering with focused 60 GHz beam in pulse operation with a 10-ms pulse duration at a 0.5Hz repetition. The densification temperature for the zirconia could be lowered by 200°C than that expected conventionally.

INTRODUCTION

Remarkable progress in the microwave sintering of ceramics have been made since 1980's following the pioneering work by W. H. Sutton [1], T. T. Meek et al. [2], and M. A. Janny and H. D. Kimrey [3]. Since then the microwave sintering technology has been widely recognized to offer a number of advantages over conventional sintering process; *i.e.*, rapid and selective heating, higher densification rates at lower temperature, achievement of sintered bodies with finer grain size and improved mechanical properties.

The microwave frequency from several GHz to several tens of GHz (millimeter wave region) has been used for ceramics sintering. The most commonly employed frequency is 2.45 GHz. But at this frequency there are problems associated with inefficiency in direct heating of low loss ceramics and difficulty in designing a multi-mode applicator with sufficient uniformity to avoid thermal runaway, which are mainly attributed to the longer wavelength of the radiation. Hybrid microwave heating [3,4] has been presented to be one of the solutions for these problems, however, limitations exist in the scale of workpiece and the choice of materials with appropriate coupling, and so on. One possible solution for these problems is to employ the millimeter-wave radiation, where the power absorption per unit volume increases linearly with increasing frequency and the sufficient uniformity of the wave field can be obtainable in a feasible and/or practical applicator size. Experiments with millimeter-wave radiations from Gyrotron tubes have been carried out in USA [3] and followed by the group at Institute of Applied Physics [5] in Russia. In Japan our group [6,7] is also extensively studying this process.

In this paper we present the results of sintering experiments performed with a 10-kW 28-GHz gyrotron in CW operation and a 100-kW 60-GHz gyrotron in pulse operation. The experiments at 28 GHz have been performed in a multi-mode applicator to sinter silicon nitride. The experiments with the 60GHz pulsed millimeter-wave radiation have been aimed at fine focusing of the radiation using a quasi-optical antenna system to obtain high power density for the efficient coupling of the wave energy to the samples. This method is convenient to heat the samples in a small reactor and has a potential for the rapid heating with actively controlled temperature distribution over the workpiece by appropriately scanning the focused beam.

Mat. Res. Soc. Symp. Proc. Vol. 430 ⊚ 1996 Materials Research Society

EXPERIMENTAL PROCEDURES

Millimeter-wave sintering of silicon nitride was performed in a multi-mode applicator operating at 28 GHz. The silicon nitride used in this work was grade Ube-SN-COA (Ube Industries, Ltd.; E-10, with inclusion of $5\%Al_2O_3$ and $5\%Y_2O_3$) with a mean particle size of ~200 nm. Green bodies of silicon nitride were prepared by the slip casting method.

The multi-mode applicator used for the sintering of the silicon nitride is schematically illustrated in **Fig. 1**. The applicator was connected to the 28 GHz gyrotron in CW operation. Typical dimension of the applicator was 600 mm in diameter and 900 mm in length (ratio L/λ ~ 90; L - scale length of the applicator, λ - wavelength of radiation), which allowed a volume of 300mm(diameter) x 300mm(height)

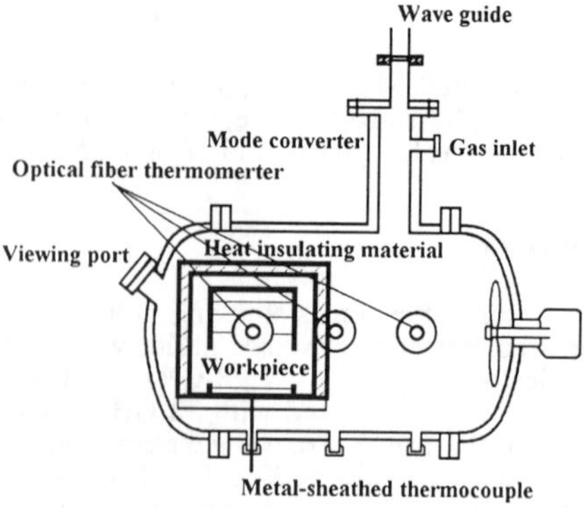

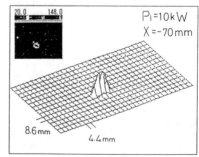

Fig. 1. Schematic illustration of multi-mode applicator used for sintering at 28 GHz.

in the applicator to offer spatially uniform distribution of the wave field.

The quasi-optical antenna system designed for the fine focusing of the 60GHz millimeter-wave radiation, described elsewhere [8] in detail, is schematically illustrated in **Fig. 2**. The millimeter-wave radiation (λ=5mm) in TE_{02} mode was irradiated from a high-power pulsed Gyrotron tube (VGE8060, Varian) with the maximum output power of 200kW and the maximum pulse width of 100ms. The pulse repetition is 0.5Hz, and the average power of the radiation was limited to about 1kW by the capability of the charging DC power supply to the high-voltage condenser bank.

The 60-GHz millimeter wave was radiated through the cut-in-half circular waveguide of 25.6mm in diameter and reflected by the elliptic mirror to the focal point O' on the O_2 axis. When the parabolic reflector additionally intercepts the radiation, the radiation was again reflected to the focal point F to form nearly Gaussian profile. The power distributions of the radiation were measured at various x-positions on the y-z plane. The radiation was injected into the microwave absorber sheet (Eccosorb AN, Emerson and Cuming) located parallel with the y-z plane. The temperature distributions on the sheet were measured using a thermal video system (TVS-3000, Nippon Avionics).

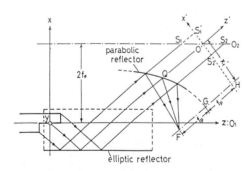

Fig. 2. Schematic diagram of quasi-optical antenna system.

Fig. 3. Beam profile of the focused 60GHz radiation measured at the focal point.

The beam profile observed at around the focal plane (x= -70mm), shown in **Fig. 3**, clearly demonstrates a finely focused and nearly Gaussian beam profile with a full width at half maximum of ~ 10mm. The focused power distribution showed an excellent agreement with the predicted profile using a simulation code[9]. These results ensure the feasibility of a high energy-density beam with 100 kW/cm^2 in the case of 100kW operation. This method of using the focused millimeter-wave radiation may be convenient to heat the samples in a small reactor and have a potential for the rapid heating with an actively controlled temperature distribution over the workpiece by appropriately scanning the radiation beam. Furthermore the focused millimeter-wave beam can be employed in advanced processing of ceramics such as sintering and joining of functionally-gradient ceramics-composite structure, which requires a technology capable of selective heating with the precisely controlled temperature gradient.

The heating characteristics have been studied in terms of the pulse duration and the power of the irradiated beams since they are directly related to the absorbed radiation energy. The focused beam was irradiated onto alumina and alumina with 5 wt% SiC samples using the experimental setup schematically illustrated in **Fig. 4**; 74 mm in diameter and 80 mm in length with a beam entrance of 39 mm in diameter and a viewing port of 40 mm in diameter for temperature measurement. The sample temperature was measured using the thermal video recorder which was separately calibrated to thermocouple readings. Here the cavity was located so that the sample was positioned at the focal point of the beam, and the pulse irradiation was performed in air at a room temperature. The alumina powder used in this work was Sumitomo AKP-50 (α-Al$_2$O$_3$ > 99.995%) with a mean particle size of ~200 nm. The silicon carbide powder used in this work was Showa Densic Ultrafine (SiC > 98 %) with a mean particle size of ~370 nm. Green bodies of alumina and alumina-SiC mixture were formed by cold press to obtain the green density of 54 and 57 %TD, respectively.

Sintering of yttria-stabilized zirconia was carried out using a cylindrical applicator with an axially inserted plunger. The zirconia powder used in this experiment was grade TZ-8Y (Tosoh; ZrO$_2$-8mol% Y$_2$O$_3$) with an average particle size of ~200nm, which was formed to obtain green bodies by slip casting method. The applicator used in the experiment had a dimension of 109 mm in diameter and 420 mm in length (L/λ~84; L-scale length of the applicator, λ-wavelength of radiation). The focused 60GHz beam was injected from the axial direction of the applicator through a 1-mm thick quartz window. The workpiece covered with alumina fiber for thermal insulation was fired with the focused beam in pulse operation with a 10-ms duration at a 0.5-Hz repetition rate. Heating and cooling rate of the samples were controlled to be 40-70°C/min and 60°C/min, respectively, by appropriately selecting the pulse duration and the output power of 60GHz gyrotron.

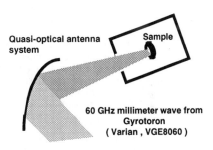

Fig. 4. Experimental setup used for the investigations of the ceramics heating with the focused 60 GHz beam.

RESULTS

Millimeter-wave sintering at 28 GHz in Multi-Mode Applicator

The results summarized in **Fig. 5** clearly demonstrate the accelerated densification of the silicon nitride samples at lower temperature in the 28-GHz sintering experiments performed using the multi-mode applicator shown in Fig. 1. When a green body with a size of 200mm in diameter and 200mm in height was sintered in this applicator, the radiation energy consumed for heating the specimen was ~93% of the incident energy from the 28GHz gyrotron; i. e., a reflected energy of 4% and a wall loss of 3%. The silicon nitride samples sintered with 28 GHz radiation at 1650°C for 30 min reached to nearly full density, >99% theoretical density (TD). In contrast, to achieve the density of ~98% TD conventionally required a temperature of > 1800°C even with specimens sintered for 60 min.

Furthermore, it was found that the achieved density of the 28GHz sintered samples was significantly dependent on the method to form the green bodies especially in terms of the initial pore size. The green bodies with initial pore sizes specifically controlled to be much less than 2000nm was found to be sintered to attain nearly full density, >99% TD, at 1600°C for 30 min. Under this sintering condition, nearly full densification was achieved without suffering appreciable grain growth; *i.e.*, the average grain size of this specifically prepared samples was as small as that of the green body (~200nm). Due to the super fine structure of these specifically prepared samples with nearly full densification, many of the workpieces demonstrated extremely high bending strength exceeding 5GPa.

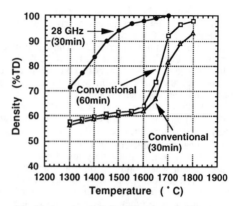

Fig. 5. Accelerated densification of silicon nitride was demonstrated with 28-GHz firing in the multi-mode applicator.

Heating and Sintering of Ceramics Using Focused 60-GHz Radiation

The heating characteristics of alumina and alumina with 5 wt% SiC samples was investigated using the finely focused 60 GHz millimeter-wave radiation in pulse operation. The increased temperature per a single pulse irradiation was measured as a function of pulse duration. The finely focused beam with an output power of 30 kW was irradiated onto the sample placed at the center of the cylindrical metal applicator schematically illustrated in Fig. 4. The results summarized in **Fig. 6** shows the linear heating characteristics of the sample as a function of the pulse duration. The inclusion of the small amount of SiC additive (5 wt%) resulted in 2.5 times higher energy absorption than that for pure alumina. This suggests that at a room temperature the dielectric loss of SiC for the 60-GHz radiation is approximately 50 times higher than that of Al_2O_3.

The results summarized in **Fig. 7** clearly demonstrate the accelerated densification of the yttria-stabilized zirconia samples at lower temperature in the 60-GHz sintering for 60 min. The zirconia sample sintered with 60 GHz radiation at 1120°C for 60 min reached to 97 %TD. In contrast, to achieve the same density conventionally required a temperature of ~ 1300°C.

The fracture surfaces of the green body and the sample densified to 97%TD are compared in **Fig. 8**. In this example the zirconia workpiece was kept at 1120°C for 60 min for sintering. These micrographs indicate that the densification of the yttria-stabilized zirconia has been

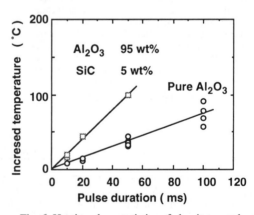

Fig. 6. Heating characteristics of alumina samples as a function of pulse duration of the focused 60-GHz beam with 30 kW.

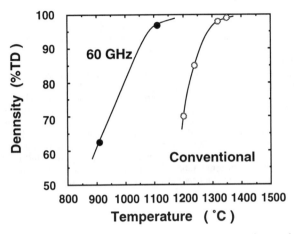

Fig. 7. Accelerated densification of yttria-stabilized zirconia was demonstrated with 60-GHz firing in pulse operation.

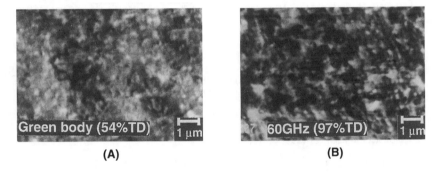

Fig. 8. Fracture surfaces of yttria-stabilized zirconia (ZrO2 - 8mol% Y2O3); (A) green body (54%TD) and (B) millimeter-wave firing at 60 GHz (97%TD).

achieved without suffering appreciable grain growth. The grain size of 60GHz sintered zirconia was approximately equal to that of the green body (~200nm), which is in contrast to the grain size of the conventionally sintered zirconia (~3000nm). At higher temperature (1360°C), however, a significant grain growth (~2000nm) was observed in the sample (96 %TD) sintered for 30 min using the focused 60 GHz radiation in pulse operateion, which was still smaller than the grain size obtained from the conventionally sintered zirconia. It is remarkable that the 10-ms pulse irradiation of 60GHz beam at a small repetition rate of 0.5 Hz demonstrated a significant effect to achieve the densification of zirconia with finer grain size at lower temperature.

SUMMARY

The results of ceramics sintering experiments performed with a 10-kW 28-GHz gyrotron in CW operation and a 100-kW 60-GHz gyrotron in pulse operation has been presented together with the fine focusing of high-power pulsed 60GHz radiation with quasi-optical nature of the millimeter-wave radiation.

Accelerated densification of the silicon nitride samples was observed at lower temperature in the 28-GHz sintering. The achieved density of the 28-GHz sintered silicon nitride was significantly dependent on the initial pore size in preparation of the green body.

Fine focusing of a high-power pulsed 60GHz millimeter-wave radiation was performed using the two-dimensional ellipso-parabolic antenna system for application of this radiation to advanced processing of ceramics. Feasibility of the high energy density radiation (~100kW/cm^2) was experimentally demonstrated. The heating characteristics of alumina using the focused beam in pulse operation was clarified. The efficient sintering of yttria-stabilized zirconia was demonstrated at about 200°C lower temperature than that conventionally required. Millimeter-wave firing in pulse operation resulted in the production of densified zirconia with finer grain size.

REFERENCES

1. W. H. Sutton, Am. Ceram. Soc. Bull. **68** (2), 376 (1989).

2. T. T. Meek, R. D. Blake and J. J. Petrovic, Ceram. Eng. Sci. Proc. **8**, 861 (1987).

3. M. A. Janny and H. D. Kimery, in Microwave Processing of Materals II, edited by W. B. Snyder, Jr., W. H. Sutton, M. F. Iskander and D. L. Johnson (Mater. Res. Soc. Proc. 189, San Francisco, CA, 1990) pp.215-227.

4. M. A. Janney, C. L. Calhoun, and H. D. Kimrey, J. Am. Ceram. Soc. **75** (2), 341 (1992).

5. Yu. V. Bykov, A. F. L. Gol 'denberg and V. A. Flyagin, in Microwave Processing of Materals II, edited by W. B. Snyder, Jr., W. H. Sutton, M. F. Iskander and D. L. Johnson (Mater. Res. Soc. Proc. 189, San Francisco, CA, 1990) pp.41-42.

6. Y. Setsuhara, Y. Tabata, R. Ohnishi and S. Miyake, Trans. JWRI **21**, 181 (1992).

7. T. Saji, New Ceramics **8**, 21 (1995) (in Japanese).

8. S. Miyake, O. Wada, M. Nakajima, T. Idehara and G. F. Brand, Int. J. Electronics **70**, 979 (1991).

9. O. Wada and M. Nakajima, Space Power **6**, 3 (1987).

Part X

Remediation of Hazardous Waste

DESIGN OF MICROWAVE VITRIFICATION SYSTEMS
FOR RADIOACTIVE WASTE

T. L. White*, W. D. Bostick**, C. T. Wilson*, C. R. Schaich*

*Oak Ridge National Laboratory, Oak Ridge, Tennessee 37831-8071, whitetl@ornl.gov
**Oak Ridge K-25 Site, Oak Ridge, Tennessee 37831-7274, bostickwd@ornl.gov

ABSTRACT

Oak Ridge National Laboratory (ORNL) is involved in the research and development of high-power microwave heating systems for the vitrification of Department of Energy (DOE) radioactive sludges. Design criteria for a continuous microwave vitrification system capable of processing a surrogate filtercake sludge representative of a typical waste-water treatment operation are discussed. A prototype 915-MHz, 75-kW microwave vitrification system or "microwave melter" is described along with some early experimental results that demonstrate a 4 to 1 volume reduction of a surrogate ORNL filtercake sludge.

INTRODUCTION

Most DOE industrial sites operate process waste-water treatment plants that produce large volumes of hazardous or mixed waste sludges. In an overview of DOE mixed waste inventories[1], the subcategory of "sludges, filtercakes, and residues", generated primarily from waste-water treatment, represented the largest volume of waste that must be treated for compliance with U.S. Environmental Protection Agency (EPA) regulations. The ORNL Process Water Treatment Plant produces a low-level radiologically contaminated water softening sludge typical of many other DOE and private sector facilities. Microwave sintering or vitrification is one of several processing options being considered for this waste. The goal of treatment is to minimize the volume of waste being stored on-site, while producing a stable waste form that may be qualified for eventual off-site disposal[2].

ORNL has been involved for several years in the research and development of high-power microwave heating systems for vitrification of DOE radioactive sludges. Microwave vitrification of radioactive waste has many advantages over conventional joule-heated melters[3]. Conventional joule heating requires electrodes in direct contact with the waste glass as well as radiant dome heater electrodes exposed to the acid off-gasses. Electrode corrosion can limit the service life of these melters. Microwave vitrification is electrodeless because microwave energy is absorbed directly by the glass. Microwave energy is absorbed by a large class of waste materials commonly found in many low-level waste streams such as fabrics, rubber, concrete, metal powders, oxides, nitrates, sulfates, nonmetallic filtering media, water, carbon, glass and may other dielectric materials. High temperature microwave waste forms[4] are refractory materials with no free liquids, particulates, or organics, and they have a high volume reduction compared to conventional grouting technology, as well as low leachability. Because microwave energy is absorbed directly, it has the considerable advantage of much higher efficiency and faster temperature control compared to conventional radiant heating. For joule-heating, the entire melter must be brought to a working temperature over a period of several days to avoid cracking the melter refractory insulation. Microwave melters can be brought to temperature in minutes because internal refractory insulation is not required. Microwave power can be transmitted through waveguides from generators that can be located safely outside a radioactive processing area where routine system maintenance can be easily performed. Waveguide windows can effectively isolate the generator from the radioactive vitrification cell. Microwave heating is extremely flexible in that a wide range of processing temperatures are available in a single system. Microwave vitrification systems can be designed to be small enough to be mobilized for on-site treatment of radioactive wastes. A microwave vitrification system can be either an "in-drum" system for vitrification in the final storage container[5], or an "overflow" type melter system[6] to vitrify the wastes in a dedicated processing chamber. In-drum vitrification of wastes eliminates transferring molten wastes to secondary containers, therefore, wastes with a wide range of viscosities can be produced. Overflow-type microwave melters are smaller, somewhat simpler to tune, and can fill any suitable container shape compared to in-drum melters. Heating profiles can be custom-tailored for each application by exciting the appropriate microwave field configuration or "mode," choosing the proper heating frequency and material geometry, and understanding how the microwave penetration depth is controlled by temperature and material composition of the waste form.

EXPERIMENT

Continuous Microwave Melter Design Criteria

The microwave melter is a design based on the principle of "Brewsters Angle". That is, for microwave fields polarized perpendicular to the surface of the glass, the microwave energy is completely transmitted into the surface without reflection when the angle of incidence is equal to Brewsters angle[7], θ_B, where

$$\theta_B = \tan^{-1} \sqrt{\varepsilon} \ , \tag{1}$$

where θ_B is measured from a perpendicular to the glass surface and

$$\varepsilon = \varepsilon' - j\varepsilon'' \ , \tag{2}$$

where ε' and ε'' are the real and imaginary parts of the dielectric constants for the dielectric material, respectively. Both ε' and ε'' are temperature dependent and some data exists on how they vary with temperature[8]. A wide range of temperatures are encountered from start-up at room temperature (~25°C) to vitrification (~900-1200°C), and therefore some empirical testing is required to optimize the microwave performance. For typical dielectric parameters, θ_B varies between 65-80° from the normal to the glass surface, depending upon temperature. The amount of microwave attenuation per unit length, α, is given by

$$\alpha = \frac{\sqrt{2}\pi}{\lambda_o} \sqrt{\sqrt{1 + \left(\frac{\varepsilon''}{\varepsilon'}\right)^2} - 1} \ , \tag{3}$$

where λ_o is the free space wavelength, and α is primarily controlled by the ratio $\varepsilon''/\varepsilon'$, called the "loss tangent" or "tan δ". The microwave fields decay exponentially as

$$E(z) = E_o e^{-az} , \qquad\qquad (4)$$

where $E(z)$ is the microwave electric field as a function of distance, z, and E_o is the electric field intensity at $z = 0$. We expect the surface attenuation to be high enough to limit microwave penetration to a few cm when the melt temperature is above 900°C. Molten glass has a moderately low electrical resistivity (.1-40 Ω-cm). In order to get good penetration by the microwave energy, a low microwave frequency of 915 MHz was chosen. The pool of glass is inclined to the axis of the waveguide by an angle of 20° to taper the microwave absorption in the glass, thus minimizing the reflected microwave power. The melter design was tested at low microwave power levels using a microwave network analyzer to determine the optimum angles for minimizing reflections. From these initial tests the drain holes were located. After initial high power testing the level of glass in the melter was reduced by using a drain located lower into the glass pool. It was desirable to deposit the microwave energy in as shallow a volume as possible for good microwave penetration and yet have enough glass to minimize the reflected power. To reduce the likelihood of arcing, a gap between the top wall of the waveguide and the melt is desirable.

Melter Description

The microwave melter was powered by a 75-kW, 915-MHz microwave generator. The waveguide transmission system connecting the generator to the melter consists of a dual-directional coupler, a fiberglass window and several waveguide bends. Air and dry nitrogen were introduced between the window and the melter to keep the waveguide clean. Forward and reflected power were measured by a dual-directional coupler with forward and reflected power monitors. The difference in forward and reflected power is simply the net power (Pn) absorbed by the melter. Since all the internal metal surfaces of the melter have very low microwave losses compared to glass, almost all of the net microwave power is absorbed by the glass. The melter is shown schematically in Fig. 1.

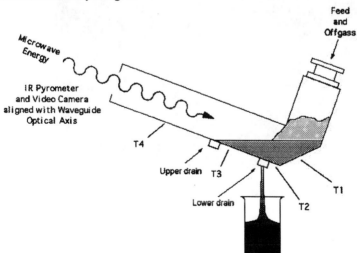

Fig. 1. A cut-away side view of the melter.

The melter is standard WR 975 rectangular waveguide with inside dimensions of 24.7-cm wide by 12.4-cm high. The waveguide is fabricated from 6.4-mm thick Inconel alloy and the waveguide axis is inclined 20° to the horizontal. The glass is in direct contact with the Inconel walls. The exterior of the melter is covered with 1.27-cm thick fiberboard insulation. Unlike a conventional joule-heated melter, there are no dome heater elements required to jump start the glass heating process since the microwave heating is direct. Also there is no internal refractory insulation. The melter is fed manually through a 10-cm diameter port on the top of the melter through which microwave energy cannot propagate (cut off). The molten glass drains out through a 1.9-cm drain hole in the bottom of the waveguide. The off-gas system has a condenser and a ventilation system capable of producing a slight negative pressure in the melter. The temperature along the outside of the bottom of the waveguide was measured by an array of four type K thermocouples (T1, T2, T3 and T4) located along the centerline of the waveguide. An infrared sensor was mounted along the axis of the waveguide looking at the surface temperature (Tir) of the glass. A video camera and light source also looked along the axis to record the vitrification process. Data from the forward and reflected power and the temperature measurements were acquired by a windows-based personal computer running the Labview data acquisition and control software. An overall view of the microwave melter and video monitoring system are shown in Fig. 2. A metal table covered with insulating firebrick supports the melter. A section of flexible bellows waveguide (shown in the upper left hand corner of Fig. 2) allows ± 5° of tilt about the 20° melter angle to adjust glass level.

Fig. 2. Overall view of the Microwave Melter and video monitoring system.

RESULTS

The authentic sludge is predominantly calcium carbonate, with trace levels of Mg, Si, Al, and Fe; the material is nonhazardous by EPA criteria, but is contaminated by low levels of Sr^{90} and Cs^{137}. A non-radiological surrogate has been formulated according to Table I for treatability studies[9].

TABLE I . Preparation of Simulated ORNL filtercake sludge (as dry solids) and frit

Compound		Weight %
Surrogate	Frit	
$CaCO_3$		34.8
SiO_2		6.8
MgO		4.2
Fe_2O_3		2.4
$Al(OH)_3$		1.8
	SiO_2 (sand) or fly ash*	33.5
	$Na_2B_4O_7\cdot10H_2O$ (Borax)	16.5

* Tennessee Valley Authority Kingston, Tennessee Steam Plant

The amount of frit (either sand or fly ash) required to produce a suitable glass waste form is also shown. Borax is added as a fluxing agent to lower the melting temperature of the simulated filtercake sludge and glass frit mixture. One gram of surrogate dry solids yields 0.34 g of calcined waste solids. For the surrogate and frit formulated in the table, the waste loading (calcined oxide basis) is approximately 44.7 wt %. Compared to a composite sample of as-stored wet filtercake, 1 g of calcined waste (or 2.24 g of glass product) is equivalent to an original volume of 3.34 cm^3 wet sludge. These 'figures of merit', together with measurement of the glass product density, allow estimation of achievable volume reduction for the treated waste. The volume reduction for the output waste form over the volume of the input filtercake sludge is estimated to be ~ 4.1 to 1.

Fig. 3 shows the operational microwave melter with glass pouring into a 400 ml beaker.

Fig. 3. Side view of the Microwave Melter.

No freeze valves were used on the melter since the glass flow can be stopped in a matter of seconds by removing the microwave power. This is due to the small thermal mass of the microwave melter. A plot of the data for this pour is shown in Fig. 4. Because of the intense

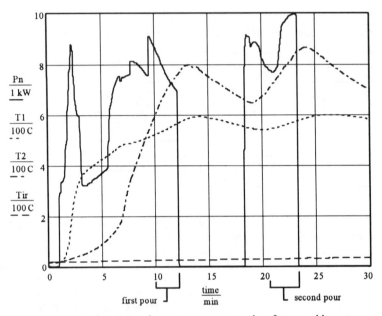

Fig. 4. Plot of absorbed power and temperatures versus time from a cold start.

direct microwave heating of the waste, the outside temperature of the melter never exceeded 900°C for this experimental run, even though the pouring temperature of the molten glass is about 950°C. The average glass output rate per unit of absorbed microwave power was measured to be 2.8 kg/kW-hr for the 275 ml pour. This figure is in fair agreement with simple heat capacity calculations which neglect heat losses. The melter has an unusually small thermal mass and can be cycled on and off much more quickly than a conventional joule-heated melter. The internal construction of the microwave melter is simple and inexpensive when compared to the complex insulation and electrode structure of joule-heated melters. This should result in much lower melter replacement costs. The infrared surface temperature is very low because the cold frit has covered the surface of the molten glass. A second pour was made at 20 minutes into the run. Fig. 5. shows some samples of the vitrified surrogate waste forms (left) and input waste form (right). The wastes have a glassy appearance with a final density of approximately 2.7 g/cm^3. Fig. 6 shows the results of an longer run where 5 L of glass was poured into a 18.4-cm diameter 304-L stainless steel beaker. Further characterization of the microwave glass waste forms is in progress.

Fig. 5. Vitrified surrogate waste forms (left) and input waste form (right).

Fig. 6. A 5-L volume of waste glass poured into a stainless steel beaker (left) and input waste form (right).

CONCLUSIONS

ORNL is involved in the development of high-power microwave heating systems for the vitrification of DOE radioactive sludges and solids. Compared to conventional joule-heated melters, microwave melters offer very simple design, fast startup and shutdown, potentially less expensive melter replacement, and small equipment size. Design criteria for a continuous microwave vitrification system capable of processing a surrogate filtercake sludge representative of a typical waste-water treatment operation are discussed. An operational 915-MHz, 75-kW prototype microwave melter is described along with some early experimental results that demonstrate a 4 to 1 volume reduction of a surrogate ORNL filtercake sludge.

REFERENCES

1. W. A. Ross, M. R. Elmore, C. L. Warner, L. J. Wachter, W. L. Carlson, and R. L. Devries, Proc. 18th Symp. on Waste Management (Waste Management '92), Tucson, AZ, March 1-5, 1992-*Am. Nucl. Soc.*, pp. 1127-1135.

2. H. T. Lee, W. D. Bostick, et al., Report No. DOE-000, to be published 1996.

3. R. G. Baxter, Proc. 15th Symp. on Waste Management (Waste Management '89), Tucson, AZ, Feb.26-*Am. Nucl. Soc.*, Mar. 2, 1989.

4. G. D. DelCul, W. D. Bostick, R. E. Adamski, W. A. Slover, P. E. Osborne, R. I. Fellows and T. L. White, 1994 Int. Incineration Conf., Houston, TX, pp. 615-620, 1994.

5. T. L. White, U. S. Patent No. 5 324 485, (28 June 1994).

6. M. S. Morrell, W. H. Hardwick, V. Murphy, and P. F. Wace, Nuclear and Chemical Waste Management, Vol. 6, pp. 193-195,1986.

7. S. Ramo, J. R.. Whinnery and T. V. Van Duzer, Fields and Waves in Communication Electronics, John Wiley & Sons, New York, pp. 363-364, 1967.

8. A. R.. Von Hippel, Dielectric Materials and Applications, John Wiley & Sons, New York, pp. 400-404, 1954.

9. W. D. Bostick, D. P. Hoffmann, R. J. Stevenson, A. A. Richmond, Report DOE/MWIP-18, Martin Marietta Energy Systems, Oak Ridge, Tennessee, January 1994.

MICROWAVE TREATMENT OF EMISSIONS FROM WASTE MATERIALS

R.L. Schulz*, D.C. Folz*, D.E. Clark*, C.J. Schmidt** and G.G. Wicks+
*University of Florida, Dept. Materials Science & Engineering, Gainesville, FL, rschu@mse.ufl.edu
**University of Florida, Dept. Environmental Engineering, Gainesville, FL
+ Westinghouse Savannah River Technology Center, Aiken, SC

ABSTRACT

A microwave off-gas system has been designed and fabricated to destroy or decompose emissions that arise from microwave treatment of printed circuit boards (PCBs). Preliminary gas chromatography data on emissions that resulted from combustion of PCBs showed significant decreases in the concentration and number of hazardous compounds detected in the off-gases after treatment. Investigations have continued in this area to demonstrate the reproducibility of the initial results and to determine critical control parameters for optimization of the system. The results of these studies are presented.

INTRODUCTION

Microwave energy has been studied by several researchers for waste remediation applications [1-13]. As most of these applications involve heating waste either for the purpose of combustion, vitrification or de-watering, it is reasonable to expect the evolution of some type of emissions from the waste material. The waste form discussed in this paper, printed circuit boards (PCBs), evolves organic compounds during combustion, some of which are hazardous to the population if released to the environment (depending on concentration and exposure level) [14]. Several attempts were made to overcome this problem. The first system used a vacuum pump to evacuate the combustion chamber and passed the emissions through a liquid nitrogen cold finger to condense the gases [4]. While this system prevented release of the hazardous emissions to the atmosphere, it did nothing to treat the emissions and created a liquid waste stream that would require remediation. Also, the system was not suitable for industrial scale-up. Therefore, another method for treating the gases was required. A second microwave off-gas system was designed and constructed to heat and decompose the gases as they were produced. Preliminary results from gas chromatography-mass spectroscopy (GC) data showed that the microwave off-gas treatment system was effective in reducing the concentration and total number of hazardous emissions [2]. Because the initial data was promising, research has continued in this area.

EXPERIMENTAL

The experimental steps for this work are summarized in Figure 1. For this study, boards were recovered from discarded electronic equipment, specifically, two PCBs from a Sears Beta videocassette recorder (Manufactured by Sanyo, 1983-84, VCR 4650) and five PCBs recovered from a Regency Scanner (1978). The boards consisted of a variety of chips, resistors and capacitors soldered to epoxy resin bases. The boards were sectioned, crushed and mixed together to simulate the random feed material that could be recovered from a landfill or scrap from electronic manufacturing operations. The experimental parameters for the individual sample runs are summarized in Table 1.

Mat. Res. Soc. Symp. Proc. Vol. 430 © 1996 Materials Research Society

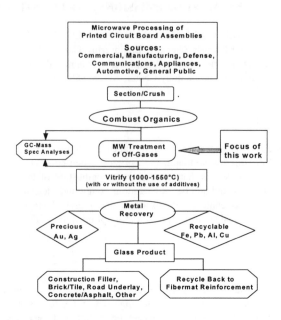

Figure 1. Experimental flowchart for printed circuit boards.

Table 1. Experimental Parameters for Printed Circuit Board Combustion/Off-Gas Study.

Sample ID	Initial Weight (g)	Final Weight (g)	%Wt Loss	Processing/ Off-gas Collection Time (min)	Duty Cycle* (%)
SR1	69.96	41.15	41.2	30	50
SR2	70.09	40.66	41.9	30	50
SR3	69.99	45.75	34.6	30	50
SR4	70.05	41.16	41.2	30	100
SR5	70.01	42.27	39.6	30	50
SR6	70.00	40.85	41.6	30	50
SR7	70.03	44.49	36.4	30	50

*A duty cycle of 50% indicates that, for a specified time interval, the magnetron is only activated 50% of the time.

The apparatus used to process the PCBs and resulting emissions has been previously described [2]. Briefly, the combustion chamber consists of an 850W, 2.45GHz Goldstar Multiwave microwave oven, lined with refractory materials. This unit is connected in tandem to a second microwave oven (Amana Model #RFS10B) that houses the off-gas system. The off-gas system consists of an Alundum (Saint-Gobain/Norton Industrial Ceramics Corp.) tube partially filled with 16 grit β-SiC. Reticulated phosphate bonded alumina (PBA) filters were placed at either end of the tube to help maintain the stability of the SiC bed and to increase emission dwell time. Strategically positioned sampling ports allow collection of gases prior to and after microwave treatment.

During processing, emissions were collected both before and after passing through the microwave off-gas system using Tenax-TA filled glass air traps (6mm OD x 4.5"; -80+60, OI Analytical, College Station, TX). Tenax-TA is a 2,6 diphenyl p-phenylene oxide porous polymer that is highly adsorbent for C6-C20 compounds and is commonly used to collect air samples. Prior to use, the air traps were conditioned in a nitrogen atmosphere for 1 hour at 210°C to dry and to remove any binders that may have been retained after manufacturing. Once the sample run was completed, the air traps were submitted for GC-mass spectroscopy analyses.

RESULTS

The results from GC-mass spectroscopy analyses are summarized in Table 2. This data shows that, while the composition and concentration of species in the off-gas emissions vary from run-to-run, there is a consistent reduction of at least one and often two order(s) of magnitude in concentration of organic species detected after passing through the microwave off-gas treatment system. The differences in pre-treatment concentrations are believed to be a result of variability in feed and air flow rate.

During the experiments, an attempt was made to keep most parameters constant (i.e., sample weight, processing time). However, the compressed air that was used to push the gases through the system was not precisely controlled and may have been somewhat slower for the first three runs (SR1-SR3). From this data and data collected in previous experiments [2], it is believed that air flow rate may be a critical parameter and that even minor changes in flow rate may cause differences in concentration of organic compounds detected before microwave off-gas treatment.

In previous experiments, compressed air was allowed to flow through the system at a much faster rate and emission concentrations before treatment were much higher for samples collected using the Tenax air traps. However, as a different feed material was used (reinforced PCB from a 286 computer), a direct comparison cannot be made. For the experiments described in this paper, the air flow rate was substantially reduced. This was done for several reasons. First, a reduction in air flow rate would increase the dwell time in the off-gas chamber. It was speculated that by increasing dwell time, complete decomposition of emissions was possible (SR2 and SR3). Secondly, faster flow rates created an over-pressure in the combustion chamber which exceeded the containment capabilities of the chamber.

From these and previous experiments, it appears that the off-gas system is capable of reducing the concentrations of emissions over a range of air flow rates and feed materials. However, additional modifications to the combustion chamber are required in order to optimize the process. Both air flow control and containment of emissions are priorities in the design of the third-generation microwave waste treatment center.

Table 2. A Summary of the GC Mass Spectroscopy Results of Emissions Resulting from Combustion of Printed Circuit Boards. (A = before microwave off-gas treatment; B = after microwave off-gas treatment).

Compound	SR-1 (ppb)		SR-2 (ppb)		SR-3 (ppb)		SR-4 (ppb)		SR-5 (ppb)		SR-6 (ppb)		SR-7 (ppb)	
	A	B	A	B	A	B	A	B	A	B	A	B	A	B
Benzene*	16.9	1.1	14.2	nd	19.8	nd	115.3	5.2	119.6	8.1	176.6	14.7	165.4	13.5
Toluene*	28.7	2.7	24.4	nd	32.6	nd	67.5	6.1	78.7	6.9	159.1	18.1	115.7	5.9
Ethylbenzene*	18.7	nd**	19.0	nd	7.8	nd	13.9	nd	26.7	nd	142.9	5.0	91.8	nd
Styrene*	38.7	1.2	66.6	nd	15.0	nd	165.2	2.9	167.7	2.6	472.3	27.2	482.9	6.5
Napthalene*	1.2	nd	11.0	nd	nd	nd	75.1	1.3	35.2	1.3	6.8	3.4	47.6	2.4
m/p Xylene*	17.5	nd	1.9	nd	nd	nd	27.5	nd	23.8	nd	53.3	1.6	60.0	nd
1,3,5 Trimethylbenzene	9.5	nd	12.4	nd	1.3	nd	15.6	1.6	18.4	nd	12.8	2.4	46.2	1.7
1,2,4 Trimethylbenzene	17.5	nd	1.7	nd	nd	nd	nd	nd	nd	nd	15.1	nd	6.1	1.8

* Listed in the Clean Air Act (as amended, 1990) as hazardous air pollutants [14].
** nd = not detected (<1ppb)

SUMMARY

Results from GC mass spectroscopy analyses of samples collected prior to and after microwave off-gas treatment verify the effectiveness of the treatment system. It also appears that while the air flow rate is a critical parameter in determining (controlling) the concentration of the emissions prior to treatment, the off-gas system is capable of a range of air flow rates and a variety of different waste feeds. Construction of a third-generation system is already underway. Modifications to optimize the system are expected to include computer control over critical processing parameters such as duty cycle (power level), temperature and air flow rate. Several changes to the basic geometry of the system have also been incorporated.

ACKNOWLEDGMENTS

The authors thank Ben Rossie for his assistance during experimental work and his input into the modifications for the next generation off-gas system. The authors also thank Westinghouse Savannah River Technology Center for partial financial support during this work under subcontract AB46395-0.

REFERENCES

1. G.G. Wicks, D.E. Clark, R.L. Schulz and D.C. Folz in Microwaves: Theory and Application in Materials Processing III (D.E. Clark, D.C. Folz, S.J. Oda and R. Silberglitt, eds), Ceramic Transactions, Vol. 59, The American Ceramic Society, Westerville, OH, pp. 79-90 (1995).
2. R.L. Schulz, D.C. Folz, D.E. Clark, C.J. Schmidt and G.G. Wicks in Microwaves: Theory and Application in Materials Processing III (D.E. Clark, D.C. Folz, S.J. Oda and R. Silberglitt, eds), Ceramic Transactions, Vol. 59, The American Ceramic Society, Westerville, OH, pp. 107-114 (1995).
3. R.L. Schulz, D.C. Folz, D.E. Clark and G.G. Wicks in, Microwave Processing of Materials IV, Vol. 347 (M.F. Iskander, ed.), Materials Research Society, Pittsburgh, PA, pp. 401-406 (1994).
4. R.L. Schulz, D.C. Folz, D.E. Clark and G.G. Wicks in, Microwaves: Theory and Application in Materials Processing II (D.E. Clark, W.R. Tinga and J.R. Laia, eds.), Ceramic Transactions, Vol. 36, The American Ceramic Society, Westerville, OH, pp. 81-88 (1993).
5. Chemistry & Industry (anonymous article), p. 440, June 21 (1993).
6. L. Dauerman, G. Windgasse, N. Zhu and Y. He in, Microwave Processing of Materials III (R.L. Beatty, W.H. Sutton and M.F. Iskander, eds.), Proceedings of The Materials Research Society, Vol. 269, The Materials Research Society, Pittsburgh, PA pp. 465-469 (1992).
7. L. Dauerman, G. Windgasse, N. Zhu and Y. He in, Microwave Processing of Materials III (R.L. Beatty, W.H. Sutton and M.F. Iskander, eds.), Proceedings of The Materials Research Society, Vol. 269, The Materials Research Society, Pittsburgh, PA pp. 465-469 (1992).
8. L. Dauerman, G. Windgasse, H. Gu, N. Ibrahim and E-H. Sedhom in, Microwave Processing of Materials II, Materials Research Society Proceedings, Vol. 189 (W.B Snyder, Jr., W.H. Sutton, M.F. Iskander and D. L. Johnson, eds.), Materials Research Society, Pittsburgh, PA, pp. 61-67 (1991).

9. R. Varma, S.P. Nandi and J.D. Katz in, **Microwave Processing of Materials II**, Materials Research Society Proceedings, Vol. 189 (W.B Snyder, Jr., W.H. Sutton, M.F. Iskander and D. L. Johnson, eds.), Materials Research Society, Pittsburgh, PA, pp. 67-68 (1991).

10. Mechanical Engineering (anonymous article), 113[11], p. 18, November (1991).

11. Design News (anonymous article), p. 30, July 4 (1988) .

12. Science (anonymous article) Vol. 257, No. 11, p.1479 (1992).

13. Process Engineering, anonymous article, p. 27, February (1991).

14. Clean Air Act, as amended, Section 112, Hazardous Air Pollutants (1990).

Part XI

Temperature Modeling and Measurements

APPLICATION OF MULTIWAVELENGTH PYROMETRY
IN MICROWAVE PROCESSING OF MATERIALS

R. S. DONNAN and M. SAMANDI
Surface Engineering Research Centre, Department of Materials Engineering,
University of Wollongong, NSW 2522 AUSTRALIA

ABSTRACT

Over the past decade microwave energy has been increasingly used in materials processing, especially for sintering and more recently for the joining of advanced ceramics. However the hostile electromagnetic and plasma environment within a high power (1-6 kW) microwave applicator poses serious problems for very accurate high temperature measurement by precluding the use of existing classes of thermometry. For instance, conventional probe-based thermometry, multiple-wavelength ratio pyrometry and even the more recently developed technologies of optical fibre thermometry by fluoroptics and radiometry, are either incompatible or of restricted application.

The main aim of this paper is to propose multiwavelength pyrometry as a viable technique for wide range (500-5000 K) thermometry in hostile electromagnetic and plasma environments. After briefly reviewing the physical basis of its operation, the experimental set up of the multiwavelength pyrometer is outlined, and consists of a comparatively inexpensive low resolving power grating monochromator and a PbS infrared single element detector. Results are presented that compare the measurements during conventional/microwave heating trials, from this multiwavelength pyrometer and from a K-type thermocouple, a double-wavelength ratio pyrometer and a single wavelength pyrometer aimed at a dummy target (carbon/metal).

INTRODUCTION

Accurate thermometry is fundamental for both reliable material characterisation and control of thermal processing of materials. It is sometimes taken for granted that thermometry is always possible and meaningful!

However, since the advent of the application of microwave energy for advanced materials processing there has been a profound and enduring difficulty in reliably performing accurate thermometry. Kilowatt microwave powers and associated induced plasmas are electromagnetically hostile systems. In the following sections the pyrometric context of multiwavelength pyrometry is delineated and its linear least squares basis of operation is described in particular and then applied.

Multiwavelength Pyrometry

Multiwavelength pyrometry is a comparatively recent technique (10-15 years) that belongs in the class of partial radiation pyrometers. Theoretical studies of multiwavelength pyrometry, [1-4] and only a few non-microwave experimental applications, [5-6] of its principles have been reported in the literature.

The principal advantage of multiwavelength pyrometry is its capacity to estimate with high precision (0.05%) an accurate temperature over a wide range (500, 5000)K without prior detailed knowledge of the material emissivity. In fact estimates of spectral emissivity are returned with their respective error estimates as a coupled solution with the associated temperature. The

557

diagram (Figure 1) below shows the wider context of multiwavelength pyrometry and the various avenues by which it may be pursued.

The seminal theoretical study by Gardner [2] and one of the more recent experimental applications of multiwavelength pyrometry by Khan [5] demonstrated that the non-linear least squares model generates more accurate temperature estimates while the linear model was faster. As with ratio pyrometry, little advantage was found by performing multiwavelength pyrometry at many numbers of wavelengths in order to gain an advantage by working with a high order expression for the emissivity.

This paper describes some initial experiments to study the applicability of multiwavelength pyrometry to the thermometry of ceramic and metal loads that are particularly heated by microwaves and microwave induced plasmas respectively. The less accurate of the least squares developments is used to study the stability and precision of the temperature estimate. Three spectral radiometric measurements are taken to allow the use of a linear exponential polynomial to model the emissivity.

THEORETICAL OVERVIEW

The principle of multiwavelength pyrometry is the minimisation of the difference between experimental radiance measurements and theoretically generated radiance data by using a fitting technique together with statistical averaging. The theoretical radiance data incorporates a spectrally dependent analytic function for the emissivity that is empirically based [2]. A statistical treatment then applies the method of maximum likelihood [7] to the unknown coefficients that together model the temperature and emissivity. In this treatment the measured radiance spectrum is regarded as being a sample data set from a parent distribution. The analytic form of the parent distribution is built upon modified coefficients from the emissivity model. A Gaussian probability function may then be generated based upon these coefficients and the associated radiometric data.

For the special case of linear least squares fitting, the emissivity expression for multiplying the Wien approximation of blackbody radiance is

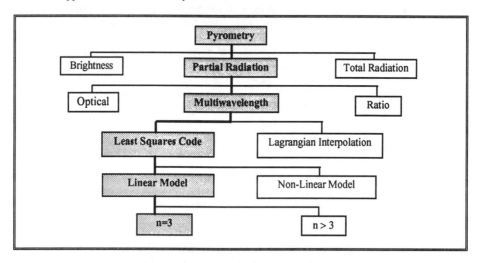

Figure 1. The context of multiwavelength pyrometry

$$\varepsilon\left(\lambda_i\right) = \exp\left\{a_0 + a_1\lambda_i\right\} \tag{1}$$

It may be shown [4] that this leads to a parent distribution expression of the form

$$\Phi\left(\lambda_i, \lambda_j\right) = a_1 + \frac{a_2}{\left(\lambda_i\lambda_j\right)} \tag{2}$$

where $\Phi\left(\lambda_i, \lambda_j\right)$ denotes the relative difference between a given pair of spectral radiances. Equation (2) now forms the basis for the Gaussian probability expression

$$P(a_1, a_2) = \Pi\left(\frac{1}{\sigma_{ij}\sqrt{2\pi}}\right)\exp\left\{-\frac{1}{2}\Sigma\left[\frac{\left(\Phi_{ij}^* - \Phi_{ij}\right)}{\sigma_{ij}}\right]^2\right\} \tag{3}$$

that descibes the likelihood of having observed a particular radiometric voltage spectrum (Φ_{ij}^*). The linear least squares fitting follows as (3) is sought to be maximised by minimising

$$\chi^2 = \Sigma\left[\frac{1}{\sigma_{ij}}\left(\Phi_{ij}^* - a_1 - \frac{a_2}{\left(\lambda_i\lambda_j\right)}\right)\right] \tag{4}$$

with respect to the coefficients a_1 and a_2. The result of the minimisation leads to an algebraic matrix equation of the form

$$A\tilde{x} = \tilde{b} \tag{5}$$

such that A contains a matrix of coefficients having spectral dependence and $\tilde{b}$ is a modified vector of the radiometric data set. The solution vector $\tilde{x}$ contains the probability coefficients a_1 and a_2. The temperature in particular is determined as

$$T = \frac{C_2}{a_2} \tag{6}$$

where C_2 is the second radiation constant in the Planck radiance formulation.

EXPERIMENTAL ARRANGEMENT

The radiance from either microwave or incandescently heated objects was collected by a visible-to-near infrared, water-free silica, fibre optic bundle (Figure 2). Signal from the fibre was

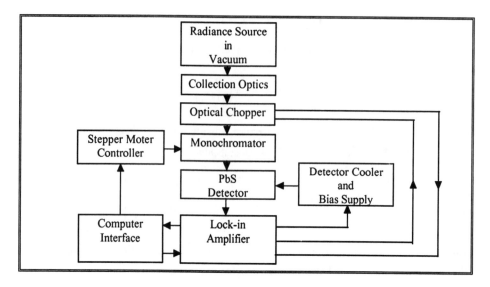

Figure 2: The experimental arrangement

modulated at a fixed frequency of 950 Hz by a mechanical chopper wheel. The modulated signal entered a quarter metre focal length monochromator configured in an asymmetric Czerny-Turner arrangement. The entrance slit image is then focused onto a 300 grooves per millimetre ruled grating blazed at 2μm. With the entrance slit width set for the fastest optics (3mm) and a reciprocal dispersion of 11.4 nm.mm^{-1} an effective bandwith of approximately 33nm. was set. The diffracted signal was imaged onto a PbS detector through the exit slit which had also been set to a 3mm width. An Oriel digital lock-in amplifier (Merlin™) performed digital signal processing and demodulation related to the chopper frequency. This frequency was selected in compliance with the detectivity of the PbS detector which peaked at 950 Hz. The lock-in amplifier is programmed to optimally co-ordinate operations to ensure a stable signal modulation, a constant rate of grating rotation and synchronous reading of the detector through the computer interface.

EXPERIMENTAL METHOD

Measurements with Non-Microwave Heating

To obtain an estimate of the reliability of the multiwavelength pyrometer, a series of measurements were made using radiation to heat the samples. Blocks (80x80x20mm^3) of steel, copper and carbon were heated under vacuum (~10mTorr). A comparison was made between the temperatures obtained with a K-type thermocouple, a single wavelength infrared pyrometer and a multiwavelength pyrometer. Two temperature regions were investigated, 250 to 600°C and 900 to 2400°C corresponding to the operating ranges of the single wavelength pyrometer (IRCON Mirage® Series 60 sensing from 2.0 to 2.6μm) and the two colour ratio pyrometer (IRCON Mirage® Series OR sensing from 0.70 to 1.08μm) respectively. The results are displayed in Tables I and II.

Measurements with Microwave Heating

A mild steel plate (~50x50x3)mm^3, which was heated by a low pressure (5-10mTorr), microwave induced nitrogen plasma, was used for the low temperature range measurements. The results are shown in Table III. A high purity sample of zirconia was used for the measurements in the high temperature range. It was initially heated by the hybrid technique of plasma assisted microwave heating [8] until it reached a sufficiently high temperature to permit further heating by direct microwave coupling alone. These results are shown in Table IV.

Note that thermocouple readings could not be obtained for Table IV due to physical damage that was sustained by the probe following overheating during arcing and antenna reception of the microwave field. In all instances where the thermocouple was used, its tip was located within the given block of material (central to a narrow face and to a depth of about 30mm). The position of the thermocouple tip generally lay near to, or within, the solid angle of view with which each pyrometer was operated.

Table I: Low Temperature Range (Incandescent Heating)

Material	Run	K-Type [°C]±0.75%	Single λ pyrometer [°C]	MWP [°C]
steel	2	552	552 ($\varepsilon = 0.92$)	567
copper	1	600	622 ($\varepsilon = 0.18$)	653
	2	563	563 ($\varepsilon = 0.20$)	584
carbon	1	600	-	616
	2	587	622 ($\varepsilon = 0.74$)	604
			610 ($\varepsilon = 0.80$)	
			588 ($\varepsilon = 0.92$)	

Table II: High Temperature Range (Incandescent Heating)

Material	Run	K-Type [°C]±0.75%	Double λ pyrometer [°C]	MWP [°C]
steel	1	740	-	-
	1	904	-	898
	1	1043	-	1080
	2	1056	1045 (greybody)	1040
copper	1	-	-	-
	2	1000	1000 (m = 1.04)	1007
			1060 (greybody)	
carbon	1	1011	1020 (greybody)	1023
	2	1216	1214 (greybody)	1220
	3	1170	1085 (greybody)	1189

Table III: Low Temperature Range (Microwave Induced Plasma Heating)

Material	Run	K-Type [°C]±0.75%	Single λ Pyrometer	MWP [°C]
Mild steel	1	436	509($\varepsilon = 0.55$)	455

Table IV: High Temperature Range (Direct Microwave Heating)

Material	Run	K-Type [°C]	Double λ pyrometer [°C]			MWP [°C]
		±0.75%	m = 0.85	greybody	m = 1.15	
zirconia	1	-	1560	1310	1090	1080
	2	-	1545	1295	1115	1159
	3	-	1596	1340	1150	1202
	4	-	1725	1445	1245	1359
	5	-	1825	1525	1322	1562
	6	-	1905	1521	1376	1632
	7	-	1816	1595	1305	1524
	8	-	1245	1051	893	1036
	9	-	1721	1443	1236	1503

CONCLUSION

When the samples were heated by radiation, the measurements with the single wavelength pyrometer and the thermocouple agreed well in the low temperature range, but the pyrometer readings are dependent on the surface emissivity. This can only be obtained by some form of calibration, as for example with a thermocouple. A change in the emissivity setting of 0.02 gave rise to a temperature change of 10°C. The multiwavelength pyrometry results are in reasonable agreement with the thermocouple measurements and require no knowledge of the emissivity. In the high temperature regime the two colour ratio pyrometer gave accurate measurements for greybody emitters but requires thermocouple calibration of the emissivity slope in other cases. For example, an emissivity slope (m) of 1.04 was required for copper. Again the multiwavelength pyrometer gave results that were in close agreement with the other two methods. It is particularly mentioned that the reflectivity of the steel block was observed to vary by approximately 30% during an experiment as the single wavelength pyrometer was calibrated against the thermocouple.

The errors in measurement for the single wavelength and ratio pyrometers become much greater in the presence of a microwave induced plasma. At low temperature the single wavelength pyrometer gave a questionable result for a suitable emissivity. At high temperatures no consistant emissivity slope can be found for the whole temperature range of the ratio pyrometer. A major difficulty lies in obtaining comparative measurements for calibration because the higher power microwave fields lead to destruction of the thermocouple. Without some prior information about the emissivity, the uncertainty in the temperature can approach 500°C. The measurements with the multiwavelength pyrometer were not disturbed by the presence of the plasma. It is also to be noted that no measurement uncertainties have been given for multiwavelength temperature estimates. This is owing to the difficult nature of obtaing a satisfactorily integrated noise model which was measured to have a dominating statistical component.

Linear least squares multiwavelength pyrometry is capable of providing a robust and accurate wide range (500-5000 K) thermometry in a variety of benign and hostile environments without the need of detailed information about the surface emissivity. It consists of conventional and comparatively low cost instrumentation and is especially useful in circumstances where other methods fail. Continuing work will test the claim that the non-linear least squares method offers greater precision and of rapid response thermometry by spectrographic array detection.

ACKNOWLEDGMENTS

The invaluable and timely assistance of Dr John Tendys of the Plasma Surface Engineering Group, ANSTO is duly acknowledged in aspects of Labview coding and in discussions of theoretical concern. This work has been sponsored by the CRC-for Welding and Joining, Project 93-15A.

REFERENCES

1. D. Ya. Svet, Sov. Phys. Dokl., **20**, 214-215 (1975).

2. J. L. Gardner, High Temperatures.HighPressure **12**, 699-705 (1980).

3. G. R. Gathers, International Journal of Thermophysics **13** (2), 361-382 (1992).

4. E. R. Spjut, Optical Engineering **32** (5), 1068-1072 (1993).

5. M. A. Khan, C. Allemand, and T. W. Eager, Rev. Sci. Instrum. **62** (2), 403-409 (1991).

6. J. Hiernaut, R. Beuker, W. Heinz, R. Selfslag, and M. Hoch, High Temperatures.HighPressure **18**, 617 (1986).

7. P. R. Bevington, Data Reduction and Error Analysis for the Physical Sciences, McGraw Hill, New York, 1969, pp. 100-101.

8. M. Samandi and M. Doroudian in Plasma Assisted Microwave Sintering and Joining of Ceramics, edited by M. F. Iskander, R. J. Lauf, and W. H. Sutton (Mater. Res. Soc. Proc. **347**, Pittsburgh, Penn., 1994) pp. 605-615.

TEMPERATURE DISTRIBUTION IN A FLOWING FLUID HEATED IN A MICROWAVE RESONANT CAVITY

J. R. THOMAS, JR†, ERIC M. NELSON‡ ROBERT J. KARES‡ RAY M. STRINGFIELD‡
† Mechanical Engineering Department,
  Virginia Polytechnic Institute and State University, Blacksburg, VA 24061
‡ Los Alamos National Laboratory, Los Alamos, NM 87545

ABSTRACT

This paper presents results of an analytical study of microwave heating of a fluid flowing through a tube situated along the axis of a cylindrical microwave applicator. The interaction of the microwave field pattern and the fluid velocity profiles is illustrated for both laminar and turbulent flow. Resulting temperature profiles are compared with those generated by conventional heating through a surface heat flux. It is found that microwave heating offers several advantages over conventional heating.

INTRODUCTION

A logical industrial application for microwave energy is that of heating a process fluid. This can be accomplished by passing the fluid through a microwave applicator at an appropriate rate to heat the fluid to the desired temperature. For many process objectives, the temperature distribution in the fluid would be of primary interest. The temperature distribution in the fluid depends on both the velocity profile of the flowing fluid and the distribution of electromagnetic field strength in the cavity. In the present work, we consider a lossy fluid flowing through a cylindrical resonant cavity and show that the presence of the fluid strongly influences the field distribution, but that the fluid temperature distribution is more sensitive to the flow regime than to the shape of the electromagnetic field. Microwave heating of flowing fluids is shown to have significant advantages over conventional heating if uniform fluid temperatures are desired.

ELECTROMAGNETIC FIELD DISTRIBUTION

A right circular cylindrical resonant cavity is a logical choice for the applicator. Typically, such a cavity would be excited in the TM_{010} mode to heat a cylinder of product on the axis. The presence of the load causes the driven fields to differ significantly from a single mode pattern, however.

The field calculation procedure has been described in detail elsewhere [1], and will be only briefly summarized. The system is modeled as a right circular cylinder consisting of two regions: load and air. The product is assumed to have an hypothetical complex permittivity $\epsilon = 10 + 2i$. The microwave source is a small aperture at the midplane of the applicator fed by a rectangular waveguide. The field is determined as a superposition of modes, all of which have the form of Bessel functions of complex argument. We considered a cavity of diameter $D = 10.86$ cm, height $d = 62.59$ cm, with a cylindrical tube on the axis having diameter $2r_0 = 2.086$ cm. It was determined that a total of 34 modes are necessary to adequately represent the field distribution in the load. The resulting field pattern on an r-z slice through the drive plane is compared to the pure TM_{010} mode in Fig. 1. It is apparent that the field is concentrated to one side of the load centerline, opposite the feed aperture. Clearly this would lead to a hot spot in a stationary product. Below, we consider the effect of this field on an electromagnetically lossy flowing fluid.

565

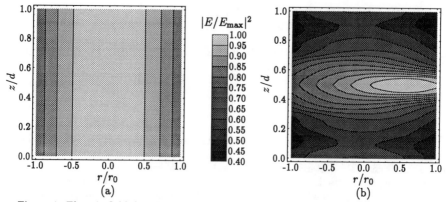

Figure 1. Electric field distribution in the tube portion of the microwave applicator. Plot (a) shows a pure TM$_{010}$ mode. Plot (b) is a more realistic picture of the field driven by an aperture at the midplane on the cavity wall (off of plot to right).

TEMPERATURE DISTRIBUTION

We assume steady flow and steady-state conditions for a fluid flowing through a cylindrical tube on the axis of the cylindrical cavity described above. Temperature distributions were computed using the thermal properties of water for both laminar and turbulent flow. The equations describing the velocity and temperature distributions are

$$\frac{1}{\rho}\frac{dP}{dz} = \frac{1}{r}\frac{\partial}{\partial r}[r(\nu + \epsilon_M)\frac{\partial u}{\partial r}] + \frac{1}{r}\frac{\partial}{\partial \theta}[\frac{1}{r}(\nu + \epsilon_M)\frac{\partial u}{\partial \theta}]; \tag{1}$$

and

$$u\frac{\partial T}{\partial z} = \frac{\dot{q}}{\rho C_p} + \frac{1}{r}\frac{\partial}{\partial r}[r(\alpha + \epsilon_H)\frac{\partial T}{\partial r}] + \frac{1}{r^2}\frac{\partial}{\partial \theta}[r(\alpha + \epsilon_H)\frac{\partial T}{\partial \theta}]. \tag{2}$$

Eqs. (1) and (2) represent conservation of momentum and energy of the fluid, respectively. The pressure gradients in the r and θ directions are neglected because they are small compared to the gradient in the downstream (z) direction. In these equations, P represents the pressure, ρ the density, and u the local fluid velocity parallel to the z-axis. Also, T is the local temperature, C_p the specific heat at constant pressure, and $\dot{q}$ represents the local energy generation density caused by absorption of microwave energy. Finally, ν represents the kinematic viscosity and α the thermal diffusivity; ϵ_M and ϵ_H are turbulence parameters.

These equations were solved by a finite-difference technique, using literature data for the temperature dependence of all physical properties. Ordinary water was used as the working fluid in these calculations. For values of Reynolds number $Re = \rho u_m D/\mu < 2300$, the flow was assumed to be laminar, and ϵ_M and ϵ_H were set to zero. For turbulent flow, these parameters were determined according to the Prandtl-Von Karmann turbulence model [2].

Typical velocity profiles for laminar and turbulent flow are shown in Fig. 2. These profiles differ from the standard shapes shown in textbooks because of the inclusion of temperature-dependent viscosity and volumetric heating of the fluid. Nevertheless, laminar flow displays a much more peaked distribution than turbulent flow, where mixing smooths the velocity distribution. We would expect these velocity distributions to strongly affect temperature distributions resulting from fluid heating. Clearly the fluid in the center of the tube, because of its higher velocity, will have less time to be heated by a microwave field.

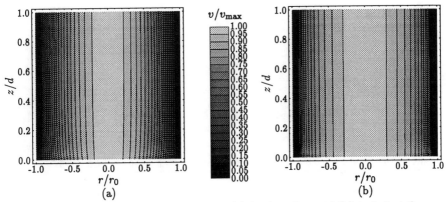

Figure 2. Normalized velocity profiles for (a) laminar flow and (b) turbulent flow. The velocity distribution is more uniform for the turbulent case, which leads to the more uniform temperature distributions shown below.

Thus for a uniform field in the tube we would expect the temperature profile to approximate the inverse of the velocity profile.

The TM_{010} mode produces a nearly uniform field across the tube cross section, leading to the temperature distributions shown on a r-z slice through the drive plane in Fig. 3(a) and Fig. 3(b). This field pattern produces a 30°C temperature difference between fluid in the center of the tube and that near the tube wall in laminar flow. Such a temperature distribution would be very undesirable for many process applications. Turbulent flow, however, produces very uniform temperatures as is apparent in Fig. 3(b).

Similar results for the realistic field pattern of Fig. 1(b) are shown in Fig. 3(c) and Fig. 3(d). Surprisingly, the rather nonuniform field pattern has little effect on the fluid temperature distribution. The field concentration seen in Fig. 1(b) is reflected in the fluid temperatures as a slight constriction in the temperature contours on the right side of the figure. Thus it appears that local nonuniformities in the microwave field are greatly overshadowed by the effects of the flow regime in determining the temperature profile in a flowing fluid heated by microwaves.

CONVENTIONAL HEATING

To provide a basis for comparison, fluid temperature profiles were computed for conventional heating of water flowing in a tube of identical size and fluid mass flow rate as that considered for microwave heating. The conventional heating was provided by a uniform surface heat flux applied at the tube surface of an appropriate magnitude to produce an exit temperature just below 100°C. The temperature distributions are shown in Fig. 3(e) for laminar flow and Fig. 3(f) for turbulent flow. In these figures, darker areas are cooler; it is seen that for laminar flow (worst case), the entire center of the flow is still at the entry temperature at the exit. In fact, the velocity-weighted mean temperature at the exit is only 32.1°C, while the fluid near the wall has reached 87.3°C. To heat the flow uniformly to 87°C would require a very long tube and a much smaller surface heat flux. Comparing this with Fig. 3(c) reveals the distinct advantage of microwave heating.

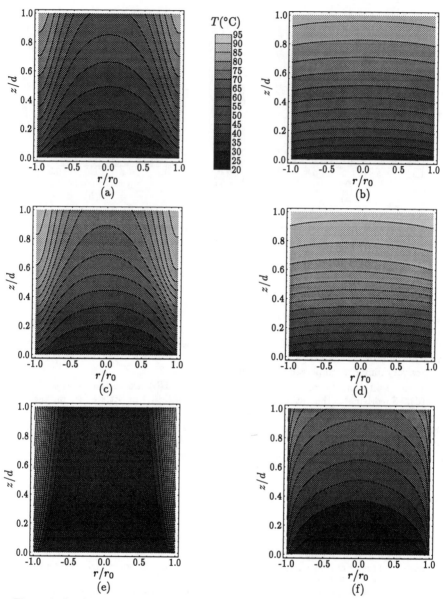

Figure 3. Product temperature on an r-z slice. The top two cases, (a) and (b), are heated by a pure TM_{010} mode field. The middle two cases, (c) and (d), are heated by a more realistic electric field which is driven by an aperture at the cavity wall to the right of the plot. The bottom two cases, (e) and (f), are conventionally heated. The cases on the left, (a), (c) and (e), are for laminar flow. The cases on the right, (b), (d) and (f), are for turbulent flow.

CONCLUSIONS

A cylindrical resonant cavity may be used as an applicator for heating a flowing fluid. The presence of the lossy fluid significantly alters the field distribution in the cavity from the design mode based on an empty cavity. However, the flow regime of the fluid is much more important in determining the uniformity of the temperature distribution in the fluid, with turbulent flow being greatly preferred if uniform temperatures are desired. Microwave heating in a properly designed applicator can produce much more uniform fluid temperatures than conventional heating.

ACKNOWLEDGEMENTS

Work supported by DOE contract W-7405-ENG-36.

REFERENCES

1. Nelson, E. M., R. J. Kares, and R. M. Stringfield, "Semi-Analytic Computation of the Driven Fields in Right Circular Cylinder Microwave Applicators", Proceedings, 30th Microwave Power Symposium, 1995.

2. Bejan, A., **Convection Heat Transfer**, John Wiley & Sons, New York, 1984.

TRANSIENT TEMPERATURE DISTRIBUTIONS IN A CYLINDER HEATED BY MICROWAVES

H. W. JACKSON, M. BARMATZ, AND P. WAGNER
Jet Propulsion Laboratory, California Institute of Technology, Pasadena, CA 91109

ABSTRACT

Transient temperature distributions were calculated for a lossy dielectric cylinder coaxially aligned in a cylindrical microwave cavity excited in a single mode. Results were obtained for sample sizes that range from fibers to large cylinders. Realistic values for temperature dependent complex dielectric constants and thermophysical properties of the samples were used. Losses in cavity walls were taken into account as were realistic thermal emissivities at all surfaces. For a fine mesh of points in time, normal mode properties and microwave power absorption profiles were evaluated using analytic expressions. Those expressions correspond to exact solutions of Maxwell's equations within the framework of a cylindrical shell model. Heating produced by the microwave absorption was included in self-consistent numerical solutions of thermal equations. In this model, both direct microwave heating and radiant heating of the sample (hybrid heating) were studied by including a lossy dielectric tube surrounding the sample. Calculated results are discussed within the context of two parametric studies. One is concerned with relative merits of microwave and hybrid heating of fibers, rods, and larger cylinders. The other is concerned with thermal runaway.

INTRODUCTION

Time dependent temperature behavior of a cylinder heated by microwaves was calculated with the aid of a realistic model of a single mode cylindrical cavity for problems relevant to materials processing. The model has been described in previous articles [1,2] where it was applied to calculate the microwave absorption and steady state temperature profiles [2] for a cylindrical sample. This new treatment extends those cylindrical sample studies along with other earlier work concerned with spherical samples in single mode cavities [3 - 5].

In this work, as well as our earlier studies, electromagnetic properties of a cavity partly filled with a sample are treated analytically using a shell model. The results provide formulas for evaluating microwave power absorption per unit volume. The absorption is included as a heat source in thermal equations that are solved using a combination of analytic and finite difference methods while taking into account the temperature dependence of thermophysical properties of the sample [6]. So far our models have not included heat of chemical reactions, changes of material properties as the sample is processed and undergoes densification, changes in chemical composition, or changes in phase. However, we plan to include some of those effects in advanced models in the future.

The present treatment of a cylindrical shell model of a microwave processing reactor has several noteworthy features. First, Maxwell's equations are solved exactly and the electrical conductivity of cavity walls is taken into account accurately. Second, realistic values of thermal emissivity are taken into account at solid boundaries inside the cavity and at the cavity walls. Third, the model addresses both direct microwave heating of the sample and hybrid heating, which additionally involves radiant heating of the sample by a microwave-heated tube that surrounds it.

A variety of applications can benefit from calculated time-dependent temperature profiles. Two important examples are the prediction and control of preheating and processing conditions, and design and optimization of new reactors. In this article, the usefulness of the calculated results will be illustrated first by investigating the relative merits of processing fibers and large samples using microwave heating alone or hybrid heating. A second illustration will focus on parametric studies that search for evidence of thermal runaway during materials processing. The theory is described next. Then calculated results are presented and discussed. Our conclusions are presented in the final section.

571

Mat. Res. Soc. Symp. Proc. Vol. 430 ©1996 Materials Research Society

THEORY

The theory is based on a shell model represented in Fig. 1. The sample is a lossy dielectric rod aligned along the axis of the cavity. A lossy dielectric tube surrounding the rod is included when hybrid heating of the sample is studied. Otherwise the tube is absent and the rod is heated by microwaves alone. Heat transfer occurs by thermal conduction alone inside the material of the rod and tube. Heat transfer by thermal radiation occurs in all vacuum spaces. This radiation is treated in a gray body approximation with realistic thermal emissivities at all curved solid boundaries in the cavity. The flat end plates are treated as perfect reflectors for thermal radiation. In practice this can be approximated by highly polished copper end plates. There is no thermal path from rod or tube to the end plates because the path is interrupted by a very small gap at each end. This model then permits the thermal radiation problem to be treated analytically, for it is as if the cylindrical surfaces were infinitely long [7].

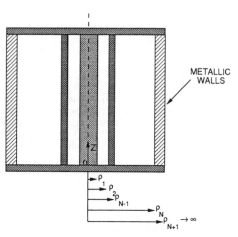

Fig. 1. Geometry of cylindrical rod and cavity.

The sample and the tube are partitioned into many thin cylindrical zones or shells, however the vacuum spaces are not subdivided. The curved cavity wall is treated as one of the zones. The complex dielectric constant is taken to be uniform in each zone, but it may vary from one zone to another, thus allowing for the temperature dependence of dielectric constant in the rod and tube. The cavity walls are taken to be at some uniform temperature. Maxwell's equations can be solved exactly for this shell model, where boundary conditions are matched at interfaces between shells and assuming perfectly electrically conducting end plates. The normal mode properties of this system of shells can be found for any assigned distribution of complex dielectric constant using a known 4 x 4 matrix formalism [8]. Any normal mode can be treated analytically by that technique, but here we will focus on TM_{0n0} modes, where there is no angular dependence or z-dependence in the fields. For these modes, microwave power absorption per unit volume varies spatially only in the radial direction. That absorbed power distribution varies as the temperature profile in the cavity reactor changes with time. Microwave absorption by the cavity end plates is calculated using a surface resistance approximation.

The solution of the electromagnetic problem was combined with thermal equations to calculate temperature distributions with microwave absorbed power density as a heat source. Under transient conditions, the general forms of the thermal equations inside the sample or tube are

$$q(r,t) = -\kappa(r,t)\,\nabla T(r,t) \tag{1}$$

$$\nabla \bullet q(r,t) = P(r,t) - C(r,t)\dot{T}(r,t) . \tag{2}$$

$C(r,t)$ is the heat capacity per unit volume and all other symbols have their usual meanings. Heat transfer through vacuum by thermal radiation between curved solid boundaries, say 1 and 2, can be evaluated exactly in a gray body context for this model using the following formula for net radiation exchange, Q, between two infinitely long cylinders:

$$Q = \frac{\sigma_{SB} A_1 \left(T_1^4 - T_2^4 \right)}{\dfrac{1}{\varepsilon_1} + \dfrac{\rho_1}{\rho_2} \left(\dfrac{1}{\varepsilon_2} - 1 \right)} . \tag{3}$$

The index 1 refers to the inner surface, where A_1, l_1, ε_1, and ρ_1 are respectively surface area, temperature, emissivity and radius. A combination of analytic and finite difference methods applied to the shell model provides an efficient means for treating the thermal problem self-consistently with the electromagnetic equations. This involves a two-step procedure, applied alternately for as many repetitions as desired, to propagate the system forward in time. First, for an assigned set of complex dielectric constants in the shells, the electromagnetic problem is solved and a microwave absorbed power density distribution is calculated. Then the thermal equations are used to calculate an updated temperature profile for a slightly later time. This is used to update the complex dielectric constant distribution and the procedure is repeated. Some results calculated with this self-consistent procedure will be presented and discussed next.

DISCUSSION

This transient model is an extension of our previous models for calculating the absorption [1] and steady state temperature profile [2] for a cylindrical rod aligned along the axis of a cylindrical microwave cavity. The present calculations correspond to the same nominal set of experimental parameters used previously [2], unless stated otherwise. The cylindrical cavity had a radius $\rho_c = 4.69$ cm and a length L = 6.63 cm corresponding to an empty cavity TM$_{010}$ mode resonant frequency of 2.45 GHz. The nominal alumina rod had radius and emissivity values of a = 0.2 cm and $\varepsilon_s = 0.31$ respectively. The cylindrical cavity walls were copper with an emissivity of $\varepsilon_w = 0.025$. Hybrid heating studies were performed using an alumina tube surrounding the rod that had a thickness, mid-radius and inner and outer surface emissivities of d = 0.1 cm, $r_{mid} = 0.4$ cm and $\varepsilon_{t_i} = \varepsilon_{t_o} = 0.31$ respectively. The temperature dependence of the real and imaginary dielectric constant for a 92% pure alumina rod [9] was used in these calculations. A Y-MP2E Cray computer performed the calculations.

The time-temperature heating response for several power levels, using the nominal set of parameters without the presence of a surrounding tube, is shown in Fig. 2. Results are shown for the center and surface of the 0.2 cm radius alumina rod. The time required to reach steady state decreases with increasing power even though the steady state temperature is higher. For a power level of 150 watts, the center temperature of the rod was 1756°C while the surface temperature was only 46°C lower. This small temperature difference is in sharp contrast to large temperature differences previously obtained for spherical samples [5].

The addition of an alumina tube around the rod leads to hybrid heating effects. Figure 3 shows the temperature profiles within the rod and tube at various times during heating with 100 watts of input power. After 300 seconds, the rod and tube have essentially reached steady state profiles. At steady state, the 100 watts is distributed between the cavity walls (0.4 watts), the tube (41.9 watts) and the rod (57.7 watts). The rod center and surface temperatures reach 1611.3°C and 1595.2°C respectively. This temperature difference of 16.1°C is almost half the 26.8°C variation obtained with no tube from the 100 watt data in Fig. 2. This reduced variation is caused by the radiant heating of the rod surface due to the surrounding tube. The hybrid heating also leads to a 67°C higher rod steady state temperature. The tube inner and outer surfaces reach steady state temperatures of 1335.4°C and 1325.5°C respectively after 300 seconds which is ≈ 270°C lower than the rod.

The calculations from this new transient model were compared to the results obtained from the previous steady state model [2]. The near steady state 300 second profile in Fig. 3 was less than 0.2% below the steady state model results for both rod and tube temperatures. This excellent agreement validates the accuracy attainable with the transient model.

In a previous study, the steady state temperatures at the center and surface of rods of various radii were calculated with and without a surrounding tube for a TM$_{010}$ mode power level of 30

573

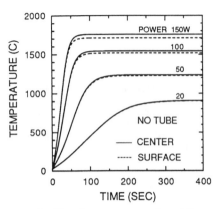

Fig. 2. Transient response of a 0.2 cm radius alumina rod for various power levels.

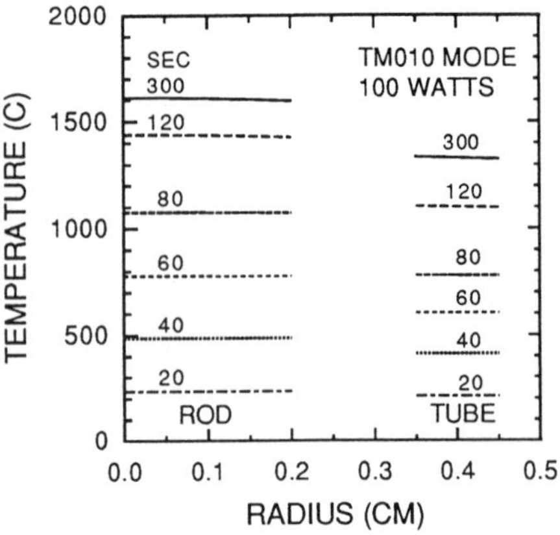

Fig. 3. Temperature profiles in rod and tube during heating with 100 watts.

watts (see Fig. 7 of ref. 2). That study found a maximum sample temperature of 1470°C with no tube (and 919°C with a tube) for a rod radius of 0.025 cm. Figure 4 shows the time-temperature transient behavior for the center of the 0.025 cm radius rod with and without the tube present for the same experimental conditions.

A parametric study was performed to compare the transient time required to reach steady state for various sample sizes ranging from fibers to large rods. Figure 5 shows the center sample temperature versus time for four rod radii. As expected, the smaller fiber-like samples were first to reached steady state (in ≈ 100 seconds) while the large 2 cm radius rod took over 10^4 seconds.

The effects of hybrid heating, using the surrounding tube, on the transient response is shown in Fig. 6. The radiation contribution from the surrounding tube has a significant

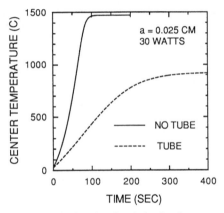

Fig. 4. Transient heating behavior for a 0.025 cm radius rod with and without a surrounding tube.

effect on the small fiber-like samples. The transient response is essentially the same for the 0.0125 and 0.025 cm radius samples. This result is due to the fact that the tube radiation dominates the heating of these samples. When the tube is inserted the steady state temperature of the 0.0125 cm radius sample is increased but the steady state temperature of the 0.025 cm radius sample is decreased. The time to reach steady state conditions is also slightly increased in the presence of the tube. An r_{mid} of 3.0 cm was used in the calculation for the rod with 2.0 cm radius. In the case of the larger 0.2 and 2.0 cm radius rods the tube slightly increases the final steady state temperatures, however it takes slightly longer to reach equilibrium conditions.

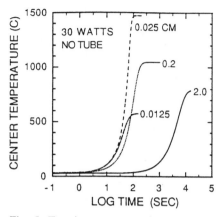

Fig. 5. Transient response of samples ranging from fibers to large rods with no surrounding tube.

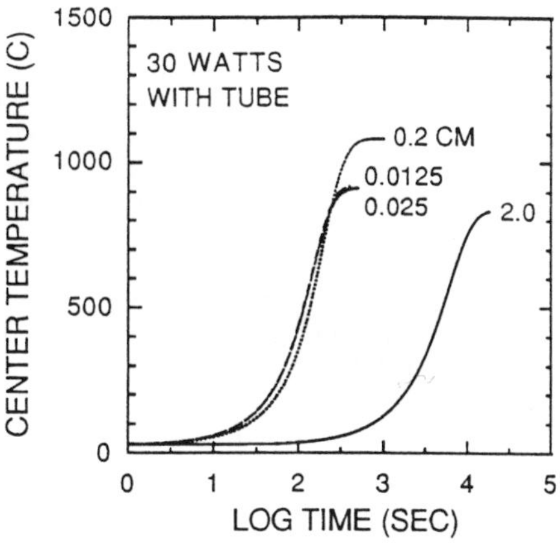

Fig. 6. Transient response of samples ranging from fibers to large rods including a surrounding tube.

THERMAL RUNAWAY

One of the main concerns regarding the application of microwave heating techniques to commercial applications is the possibility of thermal runaway. Thermal runaway can be interpreted as follows. For temperatures below some threshold, the sample attains steady state for a constant power input to the cavity. For an incremental power increase above the threshold the steady state condition is not attained and in some portion of the sample the temperature increases to the melting point. Under these conditions, the microwave power absorbed in some region of the sample becomes greater than the ability to transfer the heat away to the surrounding regions. The present transient model is ideally suited to test for thermal runaway in cylindrical samples.

To investigate thermal runaway, we first evaluated the transient response for rods of various radii using the experimentally determined temperature dependent complex dielectric constant for 92% pure alumina [9]. Figure 7 shows the transient response for three rods ranging over ≈ two orders of magnitude in radius. A typical experimental wall emissivity of 0.3 was used for these calculations. The power level was continually increased for each rod in an attempt to obtain a steady state value near the melting point of alumina (≈ 2000°C). The present transient model assumes that the normal mode resonant frequency is continuously tracked during heating. The curves in Fig. 7 for the 0.025, 0.2, and 2.0 cm radius rods correspond to power levels of 55, 300, and 950 watts respectively. While the slopes of the time-temperature curves are steep, there is no indication that thermal runaway occurs.

It has been speculated that thermal runaway is associated with a significant increase in the temperature dependence of the loss tangent at higher temperatures. To test this hypothesis, we defined two hypothetical expressions for the temperature dependence of the complex dielectric constant ε_r'' that are modifications of the experimentally determined exponential expression. The hypothetical $\varepsilon_r''(T)$ expressions are given by

$$\varepsilon_r''(T,B) = Ae^{\left[T/B\tau\right]} \tag{4}$$

$$\varepsilon_r''(T,C) = Ae^{\left[(T+C)/\tau\right]} , \tag{5}$$

where $A = 0.00438$ and $\tau = 309.49$.

The parameters B and C in Eqs. 4 and 5 can be used to amplify the exponential temperature dependence. The actual experimental exponential expression can be obtained by setting $B = 1$ in Eq. 4 or $C = 0$ in Eq. 5. The enhanced temperature dependence of the imaginary dielectric constant using Eqs. 4 and 5 is shown in Fig. 8. Values of the parameters B and C were chosen to cover the range $0 < \varepsilon_r'' < 100$. A transient response investigation of thermal runaway using these was performed for the B and C temperature dependencies shown in Fig. 8. For all cases, as the input power was increased, steady state conditions were obtained all the way up to melting.

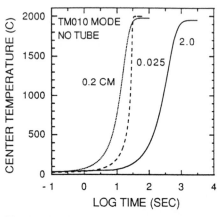

Fig. 7. Transient response for various rod radii using experimental temperature dependence of complex dielectric constant [9].

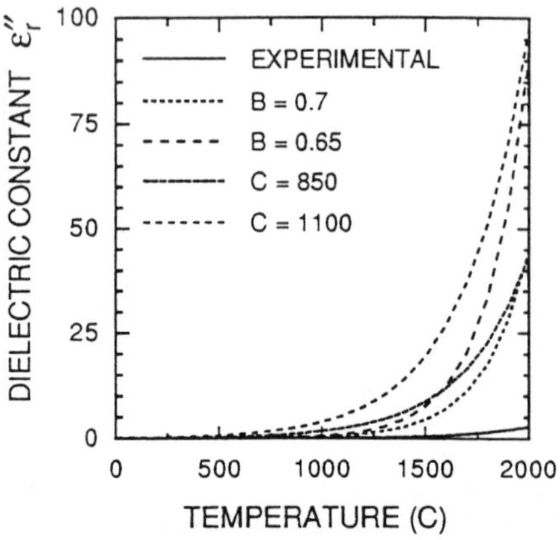

Fig. 8. Temperature dependence of the imaginary dielectric constant using the expressions in Eqs. 4 and 5.

CONCLUSIONS

The most striking feature of the parametric study of microwave and hybrid heating is that for very fine fibers where the radius $a = 0.0125$ cm the steady state temperature attained is about $400°C$ higher when hybrid heating is used. The results of another parametric study also suggest that there is no thermal runaway phenomenon associated with the microwave heating of a cylindrical rod aligned along the axis of a cylindrical cavity. How can this conclusion be reconciled with experiment reports of thermal runaway in microwave heated rods? A possible explanation is based on three important facts: (1) the slope of the temperature versus power curves can become quite steep, (2) the excitation mode resonant frequency is not being tracked in most experimental situations and (3) it can take considerable time to reach steady state conditions. In practical experimental situations, the microwave power is usually continually increased until a desired temperature is reached. If researchers are not patient, the power may be increased too quickly to a level that causes the sample to inadvertently reach the melting temperature. This scenario should be tested using controlled experimental studies.

ACKNOWLEDGMENT

The research described in this article was carried out at the Jet Propulsion Laboratory, California Institute of Technology, under contract with the National Aeronautics and Space Administration.

REFERENCES

1. H. W. Jackson, M. Barmatz, and P. Wagner, MRS Symp. Proc., **347**, pp. 317-323 (1994).
2. H. W. Jackson, M. Barmatz, and P. Wagner, Ceramic Transactions, **59**, pp. 279-287 (1995).
3. H. W. Jackson and M. Barmatz, J. Appl. Phys. **70**, pp. 5193-5204 (1991).
4. M. Barmatz and H. W. Jackson, MRS Symp. Proc., **269**, pp. 97-103 (1992).
5. H. W. Jackson, M. Barmatz and P. Wagner, Ceramic Transactions, **36**, pp. 189-199 (1993).
6. Y. S. Touloukian, *Thermophysical Properties of Matter*, (IFI/Plenum, New York Washington), **8**, p. 98 (1972).
7. J. P. Holman, *Heat Transfer*, (McGraw-Hill, New York), p. 193 (1972).
8. T. Sphicopoules, L.-G. Bernier, and F. Gardiol, IEE Proceedings, **131**, Pt. H, No.2, p. 94 (1984).
9. H. Fukushima, T. Yamanaka and M. Matsui, J. of Japan Soc. of Prec. Eng. **53**, pp. 743-748 (1987).

TEMPERATURE GRADIENTS AND RESIDUAL POROSITY IN MICROWAVE SINTERED ZINC OXIDE

L. P. MARTIN[1], D. DADON[1], M. ROSEN[1]
and
A. BIRMAN[2], D. GERSHON[2], J. P. CALAME[2], B. LEVUSH[2], Y. CARMEL[2]

[1] Department of Materials Science and Engineering, Johns Hopkins University, Baltimore, MD.
[2] Laboratory for Plasma Research, University of Maryland, College Park, MD.

ABSTRACT

ZnO samples were sintered in an overmoded 2.45 GHz microwave applicator. In-situ differential temperature measurements were made to allow comparison of surface and core temperatures during heating. At intermediate temperatures, near 600 °C, the sample core was measured to be more than 250 °C hotter than the sample surface. As the core temperature approached 1100 °C, however, the difference between the surface and core temperatures diminished. Post-sintering scanning electron microscopy (SEM) showed spatial variations in the residual porosity which were consistent with the measured temperature differential. For samples sintered to intermediate temperatures, where large temperature differences persisted, there were significant gradients in the residual porosity. For samples sintered to higher temperatures, there was little residual porosity and no observable porosity gradient. Local density versus temperature behavior was obtained by correlating porosity levels measured from the micrographs with temperature measurements made during sintering. These data demonstrate a significantly lower activation energy for microwave sintering than for conventional sintering.

INTRODUCTION

The late 1980's and early 1990's witnessed a substantial increase in research activity in the field of microwave processing of materials. Many of the issues relevant to microwave processing, including potential industrial applications, have been discussed in the literature [1-4]. The novel aspects of using microwaves for materials processing generally result from the very different ways in which materials are heated by microwave and conventional furnaces. During microwave heating the interior of the sample is heated directly due to the coupling between the electromagnetic field and the sample material (volumetric heating). As a result of the dynamic balance between the rate of electromagnetic energy absorbed within the bulk of the sample and the rate of energy lost from the sample surface, a stable temperature gradient may exist within the sample. This is supported by experimental observation of the temperature gradients in microwave heated ZnO samples subjected to isothermal (core temperature) dwells. During such a dwell, the temperature gradient through the sample achieves an equilibrium value, with the core temperature exceeding that of the surface. The temporal evolution of such a temperature gradient is a complex function of the dielectric and thermal properties of the sample material, which change with temperature and density, and of the thermal boundary conditions.

In the present investigation, the temperature gradients developed in ZnO samples during non-isothermal microwave sintering were evaluated by in-situ temperature measurements made at the surface and core of the samples. The spatial variation in the residual porosity within the sample was determined from post-sintering SEM micrographs. The residual porosity gradients were correlated with the temperature gradients measured in-situ during sintering. The results are in agreement with previous studies performed on Al_2O_3 [5, 6]. In the present work, the local temperature and porosity measurements were used to generate a real densification curve, $\rho = \rho(T)$, during microwave heating at 15 °C/min. Since the uncertainty in sample (bulk) porosity resulting from processing temperature gradients is eliminated, this method for evaluating the densification is more accurate than evaluation by bulk density measurements. The resulting density versus sintering temperature behavior was compared with that for samples sintered with similar heating and cooling schedules in a conventional furnace. It is clear from the data that significant sintering occurs during microwave heating at a lower temperature than during conventional heating.

Mat. Res. Soc. Symp. Proc. Vol. 430 © 1996 Materials Research Society

The measured densification rates were analyzed in the context of existing sintering models to yield an apparent activation energy for both microwave and conventionally sintered samples. Finally, the densification rate equation was solved in conjunction with a one-dimensional heat transport equation and one-dimensional Maxwell equations for the electromagnetic field. This allows calculation of the core temperature evolution as a function of electromagnetic power input during microwave sintering. Good correlation was found between the measured and calculated core temperature as a function of time during the sintering cycle.

EXPERIMENTAL PROCEDURE

Cylindrical ZnO green samples were uniaxially pressed without binder to 37 MPa from a commercial zinc oxide powder with a -200 mesh particle size (Z-1012, CERAC/PURE Division, Milwaukee, WI). Typical sample dimensions of 31 mm diameter and 17 mm thickness were obtained. SEM analysis of the ZnO powder determined that the individual powder particles were porous agglomerates, with a grain size of the order of 1 μm or less. For both microwave and conventional sintering, heating schedules included a slow ramp (1 °C/min) to 200 °C to allow time for water removal. Subsequently the samples were heated at a constant rate of 15 °C/min to the maximum sintering temperature, and then allowed to cool with no dwell time. Selected samples were prepared for SEM analysis by cutting in half along the diameter, polishing the exposed face to 1 μm alumina powder, etching with a 5% acetic acid solution, and sputter coating with Au.

Microwave sintering was performed in air in a computerized, highly overmoded 2.45 GHz applicator with a microwave feedback loop (Model 101, Microwave Materials Technology, Inc., Oak Ridge, TN). The samples were insulated in an alumina enclosure and loosely packed in ZnO powder from the same source as the samples. For all samples sintered in the microwave applicator a shielded, type K thermocouple, inserted into a hole drilled into the core of the sample, was used for temperature measurement and control. Additional temperature measurement was made using a second shielded type K thermocouple maintained in contact with the surface of the sample. Extensive experimentation was performed to insure the accuracy of the temperature readings. A schematic of the microwave sintering apparatus is shown in Figure 1.

Conventionally sintered samples were treated in air in a tube furnace. In this case, the sample temperature was measured by a type K thermocouple placed in direct contact with the sample surface. All samples were allowed to furnace cool to room temperature. Cooldown of the samples in the microwave applicator was designed to mimic the cooling rate of the samples prepared in the conventional furnace. The post sintering average bulk density was determined by dimensional measurement and weighing. The spatial gradient in the porosity fraction was determined quantitatively for each sample from SEM micrographs. Micrographs of 5,000 times magnification were taken at various points along the radius of the midplane of each sample. In each micrograph, the lineal fraction of porosity was measured along several traverses and used to determine the volume fraction of porosity as a function of position in the sample [7].

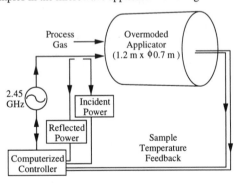

Figure 1: Schematic diagram of the microwave processing furnace.

RESULTS AND DISCUSSION

In Figure 2 the measured values of the core temperature and the difference between the core and surface temperatures are plotted as a function of elapsed time during non-isothermal microwave sintering of a ZnO sample. The results are for a single typical run, and are

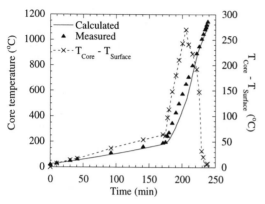

Figure 2: Core temperature, and the difference between measured core and surface temperatures, during microwave sintering of a ZnO sample.

representative of a series of sintering runs performed during this investigation. At the time when the heating rate was abruptly changed from 1°C/min to 15° C/min (at t = 178 minutes in Figure 2), the difference between the core and the surface temperatures increases dramatically. This is a strong indication that the sample is being heated internally and is consistent with theoretical predictions [8, 9]. The sintering process begins t o cause densification of the sample at around 600°C [10]. Thus, the rapid decrease in the temperature difference between the surface and the core at that temperature is attributed to the onset of densification and the resultant increase in thermal conductivity. Since ZnO is a semiconductor, the electrical conductivity increases with increasing temperature. Consequently, increasing temperature leads to increased electrical conductivity and a corresponding decrease in the depth of penetration of the electric field into the material (skin depth). This increases the surface energy deposition and explains the precipitous decrease in the difference between the core and surface temperatures for t > 220 minutes. Samples sintered to near full density showed no temperature difference between the core and the surface at the maximum sintering temperature of about 1150 °C.

Quantitative analysis of SEM micrographs was used to determine the spatial variation in the volume fraction of porosity through samples microwave sintered to core temperatures of 870°C, 960°C, and 1150°C. The porosity level was evaluated on the midplane of each sample at several points between the center and the edge. These data are shown in Figure 3, where it is clear that for samples sintered to intermediate temperatures there is a residual porosity gradient with the center of the sample being less porous than the surface. For the sample sintered to 1150°C, however, there is very little residual porosity and thus no observable porosity gradient. These observations

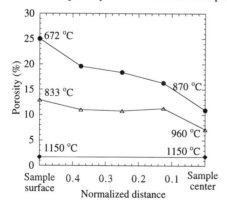

Figure 3: Porosity versus position, normalized to the radius, for ZnO samples microwave sintered to different temperatures.

support the in-situ measurements of the core and surface temperature during microwave sintering presented in Figure 2. In addition, the observed residual porosity gradients strongly indicate that the measured differential between the core and surface temperatures is a result of a continuos temperature gradient through the sample rather than the presence of isolated hot - or cold - zones.

The porosity levels shown in Figure 3 were used to evaluate the density versus sintering temperature for the sintering schedules used in this study. For each sample, the porosity determined from the SEM micrograph taken from near the center was associated with the maximum core temperature during sintering. Similarly, the porosity determined from the SEM micrograph taken from near the surface was associated with the maximum surface temperature. These data are shown in Figure 4, along with data for a set of ZnO samples conventionally sintered by similar schedules. For the conventionally sintered samples, the densities were evaluated by dimensional measurement and weighing (recall that the porosity was observed to be uniform in these samples). For the microwave sintered samples, the densities were evaluated as $100*(1-P)$ where P is the volumetric porosity determined from the lineal analysis performed on the respective SEM micrographs.

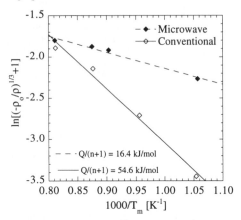

Figure 4: Ln[$(-\rho/\rho_0)^{1/3}+1$] versus T_m^{-1} for conventionally and microwave sintered ZnO samples.

The densification rate equation for initial and intermediate stage nonisothermal sintering can be determined from the general equation for isothermal sintering. The assumption made is that the isothermal and nonisothermal sintering rates are equal for any combination of shrinkage and temperature. Based on this assumption, a generalized treatment for the densification rate during nonisothermal heating has been developed [11-14]. Numerous studies of the sintering kinetics of ceramics based on in-situ dilatometric measurements have been performed for both thermal and microwave sintering [11, 15, 16]. In the present work, however, only the final dimensions of the sample were measured. Therefore, an integrated averaging approach was adopted in order to estimate the apparent activation energy for microwave sintering, and to allow modeling and comparison with conventional sintering. The resultant analysis yields [17]

$$\ln\left[\left(\frac{-\rho}{\rho_o}\right)^{1/3}+1\right] \approx -\frac{Q}{(n+1)RT_m}+\frac{1}{n+1}\left[\ln\left(T_m\left(\frac{1}{\alpha_1}-\frac{1}{\alpha_2}\right)\right)+\ln\left(\frac{AR(n+1)}{Q}\right)\right], \qquad (1)$$

where ρ is the final density, ρ_0 the theoretical density, Q the activation energy for sintering shrinkage, T_m the maximum temperature reached during sintering, α_i the heating and cooling rates, and A and n are constants. n is a geometrical constant determined by the sintering mechanism and particle contact geometry.

The quantity on the left hand side of equation (1) can be plotted as a function of $1/T_m$, and the result is very nearly a linear relation. The deviation from linearity resulting from the $ln(T_m)$ term is very small since the exponential term dominates the temperature dependence. This analysis yields, for microwave and conventionally sintered samples, $Q/(n+1)$ = 16.4 and 54.6 kJ/mol, respectively. Typical values for n are between 1 and 3 for many mechanism/geometry combinations, so that an activation energy in the range of 110 - 220 kJ/mol may be inferred from the observed data for the conventional sintering process [18]. In comparison, published values for the densification activation energy in ZnO are approximately 200 - 240 kJ/mol [19, 20]. This implies a value of $n \approx 3$. Simultaneous determination of Q and n is not possible from this treatment without explicit knowledge of the particle contact geometry or independent measurement of Q via an alternate method. However, the activation energy Q was independently measured for samples sintered conventionally at 3, 5, 7, and 11 °C/min by the method of Wang and Raj [14]. The result, $Q_{conv.}$ = 240±20 kJ/mol, is consistent with the previous discussion and supports a value of $n \approx 3$.

Direct comparison of the activation energy Q via the parameter $Q/(n+1)$ can be made for the two processes. It is clear that there is a demonstrably lower value of $Q/(n+1)$ for densification by microwave sintering. SEM micrographic evidence indicates that microwave and conventionally sintered samples exhibit similar morphologies at equal porosity [10]. This implies the same sintering mechanisms, and therefore n, for the two processes. Since n can be considered identical for the two processes, the difference of approximately a factor of 3 in this quantity results from a reduction of the activation energy Q during microwave sintering. A similar reduction in the activation energy during microwave sintering has been reported for various compositions in the alumina/zirconia system [21]. Rybakov and Semenov discuss the reduction in the activation energy and attempt to attribute it to the coupling between the electromagnetic field and the ceramic [22]. Freeman, et al., have also reported experimental evidence of microwave field enhancement (microwave effect) of charge transport in NaCl [23].

Simulation of the microwave sintering process was performed, and the results were compared with the experiment. A one dimensional (1D) slab geometry was used in the simulation. This geometry is appropriate for modeling of microwave processing in a highly overmoded applicator since, in this case, the sample is uniformly irradiated from all sides. The ZnO cylindrical sample (height 17 mm and diameter 31 mm) was therefore represented by a 31 mm thick slab sandwiched between 20 mm thick slabs of low density porous ZnO ($\rho/\rho_o = 0.12$) representing the thermal insulation. To calculate the instantaneous temperature distribution in the sample one must solve a heat transport equation with the appropriate boundary conditions. For the one-dimensional case this equation is given by

$$c_p(T)\frac{dT}{dt} = \frac{1}{\rho}\frac{d}{dx}\left[\kappa(T,\rho)\frac{dT}{dx}\right] + \frac{P}{\rho}, \tag{2}$$

where C_p is the heat-capacity, κ is the heat-conduction, ρ is the density, P is the power density absorbed in the ceramic, ω is the angular frequency of the applied field, and x is the spatial coordinate. For microwave sintering, the power P absorbed per unit volume in the sample due to the interaction with the electromagnetic field is given by the well known relation

$$P = \frac{1}{2}\omega\varepsilon_0\varepsilon''|E|^2 \tag{3}$$

where ε_o the permittivity of free space, ε'' the local imaginary part of the relative permittivity, and E the local electric field. For solution of the electric field spatial distribution, the sample and the insulation were divided into computational cells. At any time-step, each of these cells had a definite temperature and density, and the corresponding thermal and dielectric coefficients. Using previously measured values for the complex permittivity of ZnO as a function of temperature [8], the spatial distribution of the electric field E within the sample was determined by numerical solution of the Maxwell equations. Modeling of the heating process requires simultaneous solution of the heat transport equation (2), the densification rate equation (1), and the Maxwell equations for the electric field. This solution requires knowledge of all parameters appearing in equations (3) and (2). Of these, the dependence of permittivity, ε, heat capacity, C_p, and thermal conductivity, κ, on the temperature and density are discussed elsewhere [8]. The temporal profile of the injected microwave power as recorded by the control unit was used as an input to the modeling. With all these parameters, the three equations were solved simultaneously in order to calculate the temperature distribution within a sintering sample. The measured and computed core temperatures, shown in Figure 2, are in good agreement.

CONCLUSIONS

For ZnO samples sintered by microwaves to intermediate temperatures (and densities), there is a residual porosity gradient with the core being less porous than the surface. This is in agreement with the temperature gradient (as high as 260 °C) developed within the sample during the sintering process. ZnO samples heated by microwaves showed no temperature gradient at temperatures above 1150 °C. At this temperature the samples were sintered to full, uniform, density. Local temperature and porosity measurements allow evaluation of the densification rate as a function of temperature ($\rho=\rho(T)$)without the use of bulk averaging. From the experimentally measured densification behavior, $\rho(T)$, it is apparent that enhanced microwave sintering of ZnO occurs at significantly lower temperatures than conventional sintering. From these data, apparent activation energies were computed for both microwave and conventional sintering. The microwave sintering process exhibits a reduction in the apparent activation energy, compared with that of conventional sintering, of approximately a factor of three.

ACKNOWLEDGMENTS

This work was supported by the Division of Advanced Energy Projects, US Department of Energy, under contract No. DE-FG02-94ER 12140. Partial support from the Army Research Office, Research Triangle Park, and a NATO linkage grant is acknowledged. The authors are grateful for helpful discussions with V. E. Semenov, I. K. Lloyd and K. I. Rybakov.

REFERENCES

1. Several articles in *MRS Bull.*, **18** [11], (1993).
2. D. L. Johnson, in *Ceramic Transactions, Vol. 21, Microwaves: Theory and Application in Materials Processing*, D. E. Clark, F. D. Gac, and W. H. Sutton eds. (American Ceramic Society, Westerville, OH, 1991), pp. 17 - 28.
3. D. E. Clark and D. C. Folz, in *ibid*, pp. 29 - 34.
4. M. F. Iskander, O. Andrade, A. Vikar, H. Kimrey, R. Smith, S. Lamoreaux, C, Cheng, C. Tanner and K. Mhta, in *ibid*, pp. 35 - 48.
5. A. De, I. Ahmad, E. D. Whitney and D. E. Clark, in *ibid*, pp. 319 - 328.
6. A. De, I. Ahmad, E. Dow Whitney and D. E. Clark, in *Microwave Processing of Materials II*, W. B. Snyder, Jr., W.H. Sutton, M. F. Iskander and D. L. Johnson, eds. (Materials Research Society, Pittsburgh, PA, 1991), pp. 283 - 288.
7. J. E. Hilliard in *Quantitative Microscopy*, R. T. DeHoff and F. N. Rhines, eds. (McGraw Hill, Inc., New York, 1968), pp. 45 - 54.
8. A. Birman, B. Levush, Y. Carmel, D. Gershon, D. Dadon, L.P. Martin and M. Rosen, in *Ceramic Transactions, Vol. 59, Microwaves: Theory and Application in Materials Processing III*, D. E. Clark, D. C. Folz, S. J. Oda and R. Silberglitt eds. (American Ceramic Society, Westerville, OH, 1991), pp.305 - 312.
9. A. Birman, D. Gershon, J. Calame, Y. Carmel, B. Levush, Yu. V. Bykov, A. G. Eremeev, V. V. Holoptsev, V. E. Semenov, D. Dadon, L. P. Martin, M. Rosen and R. Hutcheon, submitted to *J. Appl. Phys.* (1996).
10. L.P. Martin, D. Dadon, D. Gershon, B. Levush, Y. Carmel and M. Rosen, in *Ceramic Transactions, Vol. 59, Microwaves: Theory and Application in Materials Processing III*, D. E. Clark, D. C. Folz, S. J. Oda and R. Silberglitt eds. (American Ceramic Society, Westerville, OH, 1991), pp.399 - 406.
11. D. L. Johnson, *J. Appl. Phys.*, **40**, 192 - 200 (1969).
12. W. S. Young and I. B. Cutler, *J. Am. Ceram. Soc.*, **53**, 659 - 663 (1970).
13. J. L. Woolfrey and M. J. Bannister, *J. Am. Ceram. Soc.*, **55**, 390 - 394 (1970).
14. J. Wang and R. Raj, *J. Am. Ceram. Soc.*, **73**, 1172-1175 (1990).
15. R.D. Bagley, I.B. Cutler, and D.L. Johnson, *J. Am. Ceram. Soc.*, **53**, 136 - 141 (1970).
16. J. Samuels and J. R. Brandon, *J. Mat. Sci.*, **27**, 3259 - 3265 (1992).
17. D. Dadon, L. P. Martin, M. Rosen, A. Birman, D. Gershon, J. P. Calame, B. Levush and Y. Carmel, submitted to *J. Mat. Sci. and Proc.* (1995).
18. M. J. Bannister, *J. Am. Ceram. Soc.*, **51**, 548 - 553 (1968).
19. T. K. Gupta and R.L. Coble, *J. Am. Ceram. Soc.*, **51**, 521 - 525 (1968).
20. S. K. Dutta and R. M. Spriggs, *Mater. Res. Bull.*, **4**, 797 - 806 (1969).
21. M. Janney and H. Kimrey, *Mat. Res. Symp. Proc.*, **189**, 215 - 226 (1990).
22. K.I. Rybakov and V.E. Semenov, *Phys. Rev. B*, **52**, 3030 - 3033 (1995).
23. S. A. Freeman, J. H. Booske and R. F. Cooper, *Phys. Rev. Lett.*, **74**, 2042 - 2045 (1995).

Part XII

Microwave Processing of Polymers

CONTROL OF MICROWAVE PROCESS FOR POLYMER CURING BY DIELECTRIC PROPERTY MEASUREMENTS IN A BROAD FREQUENCY RANGE, 1 MHz-10 GHz

O. MEYER, N. BELHADJ-TAHAR, A. FOURRIER-LAMER,
Laboratoire de Dispositifs Infrarouge et Micro-ondes, Université Pierre et Marie Curie
Case 92 T12 E2, 4 place Jussieu 75252 PARIS Cedex 05 FRANCE.
meyer@ccr.jussieu.fr   tel : 33-1-44 27 43 72   fax : 33-1 44 27 43 82

ABSTRACT

Let us present a new automatic system which couples a high power circuit (2.45 GHz) for heating with a low power one, for broad frequency band measurements of dielectric permittivity (1 MHz-10 GHz). The sample under microwave curing consists in an epoxy resin + hardener complex (DGEBA+DDS).
The broad band spectra shows the existence of a dielectric relaxation. This absorption is used as a marker to describe the evolution of the reaction in relation with time and temperature, in terms of conversion ratio, kinetic velocity and quality.
The phenomenum of microwave activation and interactions between permanent molecular dipoles and electromagnetic waves form the subject of this study.
We also try to establish the specificity of microwave curing from a microscopic point of view. This work emphasizes new concepts which concern microwave chemistry.

INTRODUCTION

Up to now, the dielectrical measurements carried out in order to characterize the curing phenomenon of an Epoxy-Amine system have been made under low frequencies, less than 10 kHz, or at the microwave heating frequency, i.e. 2.45 GHz. As far as conventional heating is concerned, the results obtained enabled the study of concentration variations of existing dipoles, together with the study of the molecular system during the curing process [1]. For instance, it has been observed that, at 1 kHz, the relaxation time of the DGEBA+DDM mixture increases in relation with time, during the curing process, at a constant temperature [2]. The DGEBA+DDS mixture also showed different dielectrical behaviours, even for identical curing times. These differences are due to molecule size and shape and to the distribution of loads at the level of those multiple hardener molecules including the amine function. Consequently, there are great differences between all curing kinetics [3].
In the case of microwave heating, the acceleration of the curing process is demonstrated from the very first heating steps and the quick establishment by the tridimensional network radiation observed. Dielectrical measurements at 2.45 GHz did not show fundamental differences in behaviour with system cured using a conventional method. Investigation at this frequency is not sufficient to reveal the microscopic behaviour of the molecular system. Only measurements in infrared and RMN have demonstrated that chemical processes were not similar. Infrared measurements showed the absence of etherification reaction in microwave heating. Moreover, it has been deduced that reactivity of secondary amine groups was similar to that of primary amine groups [4]. This observation recalls the effectiveness of microwave radiation in chemical reactions including saturated systems.

Mat. Res. Soc. Symp. Proc. Vol. 430 © 1996 Materials Research Society

The original instrumentation [5] presented here demonstrates that the investigated frequencies used up to now are not sufficient because they are very far from the frequencies characterising the dipoles themselves, carried out during their evolution from an ambient temperature. Conversely, the information obtained through wide-band exploration is much richer for the study of curing using a conventional method on the one hand and, on the other hand, for the demonstration of the specificity of microwave heating: it describes the evolution of both dipole behaviour before and during curing and structural behaviour of the resin+hardener mixture . This study aims to explain the role of the electromagnetic field-dipoles coupling which directs reacting entities in the case of microwave heating and to deduce its specificity at the mesoscopic level, in comparison with conventional heating, during which dipoles are oriented randomly.

This instrumentation allows the measurement of the complex permittivity $\varepsilon^*=\varepsilon'-j\varepsilon''$ of an environment submitted to heating on a very wide frequency band (1 MHz-10 GHz in conventional heating and 1 MHz-2 GHz for microwave heating). It constitutes a true process control of an environment changing according to time and temperature.

EXPERIMENT

Description of the instrumentation set up

Transformations under microwaves are usually carried out in rectangular waveguide systems. Their task consists in heating the sample as efficiently as possible using applicators optimized for heating. However, their intrinsic characteristic is that they can only propagate waves on a very narrow band. Measurements can only be made at the heating frequency. The instrumentation we have designed follows a totally different direction. In order to determine the dielectrical characteristics of the Epoxy-Amine system on the widest band via a very low power level (1 milliwatt) and to submit it alternatively to high power levels (several watts), we had to adopt the use of an aperiodic measurement cell and a circular coaxial waveguide, wide-band transmission line. The measurement and heating rack is composed of different elements :

1. The measurement cell in which the sample to be heated and qualified is placed, the SUPERMIT cell [6] is composed of a cylindrical waveguide, short-circuited at one end and excited abruptly at the other end by a circular coaxial cable in APC7 standard (Fig.1). The SUPERMIT software calculates the dielectrical characteristics from the measurement of a reflection coefficient with the resolution of inverse problem. The cell have a good precision in the 1 MHz-18 GHz frequency band with APC7 standard. Data acquistion in real time is realized with an HP-715/75 station.

The SUPERPOL cell is an adaptation of the cell presented by Kolodziej and al. [7] and J.C.Badot [8] with a dielectric crown (PTFE) to measure and heat liquids (Fig. 2). This cell have good precision in the 1 MHz-2 GHz band with General Radio GR900 standard. It is filled inhomogeneously with solid dielectrics which makes the electrical field uniform at the location of the sample. This optimization has been achieved using HFSS simulation software. This cell is optimized for homogeneous microwave heating.

2. The microwave heating device, which we will call Pump is a 50W source at 2.45 GHz with AsGa transistors manufactured by KUBIK (Fig.3).

3. The measurement and control device we call Probe, (HP 4291A for the 1 MHz-1.8 GHz band, HP 8510B for the 45 MHz-18 GHz band).

4. A coaxial line to carry energy. This aperiodic system can be easily transposed to other heating frequencies and allows wide-band measurements.

5. A system of double electro-mechanical switching for quasi-perfect insulation ($\tau_{heating}$ ≥6 s, $\tau_{measurement}$ ≤1.5 s, $\tau_{switching}$ ≈30 ms).

The measurement system (probe) and the microwave heating device (pump) are coupled alternatively to the cell (Fig. 3). This makes it possible to measure and to heat the cell alternatively. In the same configuration we can also measure the dielectrical characteristics of the same sample heated in a conventional oven.

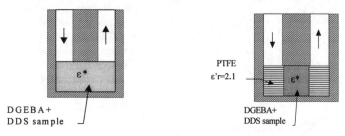

DGEBA+
DDS sample

PTFE
ε'r=2.1

DGEBA+
DDS sample

*Figure 1 : SUPERMIT cell **in APC7*** *used in conventional heating*

Figure 2 : SUPERPOL cell in GR900 *for microwave heating*

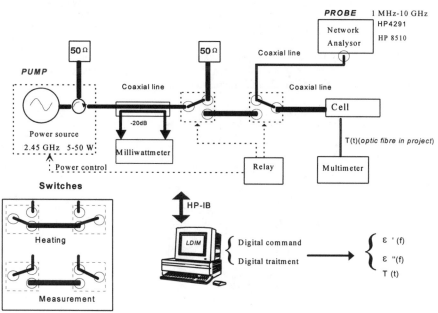

Figure 3: Instrument set for dielectric measurement under microwave heating

RESULTS

1. Results on the DGEBA+DDS mixture in conventional heating:

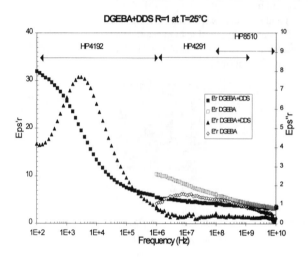

DGEBA+DDS R=1 at T=25°C

Figure 4 : Complex permittivity spectra of DGEBA+DDS ▲ and DGEBA ◆

At ambient temperature, we obtain a result from 100 Hz to 10 GHz: We can observe a high relaxation at the frequence Fr=3 kHz (Fig. 4). The relaxation is lower to the one of pure DGEBA (Fr=12 MHz). We have characterized the ε* of DGEBA+DDS mixture from an ambient temperature of 200°C in the 1 MHz-10 GHz frequency band (Fig.5). When the temperature increases, the relaxation frequency increases too. This stage is due to fluidification. The 3D representations versus frequency and time (Fig.5) describe this behaviour.

At approximatively 180°C, the frequency relaxation remains at a maximum frequency (Fig.8). Then, Fr decreases continually down to smaller frequencies. This is the curing stage. The reaction transforms the mixture into a tridimensional network.

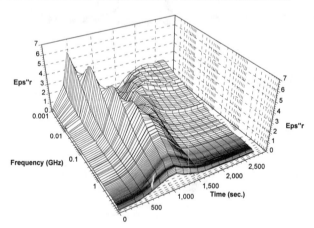

Figure 5 : Losses versus frequency and heating time in conventional heating

At the fluidification stage, (Fr increase) we can determine an activation energy by the Arhenius law, W≈2.2 eV.

2. Results on DGEBA+DDS mixture in microwave heating :

We have characterized the ε^* of DGEBA and DGEBA+DDS mixture in the 1 MHz-1.8 GHz frequency band (Fig.6). We observe a similar evolution. When heating time increases, the relaxation frequency increases to reach a maximum frequency: Frmax (Fig.8) in a short time (less than 1 minute). This stage is due to fluidification.

The slow decrease of relaxation frequency in conventional heating and in microwave should be due to the curing stage (i.e. thermal diffusion and chemical kinetics). The amplitude of the maximum absorption (ε''max) falls to an inferior value in 7 sec. (Fig.8). In conventional heating, maxima losses decrease continually (Fig.9).

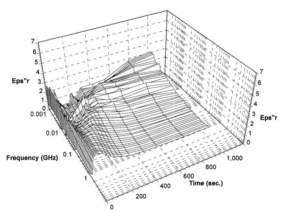

Figure 6 : Losses versus frequency and heating time in microwave heating (P=15W)

3. Results obtained using different hardener (DDS) concentrations (Fig.7).

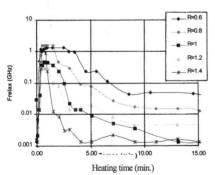

Figure 7 : Relaxation freq. Fr versus DDS concentration (R) in microwave heating

We have compared the results obtained in microwave heating and conventional heating with different concentrations of DDS. The results confirm the general evolution of the frequency relaxation. The relaxation frequency obtained depends on the concentration of cross-linking agents (R), the heating kinetics and the type of heating (microwave, conventional heating). Fr increases when the hardener rate decreases, or when dT/dt increases (Fig. 7).

4. Comparison and interpretation:

In microwave heating, Fr reaches Frmax very quickly (less than 1 minute). This appears to be the typical observation concerning the microwave process: this process induces very quick thermal. But the different results obtained as far as the amplitude of maxima losses and spectrum are concerned must be explained.

The action of power field on dipoles certainly induces a local temperature which explains the evolution of the frequency relaxation under microwave. Consequently, the frequency relaxation would be a marker of the local temperature induced by microwave interaction between dipoles and high power field.

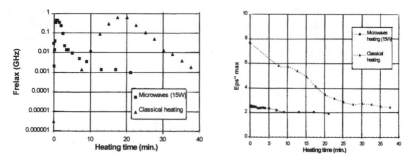

Figure 8 : Fr. versus heating time   *Figure 9 : Maxima losses versus heating time*
▲ *in conventional heating ,* ■ *in microwave heating*

CONCLUSION

Our instrumentation enabled us to follow the transformation of the DGEBA DDS mixture throughout the thermal cycle, whether using conventional or microwave heating. The evolution in temperature and over time of the mixture's characteristic relaxation frequency enabled us to describe the fluidification phase (Fr increases), followed by the curing phase (Fr decreases). We observed that the two heating types have very different behaviour at molecular level, Fr and $\varepsilon*$ being characteristics of the dipoles and the structure. The part of each remains to be determined. Among other questions, processing the signal, using an Argand diagram will make it possible to define these respective parts. The same instrumentation will also make it possible to see whether the mixture's electromagnetic response depends on the heating frequency.

ACKNOWLEDGEMENTS

The authors gratefully acknowledge Mr Delmotte from ENSAM Paris for the preparation of mixtures, Mr J.C Lacroix from ITODYS (Paris 7) for his simulations on hardeners.

REFERENCES

1. Y.Deng, C.Martin, Journal of Polymer Science: Part B,.**32**, 2115-2125, (1994).
2. M.B.M. Mangion, G.P. Johari, J. of Polym. Science: Part B, .**29**, p. 1127-1135, (1991).
3. M. Delmotte, H. Jullien, M. Ollivon, Eur. Polym. J. **27**, No 4/5, p. 371-376 (1991).
4. E.Marand, K.R. Baker,. Macromolecules, **25**, p. 2243-2252, (1992).
5. O.Meyer, N.Belhadj-Tahar, A.Fourrier-Lamer, (European Workshop on Microwave Processing of Materials, Karlsruhe 1994).
6. N. Belhadj-Tahar, A. Fourrier-Lamer, IEEE Trans. Microwave Theory Tech. **MTT-34**, No 3, p. 346-349 (1986).
7. H. Kolodziej, L. Sobscyk, Acta Phys. Pol., **A39**, p. 59 (1971).
8. J.C. Badot, Phd.Thesis from the Université Pierre et Marie Curie - Paris 6 (1988).

HIGH FREQUENCY (27.12 MHz) ACTIVATION OF THE RADICAL CURING OF UNSATURATED POLYESTERS IN STYRENE SOLUTION

P. ALAZARD, A. GOURDENNE
Laboratoire des Matériaux, ENSCT, 118 route de Narbonne, 31077 Toulouse Cedex, France

ABSTRACT

The crosslinking reaction of a divinylester resin of epoxy-acrylic type in styrene solution is activated by high frequencies at 27.12 MHz. The samples to be cured are positionned betweeen two parallel steel plates used as electrodes and an electrical voltage is applied. A parametrical study is described, where the applied electrical voltage, or power, and the concentration of benzoyl peroxide, which is the radical initiator, are taken into account. The optimization of the electromagnetic curing is performed through the determination of the glassy transition temperature of the final products.

INTRODUCTION

Microwave heating (2.45 GHz) can be used for the activation of the crosslinking reaction of various thermosetting resins such as epoxy resins, polyurethanes and unsaturated polyesters [1, 2, 3, 4, 5, 6]. The principle of the activation is based on the partial conversion as heat of the dielectric loss at 2.45 GHz in the organic medium. This electromagnetic activation process has been extended to the curing of many composites [7, 8, 9]. Another possibility of activation has recently come to light where high frequencies (27.12 MHz) are used, based on the same principle of forced dipolar relaxation. Preliminary results from our laboratory show that high frequency (HF) heating is efficient for the cure of the epoxy resins [10]. The case of unsaturated polyesters has never been considered at this frequency, in spite of its interest, although the prepolymeric polyester chains carry hydroxyl groups which are the relaxing entities [11]. So the dielectric loss depends only on the viscosity of the polymerizable medium, since the number of active species remains constant.

The present paper reports on the radical crosslinking at 27.12 MHz of a divinylester resin of epoxy-acrylic type in styrene solution.

EXPERIMENTAL

Chemicals

The divinylester used is of epoxy-acrylic type (**I**), kindly provided by Dow Chemical as Derakane D 411-C 50, with n = 2.79 and in styrene (**II**) solution (50 %). The benzoyl peroxide (**III**) which is the radical initiator (Fluka - Ref : 33851), is moistened with water (25 %) to prevent any explosive decomposition. Its concentrations will be given as weight percentages of the initial mixture. The initiator, when heated to 60 °C, gives two primary radicals (**IV**) active for the initiation of styrene (**V**) and of terminal methacrylate functions (**VI**) of the polyester chains.

These last radicals can add to their parent monomer or to each other. The initial chemical formulations are chosen to favour the copolymerization styrene-methacrylate so as to obtain a final three-dimensionnal network schematically represented by **VII**.

(I)

$\langle\!\bigcirc\!\rangle\!-\!CH\!=\!CH_2$

(II)

$\langle\!\bigcirc\!\rangle\!-\!\overset{O}{\underset{||}{C}}\!-\!O\!-\!O\!-\!\overset{O}{\underset{||}{C}}\!-\!\langle\!\bigcirc\!\rangle$

(III)

$\langle\!\bigcirc\!\rangle\!-\!\overset{O}{\underset{||}{C}}\!-\!O\cdot$

(IV)

$\langle\!\bigcirc\!\rangle\!-\!\overset{O}{\underset{||}{C}}\!-\!O\!-\!CH_2\!-\!\overset{\cdot}{\underset{H}{C}}\!-\!\langle\!\bigcirc\!\rangle$

(V)

$\sim\!\sim\!-\!O\!-\!\overset{O}{\underset{||}{C}}\!-\!\overset{CH_3}{\underset{\cdot}{C}}\!-\!CH_2\!-\!O\!-\!\overset{O}{\underset{||}{C}}\!-\!\langle\!\bigcirc\!\rangle$

(VI)

polystyrene   divinylester

(VII)

Of course various steps are involved in the synthesis of the crosslinked networks : fluidification of the initial resin-peroxide mixture ; formation of microgels ; percolation of the microgels ; vitrification of the chemical medium (fast exothermic conversion of styrene) :

Initial	Liquid	Formation of microgels	Complete gelation	Vitrification

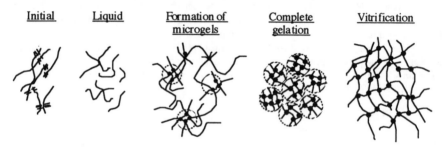

High frequency apparatus

Samples (260 g) of polymerizing matter (divinylester + styrene + benzoyl peroxide) are degassed under primary vaccum and poured inside parallelepipedal moulds with teflon walls and closed up and down by steel plates. The moulds are then set up between two stainless steel plates in a HF applicator which is connected to a Sairem system, based on the 50 Ω Technology, including a generator of HF (27.12 MHz), a phase and amplitude discriminator, and a matching box. The electrical circuit is monitored so that it works at the electrical resonance, in order to maintain the best coupling electricity-matter. The temperature of the samples is determined through a fiberoptic sensor probe immune to HF. The application of the electrical field is carried out at constant electrical voltage Vo or power Po.

The dielectric loss, P, due to dipolar relaxation in an organic medium is given by :

$$P = 2\,\pi\,.\,f\,.\,\varepsilon''\,.\,V\,.\,E^2$$

where f is the frequency, ε'' the loss factor, V the volume of matter, and E the modulus of the electrical field. When the polymerization develops, ε'' goes down. Moreover the volume remains more or less constant. So, when the power, Po, is kept constant, the field E, and consequently the potential V, have to be raised. On the other hand, at a given voltage Vo or electrical field Eo, (since Eo = Vo/d, where d is the thickness of the sample), ε'' and the power P decrease simultaneously.

The study of the variations against time (t) of the temperature T = T(t) and of the electrical power P = P(t) or voltage V = V(t), provides information on the kinetics of the polymerization reaction and on the structural changes of the chemical medium under irradiation.

The determination of the glassy transition temperature, Tg_1, of the crosslinked products is carried out by Differential Scanning Calorimetry through DSC apparatus of Setaram type (Model 111-B). Tg_1 is measured at a constant value of the heating rate q = 10 °C/min. Tg_1 is shifted towards a stabilized value Tg_∞ after many DSC analyses, because the post-cure of the resin develops as soon as the temperature becomes higher than Tg + 20°C. Tg_1 is rather linked to the extent of the polymerization, and Tg_∞ to the structural homogeneity of the crosslinked materials.

RESULTS AND DISCUSSION

Model studies

Two experiments at given electrical potential Vo = 2 000 v and power Po = 160 W have been carefully carried out from the same initial resin-peroxide (1 %) mixture.

Figure 1 displays the four curves T = T(t), P = P(t), (T)' = dT/dt and (P)' = dP/dt recorded at Vo = 2 000 v. The examination of the T curve provides information on the kinetics of polymerization based on results obtained from radical reactions thermally activated inside an oven. The temperature quickly starts increasing from 0 °C ; at the same time the Van der Waals interactions are partially broken. The following change corresponds to the production of primary radicals from the peroxide (working temperature : 60 °C), to their reactions with the anti-oxidizing agent and also with the reactants (resin and styrene) : microgels are formed and their number increases with time. When the microgels form a percolation network at 71.0 min, the fast exothermic conversion of styrene begins. The inflexion point located just before this physical step at 61.5 min is associated with the sol-gel transition which separates the fluid state, not polymerized, and the percolated gel. The vitrification of the medium develops during the exothermic step ; the coordinates of the thermal maximum are : 80.5 min, 210.6 °C. Beyond the maximum there is no more polymerization. The P curve shows that the power falls from the start ; this variation is associated with the partial break of the Van der Waals forces in which the dipolar hydroxyls are taken into account : the polarity decreases and consequently the electrical power is lowered. P continues to drop since the viscosity, due to the formation of microgels, increases. The minimum of P is located at 74.0 min, just three minutes after the start of the exothermic peak at 71.0 min. The vitrification which takes place increases the relaxation of the hydroxyls and the dielectric loss P in spite of the accompanying viscosity effect, because of the production of chemical heat. P goes through a maximum (84.5 min, 106.5 W) and then decreases to a plateau when the chemical heat is totally lost through convexion towards the external medium. The inflexion points earlier observed in the T and P curves around 1.5 min correspond to the fluidification transition.

Figure 2 shows the four curves T = T(t), V = V(t), (T)' = dT/dt and (V)' = dV/dt in the case of an irradiation at Po = 160 W. The T curve provides the same type of information as previously stated : the fluidification of the initial mixture, the sol-gel transition, the percolated gel and the vitrification step. The variations of the electrical potential, V, against time can be easily correlated with those of the temperature. V starts increasing from the origin because of the fluidification of the medium which results from a partial break of the Van der Waals interactions : the polarity goes down. Then the viscosity effect, due to the formation of microgels, lowers the relaxation of the hydroxyls : V should be raised to maintain the power at Po = 160 W. The vitrification, in spite of its viscosity effect, generates a decrease in V (or an increase in ε'') due to the associated exothermic effect. At the end, the electrical potential takes a plateau value.

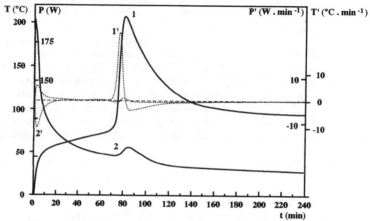

Figure 1 : High frequency crosslinking of the divinylester resin
at given electrical voltage Vo = 2 000 v
(1) : T = T(t) ; (2) : P = P(t) ; (1') : (T)' = dT/dt ; (2') : (P)' = dP/dt

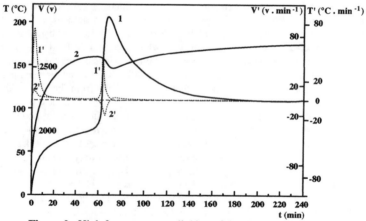

Figure 2 : High frequency crosslinking of the divinylester resin
at given electrical power Po = 160 W
(1) : T = T(t) ; (2) : V = V(t) ; (1') : (T)' = dT/dt ; (2') : (V)' = dV/dt

<u>Electrical potential Vo dependence of the kinetics of polymerization</u>

Figure 3 presents a series of T curves recorded at variable Vo values (2 800, 2 400, 2 000, 1 600 and 1 200 v) for a concentration of peroxide of 1 %. When Vo increases, the kinetics of polymerization are more and more accelerated, as expected. The same tendency is observed in the P curves.

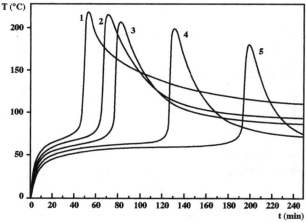

Figure 3 : Electrical voltage (Vo) dependence of the temperature $T = T(t)$
(1) : 2 800 v ; (2) : 2 400 v ; (3) : 2 000 v ; (4) : 1 600 v ; (5) : 1 200 v

The variations of P against T (Figure 4) leads to an interesting series of curves obtained at a given voltage which shows the homogeneity of the dielectric behaviour of the matter when the electrical field (E = Vo/d) varies in one sense or the other one.

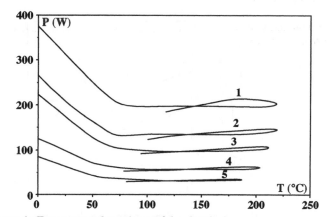

Figure 4 : Temperature dependence of the electrical power
at given electrical voltage (Vo)
(1) : 2 800 v ; (2) : 2 400 v ; (3) : 2 000 v ; (4) : 1 600 v ; (5) : 1 200 v

The broadening of the curves could be related to the fact that the final thermal plateau value remains higher than the glassy transition temperature $Tg_1 = 101.6$ °C (Figure 5) and that the corresponding dielectric absorption is enlarged. This variation is also observed in the T curve at Vo = 2 800 v which presents a final decay different from the others recorded at low Vo (Figure 3).

Figure 5 shows the Vo dependence of Tg_1 and of Tg_∞. Tg_1, which correlates to the extent of polymerization, increases with Vo. On the other hand, Tg_∞, which expresses the structural homogeneity of the networks, is independent of Vo.

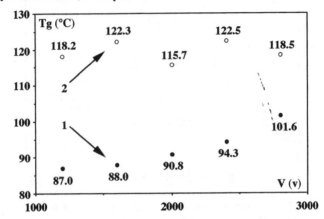

Figure 5 : Electrical voltage (Vo) dependence of the glassy transition temperature
(1) : Tg_1 ; (2) : Tg_∞

Electrical power Po dependence of the kinetics of polymerization

The Po dependence of the kinetics has been also studied. Figure 6 shows a series of T curves drawn at variable Po (200, 160, 120, 80 and 40 W).

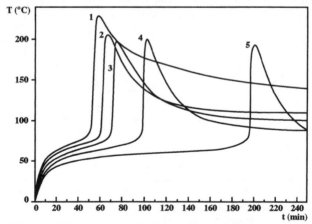

Figure 6 : Electrical power (Po) dependence of the temperature $T = T(t)$
(1) : 200 W ; (2) : 160 W ; (3) : 120 W ; (4) : 80 W ; (5) : 40 W

The polymerization is all the more accelerated when Po is high. The thermal decay is perturbed by the physical state of the crosslinked resins which is rubbery when the temperature remains higher than Tg_1 and glassy when it is lower (Figure 7). The Po dependence of the glassy transition temperature indicates that Tg_1 varies largely with Po, contrary to Tg_∞ which only rises slightly when Po decreases. This last result is in agreement with the law which states the structural homogeneity of thermosets : the final homogeneity is the best when the homogenity of the percolated gel is good. This is observed when the rate of formation of the gel is low, that it is to say when Po is small.

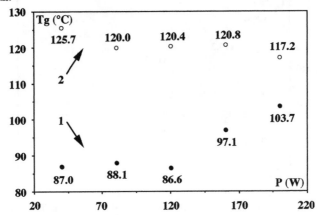

Figure 7 : Electrical power (Po) dependence of the glassy transition temperature
(1) : Tg_1 ; (2) : Tg_∞

<u>Peroxide concentration dependence of the kinetics of polymerization at given electrical voltage</u>
The mechanisms of the reactions of radical polymerization are very complicate and strongly dependent on the initial concentration of the radical initiator. This is the reason why the global effect on the polymerization of the concentration of benzoyl peroxide at Vo = 2 000 v has been studied. Five concentrations are tested : 0.25, 0.50, 0.75, 1.00 and 2.00 % (Figure 8).

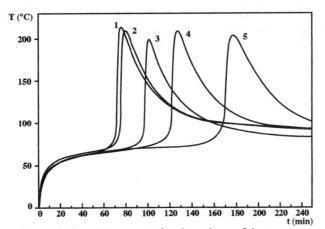

Figure 8 : Peroxide concentration dependence of the temperature
T = T(t) at Vo = 2 000 v
(1) : 2 % ; (2) : 1 % ; (3) : 0.75 % ; (4) : 0.5 % ; (5) : 0.25 %

The T curves show that the kinetics are accelerated when the concentration of the initiator is increased. The best way to illustrate the dependence of the process on the benzoyl peroxide amount is to consider the values of Tg_1 and Tg_∞ (Figure 9). The optimized concentration which corresponds to the highest value of Tg_1, is between 0.50 and 0.75 %, even when the maximum of Tg_∞ is obtained at the lowest amount (0.25 %).

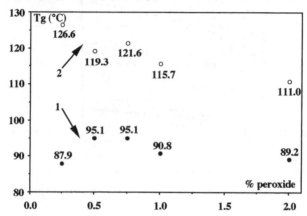

Figure 9 : Peroxide concentration dependence of the glassy transition temperature
(1) : Tg_1 ; (2) : Tg_∞

CONCLUSION

The study, which has been described, shows that High Frequencies working at 27.12 MHz are efficient to activate the crosslinking reaction of divinylester resins in styrene solution using organic peroxides as radical initiators.

Two characteristics of these electromagnetic waves give them the advantage over microwaves (2.45 GHz) for polymerization :

- the penetration by the waves, into the polymerizable matter, is much easier because the applied electrical field crosses automatically the samples to be treated.
- unlimited possibilities of shapes and dimensions of the final polymeric pieces are offered.

ACKNOWLEDGMENTS

Electricité de France (Direction des Etudes et Recherches) is gratefully acknowledged for the financial support of the present work.

REFERENCES

1 . Q. Le Van and A. Gourdenne, Eur. Polym. J. **23(10)**, p. 778 (1987)
2 . N. Beldjoudi, A. Bouazizi, D. Douibi and A. Gourdenne, Eur. Polym. J. **24(1)**, p. 49 (1988)
3 . N. Beldjoudi and A. Gourdenne, Eur. Polym. J. **24(1)**, p. 53 (1988)
4 . N. Beldjoudi and A. Gourdenne, Eur. Polym. J. **24(3)**, p. 265 (1988)
5 . A. Silinski and A. Gourdenne, Eur. Polym. J. **23(4)**, p. 273 (1987)
6 . L. Chan and A. Gourdenne, Proceedings PMSE **66**, p. 382 (1992)
7 . Y. Baziard and A. Gourdenne, Eur. Polym. J. **24(9)**, p. 881 (1988)
8 . A. Bouazizi and A. Gourdenne, Eur. Polym. J. **24(9)**, p. 889 (1988)
9 . A. Gourdenne, Proceedings PMSE **66**, p. 430 (1992)
10 . M. Palumbo, Y. Vallet, P. Alazard and A. Gourdenne, Proceedings PMSE **72**, p. 44 (1994)
11 . F. Legros, A. Fourrier-Lamer, D. Le Pen and A. Gourdenne, Eur. Polym. J. **20**, p. 1057 (1984)

MONITORING RESIN CURE OF MEDIUM DENSITY
FIBERBOARD USING DIELECTRIC SENSORS

R. J. KING*, R. W. RICE**
*KDC Technology Corp., 2011 Research Dr., Livermore, CA 94550
**Wood Science and Technology Dept., Univ. of Maine, Orono, ME, 04469

ABSTRACT

Flush mounted, in-press microwave (600-800 MHz) sensors have been developed for monitoring the complex permittivity in real time during the cure of medium density fiberboard. The measured dielectric constant (ε') and loss factor (ε'') are independent diagnostic indicators of dynamic cure events that are catalyzed by heat, pressure and moisture. In particular, with this technique the instantaneous effects of resin viscosity, rate and degree of adhesive cure, the wood density, the changes in phase of the moisture and the rate of moisture depletion can be monitored during the entire curing process.

The comparative roles of moisture and adhesive content are discussed, along with the comparative modulus of rupture with cure duration. Results are presented comparing the dynamics of phenol-formaldehyde and isocyanate resins.

INTRODUCTION

Real-time, in-press sensors for monitoring the dynamic cure processes during the manufacture of wood-based composite products (flakeboard, fiberboard, particleboard, oriented strandboard, etc.) are lacking. The process typically involves pressing a low moisture mat of wood flakes, fibers or particles which have been sprayed with thermoset resin between two high temperature platens. Pressures can exceed 1000 psi and temperatures can exceed 200°C (392°F) in the panel over an extended period of 6 to 15 minutes. Inside the panel, heat transfer occurs by conduction and by convection, with steam and wood as the heat transfer media. During the initial cure phase, heat reduces the resin viscosity to improve resin penetration into the wood fibers, and to activate the resin's chemical reaction.

Obviously, the curing process is complex and dynamic, involving simultaneous densification, heat and mass transfer, compounded by the concurrent chemical reaction. To date, the only real-time diagnostic tools that the wood technologist has are monitoring the pressing time, the pressure and the temperature, combined with after-the-fact destructive and nondestructive testing of the fabricated product. No diagnostic tools are currently available for *in situ* monitoring of the curing reaction through the entire panel thickness. Such information provides new insight into the interacting roles that moisture, adhesive content, density, pressure, and temperature play in determining properties of the manufactured product. When installed in a laboratory press, real-time displays of the sensor outputs can serve as the basis for an intelligent process monitoring system. As a result, the overall quality and uniformity of the manufactured product can be improved while decreasing the pressing time and allowing the press dynamics and critical timing points to be monitored directly.

This new technology is useful for research and development of resins, resin systems and pressing schedules. It can be used to optimize manufacturing processes of existing and new wood-based products, and for fine tuning of adhesive systems. Rapid and efficient screening of resin systems, timing of critical events through the entire pressing process from start to finish, and estimates of 'pre cure' conditions become possible.

Besides uses for improving conventional wood-based composites, this technology can also play a key role in the development of the continually expanding list of base materials and manufacturing processes. This list includes under-utilized wood species not suitable for other purposes, recycled plastics and cellulose forms which are now agricultural waste products.

This paper briefly describes the use and operational principles of Microwave Dynamic Dielectric Analyses (MDDA) to fill this need. Typical illustrative results are given and interpreted.

BASIC PRINCIPLES OF MDDA

Figure 1. Photograph of a resonant sensor that is flush mounted in a caul plate between the platens of a laboratory press.

In use, a thin and flat microwave sensor is flush mounted in a caul plate that rests on the lower platen of a laboratory press, as shown in Figure 1. In this way, the electromagnetic fields emanating from the sensor penetrate into the wood test material between the upper platen and the caul plate. The sensor thus responds to two fundamentally independent electrical properties of the wood, namely its dielectric constant (ε') and its loss factor (ε''). In the present application these properties depend, to different degrees, on such important physical parameters as:

[1] wood species, form factor (fibers, particles, chips, strands) and bulk fiber orientation (as for oriented stand board)
[2] type and amount of adhesive

When pressing a given mat, the above parameters are constant. The following parameters, however, vary during the pressing process:

[3] bulk moisture content and its phase (bound moisture, free moisture and vapor)
[4] mat bulk density through consolidation and relaxation
[5] adhesive viscosity and rate and degree of cure through chemical reaction
[6] mat temperature

It is important to realize that while the electrical constitutive parameters (ε' and ε'') depend on the above physical properties, their dependencies are in fact independent of each other. That is to say, the above properties affect ε' and ε'' in different ways and for different reasons. For example, densification of a bone dry mat has a strong effect on ε' while the affect on ε'' is only slight.

Of immediate and practical interest here is the use of MDDA to make comparative tests, e.g., to compare the relative cure profiles among cure cycles having different types or amounts of adhesive, different wood species, different moisture contents, different pressures or temperatures, etc. In this context, ε' and ε'' merely become convenient vehicles or stepping stones for interpreting phenomenological effects in terms of the behavior of the physical parameters listed above.

More practically then, we turn attention to the two parameters that are actually measured at the output of the MDDA sensor, and their functional relationship to ε' and ε''. Without delving into the technicalities of these patented sensors [1-5], it is sufficient to note that they are electromagnetically resonant at a frequency of about 750 MHz, and that this resonant frequency, f_r, decreases linearly with increasing bulk dielectric constant, ε', of the test material (the mat or panel). Thus, in relative terms the change in f_r due to a change in ε' is

$$\Delta f_r = -K_1 \Delta\varepsilon'. \tag{1}$$

where K_1 is a calibration constant. In this way, simply monitoring f_r provides a relative measure of changes in ε'. Calibration to determine the absolute values of ε' is accomplished using one or more dielectric standards.

The second measured sensor parameter is its input resistance r_o ($=R_o/R_c$), normalized to the characteristic resistance of the sensor input cable, R_c ($=50$ ohms). By means of a patented adjustment mechanism, the sensor is matched to the test material such that r_o varies within the range $0.5 < r_o < 1.0$ during a typical panel pressing cycle. For practical use, the measured r_o is inversely related to the loss or dissipation factor of the test material, ε''. That is,

$$r_o^{-1} = K_2\,\varepsilon''. \qquad\qquad (2)$$

where K_2 is a calibration constant that can be determined using a material having a known ε''.

SENSOR AND INSTRUMENTATION

The MDDA sensor is thin and flat (3 x 6 x 5/8 inch), suitable for custom mounting by the user in a 0.75 inch thick caul plate (see Figure 1.) A single coaxial cable extends from the sensor to the edge of the caul plate for connection to the microwave instrumentation via another coaxial cable. A thin Teflon™ sheet is bonded to the sensor face to prevent adhesion to the test wood panel during pressing. The sensor and its coaxial cable can operate up to 500°F continuously. Thermocouples can be added to monitor the sensor and caul plate temperatures.

The ancillary microwave and electronics instrumentation package is bascially a reflectometer that excites the sensor with low power (20 mw) electromagnetic waves over a stepped spectrum of frequencies ranging from 600 to 800 MHz, and monitors the spectrum of the signal reflected from the sensor [2-5]. As the microwave source frequency is stepped through its range in increments of 0.1 MHz, the amplitude of the reflected signal dips to a sharp minimum at the sensor's resonant frequency, f_r. Further, the standing wave ratio (SWR) at f_r is equal to r_o. Typically, each sweep takes about 0.5s which permits tracking of rapid dynamic changes in the sensor's response.

Besides the microwave source and reflected signal detection and processing electronics, the system includes a 386 computer with solid state memory, a DOS-based operating system and custom software, and a handheld display (4 lines x 20 character) and control terminal. Real time plots of the sensor's response parameters are displayed on a color monitor, and a floppy drive is used for storing and downloading data. A remote PC can communicate with the system over an RS232 link.

TYPICAL CURE PROFILES

This section gives example profiles of ε' and ε'' during cures of Medium Density Fiberboard (MDF) panels. All results given here were obtained using a laboratory press set at 375°F and Douglas Fir fibers that were conditioned to 6.7% moisture content, prior to addition of the adhesive. One square foot panels were pressed to 0.5 inch stops to yield a nomional density of 41-44 lbs/ft^3. Figure 2 compares cure profiles of (a) sensor resonant frequency ($f_r(t)$) and (b) the corresponding real dielectric constant ($\varepsilon'(t)$) for three cases:

•Oven dried fibers (start mc = 0.66%, end mc = 0%; dry basis). The effects of ligneous bonding, which is generally aided by moisture losses that free bonding sites, is not evident. The changes in f_r and ε' are small and monotonic, indicating only a small change in the polar character of the molecules. The completed samples crumbled easily with little or no evidence of bonding.

•Normal fibers (start mc = 6.7%, end mc = 0.19%) without resin. Here we see the effects of moisture loss and ligneous bonding. The dielectric constant ε' first peaks during consolidation, relaxation and moisture loss, then gradually decreases to a level somewhat below its starting value. This type of profile (i.e., broad peak and small swing in ε') is typical of a poorly bonded composite. The completed board was firm and appeared to be well-bonded on the surface, although friable and delaminated in the interior.

•Normal fibers (start mc = 6.7%, end mc = 0.61%) with 7% phenol-formaldehyde resin. This profile is typical of a well-bonded board. The dielectric constant has a well-defined peak followed by a rapid decrease, finally ending at a point that is well below the starting level.

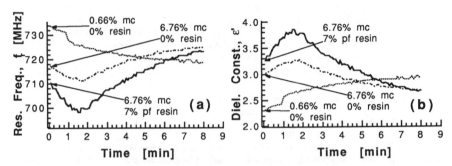

Figure 2. Comparitive cure profiles of (a) sensor resonant frequency and (b) dielectric constant for MDF panels with and without moisture, and with moisture and resin.

The starting values of ε' in (b) are chiefly determined by the initial total moisture contents, including the moisture in the resin. Between the start and the finish, moisture is vaporized (steam) and catalyzes the resin. In addition, moisture is a curing reaction by-product of phenol-formaldehyde resin. The dynamics of this reaction is a more pronounced peak in ε' than in the case of no resin.

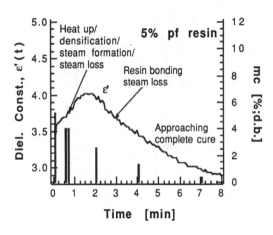

Figure 3. Moisture depletion with cure time and annotated cure profile of ε'.

Throughout the cure, the moisture is being depleted until it is essentially zero at the end, as shown in Figure 3. Here, six boards were pressed and removed from the press at successive times into the cure cycle. The moisture in each board was then determined by weighing and drying.

To quantify board strength with cure time, modulus of rupture (MOR) tests were made on 1 x 12 inch strips of partially cured MDF, immediately upon removal from the press. In these tests, a load was applied at the center of each strip, bending it until failure.

Resin (%)	Duration (min)	MOR (psi)
5	3	315
5	5	2059
5	8	3003

This shows that the major strength is formed between three and five minutes in the cure cycle, i.e., to the right of the peak in ε'.

Figure 4 compares three profiles of (a) ε' and (b) ε" for boards that were conditioned to 6.7% moisture: no resin; 5% phenol-formaldehyde resin; and 5% isocyanate resin. The quite different reactions are clearly evident. For the phenol-formaldehyde resin, both ε' and ε" sharply peak earlier than in the no resin case, presumably because moisture is a byproduct of the curing action. In addition, phenol-formaldehyde resin is somewhat resistive. It is believed that progressive bonding of the now more resistive resin laden fibers causes the large swing in ε". As a result, the loss factor ε" for this case is probably indicative of the rate of reaction.

In contrast, isocynate resin absorbs moisture during its reaction and it is non-resistive. As a result, the swings in both ε' and ε" are substantially less than for the phenol-formaldehyde case. Moreover, moisture absorption is believed to cause both ε' and ε" to peak earlier in the cycle. Interestingly, the final ε" for all three cases asymptotically converge to nearly the same values, as chiefly determined by zero moisture.

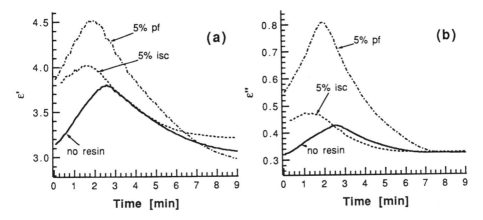

Figure 4. Comparitive cure profiles of ε' and ε" for MDF panels with no resin, with 5% phenol-formaldehyde resin and with 5% isocyanate resin.

CONCLUSIONS

The preliminary results presented here are rich in new information about the curing process of wood-based composites. In particular, it now becomes possible to associate certain features of cure profiles with consolidation and relaxation of the pressing process, moisture generation (or absorption, depending on the type of resin), steam loss, resin bonding, degree of cure, and final strength of the product. More research in the coming months will shed more insight, and hopefully quantify these and other parameters of the curing process.

ACKNOWLEDGEMENTS

This material is based upon work supported by the Cooperative State Research Service, U.S. Dept. of Agriculture, under Agreement Nos. 93-33610-8589 and -0894. Any opinions, findings, conclusions or recommendations expresSed in this publication are those of the authors and do not necesarily reflect the view of the U.S. Dept. of Agriculture.

REFERENCES

1. US Patent No. 5,334,941
2. R. King, SENSORS, 9(10), p. 25 (1992)
3. M. Werner and R. King, MRS Proc. 347, p. 253, (1994)
4. R. King, M. Werner and G. Mayorga, J. Reinforced Plastics, 12, p. 173 (1993)
5. R. King in Electromagnetic Wave Interactions With Water and Moist Substances, edited by A. Kraszewski, (Proc. 1993 IEEE MTT-S Int'l. Microwave Symp. Workshop, IEEE Press, 1996).

MECHANICAL BEHAVIOR OF MICROWAVE PROCESSED POLYMER MATRIX COMPOSITES: THE EFFECT OF THE TEMPERATURE INCREASE RATE

M. DELMOTTE, J. FITOUSSI, J. TOFTEGAARD-HANSEN, C. MORE, H. JULLIEN*, D. BAPTISTE.
Laboratoire Microstructure et Mécanique des Matériaux, CNRS URA 1219, ENSAM, 151 blvd. de l'Hôpital, 75013 Paris, France.

ABSTRACT

Microwave processed glass reinforced epoxies or glass reinforced polyesters exhibit mechanical behaviors different from conventionally cured materials, relatively to tensile tests.

The faster increases of temperature due to microwaves cause a competition between the matrix flow and the crosslinking reaction which can be estimated by porosity variations. Higher mechanical moduli are also obtained, because of both an effect on chemical kinetics and a more homogenous distribution of temperature in materials.

Nevertheless, to provide such specific mechanical behaviors in microwave processed composite materials, a best control of the experimental pressure parameters is requested.

INTRODUCTION.

The microwave processing of polymer composites was first considered on a molecular level. Many authors have described the effects of microwaves on the crosslinking chemical reaction of thermoset resins [1-7]. The second step was to approach the process on a macroscopic level. The question was how to make the microwave curing of large size epoxy composite samples homogeneous? This approach involved considerations about the wave propagation through molded samples: Outifa et al [8, 9] proposed a solution to this problem, using the dielectric matching of the applicator by tapered transitions, and the focusing of the energy inside samples by means of appropriated values for the permittivity of the dielectric molds.

The third step now is to consider the microscopic levels and to examine if the microstructures are changed by microwave processing. And if yes, does it induce significant changes in the mechanical properties of samples? In a previous work [10] rheological properties in microwave or conventionally cured epoxies were correlated with molecular data as conversion extents: no significant difference was found. Inversely, Bai Shu Lin found better mechanical properties in unidirectional epoxy composites, and despite a higher porosity due to a lower external pressure and a shorter curing time, inducing a delamination of the composites under mechanical load [11].

The present paper describes the microwave processing of tensile test samples, and exhibits the results of tensile tests and micrograph analysis, as a first step in the study of microwave induced microstructures in bi-directionnal epoxy composites.

EXPERIMENT.

Microwave equipment.

Figure 1 shows an overview of the experimental set-up, equipped with two facing 800W, 2.45GHz magnetron generators. Each of the magnetrons is protected from reflected power by an insulator. The control of the applied power is obtained from a cross coupling guide fitted

---

*To whom correspondence should be addressed.

with a power sensor connected with a wattmeter (Boonton RF Micro-Wattmeter Model 4200). It is assumed that the two generators deliver the same output power with the same setting.

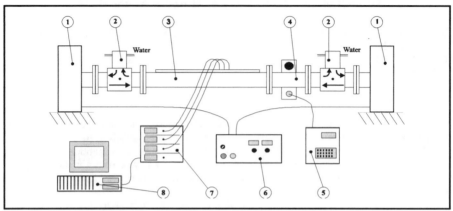

Figure 1. General scheme of the microwave processing equipment.

1. Microwave generators, 2.45GHz, 800W. 2. Insulators. 3. Waveguide applicator. 4. Directive coupler with power sensor. 5. Power meter. 6. Power supply. 7. Fiber optic thermometer. 8. PC computer.

Microwave applicator.

The applicator is a standard WR340 waveguide, filled with a dielectric mold containing two tensile test samples. The mold is microwave matched as indicated on Figure 2, so that only the fundamental single mode TE_{01} is allowed to propagate [9].

The dielectric mold is made of two parts: an external polyimide coated paper honeycomb dielectric, transparent to microwaves, the aim of which being to transfer the pressure from the cover to the mold and to assure a symmetrical placement of the mold in the applicator, and an internal dielectric mold made of a silicone-glass composite (Silirite Silicone™), also transparent to microwaves.

In these applicator and mold a measurement of the pressure applied to sample was not possible. But the applicator and the mold were made in such a way that the pressure on samples is carried out by applying to the mold a deformation of 1mm. The pressure is assumed to be constant for every experiment.

Samples

Samples were made of a prepreg of diglycidylether of bisphenol-A and dicyandiamide (DGEBA/DDA) based epoxy resin, reinforced with glass E fibers. Both sides of the samples were isolated by a 10mm thick silicon-glass composite, to prevent heat transfer to the guide walls. Above and underneath the central area of each sample waste material of the same resin as the test material were added to prevent transversal temperature gradient in curing samples.

Three fiber optic temperature probes were placed into quartz tubes that were inserted into one of the samples, through holes in applicator and mold. This sample became unsuitable for tensile tests. Temperature was measured by means of a Luxtron 755 Fluoroptic™ Thermometer, and a PC computer was used for the acquisition of temperature-time profiles. The middle of both samples is assumed to be cured identically, but because of the wave attenuation there is a small difference between the ends of each sample. This is assumed to have no influence upon the mechanical tests which will be carried out with reference to the central area of each sample.

Thermosetting materials have a limited range of temperature within which curing can occur. The lowest temperature necessary to initiate the crosslinking reaction for Glass E/Epoxy DGEBA-DDA is 120°C and the highest limit where degradation occurs is 220-230°C. The lowest limit determines the minimum power necessary to reach the minimum temperature of curing for the resin and to stay over this temperature within the whole curing duration. The highest limit determines the maximum power to avoid heat degradation of the material. The three series of samples were cured within this range. Three microwave power values were used: 200W, 320W and 450W. The treatment was carried out with a fixed time starting from the time when the temperature of cross-linking occurs, resulting in a fixed interval of curing but a variable time of treatment. A fixed time of curing is preferred because it gives a time independent curing interval. This interval was 10 min starting from the time the measurement of the central temperature probe reaches 120°C: the total curing duration were 23.00, 16.40 and 13.00 minutes for 200W, 320W and 450W respectively.

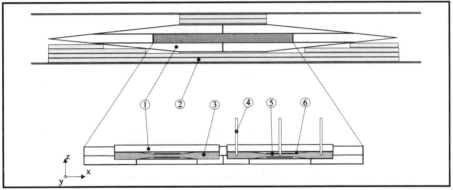

Figure 2. Design of the mold. Top: external mold; bottom: internal mold.
1. Silicon-glass mold. 2. Honeycomb external dielectric. 3. Tensile test sample to cure. 4. Quartz tubes for temperature sensors. 5. Silicon-glass composite. 6. Waste epoxy material.

For mechanical tensile tests, raw samples are preferred because of the influence of a further treatment on mechanical properties: it could initiate cracking of samples, that would be falsely attributed to the process. For this reason the design described by Sih and Skudra [11] was used, as shown in Figure 3.

The central part of samples was a bi-directional prepreg, based on symmetric angle-ply laminates which were symmetric about the center plane to avoid coupling stresses. The coupling stress causes a radical change in the shape of the laminate under load. In such a way, samples were made of 16 layers, symmetrically crossed relatively to the directions of both the wave propagation and the direction of the mechanical test load. Three series of samples were made, with three ply angles (±40°, ±45°, ±50°). The ends of samples were thicker than the central part, with 6 extra unidirectional layers to reinforce each end and to assure that cracking will not happen between the jaws.

Test equipment.

After microwave processing the samples were characterized by tensile tests and optical micrograph analysis. For the tensile tests a 10 ton hydraulic Instron Model 1185 tension machine was used, with regulated velocity and force, and load simulation facilities. Tensile tests of were carried out with a displacement of 2mm/min (static load). The samples were prepared by clueing strain gauges parallel to the direction of load.

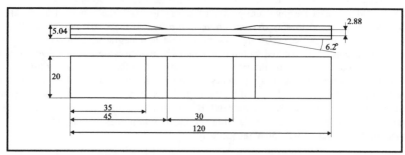

Figure 3. Geometry of the samples (sizes in mm).

For the optical analysis cross sections were cut in samples; optical microscopes Olympus Tokyo and Olympus PMG3 with enlargement in the ranges of 25-500 and 50-1000 respectively were used.

RESULTS.

Temperature-time profiles.

Typical temperature-time variations are given in Figure 4, for 200, 320 and 420W treated samples, exhibiting temperature increase rates raised with microwave power. The exothermic peak almost fails to appear because of the individual isolation of sample which regulated heat transfer.

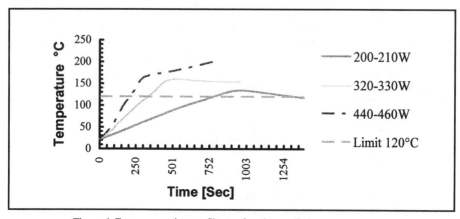

Figure 4. Temperature-time profile as a function applied microwave power.

The treatment of 200W has a negative heat transfer with the surroundings and the temperature decreases after the reaction occurred. The treatment of 320W is almost neutral which gives a neutral evolution of the temperature after the exothermy occurred and the treatment of 450W has a positive heat transfer with its surrounding: temperature was increasing during the whole curing time.

Load-unload tensile tests.

Load-unload tests were used to determine the limit of elasticity, which is used as a macroscopic criterion of the starting damage. The load was incremented by 60daN until cracking, thus providing an evaluation of the evolution of the elastic modulus and of the anelastic strain as a function of stress.

These tests have shown that the material is heavily dependent on the orientation of the angle with the direction of load. Especially the 50° laminate do have a catastrophic evolution from the moment when the load exceeded the limit of elasticity.

Figure 5 shows histograms of the variations of the maximum stress and of the cracking strain as functions of both the microwave power and the ply angle. The most important results are the decreasing maximum stress for 450W, and the maximum of cracking strain exhibited for 320W and 45°.

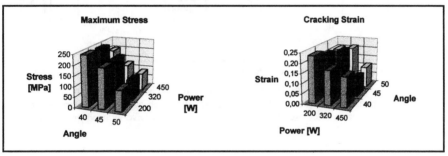

Figure 5. Histograms of the variations of the maximum stress and of the cracking strain as functions of both the microwave power and the ply angle.

Micrograph analysis.

Microwave have several effects on morphology: orientation, crystallization, spatial heterogeneity such as micro-gels and changes in microstructure as functions of the applied power. Mechanical properties are especially affected by voids, due to: 1. an incomplete wetting out of the fibers by the resin resulting in an entrapment of either air or water vapor; 2. the presence of gas produced during the curing: residual solvents, products of chemical reactions or low molecular weight fractions.

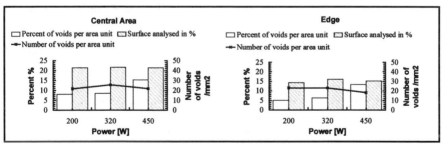

Figure 6. Percentage of voids per area unit and percentage of measured area in function of the total area. (left axis). Number of voids per area unit (squared millimeter) (right axis).

The volume fraction of voids were determined by point counting on micrographs, one in the center and another at the edge of sample cross sections.

Figure 6 exhibits the percentage of voids per area unit, increasing with the microwave power, while the number of voids was relatively constant with a slight maximum for 320W and a decrease tendency for 420W. Figure 7 shows that the mean size of voids was largely growing with the applied microwave power. These results are easily correlated with the histograms of Figure 5.

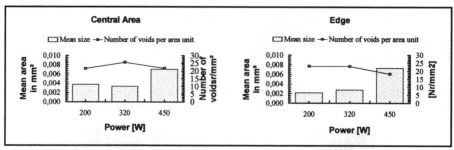

Figure 7. The evolution of mean size and number of voids as a function of the applied power.

The volume fraction of voids was increasing with the applied microwave power. Most of voids were interlaminar, due principally to the uncompleted wetting between layers, and not to resin flow. Moreover the increase chemical reaction rate might have induced gas formation in high power processing.

CONCLUSION.

A good correspondence was found between the mechanical behavior and the structural voids in a glass epoxy composite, as functions of microwave processing parameters such as the temperature increase rate. The excessive number of voids could be reduced by optimizing the process including a better control of pressure. Further investigations about microstructures might help this optimization of the process.

REFERENCES.

1. W.I. Lee and G.S. Springer, J. Compos. Mater. **18**, p. 357 (1984).
2. A. Gourdenne, A.H. Maassarani, P. Monchaux, S. Aussudre and P. Thourel, Polym. Prepr. **20***(2)*, p. 471 (1979).
3. F.M. Thuillier, H. Jullien and M.F. Grenier-Loustalot, Polymer Communic. **25**, p. 206 (1986).
4. M. Delmotte, H. Jullien and M. Ollivon, Eur. Polymer J. **27**, p. 371 (1991).
5. D.A. Lewis, J.D. Summers, T.C. Ward and J.E. MacGrath, J. Polymer Sci., part A: Polymer Chemistry **30**, p. 1647 (1992).
6. J. Mijovic and J. Wijaya, Macromolecules **23**, p. 3671 (1990).
7. J. Wei, MC. Hawley, MT DeMeuse, Polym. Mater. Sci. Eng. **66**, p. 478 (1992).
8. L. Outifa, H. Jullien and M. Delmotte, Polym. Mater. Sci. Eng. **66**, p. 424 (1992).
9. L. Outifa, C. Moré, H. Jullien, M. Delmotte, Ind. Eng. Chem. Res. **34**, p. 688 (1995).
10. C. Jordan, J. Galy, J.P. Pascault, C. Moré, M. Delmotte and H. Jullien, Polym. Eng. Sci. **35**, p. 240 (1995).
11. Bai Shu-Lin, PhD Thesis, Ecole Centrale de Paris, Paris, 1993.
12. G.C. Sih and A.M. Skudra, Failure mechanics of composites (North Holland, 1983).

COMPARATIVE STUDY OF Si$_3$N$_4$ - BASED CERAMICS SINTERING AT FREQUENCIES 30 AND 83GHz

Y. BYKOV, A. EREMEEV, V. HOLOPTSEV
Institute of Applied Physics, Russian Academy of Sciences, 46 Ulyanov St., Nizhny Novgorod 603600 Russia.

ABSTRACT

The study of microwave ceramics sintering at different frequencies of radiation has important implication for clarifying a specific effect inherent to microwave-based processes and determining the optimal conditions of sintering which yield a high-quality final product. Comparative study of Si$_3$N$_4$-based ceramics sintering at the frequencies of 30 and 83GHz was undertaken on retention of all the conditions but a frequency and microwave power. It is found that an increase in microwave frequency results in reduction in silicon nitride decomposition during sintering.

INTRODUCTION

Features peculiar to microwave heating have aroused considerable interest in ceramics sintering and synthesis of new materials using microwave energy. However, for the volumetric, inertialess, selective microwave heating to become practical it is necessary to gain a thorough knowledge how these specific properties affect the process of sintering and how they can be intelligently used to yield materials of new qualities. One of the approaches to handling the problem is comparative study of the microwave and conventional sintering. This approach has gained a wide-spread acceptance and proved its fruitfulness in the researchs in this field. However, owing to the fundamental distinction between mechanisms of conventional and microwave heating, too many factors may be responsible for the difference between the results of processing. The first and most crucial distinction is in radically different temperature profiles inside heated bodies. This factor frequently causes serious doubt upon the validity of such a comparison even when an identity of the temperature measurements in both cases is out of question. More reliable information can be obtained when microwave powers of two rather distinct frequencies are used. This method of comparison allows to deal with the same mechanism of heating, and only particular characteristics of heating vary with the parameters of applied powers. Two-frequency comparative study makes it possible to obtain the dependence of the specific microwave effect, if any, upon frequency and amplitude of the electromagnetic field and thereby to elucidate the physical mechanism involved. Additionally, as long as the microwave absorption is frequency dependent, such a comparative study is of great importance for the development of the two-frequency microwave processing technique [1] which is supposed to alleviate the problem of non-zero temperature gradients attributed to any heating by volumetrically spread thermal sources.

Two-frequency comparative study of ceramics sintering is not a new technique. It was used, for example, by the Oak-Ridge researchers [2, 3]. However, as a rule, such investigations are limited to displaying the ultimate results of processing only, but do not deal with experimental revealing of the frequency dependence of these results. In the present work, an attempt has been made to compare the behavior of the main characteristics of sintering of Si$_3$N$_4$ - based ceramics by using microwaves with frequencies 30 and 83GHz. The kinetics of densification at its initial stage are compared, as well as weight loss and $\alpha \rightarrow \beta$ phase transformation. Based on the

613

experimental data, the main advantages of high frequency sintering of Si_3N_4 - based materials are discussed.

EXPERIMENTAL PROCEDURES

The powder compacts of the compositions α- Si_3N_4 + 2% Al_2O_3 + 9% Y_2O_3 (2A9Y) and α-Si_3N_4 + 1% Al_2O_3 + 5% Y_2O_3 (1A5Y) were used in the present study. The green samples were obtained from the company "Teknologia" (Obninsk, Russia). The starting technical grade powders were mixed in a mill with Si_3N_4 - ceramic balls for 100 hours. The particle size after milling was less then 5mkm (>93%). Green samples in the form of $10 \times 10 \times 50$ mm bars were prepared by pressure molding with a paraffin-based organic mixture as a binder. The binder was burnt out and the samples were prefired conventionally in air at $T \approx 400°C$ for 100 hours. Green density of the samples was about 55% of the theoretical density. For this study, the samples of the length about 10mm were cut of the bars. For microwave processing, the samples were housed in the center of a 60mm dia × 60mm height fused quartz vessel and covered with Si_3N_4 powder mixed with Al_2O_3 - Y_2O_3 powder additives in the proportion corresponding to the composition 2A9Y. The quartz vessel with a sample was placed inside of multimode unturned cavities. Two kinds of microwave set-up were used in the study. One of them, a 83GHz gyrotron-based Heating System had the cylindrical cavity with the ratio of $L/\lambda \sim 80$ (L is the dimension of the cavity, λ is the wavelength). Another Heating System used a 30GHz gyrotron for feeding the multimode cylindrical cavity with $L/\lambda \sim 50$. The cavities have been evacuated and then filled up with dry nitrogen up to 1atm pressure. The sample temperature was measured by (W + 5%Re - W + 20%Re) thermocouple. The thermocouple head was inserted into a narrow bore drilled to the center of sample, which is assumed to give no coupling with the electric field. In each run, microwave heating was done in a computer control mode. The constant heating rate of 30°C per minute was maintained by controlling the output power of the gyrotron. Having reached the maximal prescribed temperature, microwave power was switched off and the sample cooled down together with the thermal insulation. All samples were weighed and their sizes were measured before and after heating to determine the weight loss during processing and the linear shrinkage. Densities were determined by the Archimedes method after one-hour boiling of samples in distilled water. Microstructure of samples was characterized using JEOL-200CX electron microscope and the phase composition was identified by X-ray diffraction analysis.

RESULTS AND DISCUSSION

Figures 1, 2, 3 show the temperature dependencies of the main parameters measured in every experimental run. Each point in these plots relates to the sample heated to the corresponding maximal temperature T_{max}. Figure 4 shows the sample cooling rates, a parameter which will be essential for further analysis of the experimental results. As it seen from Figures 1 and 2, only the kinetics of the weight loss differ from one another. Based on the experimental data it may be inferred whether this difference comes from distinctions in the frequency or the strength of the electromagnetic field used for heating. For a comparison between the data corresponding to 30 and 83GHz heating we will use the equation for the conservation of energy for a sample being heated. For the aim of this comparison it is sufficient to consider this equation written for the volume-average parameters, not taking into account, at this stage, non-uniformity of the

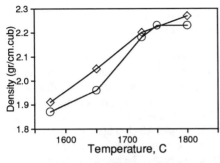

Fig. 1. Sample density vs. sintering temperature.

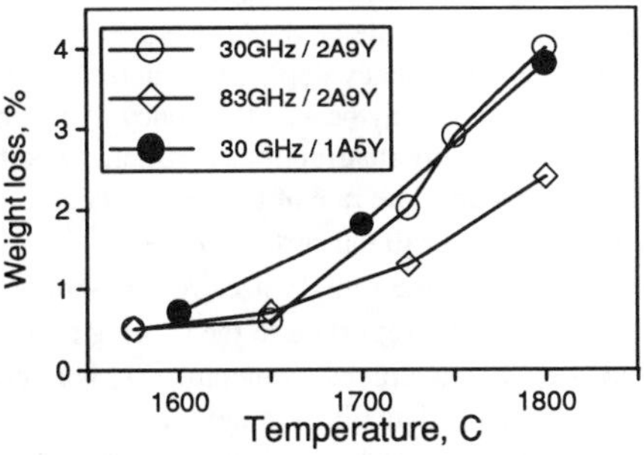

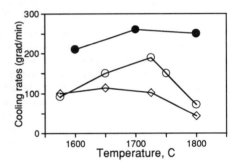

Fig.2. Weight loss vs. sintering temperature.

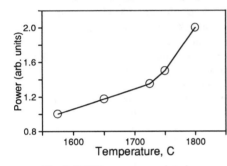

Fig. 3. 30GHz power vs. sintering temperature.

Fig. 4. The cooling rates vs. sintering temperature

temperature within a sample:

$$cp(dT/dt)_+ = \sigma_{eff}E^2 - Q_- = \sigma_{eff}E^2 - cp(dT/dt)_- \qquad (1)$$

Here c, p are specific heat capacity and density, respectively; σ_{eff} is effective microwave conductivity; E is the electric field in the sample; Q_- is the specific thermal loss power; $(dT/dt)_+$ and $(dT/dt)_-$ are the rates of the sample heating and cooling, respectively. It is assumed in (1) that at temperatures close to $T_{max.}$ parameters c and p for the cooling stage are the same as for the heating stage. We will further operate with the specific microwave power absorbed in a sample $P_{abs} = \sigma_{eff}E^2$ which is directly proportional to the sum of the absolute values of $(dT/dt)_+$ and $(dT/dt)_-$ registered in the experiments.

The results relating to the fixed temperature of the samples center will be further compared. As it follows from the energy balance equation (1), the distinction in the cooling rates of samples, providing for their constant heating rate, implies that P_{abs} is higher for 30GHz heated samples, at any given maximal temperature. The experimental results of this study do not allow to indicate unambiguously the reason for the difference in the values of P_{abs} pertinent to 30 and 83 GHz heating. The data reflect only the fact that under given experimental conditions more microwave power needs to be absorbed at the frequency of 30GHz than at 83GHz for heating the samples to the same T_{max} of their centers. Since the maximal temperature of the compared sample is fixed, any difference in behavior of the registered parameters for the samples heated with 30 and 83GHz powers can be attributed either to the difference in the temperature distribution within them, caused by the frequency-dependent microwave absorption, or to some specific non-thermal effect which potentially may be also frequency - dependent.

The temperature distributions pertinent to both cases of heating may be compared using the cooling rate data. The rate of the sample central part cooling is governed by the temperature drop ΔT from its center to the surface: $\Delta T \sim (r^2/k) \cdot Q_- = (r^2/k) \cdot c\rho (dT/dt)_-$, here k is the thermal conductivity, r is the sample size. For the samples of $r = const$ and nearly equal density, the value of ΔT is in direct proportion to the cooling rate and when the temperature of the sample center is fixed, a volume-averaged temperature $\overline{T}$ is larger for the samples which cool down slower. As it follows from the data given in the Figure 4, a volume-average temperature for the samples heated with 83GHz power is larger then those of their 30GHz counterparts. The actual magnitude of ΔT was measured with two thermocouples placed in the center and on the surface of the sample and reached approximately 100°C for samples heated with 30GHz power. According to the measured value of ΔT, the difference between the value of $(\overline{T})_{30}$ (samples heated with 30GHz power) and $(\overline{T})_{83}$ (samples heated with 83GHz power) may be around 50°C in these experiments.

As is seen from the Figure 1, sample densities increase in an identical manner for both microwave frequencies, with a due regard to the experimental errors. However, since the volume-averaged temperatures are different for the samples heated with 30 and 83GHz powers, this fact signifies that densification under microwave heating depends not only on the local temperature, but also on other factors intrinsic to this heating technique, such as temperature gradient, specific microwave absorbed power P_{abs}, or the amplitude E of the electromagnetic field. The following chain may be derived from the experimental data taking into account the considerations on the differences in $\overline{T}$ and P_{abs} relative to heating with powers of those two frequencies.

$$(gradT)_{30} > (gradT)_{83}$$
$$(A)$$

$$(\dot{T}_-)_{30} > (\dot{T}_-)_{83} \quad \rightarrow \quad \begin{matrix} \uparrow \\ \downarrow \end{matrix} \qquad\qquad\qquad (2)$$

$$\overline{T}_{30} < \overline{T}_{83} \qquad (\sigma_{eff} E^2)_{30} > (\sigma_{eff} E^2)_{83} \xrightarrow{\ \sigma_{eff}=const\ } E_{30}^2 > E_{83}^2$$
$$(B) \qquad\qquad\qquad\qquad\qquad (C)$$

Previously, only the "$grad\,T$" influence (scenario (A)) was verified in sintering experiments but with another, piezoceramic materials [4]. In that work, no appreciable "$grad\,T$" effects at microwave sintering of piezoceramics was detected. The absence of the elastic stress influence on the results of solid state sintering allow to suggest that "$grad\,T$" enhancement of densification is even less plausible during the liquid phase sintering of Si_3N_4 - ceramics, because of much faster relaxation of stresses.

As shown in Figure 2, a heavier weight loss is observed when samples heated with 30GHz power. This fact indicates that weight losses of the microwave heated samples are determined not only by the value of the local temperature resulting by averaging the temperature over a length much larger then the grain size. It is unknown to the authors, whether any data exist for Si_3N_4 microwave absorption in the frequency range of 30 - 90GHz and at the temperatures as high as 1800° C. Some tendency for an increase in loss tangent with a frequency is observed when comparing the experimental data for the dielectric properties of Si_3N_4-ceramics obtained at 35GHz [5] and 132 - 145GHz [6].

Following [7], we assume that at elevated temperatures, an effective microwave conductivity of the polycrystalline materials coincides with the direct current conductivity. If the effective microwave conductivity is not frequency dependent, then the data on the specific microwave power absorbed in a sample may be directly used for treating the results of the two-frequency comparative study in the terms of the strength of the electromagnetic field. This, in principle, makes it possible to link mass-transport processes and the electromagnetic field existing in processed materials. Under the experimental conditions corresponding to the scenarios (B)-(C) of the scheme (2), an enhanced weight loss is indicative of a direct influence of the electric field strength E on the rates of mass-transport processes - dissolution, diffusion, precipitation, chemical reaction, etc., which are singly or in combination responsible for the weight loss in Si_3N_4 - ceramics heating. It may appear that this statement is rather speculative when it is given without any supporting evidence obtained through the direct experimental observation of the connection between mass-transport rates and value of E^2. At the same time, without this assumption it is not clear how to explain that an increase in the weight loss occurs while the average temperature of the samples goes down. An additional argument in favor of the direct influence of the electric field strength on mass - transport rates follows from a comparison between the data on weight loss and cooling rates while sintering Si_3N_4 - samples with different content of oxide additives. As it seen from Figure 4, the cooling rates are larger for the low - additive A1Y5 samples. But as a result of an adequate competition of lower average temperature and larger value of $\sigma_{eff}E^2$, the low-additive samples 1A5Y have practically the same weight loss as their rich-additive counterparts 2A9Y (Figure 2).

A comparison between the results of experiments at the two, quite distinct frequencies of the microwaves shows that an increase in frequency is beneficial for sintering of Si_3N_4-based ceramics owing to reduced weight loss.

When samples are heated with the power of rather high frequency, microwave absorption in thermal insulation leads to a decrease in microwave power needed to be absorbed in the sample itself. Moreover, the experiments show that conditions can be realized when the temperature of the thermal insulation in the region nearly touching the sample may be even higher then the sample temperature. Such a temperature profile which is reverse of that ordinary for microwave heating can be obtained at temperatures close to the temperature of sintering by a proper choice of the insulation geometry. That is, the technique which is similar in function to a hybrid heating is realized. What is essential, this technique is achieved while using only one means of heating - absorption of the microwave energy of a rather high frequency. Due to the strong absorption of high frequency microwave power, an efficient heating at room temperature is provided. Then, at elevated temperatures, predominant heating of thermal insulation leads to a decrease in both the thermal loss from the sample and the microwave power needed to be absorbed in the sample. The former factor has a beneficial effect on the microstructure homogeneity within sintered material, the latter is favorable to a reduction in Si_3N_4 -decomposition and weight loss. When an arrangement of thermal insulation is chosen with due regard to its geometry, composition, and microwave absorption, an optimal ratio of powers absorbed in a sample and within thermal insulation may be maintained and the hybrid heating may occur in a self-regulating manner.

CONCLUSION

A comparative study of materials processing with microwave powers of the two, distinctive frequencies may be even more informative than the traditional method of a comparison between conventional and microwave processes. The two-frequency technique allows to deal with the same mechanism of heating and only particular characteristics of heating vary with the parameters of applied powers. The results of investigation of the Si_3N_4- powder compacts sintering at the frequencies of 30 and 83GHz shows that the densification and weight-loss at heating are associated with not only the local temperature within the sample, but with the specific absorbed microwave power and the amplitude of the electric field in the sample. Less weight loss occurs while sintering with 83GHz power. Applying the higher frequency power for Si_3N_4- ceramics sintering, a technique similar in function to a hybrid heating can be realized. The hybrid heating features are provided by easy coupling of the high-frequency microwaves with the material at low temperature and by partial electromagnetic shielding of the sample at the sintering temperature due to strong absorption of microwave power in the thermal insulation parts adjacent to the sample.

ACKNOWLEDGMENT

This work was supported in part by the Russian Basic Science Foundation through grant No95-02-05000-0. A. Sorokin and T. Borodacheva contributed valuable assistance to the conduct of this research.

REFERENCES

1.   Yu.V. Bykov, A.G. Eremeev, V.V. Holoptsev, V.E. Semenov, A. Birman, J. Calame, Y. Carmel, D. Gershon, B. Levush, D. Dadon, D. Martin, M. Rosen, and R Hutcheon, "Comparative studies of microwave sintering of zinc oxide ceramics at frequencies of 2.45, 30, and 84 GHz", this Proceedings.

2.   M.A. Janney and H.D. Kimrey, in Microwave Processing of Materials II, edited by W.B. Snyder, W.H. Sutton, M.F. Iskander and D.L. Johnson (Mater. Res. Soc. Proc. **189**, Pittsburgh, PA 1991), pp. 215 - 227.

3.   T.N. Tiegs, J.O. Kiggans and H.D. Kimrey, in Microwave Processing of Materials II, edited by W.B. Snyder, W.H. Sutton, M.F. Iskander and D.L. Johnson (Mater. Res. Soc. Proc. **189**, Pittsburgh, PA 1991),   pp. 267 - 272.

4.   Yu. Bykov, A. Eremeev, V. Holoptsev, in Microwave Processing of Materials IV, edited by M.F. Iskander, R.J. Lauf and W.H. Sutton (Mater. Res. Soc. Proc. **347**, Pittsburgh, PA 1994), pp. 585 - 590.

5.   D.R. Clarke and W.W. Ho, Adv. Ceram., 7, 246 (1983).

6.   V.V. Parshin, private communication.

7.   W.W. Ho, in Microwave Processing of Materials, edited by W.H. Sutton, M.H. Brooks and I.J. Chabinsky (Mater. Res. Soc. Proc. **124**, Pittsburgh, PA 1988), pp. 137 - 148. .

MICROWAVE AND MILLIMETER WAVE PROCESSING OF POLYMER-DERIVED SILICON NITRIDE

S.T. SCHWAB, S.F. TIMMONS, C.R. BLANCHARD, M.D. GRIMES, R.C. GRAEF
Materials & Structures Division, Southwest Research Institute, San Antonio, TX 78228-0510
J.D. KATZ, D.E. REES, T.W. HARDEK
Los Alamos National Laboratory, Los Alamos, NM 87545

ABSTRACT

Chemical methods of processing ceramics have the potential to overcome many of the processing-related obstacles that have hindered widespread commercialization. The Southwest Research Institute (SwRI) has focused on the development of polymeric precursors to silicon nitride (Si_3N_4). One such precursor, perhydropolysilazane (or PHPS), has been shown to be a useful binder for Si_3N_4 powder processing, a useful matrix precursor for the polymer infiltration/pyrolysis (PIP) processing of fiber-reinforced Si_3N_4 , and a useful ceramic coating precursor for the repair of oxidation protection coatings on carbon-carbon composites. While conventional, thermal pyrolyses of these preceramics has been sufficient to demonstrate their potential, substantial cost savings could be realized if the polymer-to-ceramic conversion could be instigated with electromagnetic energy. We have investigated the use of millimeter wave heating as a means of converting PHPS into Si_3N_4, and report here the results of our efforts to produce bulk compacts, coatings, and fiber-reinforced ceramics.

INTRODUCTION

The market for advanced ceramics and composites has been predicted to enjoy explosive growth for the past several years.[1] This growth has not taken place at the predicted pace partly because high cost has not yet been overcome for structural ceramics.[1] For many years, research in the area has been directed towards meeting the needs of the cold war aerospace industry, in which performance often outweighed cost. With the paradigm shift that has followed the end of the cold war, advanced ceramics increasingly find themselves competing against metals and polymers for applications—and losing because cost now often outweighs performance. Entirely new manufacturing methods are needed if ceramics are to compete with metals and polymers on a cost basis without sacrificing their performance advantages.

Difficulties inherent to traditional, powder-based methods for manufacturing advanced ceramics prompted the development of chemical techniques of ceramic fabrication, including "sol-gel" and "preceramic polymers."[2] While the sol-gel technique is most readily applied to the fabrication of oxide glasses and ceramics, preceramic polymers are generally applied to the fabrication of non-oxide ceramics, such as silicon nitride (Si_3N_4). Si_3N_4 is being considered for a wide variety of high temperature applications because of its resistance to thermal shock and other properties.[3] A particularly useful precursor to Si_3N_4 has been developed at SwRI. This material, known as perhydropolysilazane (or PHPS), is isolated as a low-viscosity, thermosetting liquid that exhibits a high ceramic yield. Although originally developed as a binder for Si_3N_4 powder processing,[4] PHPS has been shown to be an effective repair material for damaged oxidation protection coatings on carbon-carbon composites,[5] and a useful matrix precursor for the polymer infiltration/pyrolysis (PIP) processing of fiber-reinforced Si_3N_4 composites.

While chemical fabrication techniques do offer substantial advantages over traditional powder methodologies, they still rely on conventional heating methods to convert the polymer into ceramic. Typical oven or autoclave process cycles require many hours to complete, and are not energy efficient. Recent advances have made electromagnetic sources, such as microwave or millimeter wave, attractive alternatives to conventional heating. A series of experiments was conducted at Los Alamos National Laboratories to determine if electromagnetic radiation was effective in converting PHPS into Si_3N_4.

EXPERIMENTAL

All manipulations of uncured polysilazanes were carried out under anhydrous and anaerobic conditions using common synthetic techniques in combination with an inert atmosphere/vacuum manifold system or an argon-filled drybox (Vacuum Atmospheres HE-43-2 with HE-493 Dritrain).[6] Irradiation experiments were conducted using the quasi-optical gyrotron at Los Alamos National Laboratory, operating at 37 GHz.[7] Samples were irradiated with approximately 4.5kW; a typical power curve is presented in Figure 1. The PHPS/powder mixtures were irradiated in a quartz tube under flowing Ar/H_2 to minimize oxygen exposure and plasma formation. Prior to irradiation, the PHPS/powder mixtures were cured in quartz crucibles under flowing nitrogen in a quartz tube furnace at 180°C for three hours.

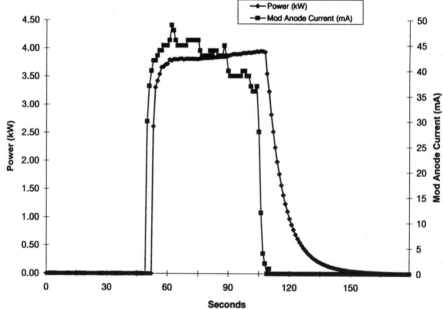

FIGURE 1. Typical Power Curve for Gyrotron Heating Experiment.

The synthesis of PHPS is described elsewhere.[8] Silicon carbide (SiC) powder was purchased from H.C. Starck. Graphite-coated Nicalon fabric (Ceramic Grade, 8-harness satin weave, carbon coated with poly(vinylalcohol) sizing) was purchased from Dow Corning. Poly(vinylalcohol) (PVA) sizing was removed by soaking the fabric in hot, deionized water, rinsing with acetone, and drying at 180°C for 24 hr. The Nicalon fiber-reinforced PHPS was

prepared as described elsewhere for composites.[9]   RCC-3 carbon-carbon specimens were
provided by NASA-Johnson Space Center.

RESULTS & DISCUSSION

Depending on the curing and firing conditions, PHPS converts to ceramic in yields of 80%
by weight or greater. X-ray Diffraction (XRD) analysis indicate that the polymer-derived ceramic
obtained at temperatures above 1270°C is crystalline, and is composed of α-Si_3N_4, β- Si_3N_4, and
elemental silicon with the proportions varying with temperature and pyrolysis atmosphere.[10]
At temperatures below 1270°C, the char product is largely amorphous, with elemental silicon the
only crystalline species identified.

Preliminary experiments revealed that PHPS does not couple with the 37 GHz radiation;
however, irradiation of PHPS loaded with SiC powder resulted in rapid, visible pyrolysis.
Experiments were then conducted to determine the effect of SiC powder-loading and irradiation
time on the polymer-derived product. The experimental matrix is presented in Table 1.

TABLE 1: Experimental Matrix for Gyrotron Heating

Sample ID	% SiC Powder	Target Beam Time (s)	Actual Beam Time (s)
A	2.5	20	19
B	2.5	90	58
C	10	20	19
D	10	90	58
E	30	20	19
F	30	90	58
G	50	10	10
H	50	20	19
I	50	45	37
J	50	90	58
K	50	180	112

In general, the samples emitted a red glow and appeared to be evolving gas shortly after
power was applied to the gyrotron. With most samples, a blue-purple plasma appeared after the
gas evolution had neared completion; however, with some specimens, particularly those with less
SiC powder, the plasma appeared shortly after power was applied. We believe the plasma tended
to shield the specimen from the beam. While the "beam on" time was known fairly accurately,
it was not possible to measure the energy density at the sample, and impossible to correct for the
effects of the plasma.

Following irradiation, the samples were ground and classified under inert conditions, and then
examined by XRD. A typical XRD pattern is presented in Figure 2. Preliminary analysis of the
XRD data has been conducted, using the method described previously,[10] to identify the relative
proportions of the observed crystalline constituents. The data has not been corrected for the
amorphous content of the chars, nor for the SiC powder loading. Nonetheless, the millimeter
wave induced conversion process appears qualitatively similar to that observed under thermal
conditions.[10] Figures 3 and 4 present the crystalline phase composition of the chars versus SiC
powder loading for two irradiation times, while Figure 5 presents the crystalline phase
composition versus irradiation time for a fixed SiC powder loading (50%).

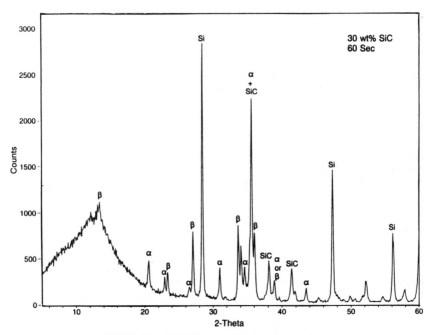

FIGURE 2. XRD Pattern Obtained From Char.

For both the 20 second and 60 second irradiation, no crystalline phases were observed for samples with 2.5% SiC. At 10% SiC content, elemental silicon is the major phase observed, although small amounts of α-Si_3N_4 and β-Si_3N_4 are observed. With 30% SiC content, approximately half the observed crystalline material was elemental silicon for both irradiation times. The sample irradiated for 20 seconds, however, contained slightly more beta phase than alpha, while the reverse was true for the sample irradiated for 60 seconds. At 50% SiC loading, α-Si_3N_4 was the dominant crystalline phase, with approximately equal amounts of β-Si_3N_4 and elemental silicon observed in the 20 second sample, but substantially more elemental silicon than β-Si_3N_4 observed in the 60 second sample. The phase composition of the 20 second sample is similar to that obtained under thermal conditions at 1350°C, while the phase composition of the 60 second sample more closely resembles that obtained at 1420°C.[10] The 60 second sample, however, has a higher elemental silicon to β-Si_3N_4 ratio than has ever been obtained thermally.

Increasing the SiC powder loading appears to increase the Si_3N_4 content of the char material, as does increasing the irradiation time. As seen in Figure 5, a 10 second irradiation was not sufficient to produce crystalline material. A 20 second irradiation was sufficient to produce crystalline material, with an approximate composition of 50% α-Si_3N_4, 25% β-Si_3N_4, and 25% elemental silicon. The crystalline phase content is essentially unchanged after 40 seconds. After 110 seconds, the char crystalline composition is roughly 50% α-Si_3N_4, 30% β-Si_3N_4, and 20% elemental silicon, which most closely resembles the phase composition of material obtained thermally at 1550°C. The 60 second sample, with an increased alpha content, decreased beta content and constant elemental silicon content, is not consistent with the other chars obtained thermally or through irradiation. Plasma formation could have caused the pyrolysis to pursue an anomalous pathway.

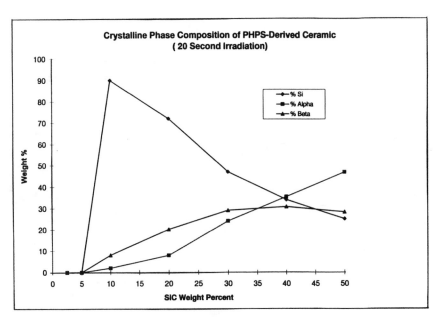

FIGURE 3. Crystalline Phase Composition of PHPS-Derived Ceramic (20 second irradiation).

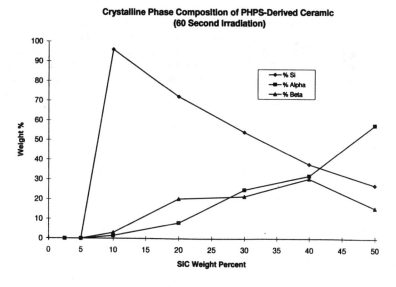

FIGURE 4. Crystalline Phase Composition of PHPS-Derived Ceramic (60 second irradiation).

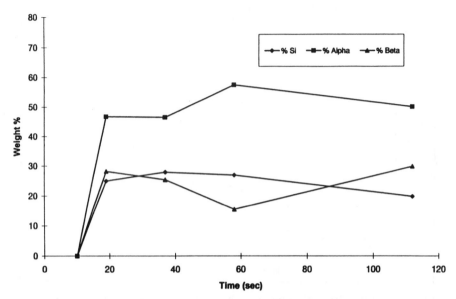

FIGURE 5. Crystalline phase composition of PHPS-Derived Ceramic (50% SiC loaded sample).

In addition to the SiC powder loaded PHPS, two Nicalon fiber-reinforced PHPS composites and a carbon-carbon (RCC-3) specimen with a damaged oxidation protection coating that had been prepared with powder-loaded PHPS were irradiated. The first composite was irradiated for 60 seconds and the second for 20 seconds. In both cases, the material glowed red, then white, and appeared to be evolving gas. Upon removal from the gyrotron, the composites appeared very similar to Stage 0 composites after pyrolysis.[9] The was no visible degradation of the fibers in either composite. The repair material on the carbon-carbon composite was irradiated for approximately 100 seconds. The repair material glowed red, then white, and appeared to be evolving gas. Upon removal from the gyrotron, the repair material appeared very similar to that obtained thermally.[5]

CONCLUSIONS

While PHPS does not couple efficiently to the gyrotron beam at 37 GHz, introduction of SiC powder enables conversion to ceramic. The degree of conversion and phase composition of the char can be controlled by the irradiation time or the SiC content. Overall, the conversion process appears qualitatively similar to that observed under thermal conditions, except that only seconds, rather than hours, are required to produce silicon nitride. Further studies are in progress.

ACKNOWLEDGEMENT

This effort was funded by the Southwest Research Institute Internal Research Program. The authors acknowledge with gratitude the contributions of Mr. Rafael J. Cordova, Mr. James F. Spencer, Mr. Stephen Salazar, Ms. Julie A. McCombs, and Ms. Rachel F. Muñoz.

REFERENCES

1.   a) T. Abraham, Ind. Ceram., **13**, p. 82 (1993); b) V. DeSapio, Chemtech, p. 46 (November, 1993).

2.   K.J. Wynne and R.W. Rice, Ann. Rev. Mat. Sci., **14**, p. 297 (1984).

3.   a) G. Ziegler, J. Heinrich, and G. Wötting, J. Mater. Sci., **22**, p. 3041 (1987); b) C. Boberski, R. Hamminger, M. Peuckert, F. Aldinger, R. Dillinger, J. Heinrich, and J. Huber, Angew. Chem. Int. Ed. Engl. Adv. Mater., **28**, p. 1560 (1989).

4.   a) S.T. Schwab and C.R. Blanchard-Ardid, Mat. Res. Soc. Symp. Proc., **121**, p. 581 (1988); b) S.T. Schwab, C.R. Blanchard, and R.C. Graef, J. Mat. Sci., **29**, p. 6320 (1994).

5.   S.T. Schwab, R.C. Graef, Y-M. Pan, D.M. Curry, V.T. Pham, J.D. Milhoan, and D.J. Tillian (Proceedings of the 39th International SAMPE Symposium (Closed Session), Society for the Advancement of Material and Process Engineering, Covina, CA, 1994) p. 68.

6.   a) D.F. Shriver and M.A. Drezdzon, *The Manipulation of Air-Sensitive Compounds*, 2nd ed. (John Wiley, New York, 1986); b) A.L. Wayda and M.Y. Darensbourg, *Experimental Organometallic Chemistry* (ACS Symposium Series 357, American Chemical Society: Washington, DC, 1987).

7.   J.D. Katz and D.E. Rees, Cer. Trans, **59**, p. 141 (1995).

8.   S.T. Schwab, U.S. Patent No. 5 294 425 (15 March 1994).

9.   S.T. Schwab, D.L. Davidson, Y-M. Pan, and R.C. Graef, *PIP Processing and Properties of Nicalon-Reinforced Silicon Nitride* (Proceedings of the 39th International SAMPE Symposium (Closed Session), Society for the Advancement of Material and Process Engineering, Covina, CA, 1994) p. 53.

10. C.R. Blanchard, S.T. Schwab, J. Am. Cer. Soc., **77**, p. 1729 (1994).

Part XIII

Plasma Processing

Microwave Plasma Sintering of Alumina

D. LYNN JOHNSON AND HUNGHAI SU
Department of Materials Science and Engineering, Northwestern University
Evanston, IL 60208-3108, dl-johnson@nwu.edu

ABSTRACT:

Microwave induced plasma provides an effective heating source for sintering of ceramics without the thermal instability that occurs in microwave sintering of some materials. In the present study, small cylindrical tubes were sintered in a microwave induced oxygen plasma inside a single mode (TM_{012}) cavity. A dilatometer and an optical fiber thermometer (OFT) black body sensor were employed to follow the changes of shrinkage and temperature during the sintering process. A similar specimen setup was built inside a conventional rapid heating furnace for comparison. The estimated activation energies were 468 ± 20 kJ/mol for plasma sintering and 488 ± 20 kJ/mol for conventional sintering. An athermal effect due to the plasma was observed. Sintering data were analyzed using the combined stage sintering model and the results suggest that the athermal effect may be ascribed to an increase of aluminum interstitial concentration during plasma sintering.

INTRODUCTION

Numerous studies[1][2][3][4][5][6][7] of enhanced plasma sintering of ceramics have been reported since the pioneering work of Bennett et al.,[8] who sintered small alumina specimens at constant temperature in a microwave (MW) plasma and compared the properties of sintered specimens with those of specimens sintered in a conventional furnace. They observed a reduction in sintering temperature on the order of 200°C, smaller grain size and higher mechanical strength than conventionally sintered specimens. Possible mechanisms of enhancement were discussed but not explored. The more recent work on MW sintering in which a "microwave effect" is reported[9][10] prompt us to ask whether the enhancement was due to the plasma, or to the microwaves. Among other things, this study answers this question.

Although significantly enhanced densification has been reported in all plasma sintering studies for a variety of ceramic materials, the mechanisms are still not clear. Young and McPherson claimed that the observed sintering rates were too great to be accounted for on the basis of capillarity-driven diffusion, and proposed that temperature gradients enhanced the sintering rate.[11][12] However, we formulated a combined stage sintering model and suggested that the observed sintering rates were within the range expected for diffusion-controlled sintering at the heating rates employed.[13][14] Hansen's[7] results and the more complete work reported below show that the rates are, indeed, enhanced above those for conventional sintering, but not by the mechanism proposed by Young and McPherson. In fact, the apparatus used for the present study was designed specifically to minimize temperature gradients in the specimens.

A critical issue in determining whether sintering rates are enhanced is the accurate measurement of temperature. The reported enhancements could be artifacts of a low bias in temperature measurements. Indeed, both Bennett et al.[8] and Hansen[7] observed such a bias and attempted to compensate for it using various calibration procedures. A thermally well-coupled black body sensor for an optical fiber thermometer (OFT) was employed in the present study to minimize the bias (see below).

The plasma sintering work reported to date has been limited to measurements of final density and other properties of individual specimens. The shrinkage rate data, which are crucial in interpreting sintering mechanisms, are very difficult to obtain from these experiments. More detailed investigations utilizing dilatometry and accurate temperature measurement are needed to further elucidate the effects of the plasma on sintering. This paper summarizes the results of such a study.[15]

EXPERIMENTAL PROCEDURES

High purity small particle size alumina powder* was formed into 10 mm ϕ x 10 mm long tubes with 1 mm walls by isostatic pressing and presintering at 650°C for a half hour to burn off the polyvinyl butyral binder (green density $\approx$ 55%).

A single mode (TM_{012}) MW cavity[16] was employed to excite the plasma because of flexibility in the selection of plasma gases and the large volume of the plasma. Ultra-high purity oxygen was used because a nitrogen plasma etches alumina[5] and neutral plasmas tend to reduce oxides.[17] The 41 mm alumina plasma tube was insulated with an alumina fiber cylinder to minimize temperature gradients.

The specimens were centered in the hot zone on a sintered high purity alumina plate. A dilatometer** was mated to the specimen with the reference sensor connected to a 1.27 mm sapphire rod resting on the base plate and the moving sensor connected to a coaxial 3.2 mm (OD) sapphire tube resting on a sintered lid on the specimen. Temperature was measured using an OFT black body sensor*** which was inserted through the base plate into the space enclosed by the specimen, base and lid. Figure 1 illustrates the arrangement of the specimen, the dilatometer probes and the OFT probe. The assembly was supported on an 8 mm diameter single crystal sapphire tube.

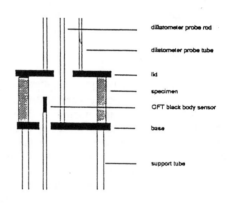

Figure 1. Schematic diagram of specimen, dilatometer probes, and OFT sensor.

Quenching the plasma by chopping the power to the magnetron showed no instantaneous drop in measured temperature, showing that the OFT probe was not biased by the luminous plasma.

A similar setup was built inside a low thermal mass furnace to compare conventional sintering with plasma sintering under identical thermal histories and oxygen pressures. An R-type thermocouple was used to measure the temperature of specimens in the position of the OFT sensor in plasma sintering. Comparison of the thermocouple and the OFT sensor in a black body cavity inside an isothermal furnace revealed differences of ±3°C.

---

*   AKP-50, Sumitomo Chemical America, Inc., New York, NY.

**   Dilaflex 5, Series 6548, Theta Industries, Inc., Port Washington, NY.

*** Accufiber Div., Luxtron Corporation, Santa Clara, CA.

Calibration runs were made under each set of conditions to correct for specimen and system thermal expansion effects. Dilatometry results agreed with caliper measurements before and after sintering within the 0.025 mm precision of the caliper.

Plasma and conventional sintering experiments were carried out at different constant heating rates under the same oxygen pressure, and under different oxygen pressures at the same heating rate. In the first set of experiments, the specimen was initially heated to 750°C at arbitrary rates and then to 1400°C at 7.5, 15, 30, and 45°C/min with no hold at the maximum temperature. The oxygen pressure was fixed at 1.7 kPa with a flow rate of 2 cc/sec (STP). For the second set of experiments, the specimens were heated to 1500°C at a constant heating rate of 30°C/min under oxygen pressures of 1.1, 1.7, and 2.4 kPa. Two or three repeated runs were performed under each sintering condition. The measured final densities at each condition differed by less than 1% of theoretical density.

The final bulk density and percent open porosity of the sintered specimen was measured by the Archimedes method with deionized water as the immersion medium. The instantaneous density during densification was obtained from the final density and the dilatometry data, assuming isotropic shrinkage. Isotropic specimen shrinkage was confirmed by comparing the final shrinkage in height and outside diameter. Also, the specimens were not tapered, indicating that they were uniformly heated. Grain size at different densities was estimated from measurements made on fracture surfaces.[15]

RESULTS AND DISCUSSION

The final bulk density of specimens sintered at constant heating rates is plotted in Fig. 2. The enhancement in densification is obvious. Figure 3 shows the density computed from dilatometry data as a function of temperature for plasma and conventional sintering at constant heating rates, again showing enhanced densification in the plasma. A reduction in sintering temperature of about 140° is indicated in the intermediate density range (75% ~ 80%).

Examination of fracture surfaces revealed no discernable difference in the character of the microstructure of plasma and conventionally sintered specimens. Likewise, the density dependence of both open pore volume fraction and grain size was similar for both processes. These observations imply that the basic mass transport processes which cause densification and grain growth are similar for both heating process.

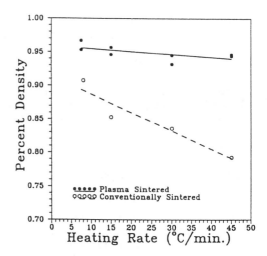

Figure 2. Final bulk density of plasma and conventionally sintered specimens heated to 1400°C with no hold at temperature.

The data were analyzed using the combined stage sintering model, and the grain boundary diffusion coefficient calculated, with the results shown in Fig. 4. The activation energy was found to be 468 ± 20 kJ/mol for plasma sintering and 488 ± 20 kJ/mol for conventional sintering. These activation energy values are consistent with those reported for conventional sintering and creep of alumina.[18][19][20][21][22] We thus conclude that plasma and conventional sintering proceed by grain boundary diffusion, and the differences in rate are due to differences in the pre-exponential term of the diffusion coefficient.

The deviation of the high temperature data (corresponding to density larger than 85%) from the straight line of the lower temperature data indicates changes the interaction of the plasma with the material. This might be expected since the plasma interacts only with the surface, and sintering is controlled by bulk properties. Although surface diffusion can rapidly equilibrate the interior with the surface at first, as the pores shrink and are closed off from the surface this becomes slower, resulting in less enhancement.

We proposed that the observed enhancement in the plasma is caused by increased concentration of aluminum interstitials by stripping of $O^=$ from the surface by bombardment of the surface by O^+ from the surface.[15] The data obtained from the series of runs made at various plasma pressures at constant heating rate gave support to this hypothesis. At lower pressures, where the oxygen would be more heavily ionized, the rate was enhanced more than at higher pressures. The enhancement in the diffusion coefficient, up to ten-fold as the pressure was decreased from 2.4 to 1.1 kPa, is far greater than the 2.25-fold increase observed when the oxygen partial pressure was reduced by seven orders of magnitude in conventional sintering of alumina.[23]

These data also demonstrated that the

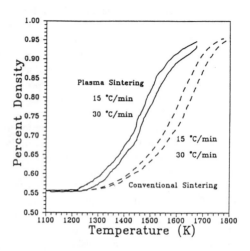

Figure 3. Density for plasma and conventional sintering at constant heating rates. Data for 7.5° and 45° per minute are not shown for clarity.

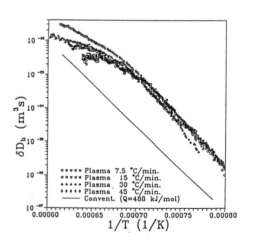

Figure 4. Diffusion coefficients for plasma and conventional sintering (the solid line represents the average value for conventional sintering).

enhancement was a plasma effect, and not a "microwave effect". As the pressure is reduced, with the resulting increase in the degree of ionization, the MW penetration depth in the plasma will be reduced. The sintering rate would have <u>decreased</u> at lower pressure if the effect were due to MW interaction with the material.

CONCLUSIONS

Significantly enhanced densification rates have been achieved in microwave induced plasma sintering of alumina compared to conventional sintering under the same temperature-time excursions. The results demonstrate that besides the ultra-rapid heating rate achievable in a plasma, an athermal effect also contributes to the enhanced sintering. The sintering is sensitive to the oxygen plasma pressure, with lower pressures yielding greater rates. The activation energy for sintering in the oxygen plasma was estimated to be 468 ± 20 kJ/mol, which was in agreement with the activation energy for conventional sintering. It is concluded that the dominant sintering mechanism for alumina in a plasma is grain boundary diffusion of aluminum ions, the same as with conventional heating, but with a significant enhancement of the concentration of aluminum interstitials caused by exposure to the plasma.

Acknowledgement: Supported by the National Science Foundation under Grant No. DMR-9200128.

REFERENCES

1. L.G. Cordone and W.E. Martinsen, "Glow-Discharge Apparatus for Rapid Sintering of Al_2O_3", J. Am. Ceram. Soc., **55** [7] 380 (1972).

2. G. Thomas, J. Freim, and W. Martinsen, "Rapid Sintering of UO_2 in a Glow Discharge", Trans. Am. Nucl. Soc., **17**, 177 (1973).

3. D. L. Johnson and R.A. Rizzo, "Plasma Sintering of β''-Alumina", Am. Ceram. Soc. Bull., **59** [4] 467-468,472 (1980).

4. J.S. Kim and D. L. Johnson, "Plasma Sintering of Alumina", Am. Ceram. Soc. Bull., **62**[5] 620-622 (1983).

5. E.L. Kemer and D.L. Johnson, "Microwave Plasma Sintering of Alumina", Am.Ceram. Soc. Bull., **64** [8] 1132-1136 (1985).

6. D. L. Johnson, W.B. Sanderson, J.M. Knowlton, E.L.Kemer, and M.Y. Chen, "Advances in Plasma Sintering of Alumina", pp. 815-820 in High Tech Ceramics, Edited by P. Vincenzini, Elsevier Science Pub. B.V., Amsterdam (1987).

7. J.D. Hansen, "Rapid Sintering Kinetic of Alumina in an Argon Induction Coupled Plasma", Ph.D. Dissertation, Northwestern University, Evanston, IL. December 1991.

8. C.E.G. Bennett, N.A. McKinnon, and L.S. Williams, "Sintering in Gas Discharges", Nature (London), **217**, 1287-88 (1968).

9. M.A. Janney and H.D. Kimrey, "Diffusion-Controlled Processes in Microwave-Fired Oxide Ceramics", pp. 215-227 in Microwave Processing of Materials II, Materials

Research Society Symposium Proceedings vol. 189, Materials Research Society, Pittsburgh, Pennsylvania, (1990).

10. I. Ahmad and D.E. Clark, "Effect of Microwave Heating on the Mass Transport in Ceramics", pp. 287-295 in Microwaves: Theory and Applications in Materials Processing II, Ceramic Transactions, vol. 36, ed. by D.E. Clark, W.R. Tinga and J.R. Laia, Jr., American Ceramic Society, Ohio, (1993).

11. R.M. Young and R. McPherson, "Temperature-Gradient-Driven Diffusion in Rapid-Rate Sintering", J. Am. Ceram. Soc., **72** [6] 1080-1081 (1989).

12. R.M. Young and R. McPherson, "Reply", J. Am. Ceram. Soc., **73** [8] 2579-80 (1990).

13. D.L. Johnson, "Comment on 'Temperature-Gradient-Driven Diffusion in Rapid-Rate Sintering'", J. Am. Ceram. Soc., **73** [8] 2576-78 (1990).

14. J.D. Hansen, R.P. Rusin, M.H. Teng, and D.L. Johnson, "Combined-Stage Sintering Model", J. Am. Ceram. Soc., **75** [5] 1129-35 (1992).

15. Hunghai-Su and D. Lynn Johnson, "Sintering of Alumina in Microwave Induced Oxygen Plasma", J. Am. Ceram. Soc., in press, 1996.

16. M.P. Sweeney and D.L. Johnson, "Microwave Plasma Sintering of Alumina", pp. 365-372 in Microwave: Theory and Application in Materials Processing, Edited by D.E. Clark, F.D. Gac and W.H. Sutton. American Ceramic Society, Inc. 1991.

17. C.A. Kotecki, "RF Plasma Sintering of TiO_2", MS Thesis, Northwestern University, Evanston, IL. June 1987.

18. W.E. Young and I.B. Cutler, "Initial Sintering with Constant Rates of Heating", J. Am. Ceram. Soc., **53** [12] 659-663 (1970).

19. J. Wang and R. Raj, " Estimation of the Activation Energies for Boundary Diffusion from Rate-Controlled Sintering of Pure Alumina, and Alumina Doped with Zirconia or Titania", J. Am. Ceram. Soc., **73** [5] 1172-75 (1990).

20. S.H. Hillman and R.M. German, "Constant Heating Rate Analysis of Simultaneous Sintering Mechanisms in Alumina", J. Mater. Sci., **27**, 2641-48 (1992).

21. J. Wang and R. Raj, " Activation Energy for the Sintering of Two-Phase Alumina/Zirconia Ceramics", J. Am. Ceram. Soc., **74** [8] 1959-63 (1991).

22. R.M. Cannon and R.L. Coble, "Review of Diffusional Creep of Al_2O_3", pp.61-100 in Deformation of ceramic Materials, Eds. R.C. Bradt and R.E. Tressler, Plenum Press, New York (1975).

23. A.M. Thompson and M.P. Harmer, "Influence of Atmosphere on the Final Stage Sintering Kinetics of Ultra-High-Purity Alumina", J. Am.Ceram. Soc., **76** [9] 2248-56 (1993).

MICROWAVE PLASMA ASSISTED CVD OF DIAMOND ON TITANIUM AND TI-6AL-4V

D.A. Tucker*, M.T. McClure**, Z. Fathi*, Z. Sitar**, B. Walden***, W.H. Sutton****, W.A. Lewis*, and J.B. Wei*

*Lambda Technologies, Inc., 8600 Jersey Ct., Suite C, Raleigh, NC 27613
**Department of Materials Science and Engineering, North Carolina State University, Campus Box 7919, Raleigh, NC 27695-7919
***Department of Physics, Trinity College, 300 Summit St., Hartford, CT 06106
****United Technologies Research Center, 411 Silver Lane-MS-24, East Hartford, CT 06108

ABSTRACT

The ultimate goal of this research was to demonstrate a Microwave Plasma assisted Chemical Vapor Deposition (MPCVD) process to coat a Ti-6Al-4V bearing shaft. Preliminary experiments were performed in an ASTeX™ system on flat chemically pure titanium and Ti-6Al-4V coupons. Diamond deposition was also attempted on a Ti-6Al-4V wedge sample which contained a curved surface that simulated the bearing. Although uniform diamond deposition was attained on the flat samples, very poor uniformity was observed on the curved sample. This lack of uniformity was attributed to the difficulty in controlling the plasma-to-substrate distance in a single mode, single frequency reactor and it was believed that by varying the frequency and using a multimode cavity one could solve this problem. Thus, depositions on a titanium rod were performed in a variable frequency MPCVD reactor. It was determined visually that the variable frequency operation provided uniform plasma distribution along the length and circumference of the rod and resulted in a fairly uniform coating. However, scanning electron microscopy revealed that the morphology of the particles was poor and micro-Raman spectroscopy showed weak, broad peaks that were attributable to amorphous carbon. It is believed that by fine-tuning parameters, a uniform diamond film of good quality can be achieved along the entire rod.

INTRODUCTION

Diamond films have unique properties (e.g. abrasion resistance, chemical inertness, thermal conductivity, and low coefficient of friction) which meet the requirements of a variety of applications including wear and thermal barrier applications. A diamond coating on Ti-6Al-4V, an alloy currently used as an aerospace and biomedical material, would certainly prolong its useful life. Although the chemical vapor deposition of diamond has made significant advances over the last ten years, there are still many key problems to be addressed, including improving coating uniformity on three-dimensional shapes. This is one of the limitations in the current usage of microwave plasmas to deposit diamond films. Because the plasma is approximately spherical in shape and plasma-to-substrate distance is critical, coating anything but approximately flat samples is difficult. The research presented in this paper attempts to resolve this issue by studying the use of variable frequency in a multimode cavity to raster plasmas over a titanium rod.

Mat. Res. Soc. Symp. Proc. Vol. 430 © 1996 Materials Research Society

Single Mode, Single Frequency Experiments

Bias-enhanced nucleation (BEN) and diamond deposition were performed in an ASTeX™ microwave (2.45 GHz) plasma CVD chamber. This system has been described in detail elsewhere [1]. BEN was performed as a pretreatment routine to promote nucleation and involved applying a negative dc bias to the substrate holder while the positive potential was connected to ground. Table I lists typical parameters used for deposition on the titanium and Ti-6Al-4V substrates.

Multimode, Variable Frequency Experiments

Diamond deposition was performed in a multimode cavity utilizing a broadband microwave source. This apparatus is schematically illustrated in Figure 1. A nominally 4.0 to 8.0 GHz Teledyne traveling wave tube (TWT) was used to amplify input from a Hewlett Packard Sweep Oscillator, which also controlled frequency range and sweep rate.

The plasma was contained within a 8" O.D. quartz bell jar that was sealed to the bottom plate of a rectangular cavity with a silicone rubber L-gasket. Hydrogen and methane gas flow rates were controlled by MKS mass flow controllers and introduced premixed into the bell jar

Table I. Typical parameters for single mode, single frequency deposition on polycrystalline Ti and Ti-6Al-4V coupons.

Process Parameters	H_2 plasma cleaning	Carburization	BEN	Growth
Power (Watts)	650	700	700	700
Pressure (Torr)	25	30	20	30
H_2 flow (sccm)	500	500	500	500
CH_4 flow (sccm)	0	10	25	4
O_2 flow (sccm)	0	0	0	0.5
Bias current (mA)	-	-	80	-
Bias voltage (V)	-	-	-220	-
Temperature (°C)	0-630	950	1000	1000
Duration	15 min.	30 min.	10 min.	4.5 hrs.

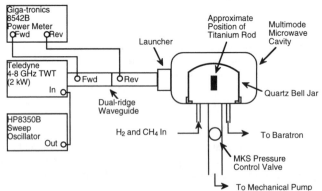

Figure 1. Schematic illustration of the multimode, variable frequency microwave plasma apparatus.

via a port in the bottom plate. The pressure within the bell jar was monitored by a MKS baratron and controlled with a MKS control valve.

A titanium rod, approximately 0.5" in diameter and 2" in length, was placed upright on a quartz plate in the center of the bell jar. The ends of the rod were masked by quartz cups which left about a 1" length of rod free for coating. A frequency range of 3.7 to 6.5 GHz was found to give visibly uniform plasmas around the circumference of the rod. Table II lists diamond deposition parameters for the variable frequency MPCVD system.

RESULTS

Single Mode, Single Frequency Experiments

The diamond particles and films deposited were examined by Scanning Electron Microscopy (SEM) for particle morphology and deposition uniformity and micro-Raman spectroscopy for qualitative composition information. A trace amount of oxygen gas was used in the growth procedure to limit twinning on the (111) faces of the diamond particles and also to provide an environment that would promote [001] texturing of the film. These attempts, however, were unsuccessful and film texture in all cases appeared to be [110]. Figure 2 presents micrographs of diamond crystallites from deposition on titanium and Ti-6Al-4V with BEN used as a nucleation mechanism. These two films are very similar in morphology and uniformity, which is expected considering that Ti-6Al-4V is composed of approximately 90 wt.% titanium. This is very useful information because titanium is a readily available material and can be used for preliminary trials instead of the more expensive alloy.

A micro-Raman spectrum from deposition on the Ti-6Al-4V alloy is presented in Figure

Table II. Typical parameters for multimode, variable frequency
deposition on polycrystalline Ti rods.

Process Parameters	Carburization	Growth
Frequency range (GHz)	3.7-6.5	3.7-6.5
Power (Watts)	1400	1400
Pressure (Torr)	30	30
H_2 flow (sccm)	800	800
CH_4 flow (sccm)	40	4
Temperature (°C)	800	800
Duration	15 min.	23 hrs.

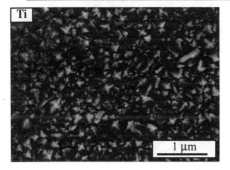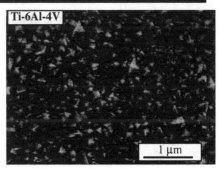

Figure 2. SEM micrographs of diamond particles
deposited on titanium and Ti-6Al-4V.

3 and shows the presence of a sharp peak at 1350 cm^{-1} and a broad peak centered at approximately 1530 cm^{-1}. Typically, a very sharp peak at 1332 cm^{-1} is ascribed to diamond. However, this peak can shift due to stress in the film from thermal expansion differences between the substrate and diamond coating [2]. The broad peak at 1530 cm^{-1} is attributed to sp^2 components within the film, which is not unusual for thin polycrystalline films where Raman scattering occurs at the interface and grain boundaries. Thus, the peaks at 1350 and 1530 cm^{-1} are diamond and non-diamond components, respectively.

Diamond deposition was also attempted on a Ti-6Al-4V wedge sample which contained a curved surface that simulated the bearing. Figures 4 and 5 show a schematic illustration of the curved wedge and SEM micrographs of resultant diamond deposition, respectively. Deposition was non-uniform with only patchy deposition of diamond particles across the curved surface. The highest nucleation density is observed at the top of the wedge. Plasma-to-substrate distance is very critical for nucleation and growth of diamond. Thus, the result on the curved sample is not unsurprising because in a single mode, single frequency MPCVD reactor it is difficult to get a plasma uniformly displaced from a curved surface.

Multimode, Variable Frequency Experiments

The capability of varying frequency to cover three-dimensional samples has been alluded to previously [3, 4]. However, to the best of our knowledge, this is the first publication where a bandwidth of frequencies (3.7 to 6.5 GHz) have been utilized to deposit diamond. As the frequency changes, the power density distribution throughout the multimode cavity changes and results in plasmas that move with the maxima in power density. The ideal method that would provide the best efficiency would be to frequency jump so that plasmas only form in the vicinity of the sample to be processed and energy would not be wasted in forming plasmas

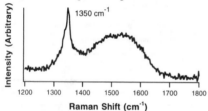

Figure 3. Micro-Raman spectrum of deposit on Ti-6Al-4V.

Figure 4. Schematic illustration of wedge (pie-shaped slice from a 0.5" diameter rod).

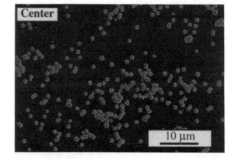

Figure 5. SEM micrographs of diamond particles deposited on the Ti-6Al-4V wedge .

away from the object to be coated. We expect to expand our capabilities to frequency jumping in the near future.

Again, samples were analyzed by SEM and micro-Raman spectroscopy. Figure 6 presents low-magnification photographs of the titanium rod taken at 0° and 180° rotations and shows that the rod has a fairly uniform coating. The photographs also show that masking with quartz cups was marginally effective at best. (Because the cups were slightly oversized to cover thermal expansion differences, small plasmas may have been initiated between the quartz and rod at the higher frequencies.) Figure 7 shows SEM micrographs obtained from different areas along the rod. (The macro photographs of the rod in Figure 6 have been marked with the approximate locations on the rod where these micrographs were acquired.) Figure 7a is a typical morphology covering the majority of the rod and suggests that there are two layers: a dark amorphous/crystalline mixture and a sparse overlayer of white crystalline clusters. This may mean that the initial carburization step left a layer which probably poisoned future diamond deposition. A micro-Raman spectrum obtained in an area of morphology similar to Figure 7a is presented in Figure 8. It shows two weak/broad peaks centered at approximately 1340 and 1590 cm^{-1} which are typically attributed to a mixture of amorphous carbon and graphite [5]. Micrographs b, c, and d (where d is a higher magnification micrograph of c) of Figure 7 show spots on the rod where there are ball shaped nuclei with multiple facets. These results again suggest that too high a methane concentration was used [6]. By fine-tuning parameters, such as sample temperature, methane concentration, and plasma position, it is believed that a uniform diamond film can be deposited on this rod.

CONCLUSIONS

Preliminary experiments for diamond deposition coating on a cylindrical titanium rod have been performed. Similar results on flat samples of titanium and Ti-6Al-4V in a single mode, single frequency MPCVD reactor have shown that titanium can be used as a substitute for the alloy in mock-up trials. Deposition on a Ti-6Al-4V wedge sample showed patchy diamond deposition. This lack of uniformity was attributed to the difficulty in controlling plasma-to-substrate distance in a single mode, single frequency reactor and it was believed that by varying the frequency and using a multimode cavity one could help solve this problem. Thus, depositions on a titanium rod were performed in a variable frequency MPCVD reactor. It was determined visually that the variable frequency operation provided a uniform plasma distribution along the length and circumference of the rod and resulted in a fairly uniform coating. However, scanning electron microscopy revealed that the morphology of the particles was poor. It is believed that by fine-tuning parameters, such as sample temperature, methane concentration, and plasma position, a uniform diamond film of good quality can be achieved along the

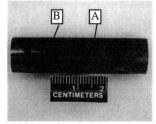

Figure 6. Low-magnification photographs of titanium rod processed with variable frequency MPCVD reactor.

639

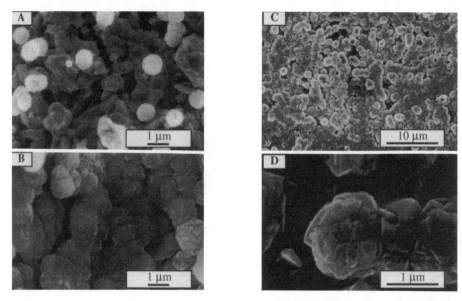

Figure 7. SEM micrographs obtained in different regions of the titanium rod.

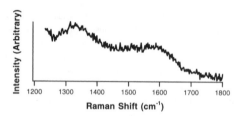

Figure 8. Micro-Raman spectrum of deposit on Ti rod.

entire rod and that this work could be extended to diamond deposition onto a Ti-6Al-4V bearing shaft.

CONCLUSIONS

1. B.R. Stoner, G.-H.M. Ma, S.D. Wolter, and J.T. Glass, Phys. Rev. B **45**, 11067 (1992).
2. J.W. Ager III and M.D. Drory, Phys. Rev. B **48**, 2601 (1993).
3. R.A. Rudder, R.C. Hendry, G.C. Hudson, R.J. Markunas, A.C. Johnson, L.T. Thigpen, R.S. Garard, and C.A. Everleigh in <u>Microwaves: Theory and Application in Materials Processing II</u>, edited by D.E. Clark, W.R. Tinga, and J.R. Laia Jr. (American Ceramic Society Proc. **36**, Cincinnati, OH, 1993) p. 377.
4. A.C. Johnson, R.A. Rudder, W.A. Lewis, and R.C. Hendry in <u>Microwave Processing of Materials IV</u>, edited by M.F. Iskander, R.J. Lauf, and W.H. Sutton (Mater. Res. Soc. Proc. **347**, San Francisco, CA, 1994) p. 617.
5. D.S. Knight and W.B. White, J. Mater. Res. **4**, 385 (1989).
6. W. Zhu, A.R. Badzian, and R. Messier in <u>SPIE Diamond Optics III</u> (SPIE Proc. **1325**, 1990) p. 187.

MICROWAVE ANNEALING OF ION IMPLANTED 6H-SiC

J.A. GARDNER*, M.V. RAO*, Y.L. TIAN*, O.W. HOLLAND**, G. KELNER*** , J.A. FREITAS, Jr*** and I. AHMAD****
* ECE Department, George Mason University, Fairfax, Virginia 22030
** Oak Ridge National Laboratory, Oak Ridge, Tennessee, 37831
*** Naval Research Laboratory, Washington, D.C. 20375
**** FM Technologies Inc., Fairfax, Virginia, 22032

ABSTRACT

Microwave rapid thermal annealing has been utilized to remove the lattice damage caused by nitrogen (N) ion-implantation as well as to activate the dopant in 6H-SiC. Samples were annealed at temperatures as high as 1400 °C, for 10 min. Van der Pauw Hall measurements indicate an implant activation of 36%, which is similar to the value obtained for the conventional furnace annealing at 1600 °C. Good lattice quality restoration was observed in the Rutherford backscattering and photoluminescence spectra.

INTRODUCTION

In the past several years SiC has received a great deal of attention for its potential use in semiconductor device applications. SiC has certain qualities that could make it extremely attractive for use in high-temperature and high-power integrated circuits [1]. The SiC devices could be used up to temperatures as high as 700 °C provided that the packaging technology can withstand such an extreme environment. This material has the additional advantage of being chemically inert, which would allow the devices to be exposed to corrosive or similarly harsh environments without degradation. Another aspect of SiC, is that the electron saturation velocity is about three times higher than that of Si [1]. This would allow the device to operate at higher speeds, as far as switching logic is concerned. The higher speed of the device would also permit circuits to operate in a higher frequency range (i.e, microwave range) as compared to Si. SiC has three main polytypes which are currently being used for discrete and integrated circuit fabrication. They consist of 3C (cubic), 4H, and 6H (hexagonal). In this study we have used only 6H-SiC.

In fabricating planar SiC devices, it is necessary to use ion implantation as the vehicle for doping, using various impurities. Thermal diffusion techniques would be highly impractical since the diffusion coefficients of impurities in SiC are extremely low thereby requiring temperatures as high as 1800 °C for long periods of time [1]. The sample would decompose at such a high temperature due to the thermal evaporation of Si. Ion implantation provides the user with the capability of selective area doping without the need of subjecting the material to extreme temperatures. However, during the process of implantation the crystal lattice of the SiC is damaged as the ions penetrate it. The damaged material must then be subjected to heating in order to restore the crystal lattice to its original structure. This process also has the effect of electrically activating the implants [2]. At present, the implanted SiC is annealed in ceramic processing furnaces at temperatures as high as 1650 °C [3]. Because the conventional annealing process can take several hours due to slow ramp up and ramp down times, the possibility of using microwaves for rapid thermal processing has been explored. This helps to prevent redistribution of impurity atoms during annealing [4].

Microwave energy can provide a fast and efficient method to high temperature anneal SiC, with the proper insulating material. Microwave energy can be directed and tuned, which will cause the material to heat up locally at an accelerated rate [5-8]. Rise rates of 300 °C/ min have been achieved, compared to 10 °C/min, in the conventional annealing techniques. This process has the possibility of being scaled up so as to anneal arrays of wafers in the same short time period, (unlike halogen lamp RTP) which is one of the requirements for commercial application. The intention is to investigate this new technology by annealing 6H-SiC samples which have been implanted with nitrogen. Electrical and lattice quality tests were performed on these samples in order to compare them with conventional annealing.

EXPERIMENT

The microwave annealing experiments were carried out in a TE_{103} single mode cavity with a variable power source of 1 kW. In order to achieve maximum energy efficiency and heat the samples to a high temperature in a short ramp time, critical tuning and coupling was achieved and maintained during the entire annealing process by adjusting the position of a moving plunger and the size of an adjustable iris. A directional coupler was used to monitor the reflected power. The temperature of the samples was measured by a two-color pyrometer and controlled by manually varying the input power.

The samples used for the annealing tests were N-implanted p-type 6H-SiC with a background carrier concentration of $\sim 10^{18}$ cm^{-3}. The samples were placed in a SiC crucible, which acted as a susceptor for absorbing microwave power and heating the 6H-SiC samples. In addition, the use of the SiC crucible protected the surface of the samples from pitting due to thermal dissociation caused by the evaporation of silicon. The crucible itself may also have some dissociation and hence maintained high Si partial pressure in the vicinity of the SiC sample during annealing. The SiC crucible was enclosed in an insulating material which minimized the heat loss from radiation and stabilized the temperature over the annealing cycle. Especially in the high temperature range of over 1300 °C, significant radiation loss would limit the maximum annealing temperature to be reached if the insulation were not used.

Arcing was one of the problems frequently encountered in the microwave annealing experiments. It was caused by an intense local electric field near a sharp corner or discontinuities of the crucible and insulation. A high and uncontrollable temperature caused by arcing could heat up the SiC samples to a temperature at which it would degrade and become useless for device fabrication. Therefore the geometry of the crucible and the insulation must be carefully designed to eliminate the arcing.

The sample was placed in the cavity and temperatures ranging from 1150 °C to 1400 °C were achieved in a few minutes. A typical temperature/time cycle obtained for 1230 °C annealing is shown in Fig 1. The temperature rose at a rate of ~300 °C/min and falls at a rate of ~200 °C/min, which is monitored by an optical pyrometer through a narrow orifice in the insulating case. Various gas ambients were tried during the annealing process. Argon and helium gases resulted in arcing at relatively low temperatures whereas nitrogen could withstand at least 1450 °C. The annealing was performed for a duration of 10-15 min after the temperature had been stabilized. After annealing, the samples were removed and subjected to a thorough cleaning process by solvents. A dip in HF is necessary to remove the nitride layer formed on the surface during annealing in the N_2 ambient. Hall measurements, Rutherford Backscattering (RBS), and photoluminescence measurements were performed on the annealed samples in order to evaluate the electrical activation of the implant as well as the lattice perfection.

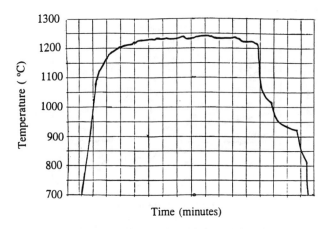

Fig.1. Typical temperature / time cycle of microwave annealing.

RESULTS

The N-implants were done at 700 °C down to a depth of 2.5 μm using a multiple ion energy schedule. The nitrogen implant schedule used was 50 keV/ 6.1x10^{13} cm^{-2}, 100 keV/ 9.2x10^{13} cm^{-2}, 250 keV/ 1.4x10^{14} cm^{-2}, 500 keV/ 1.6x10^{14} cm^{-2}, 800 keV/ 1.9x10^{14} cm^{-2}, 1 MeV/ 2.05x10^{14} cm^{-2}, 2 MeV/ 2.2x10^{14} cm^{-2}, 3 MeV/ 2.4x10^{14} cm^{-2}, and 4 MeV/ 2.67x10^{14} cm^{-2} in order to obtain a uniform nitrogen concentration of 10^{19} cm^{-3}. The as-implanted material is dark/opaque in color due to the lattice damage. Once annealed the samples regained their transparency with increasing annealing temperatures. No deterioration was observed in the surface morphology after annealing at 1400 °C.

The lattice perfection of the as-implanted and annealed samples was evaluated by performing RBS via channeling measurements. A 2.3 MeV He^{++} beam was used along with a standard solid-state detector which was positioned to record backscattered ions at 160°. Recorded spectra are shown in Fig. 2. Spectra from an aligned and randomly oriented virgin sample are also given in Fig. 2 to use as a basis for comparison with the RBS data. As shown in Fig. 2 the yield in the as-implanted sample is closer to the virgin level than to the random level. This random level indicates the scattering from a completely amorphized material. The low yield of the as-implanted sample is due to the implantation at an elevated temperature of 700 °C. Accumulated lattice damage is less at elevated implant temperatures due to dynamic self-annealing. This is in contrast to the room temperature implantation in which the as-implant yield would be at the random level [3]. After annealing at 1140 °C there is a significant reduction in the scattering yield, bringing the spectra closer to the virgin level. Nearly identical spectra were obtained for 1230 and 1320 °C anneals. For 1400 °C annealing, the spectra coincide with the virgin level indicating highly effective lattice damage removal. Similar RBS spectra were obtained on the samples annealed at 1500 and 1600 °C in the conventional Brew ceramic processing furnace.

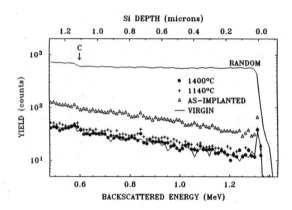

Fig. 2. RBS spectra on N-implanted SiC before and after annealing.

The electrical characteristics of the annealed material were evaluated by performing van der Pauw Hall measurements at room temperature. We were unable to perform Hall measurements on samples annealed at temperatures less than 1400 °C. This is due to the high resistance of the material associated with the residual implant damage. We observed a high resistance behavior in SiC bombarded with Si or C at levels which produce a high concentration of point defects. RBS provides a qualitative measure of gross lattice damage, which is in the form of extended defects like dislocation loops. Though the concentration of the extended defects is reduced at temperatures below 1400 °C, a large concentration of point defects, which cannot be detected by RBS, may still exist.

For the sample annealed at 1400 °C we measured a sheet resistance of 3×10^2 $\Omega/\square$, a sheet carrier concentration of 5.6×10^{14} cm^{-2}, and a carrier mobility of 37 cm^2/V.s. The percent activation, which is obtained from the ratio of sheet carrier concentration and the total implant dose of 1.57×10^{15} cm^{-2}, was found to be approximately 36%. This activation is comparable to the values obtained [3] for conventional 1600 °C annealing. However the carrier mobility measured for 1400 °C microwave annealing is less than the value of 50 cm^2/V.s obtained for the 1600 °C conventional annealing.

Photoluminescence measurements have also been taken in order to obtain addition information on the lattice quality of the sample as well as pair recombination information. Figure 3 shows the low temperature and low resolution photoluminescence (PL) spectrum of a p-type substrate, which has been implanted with N and microwave annealed at 1400 °C for 10 min. Also represented in the figure is the PL spectrum of the back-surface of the sample (un-implanted surface). The sharp lines observed in the spectral range between 3.0 and 2.8 eV, not clearly visible in the low resolution spectra shown in Fig. 3, are assigned to recombination processes involving the distribution of close N-donor and Al-acceptor pairs (close DAP recombination) present in the substrate. The broad shoulders detected between 2.85 and 2.65 eV are the zero

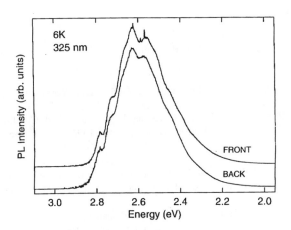

Fig. 3. Photoluminescence spectrum in implanted (front) and un-implanted (back) 6H-SiC.

phonon lines (ZPL) associated with the recombination of distant N-Al DAP. The only difference observed between the two spectra are the weak sharp lines at 2.623 eV, 2.588 eV, and 2.586 eV in the implanted sample. These lines are associated with defects induced by the ion implantation [9]. The relatively small intensity of these lines and the similarity between both spectra, suggested that the annealing treatment has recovered the crystalline quality of the substrate. Although the transport measurements performed on such layers shows conclusive evidence of N activation, we were not able to observe any emission band which could undoubtedly be assigned to the implanted N species. Since the penetration depth of the laser-line is much larger than the implanted layer thickness, we are planning to repeat this experiment with a high absorbing light source.

CONCLUSIONS

Microwave annealing seems to be an attractive method for rapid thermal annealing of implanted SiC. The electrical characteristics and lattice perfection of the annealed material are comparable to the values obtained for conventional ceramic processing furnace anneals.

ACKNOWLEDGMENT

The work at George Mason University is supported by NSF under Grant #ECS-9319885, and Oak Ridge National Laboratory, managed by Lockheed Martin Energy Research Corp for the U.S. Department of Energy, under contract number DE-AC05-96OR22464. We also wish to give special thanks to Ilene S. Manster for her assistance in editing this report.

REFERENCES

1. R.F. Davis, G. Kelner, M. Shur, J.W. Palmour, and J.A. Edmond, proc. IEEE 79, p.677 (1991).

2. S.K. Ghandhi, VLSI Fabrication Principles, John Wiley, New York, 1994, pp. 368-406.

3. J. Gardner, M.V. Rao, O.W. Holland, G. Kelner, D.S. Simons, P.H. Chi, J.M. Andrews, J. Kretchmer and M. Ghezzo, J. Electron. Mater. 25, May 1996 issue and the references therein.

4. M. R. Splinter, R. F. Palys, and M. M. Beguwala, Low Temperature Microwave Annealing of Semiconductor Devices, United States Patent, # 4,303,455 , Dec. 1, 1981.

5. P. D. Scovell, Method of Reactivating Implanted Dopants and Oxidation Semiconductor Wafers by Microwaves, United States Patent # 4,490,183 , Dec. 25,1984.

6. S.L. Zhang, R. Buchta, and D. Sigurd, Thin Solid Films 246, p. 151 (1994).

7. T. Fukano, T. Ito, and H. Ishikawa, Microwave Annealing for Low Temperature VLSI processing, IEDM Tech. Digest, p. 224 (1985).

8. H. Amada , Method and Apparatus for Microwave Heat-Treatment of a Semiconductor Wafer, United States Patent # 4,667,076 , May 19, 1987.

9. M.V. Rao, P. Griffiths, O.W. Holland, G. Kelner, J.A. Freitas, Jr., D.S. Simons, P.H. Chi, and M. Ghezzo, J. Appl. Phys. 77, p. 2479 (1995).

DEPOSITION OF THIN SiO$_2$ FILMS ON POLYMERS AS A HARD – COATING USING A MICROWAVE – ECR PLASMA

K. SANO*, H. TAMAMAKI*, M. NOMURA*, S. WICKRAMANAYAKA**,
Y. NAKANISHI** and Y. HATANAKA**
*Development Division II, Suzuki Motor Corporation, 300 Takatsuka-cho, Hamamatsu-shi
**Research Institute of Electronics, Shizuoka University, 3-5-1 Johoku, Hamamatsu-shi

ABSTRACT

SiO$_2$ thin films were deposited on automobile plastics at low temperatures using a microwave activated ECR plasma. Oxygen was used as the plasma gas while tetraethoxysilane (TEOS) was used as the source gas which was introduced into the downstream. In the present investigation high quality SiO$_2$ films were deposited on polycarbonate (PC) and polypropylene (PP) substrates with and without a mesh and the characteristics of hard coating films were studied. The film growth rate increases with the decrease of substrate temperature when a mesh is inserted into the plasma. The irregularities of polymer surfaces could be planarized by the deposition of 1.0 μm thick SiO$_2$ film. The dynamic hardness of PC and PP are increased by the deposition of SiO$_2$ film, however, films deposited on PP is seen to be cracked while that of on PC is crack-free.

INTRODUCTION

There is a current public desire for an environmental-conscious car having a high performance. Most of the automobile components have been converted into those made from polymer materials, resulting in cars of lighter body weights which enable more efficient gas mileage. Especially, thermoplastics are considered to be available materials which can be recycled easily and utilized in automobile components [1]. However, polymer materials have low mechanical strength, hardness, weatherability, heat and friction resistance as compared with metallic materials. Considering the function and quality of automobiles, the polymer application for automobile components are limited. Unless the function of polymer surface is enhanced by surface treatments, more advanced applications of polymer for automobile industry can not be expected [2]. The SiO$_2$ thin film has the potential to use as a hard and smooth protecting layer for soft and rugged-surface materials. As a response to various requirements, several SiO$_2$ deposition methods are currently being investigated, including plasma enhanced CVD, thermal CVD and sol-gel coating [3-9]. However, it has been extremely difficult to find a method that obtains high quality films at room temperature. For instance, a remote plasma CVD where TEOS/O$_2$ chemistry is used, produces high quality SiO$_2$ films at substrate temperature above 400 ℃ [4-5]. Hence, this method can not apply for the polymer materials. The ECR plasma CVD method is expected to be a better deposition method because it has a higher ionization ratio of the reaction gas, and permits efficient deposition of high quality film at low temperatures [10-11]. In the present research, we have investigated the possibility of SiO$_2$ film depositions by ECR plasma on the surface of two types of polymers. These materials are PC and PP, which have wide application in automobile industry. A recent work reports that SiO$_2$ has a good adhesive property on PC surface. PP is commonly used as a bumper while PC is used as a meter panel window in automobiles. Polymer materials have lower melting point, therefore, the SiO$_2$ fabrication should be carried out at a lower temperature. Moreover, the thermal expansion coefficients of PC (6.5×10^{-5} / ℃) and PP (8.5×10^{-5} / ℃) are higher as compared with that of SiO$_2$ (0.6×10^{-6} / ℃) [12]. Therefore, even a small temperature difference between the deposition temperature and the room temperature, where the material is expected to be used, may build up film stress or causes cracks. This latter fact limits the deposition temperature to a lower value.

This paper reports a technique for SiO$_2$ deposition on polymer substrates. The SiO$_2$ film deposited by TEOS/O$_2$-microwave ECR plasma using a mesh could be improved polymer surfaces effectively. The SiO$_2$ films were investigated for their compositions, hardness and morphology and result obtained are discussed.

EXPERIMENTAL

Figure 1 shows a schematic diagram of the ECR plasma apparatus used for the experiments. Magnetic coils were placed around the plasma chamber to supply a magnetic-field density of 875 Gauss. Oxygen gas was introduced through the upstream inlet and TEOS gas was introduced into the downstream through a ring-shape outlet orifice. Then, a microwave power was applied to the chamber to initiate the plasma. The microwave generator was used Nihon Koushyuha Corporation, MKN-103-3S (1.0 kW). A round stainless steel mesh of 150 mm diameter was placed between the TEOS outlet-orifice and the substrate. The mesh pitch is 1.5×1.75 mm and the mesh wire, 0.5 mm in diameter. The distance from the mesh to the TEOS outlet-orifice was 50 mm, and the mesh was grounded. Films were deposited on PC, PP, c-Si<100> and quartz substrates at room temperature. Table 1 lists typical ECR plasma conditions used for SiO_2 film deposition on polymers that were determined by preliminary experiments. The emission spectrum of the downstream plasma was measured by a spectrophotometer.

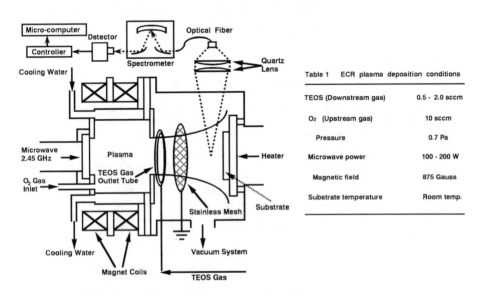

Table 1   ECR plasma deposition conditions

TEOS (Downstream gas)	0.5 - 2.0 sccm
O_2 (Upstream gas)	10 sccm
Pressure	0.7 Pa
Microwave power	100 - 200 W
Magnetic field	875 Gauss
Substrate temperature	Room temp.

Fig.1   Schematic diagram of ECR plasma CVD reactor.

Further, the film thickness was estimated by an ellipsometer. The characteristics of the films deposited were investigated by XPS, FTIR and SEM. The surface temperatures of PC, PP and Si substrates in the plasma deposition were measured by using thermocouples. The thermocouples were contacted on the substrate surface. Furthermore, the hardness of the SiO_2 deposited polymer surface was measured by a dynamic micro-hardness meter (SHIMADZU Corporation , DUH-50). In this method, a diamond tip (pyramid shape with the edge angle of $115°$) is inserted into the substrate surface with a constant speed, and the road in the direction of the depth is read directly. The dynamic hardness is calculated by the following equation.

Dynamic hardness = $\alpha \times P / D^2$

where, α is a constant (= 37.838) depending on the shape of the diamond tip, P is the testing load (gf) and D is the indentation depth (μm) [2][13]. The insertion depth of the diamond tip was adopted as 0.8 μm from the film surface.

RESULTS

TEOS Flow Rate and the Deposition Rate

Figure 2 shows the relationship between the TEOS flow rate and the deposition rate under varying microwave power at room temperature. When the mesh is inserted into the plasma, a higher deposition rate of SiO_2 was observed compared to that of without the mesh. Figure 3 shows the plasma emission spectra with and without the mesh. When the mesh was inserted, the emission peaks of alkyl groups and hydrogen atoms generated by the dissociation of TEOS molecules were decreased. This is considered due to the elimination of excess electrons and hydrogen radicals in the plasma by the mesh. The etching reaction on the substrate surface by hydrogen radicals was suppressed. The film growth rates deposited with and without the mesh rise with increasing TEOS flow rate, until the flow rate comes to a particular point. This is considered due to a large volume of oxygen radicals as compared with TEOS volume. When the TEOS flow rate was increased beyond this particular point, the both film growth rates were decreased. Also, the emission intensity of O_2^+ peak at 306 - 312 nm was observed to decrease [14]. This is attributed to the shortage of oxygen radicals. Further, the vapor-phase reaction-area shifts to the upstream with an increase of TEOS concentrations. When the microwave power is increased from 100 W to 150 W, the deposition rate increases. However, when the microwave power is further increased to 200 W, the deposition rate drops to a lower level than that of at 100 W. This reason is not clear in detail at the present.

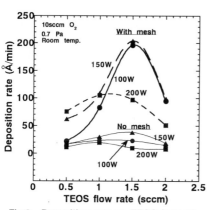

Fig.2   Deposition rate vs. TEOS flow rate.

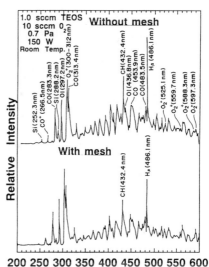

Fig.3   Optical emission spectra of plasma.

Chemical Analysis of SiO₂ Film

Figure 4 shows the relationship between the TEOS flow rate and the carbon content of SiO₂ films deposited with and without the mesh. The carbon contents were analyzed by XPS measurements. The carbon content of both films increases gradually when the TEOS flow rate is increased from 0.5 sccm to 2.0 sccm. Further, the adsorption spectra associated with -OH and H_2O stretching vibration modes are observed in the FTIR spectra when the TEOS flow rate is changed to 1.5 - 2.0 sccm [2][10]. This is due to the shortage of oxygen radicals to remove the intermediate products such as -OH, H_2O and some other hydrocarbons from the reaction site before they attached to the substrate surface. If the TEOS flow rate is excessive compared with that of oxygen radicals, a decrease of the SiO₂ film deposition rate and a degradation of the film quality are observed. The possibility for high quality SiO₂ film deposition without -OH, H_2O and hydro-carbon at room temperature suggests to be able to apply to the film deposition on polymer substrates that have low heat resistance.

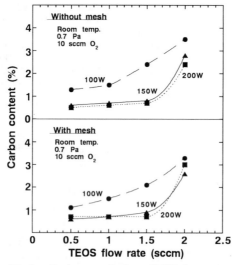

Fig.4   Carbon content vs. TEOS flow rate.

Surface Temperature of Polymers

Figure 5 shows relationship between the substrate temperature and the deposition time with and without the mesh. The surface temperature of substrates is increased by the plasma irradiance with the SiO₂ deposition. When the mesh is inserted into the plasma, the increasing of substrate temperature is suppressed. This is caused by the decrease of bombardments on the substrate surfaces by ions and electrons. The PP surface was heated up to the high temperature as compared with that of PC. Further, the surface temperature increase for PC was observed as the same level as the Si surface. This is considered to be due to the difference of thermal conductivities of substrates.

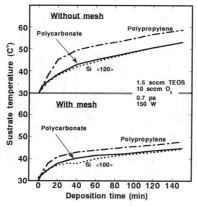

Fig.5 Substrate temperature vs. deposition time.

SEM Observation of SiO$_2$ Film

The film surfaces of PC and PP were observed by SEM after depositing SiO$_2$ films of 1.0 μm thick on polymer surfaces. Further, SiO$_2$ deposition was performed in the presence and absence of a mesh. The SEM photograph of the film deposited on PC without the mesh is shown figure 6, and that of deposited with the mesh is shown figure 7. In addition, the SEM photograph of the film deposited on PP without the mesh is shown figure 8, and that of deposited with the mesh is shown figure 9. In the case of without the mesh, the cracks were found in films deposited on both polymers. These cracks are generated due to the differences of thermal expansion coefficients of polymers and SiO$_2$ films. Because, substrate is heated up to a higher temperature by the long time-plasma irradiance since the film growth rate is lower. When the mesh is used, an uniform SiO$_2$ film without cracks is deposited on the PC. Also, the density of cracks on PP surfaces was decreased than that of without the mesh. The decrease of crack density is considered to be due to the following reason. The both reason of the higher growth rate and no-ion bombardment makes the substrate temperature suppress by low level when the mesh is inserted.

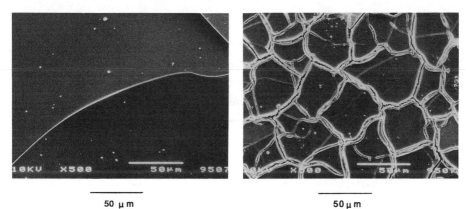

50 μm   50 μm

Fig.6 SEM micrograph of SiO2 on PC (without mesh). Fig.7 SEM micrograph of SiO2 on PP (without mesh).

651

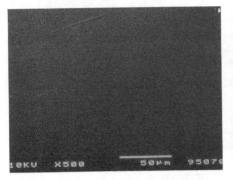

50 µm	50 µm
Fig.8 SEM micrograph of SiO2 on PC (with mesh).	**Fig.9 SEM micrograph of SiO2 on PP (with mesh).**

Furthermore, a large quantity of cracks are generated in the SiO_2 film on PP surface compared with PC. This is considered due to the higher thermal expansion of PP than that of PC. Hence, PP surface is rapidly heated up to a high temperature by the plasma irradiance, because the thermal conduction coefficient of PP is lower than that of PP.

Hardness of SiO_2 Films Deposited on Polymer

Table 2 shows the dynamic hardness values of polymer surfaces before and after SiO_2 deposition. The hardness of PC surfaces could be increased by 50% by the deposition of high quality SiO_2 film. Further, in spite of having cracks in the film, the hardness of PP surface increases. Even though the films deposited on PP and PC have the same film composition and the same film thickness, a difference of film hardness was observed. This is considered due to the influences of film cracks and the dependence on the kind of basal substrate materials.

Table 2 Dynamic haradness of polymer deposited SiO2

Substrate	Dynamic hardness (HD115°)	
	Original suraface	deposited surface
Polycarbonate	33.6	53.5
Polypropylene	12.5	22.2

CONCLUSIONS

Deposition of SiO$_2$ films on polymer surfaces by using TEOS/O$_2$ chemistry in an ECR plasma was studied in order to realize a higher deposition rate of high quality films at low substrate temperatures. If all the TEOS molecules introduced into the plasma react with oxygen, the films are observed to be free of -OH, H$_2$O and C. Therefore, a precise control of TEOS and oxygen flow rates is required to deposit high quality SiO$_2$ films on polymer surfaces. In order to realize a higher deposition rate with high quality films at the low substrate temperatures, use of the mesh is essential for ECR plasma deposition. In other words, if a mesh is attached to the reaction downstream, the increase of polymer substrate temperature is suppressed because of the shorter deposition time and no ion-bombardment, hence, the cracks in the film deposited can be decreased. To increase the hardness of the polymer surfaces, the deposition of crack-free film is of importance, therefore a low temperature deposition and no ion-bombardment by ECR plasma using the mesh is effective.

ACKNOWLEDGMENTS

The authors wish to thank Dr. H. Oikawa at the Development Division of Aftex Co., Tokyo, Japan, S. Hayashi at the Research and Development Laboratory of SUZUKI Motor Co., Yokohama, Japan, and Dr. T. Aoki at the Research Institute of Electronics, Shizuoka University, Hamamatsu, Japan, for their guidance and assistance.

REFERENCES

1. K. Sano, M. Nomura, S. Hayashi, and N. Inagaki, Trans. Sci. Auto. Eng. Jpn., **25**, 153 (1994).
2. K. Sano, S. Hayashi, M. Nomura, S. Wickramanayaka, Y. Nakanishi and Y. Hatanaka, SUZUKI Tech. Rev. Jpn., **21**, 49 (1995).
3. M. Shinoda and T. Nishide, J. Vac. Sci. Technol., **A12**, 746 (1994).
4. S. Wickramanayaka, A. Matsumoto, Y. Nakanishi, N. Hosokawa and Y. Hatanaka, Jpn. J. Appl. Phys., **33**, 3520 (1994).
5. S. Wickramanayaka, A. Matsumoto, Y. Nakanishi and Y. Hatanaka, Tech. Report of Jpn. IEICE, **93**, 81 (1993).
6. T. S. Cale, G. B. Raupp and T. H. Gandy, J. Vac. Sci. Tech., **A10**, 1128 (1992).
7. S. Nguyen, D Dobuzinsky, D. Harmon, R. Gleason and S. Fridamann, J. Electrochem. Sci., **137**, 2209 (1990).
8. Y.Furukawa, Tech. Proceedings of SEMICON Jpn., **DEC.** 339 (1992).
9. K. Fujino, Y. Nishimoto, N. Tokumasu and K. Maeda, J. Electrochem. Sci. **138**, 550 (1991).
10. K. Sano, S. Hayashi, S. Wickramanayaka and Y. Hatanaka, Tech. Report of Jpn. IEICE, **94**, 7 (1995).
11. K.Machida, H.Oikawa, J. Vac. Sci. Technol., **B4**, 818 (1986).
12. A. K. Sinha, H. J. Levinstein and T. E. Smith, J. Appl. Phys., **49**, 2423 (1978).
13. K. Kanazawa and K. Teragima, J. Surface Finishing Sci. Jpn., **40**, 41 (1989).
14. K. Sano, S. Hayashi, M. Nomura, S. Wickramanayaka, Y. Nakanishi and Y. Hatanaka, SUZUKI Tech. Rev. Jpn., **22**, 45 (1996).

AUTHOR INDEX

SUBJECT INDEX